Comparative Anatomy of Vertebrates

Comparative Anatomy of Vertebrates

Second Edition

R.K. Saxena
Sumitra Saxena

London • New Delhi

Every possible effort has been made to ensure that the information contained in this book is accurate at the time of going to press, and the publisher and author cannot accept responsibility for any errors or omissions, however caused. No responsibility for loss or damage occasioned to any person acting, or refraining from action, as a result of the material in this publication can be accepted by the editor, the publisher or the author.

Every effort has been made to trace the owners of copyright material used in this book. The author and the publisher will be grateful for any omission brought to their notice for acknowledgement in the future editions of the book.

Copyright © MV Learning, 2015

All rights reserved. No part of this book may be reproduced, stored in a retrieval system, or transmitted in any form or by any means, electronic, mechanical, photocopying, recorded or otherwise, without the written permission of the publisher.

MV Learning

3, Henrietta Street
London WC2E 8LU
UK

4737/23, Ansari Road,
Daryaganj, New Delhi 110 002
India

A Viva Books imprint

ISBN: 978-81-309-3000-8

Printed and bound in India.

SUB-0001/SSUB-0007

Contents

Preface *vii*

PART A

Origin Evolution and Descent of Metazoan Animals and Classification

1. Evolutionary Descent of Metazoan Animals — 3
2. Recent Trends in Animal Classification — 41
3. Agnathostomes — 94
4. Gnathostomes — 106
5. Amphibia — 131
6. Reptilia — 148
7. Aves — 171
8. Mammalia — 189

PART B

Comparative Anatomy

9. Integumentary System — 227
10. Skeletal System — 258

11.	Muscular System	350
12.	Digestive System	375
13.	Respiratory System	401
14.	Circulatory System	426
15.	Excretory System	457
16.	Reproductive System	470
17.	Nervous System	502
18.	Receptor System	540
19.	Endocrine System	567
	Glossary	589
	Index	651

Preface

We have developed this book around two fundamental themes. The first eight chapters of Section A of this book are devoted to a class-wise discussion of Chordates evolution and classification. Fairly detailed taxonomic synopsis for each class is included in each chapter. We have used a combination of phylogenetic trees (Cladograms) and explained narrative discussions to talk about animal evolution. The Cladograms are used where appropriate, because they provide the least ambiguous statements that can be proposed about animal relationships. Unless otherwise indicated, the classification in each chapter deals only with extant taxa. However, We have discussed molecular-based hypotheses that have received widespread attention. We hope that Section A of this book will make this seemingly overwhelming task a bit more manageable.

The Section B deals with the comparative anatomy. This section is updated and has not been modified much, because it deals with the study involving the structural pattern and the influence of evolutionary design. In this section the structure function concept at the level of organs and organ system is fundamental to our understanding of comparative morphology. It is upon these interrelated aspects—structure, function and evolution—that we have arranged and presented the diversity of vertebrate organisation of each organ system.

We are extremely grateful to those authors of the textbooks and research papers we have freely consulted, and used the available information and data in the respective chapters of the book. In spite of our best efforts and critical screening of the text in the book at all levels, there might have crept in certain unnoticed errors. We shall be obliged if the readers point out those to us for corrections in the subsequent editions.

Any sugggestions and criticism from the readers for the improvement of the text will be highly appreciated.

Authors

"Were I to await perfection, my book would never be finished."

– Tai Tung

Part A
Origin Evolution and Descent of Metazoan Animals and Classification

Chapter 1

Evolutionary Descent of Metazoan Animals

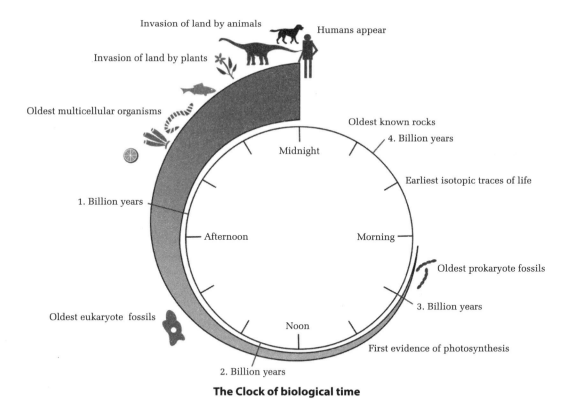

Figure 1.0 A billion seconds ago most using this text had not yet been born. A billion minutes ago the Roman empire was at its zenith. A billion hours ago Neanderthals were alive. A billion days ago the first bipedal hominids walked the earth. A billion months ago the dinosaurs were at the climax of their radiation. A billion years ago no animal had ever walked on the surface of the earth.

Origin of Metazoa

Evolutionary changes in animal body plans might be likened to a road map through a mountain range. What is most easily depicted are the starting and ending points and a few of the "attractions" along the route. What cannot be seen from this perspective are the tortuous curves and grades that must be navigated and the extra miles that must be traveled to chart back roads. Evolutionary changes do not always mean "progress" an increased complexity. Evolution results in frequent dead ends, even though the route to that dead end may be filled with grandeur. The account that follows is a look at patterns of animal organization. As far as evolutionary pathways are concerned, this account is an inexplicit road map through the animal kingdom

First cell to appear was a prokaryote cell without a definite nucleus. Later, the eukaryotic cell developed having a well-organized nucleus. The earlier unicellular or non-multicellular organisms were neither plant (Protophyta) nor animal (Protozoa) in nature. With the increase in complexity of life, multicellularity developed in plants and animals. The term Metazoa is used for multicellular animals. There is no direct proof about the ancestors of Metazoa. But all zoologists agree that metazoans have evolved from some unicellular or acellular organisms, such as Protozoa. How single-celled Protozoa, which evolved possibly 2,000 million years ago, gave rise to the multicelled Metazoa (Fig. 1.0), remains one of the fundamental mysteries of evolution shrouded in complete obscurity.

Premetazoan Ancestors

It is quite true that along with eukaryotes came large cell size, enhanced motility, and the ability to feed on their prokaryotic progenitors. Multicellularity resulted in another quantum leap in body size, enabling animals to feed on protozoa (as well as prokaryotes), but also conferring on them metabolic and other advantages over the unicellular protists. The evolution of epithelia allowed eumetazoans to physiologically regulate internal and, extracellular compartments, such as the gut and coelom. This regulation not only improved the performance of digestive and other functions, but also liberated eumetazoans to some extent from the vagaries of environmental variations, thus enabling them to sustain relatively high levels of activity. The adoption of bilateral symmetry by eumetazoans resulted in a directionally polarized body equipped with sensory organs used for pursuing food and mates or escaping from enemies. The metamerism played significant role in locomotion and cephalization.

The origin of multicellular animal like form from single-celled ancestors is perhaps the most enigmatic of all phylogenetic problems, and the least likely to be 'solved' by additional evidence.

The problem of metazoan origins has naturally exercised the imaginations of invertebrate biologists, and many theories have been advanced in the last 150 years. Most of these have been compounded with views on subsequent evolution from the earliest metazoan to the groups of existing lower Metazoa, and on relationships between these various phyla i.e., particularly sponges, cnidarians and platyhelminths. In fact it is difficult to separate these issues, as any theory of the nature of the first multicellular animals inevitably has consequences for conceptions of 'what happens next'.

Although there is a multiplicity of theories to be considered, it is fortunate that there are in fact only a very few ways in which the initial multicellularity could conceivably have been achieved. Reduced to their very simplest, all the theories are of only two kinds, and could only be of two kinds (Fig. 1.1).

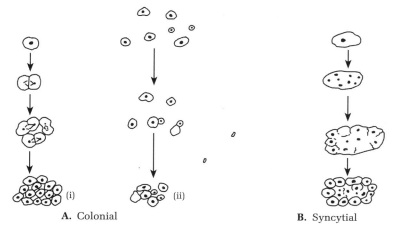

Figure 1.1
Methods for achieving *multicellularity*, by aggregation of mitotically-related (A i) or possibly unrelated (A ii) cells, or by incomplete subdivision of a single large cell (B).

a) The **first possibility** is that metazoan status was achieved by the coming together of two or more cells, forming a colony, whether by incomplete separation of daughter cells after mitosis (Fig. 1.1 A i), or by independent aggregation of formerly separate cells in a symbiotic fashion (Fig 1.1 A ii).

b) The **second major possibility** is that a single cell became multinucleate without divisions of the cytoplasm, forming a 'plasmodial' organism (Fig. 1.1 B), and that cell boundaries were established later to give a multicellular animal with many primitively syncytial tissues.

The first true eucaryote Protists may have arisen as early as 1400 MYA, (Fig. 1.2) and it is widely assumed that they represent a monophyletic entity given the uniformity of intracellular structures, notably mitochondria and microtubular complexes. Most authors have concluded that autotrophic protists with aerobic and photosynthetic metabolism

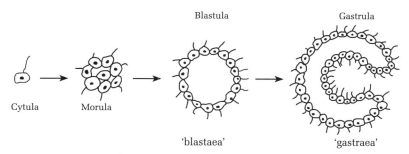

Figure 1.2
The early stages of development, of ancestral forms as 'blastaea' and 'gastraea'.

were the ancestral type, probably of an amoeboid form with a flagellar apparatus. Loss of the plastids would give primary heterotrophic protozoans (probably polyphyletically), having a naked cell membrane and a rather more complex flagellar/centriolar system. From these early single-celled 'animals' many further lines of heterotrophic protozoans would probably evolve before the first metazoans were to emerge. Which of these lines gave rise to those archemetazoans, perhaps 700-1000 MYA, is therefore inevitably a matter of pure speculation.

To qualify as a **true metazoan**, an organism must be multicellular, heterotrophic and potentially motile, with specialised gametic cells; some authors also require that it possess distinct tissue types other than the gametic tissue. An important stage in the development of metazoans from eukaryote protists must therefore have been the acquisition of meiosis and hence sexual reproduction; but the sporadic occurrence and degree of variation of this process in protists suggests that it may have had multiple convergent origins. Hence the particular sexual pattern of metazoan cells is not necessarily inherited from any one group of protists but may have evolved separately within the forerunners of the multicellular forms themselves. Following on from this argument, Salvini-Plawen suggests that the identical meiotic patterns, and diploidy, of all extant metazoans are convincing evidence for their phyletic unity – in other words, sex itself is polyphyletic, but the metazoans are not. To start with then, we should examine existing alternative theories of how this supposedly unitary evolutionary step could have occurred, and what the consequences are of each theory.

There are four main Possible theories

1. Colonial 'blastaeal gastraea' theories

A view of metazoan origins involving blastula-like and gastrula-like stages succeeding a phase of aggregation of protistan flagellates has long been influential. The version of this story proposed in the nineteenth century by Haeckel himself, (Fig. 1.2), was heavily dependent on recapitulation theories, suggesting that the blastula of modern embryos recapitulates their ancestry as a 'blastaea' organism. The basic metazoan was thus a pelagic, radially symmetrical aggregation of flagellated cells, as currently represented by forms like Volvox, which have incipient division of labour between cells (even with a separation of somatic and gametic cells) and some coordination of flagellar activities. A 'gastraea' stage, which the gastrula of modern embryos recapitulates, would arise from the blastaea by a separation of locomotory and digestive regions, the latter being posterior and gradually tucking in to form an interior endoderm. This gastraea form, a simple two-layered, sac-like organism, readily gave rise to the similarly constructed aseptate cnidarians, from which other metazoans subsequently evolved.

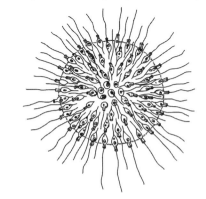

Figure 1.3

A colonial flagellate protist of the choanoflagellate, *Sphaeroeca volvox*;

Most obviously, though there are many colonial flagellate protists, all the existing blastula-like types are unequivocably plants, and all have incomplete separation of constituent cells, with fine cytoplasmic bridges connecting the colony. The phytophagous volvocid organisms (possessing) are also primarily haploid. Clearly a zooflagellate colony with prolonged diploidy is needed as the root for this theory, yet the zooflagellates do not commonly show such a tendency to form colonies of the right blastuloid type. The nearest equivalents are the colonial choanoflagellates, some of which Sphaeroeca (Fig. 1.3) are pelagic and could provide suitable models for an **archemetazoan** (Fig. 1.4); though the colonies in question are all fresh-water forms, and do not appear to show any division of labour as seen in some volvocids.

The simple blastula form must invaginate to form a diploblastic gastraea organism, and in all phyla except sponges there must be a degree of internal elaboration of gastric surfaces. Remane took the view that this would occur in a pelagic animal, and proposed four original gastric pouches in his 'cyclomerism theory' (Fig. 1.4 A) to account for archemetazoan origins. Jägersten, and subsequently Siewing, suggested that the ancestral form had become benthic before gastrulation occurred, giving a bilaterogastraea' (Fig. 1.4 B) or a rather similar 'benthogastraea' respectively, each having three pairs of gastric pouches.

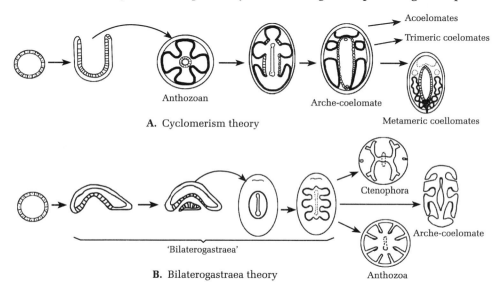

Figures 1.4 A, B
Archecoelomate theories of metazoan origins: **A.** Remane's cyclomerism theory, where the gastric pouches of a cnidarian are transformed into a trimeric coelomate condition; **B.** Jägersten's *bilaterogastraea theory*, where a bilateral and benthic organism gastrulates and acquires gastric pouches.

Even with these modified versions of gastraea theories significant controversies are raised. While some lower invertebrates do gastrulate by invagination, poriferans and cnidarians do not. Much more seriously, nearly all modern gastraea theories are inextricably linked with enterocoel theories of early coelom acquisition, for in the archemetazoans the fully formed gastric pouches of the cnidarian-like ancestor are supposed to transform rapidly into enclosed coelomic cavities (Fig. 1.4). All current acoelomates must therefore have regressed from a coelomate state; and the archemetazoan becomes synonymous with the archecoelomate. But the various gastraea-type phases that these authors support, being so closely tied to enterocoelic theories, have difficulty in accounting for the diversity of modern coelomate embryology and for the very existence of many acoelomate forms; the arguments advanced could be reiterated here. Once divorced from the enterocoelic overtones, the gastraea is a purely recapitulationist construct, and has little plausibility as a metazoan ancestor. Many authors have therefore retained the idea of a **simple choanoflagellate/blastula** colony as a plausible early metazoan, but looked elsewhere for the development of the next 'grade' of multicellular forms.

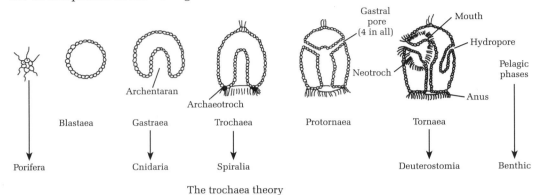

Figure 1.5
The trochaea theory, in which the pelagic trochophore larva (trochaea) succeeds the blastaea and gastraea, to form the ancestor of spiralians, and in turn gives rise to a pelagic tornea as the ancestor of deuterostomes.

One variant of gastraea theories advanced by Nielsen & Norrevang suggest that a common monociliate gastraea stock gave rise directly to Cnidaria, and also via a 'trochaea' stage, with an equatorial ring of multiciliate cells, to the remaining bilateral phyla. This occurred through trochophore larvae on the spiralian side, and through a 'tornaria' larva on the deuterostome side, the latter form losing its multiciliation again (Fig. 1.5). While avoiding explicit archecoelomate postulates, this theory does require awkward and functionally inexplicable gains and losses of ciliation patterns. It follows Jägersten in placing perhaps excessive stress on the primacy of pelago-benthic life-cycles and the primitive stature of larvae, all of which is interesting dogma but untestable; in fact, most lower Metazoa do not have larval stages. The resultant theory seems over-elaborate on the one hand, and insufficiently supported by a functional rationale on the other.

In passing, one more theory related to but somewhat different from the blastaea/gastraea theory is the **'plakula' theory,** invoking a flattened disc of cells (rather than a hollow ball) as the plakula stage to precede a gastraea. The theory was revived by Grell (1971, 1981) with particular reference to the placozoans, peculiar animals that have a dorsal layer of flagellated cells and a ventral absorptive layer with little in between . However, since they do have a certain amount of intervening parenchyma, it seems preferable to refer them to a planula grade and to do without the confusing plakula grade; especially as Trichoplax in fact passes through a perfectly good blastula stage to arrive at its adult morphology.

2. Colonial 'blastaea/planula' theories

One way to avoid the problems discussed above is to retain the blastula as a first stage in metazoan origins but combine it with a non-invaginated, solid stage as the next step in evolution of the archemetazoans. This second stage has been termed the **'parenchymula'**, or **'planaea'**, or most simply the **'planula'** The planula ancestor is taken to be a small, pelagic, ovoid but radially symmetrical solid mass of cells, the internal cells acquired by immigration from the blastula stage. From it radiated the cnidarian and ctenophoran lines, and when it became benthic and differentiated its dorsal-ventral and anterior-posterior axes it gave rise to the primitive Bilateria such as flatworms – the turbellarian acoels are therefore assumed to be derived from sexual planulae. The theory, partly foreshadowed in the late nineteenth century by Metschnikoff when he proposed a solid gastraea as a metazoan precursor. Hyman favoured it specifically as an attempt to update and render acceptable the colonial theories of Haeckel, avoiding most of the problems thereof. It has been added to and modified by Hand (1963), Ivanov (1968), Beklemishev (1969), and Reisinger (1970). A thorough review of these theories and their strengths is given by Salvini-Plawen, and a summary of likely stages in metazoan evolution based on his views and to a lesser extent those of Ivanov is given in (Fig. 1.6).

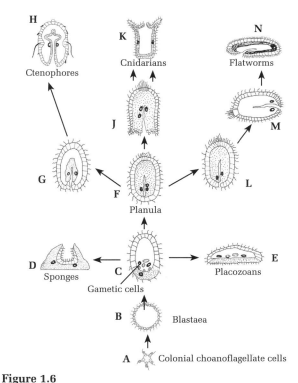

Figure 1.6
Proposed invertebrate origins in which the planula is central to all developments after acquisition of multicellularity. Colonial choanoflagellate cells **A** give rise to a blastaea, **B**. Separate gametic cells develop (black circles), and ingression **C** gradually produces an internal layer and thus a true planula **F** Sponges **D** and placozoans **E** are derived from the very early planula, with ctenophores **H**, cnidarians **K** and flatworms **N** as offshoots from the fully developed planula stage.

Several points can be made in support of this theory. It embraces both pelagic and benthic forms, and radial and bilateral forms, in a single theory, permitting all the Metazoa to be monophyletic from flagellate ancestors. It provides a good starting point for most of the 'lower' phyla, as planuloid larvae occur not only in many cnidarians but also in a few sponges and at least one ctenophore). It neatly provides a plausible status for the placozoans, as rather solid forms of blastaea forming incipient planulae (Ivanov 1973). On the whole it makes ecological and functional sense, with varieties of planulae diversifying according to their different modes of life as planktonic, surface-dwelling or thiobenthic forms (though it is not clear how the initial planula stage was supposed to feed, having no mouth or gut). The theory also establishes the platyhelminths firmly near the rootstock of the bilateral phyla, with little modification required to derive them direct from a planula; indeed, the acoel turbellarians are superficially most convincingly like large planulae. All that is required to complete their transformation is the increasing potency of the endoderm to form mesodermal parenchyma and actual gut tissues leading from a new ventral mouth; and the acquisition of reproductive system.

There are also, of course, some problems, but having as a first stage a colonial flagellate ancestor does involve certain difficulties, even if the choanoflagellates are used as a model. At the next stage, the theory in its original versions suggested that the medusa form of cnidarians would be the more primitive), as being more easily derived from essentially radial forms, so that Hydrozoa and Scyphozoa (having medusoid stages) were more primitive than the solely polypoid and more pronouncedly bilateral Anthozoa. Although this has long been the conventional view of cnidarian relations, various authors have recently taken the opposite view.

However, Salvini-Plawen's review combines a planuloid theory with a revised view of cnidarian evolution, invoking a settling planula with a single pair of tentacles as the first cnidarian and so starting with biradial polyps. This phase actually occurs in the embryology of Scyphozoa and some hydrozoan polyps; and as a theory it explains the differences between the medusae in these two groups, proposing that their pelagic medusoid phases must have arisen convergently. This seems to provide the most broadly acceptable colonial flagellate/blastaea type of theory to date.

But it is worth noting two important issues here. **Firstly the 'planula'** is an extremely simple architectural construct, and is almost an obligate stage in any metazoan's embryology given the way in which cell aggregations behave; so it may well have developed independently several times). **Secondly, the problems** with the occurrence of cilia and flagella. Monociliated cells, as derived from a flagellate, are represented amongst lower metazoans in cnidarians, placozoans, gnathostomulids and probably sponges, but multiciliated cells occur in ctenophores and platyhelminths. If the whole origin of Metazoa is monophyletic from flagellates, the presence of monosomal multiciliation in the very early planula leading to the turbellarians and other spiralian animals is somewhat difficult to account for. A number of authors have developed theories for metazoan origins that do not rely on colonial (monociliated) protistan flagellates, forming blastulas, as their starting point; can such theories avoid some of these difficulties?

3. Colonial amoeboid/acoeloid theories

Reutterer proposed an archemetazoan origin amongst the amoeboid protozoans that exhibit some tendency to aggregate, and are essentially diploid. The ancestral forms would therefore be benthic, with nutrition by ingestion of surface deposits. Lacking primary surface ciliation, such acoeloid ancestors could perhaps readily give rise to either modern multiciliated groups like the Acoela, or to monociliated lineages like the cnidarians.

Hanson also advocated a colonial amoeboid form as the likely ancestor of cnidarians, as part of his effectively polyphyletic theory on metazoan origins,

4. Syncytial ciliate/acoeloid theories

The possibility of metazoan origins occurring through growth, nuclear division, and cellularisation of a single protozoan, rather than by aggregation of several, was realised in the nineteenth century, but owes its modern formulation largely to Hadzi (Figs. 1.7 A, B). His

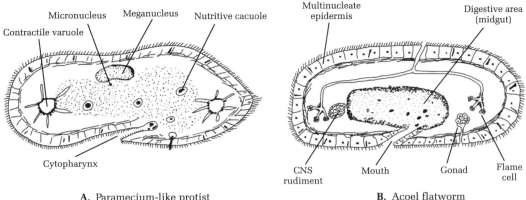

Figures 1.7 A, B
The similarities between a large ciliate protozoan such as *Paramecium* and a small acoel flatworm.

version of the theory, also advocated by Steinböck and in part by Hanson envisaged a process of cell boundary formation in a multinucleate (plasmodial) protozoan of a ciliate type (perhaps one like Paramecium), or possibly of an aberrant form such as Opalina. A bilateral ancestor of this kind could very directly give rise to the acoel turbellarians (Fig. 1.8), which lack any gut lumen. In particular, Hadzi believed the 'midgut' tissue in Acoela to be primitively syncytial in nature, standing as testimony to a genuine link with plasmodial protists. Acoels would then be ancestral to all other metazoans, except possibly sponges which could have a separate origin from flagellates.

Hadzi drew heavily on the simple fact of resemblance between ciliates and acoels, and was imprecise in considering which features could actually be described as homologous; indeed he advocated somewhat rash parallels in pursuit of this theory. All this has allowed his views to be extensively criticised by the German school, accustomed to applying the rigid rules laid down by Remane (1952) to questions of homologies. As Hadzi's views of the later evolution of the metazoans were peculiarly linear, with all phyla derived from a single stem (Fig. 1.9), his work has often been ridiculed and has certainly not been widely accepted. Remane and Ivanov further discredited it with attacks on the plausibility of cellularisation itself. Nevertheless Hadzi's ideas on archemetazoan origins have been accorded with some serious consideration. The planuloid/acoeloid form leads readily to flatworms, which are now favourite starting point for metazoan radiation, but which are in fact multiciliate (like ciliates) and not monociliate (like flagellates). The order Acoela includes animals that are of a similar size to large multinucleate ciliates, and of an appropriate bilateral form; they have a similar feeding

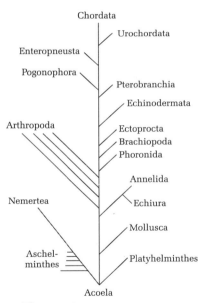

Figure 1.8
Phylogeny of the animal kingdom, above the Acoela level.

habit, ingesting bacteria and small protists; and no awkward hypothetical intermediate stages need be invoked to make the transition from protist to archemetazoan.

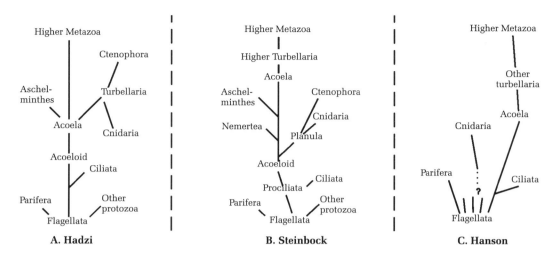

Figure 1.9
Various interpretations of the syncytial theory of metazoan origins, all of which regard the acoeloid form as primitive and derived from a ciliate like protistan stock, but differ in their views on the status of cnidarians, ctenophores and other lower groups.

However, One major stumbling block must be the gnathostomulids, a group of flatworms now given phyletic status on their own and thought to be closely allied to but primitive relative to platyhelminths. They are actually a monociliate group, so it is hard to see where they could fit into a ciliate/acoel scheme. Furthermore, the turbellarian starting point as envisaged by Hadzi requires the so-called diploblastic groups to be regressive, and bilateral anthozoans to be ancestral to the other cnidarian classes. Hadzi proposed (Fig. 1.9 A) that the ctenophores evolved from neotenous polyclad larvae, and that cnidarians derived more directly from adult rhabdocoel-type platyhelminths. This is rather difficult to envisage, necessitating a link between the flatworm gut diverticula and the septae of cnidarians. Alternatively, Steinböck suggested that cnidarians branched off from a very early acoel form (Fig. 1.9 B) similar to the planula invoked by other theories; the acoel preceded the planula, rather than vice versa.

Hanson's version is perhaps the most acceptable. He presents plausible selectionist arguments for the ciliate ancestry of turbellarians, and for cellularisation as a process, rejecting the more esoteric aspects of Hadzi's views; Overall he invokes a plasmodial/syncytial ciliate origin for platyhelminths and the spiralians, but a choanoflagellate origin for sponges and perhaps an amoeboid origin for cnidarians; his views are summarised in (Fig. 1.9).

Other possible theories

Metazoans originated from procaryotes. A theory of this type was presented by Pflug, when he proposed that animals actually arose from multicellular procaryotes that developed a nucleus. He based this view on certain fossils known as Petalo-organisms, from Pre-Cambrian (Ediacaran) rocks. Some of these seem to be plants but others have an apparent internal cavity that could have had a digestive function, and may have been cnidarian-like. Apart from the difficulties of proposing diphyletic acquisition of complex eucaryotic features, this theory also seems incompatible with any known embryological and growth patterns of animal life, which is never filamentoid or meristem-like as in the Petalo-fossils. Salvini-Plawen dismisses the petaloid organisms as an irrelevancy, mostly referred more plausibly to a plant or fungus lineage. Many other authors would concur with this view, and as was discussed Glaessner suggests that some of the more animal-like petaloid forms are in fact perfectly good anthozoan cnidarians, whilst Seilacher believes they are evidence of a totally separate and extinct 'type' of metazoan life. They certainly do not seem to be good material on which to base such a radical theory of the origins of extant Metazoa.

Metazoans originated from plants. Consideration of the classic blastaea/gastraea theories of Haeckel raised the difficulty that all existing spherical hollow flagellate colonies are actually plants and not animals at all. Most authors have therefore discounted volvocid colonies as models. But Hardy proposed instead that such colonies could plausibly have given rise to animals by a series of logical steps. The colony gradually specialised its leading cells as flagellated locomotory structures, retaining their photosynthetic capacity; while trailing cells began to invaginate, forming a saucer-shaped creature. Due to the currents created by the flagella, debris including food particles would collect in the depression of the saucer, and cells bordering this region would lose their plastids and resort to heterotrophy instead. Thus a simple gut was formed and eventually photosynthesis was no longer required. Though less implausible than some, this theory has received little support, and does seem to propose somewhat superfluous steps given the presence of so many already heterotrophy single-celled protists as possible metazoan ancestors.

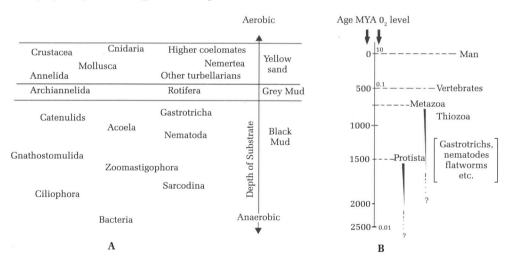

Figures 1.10 A, B
The *thiozoon* theory of metazoan origins proposing the modern inhabitants of anoxic muds to be relic representatives of an early anoxic metazoan radiation: **A.** The presentday distribution with depth of various protist and invertebrate groups, based on **Boaden** (1975); **B.** The geological sequence, setting metazoan origins very much further back in time than most other theories (From **Platt** 1981).

Metazoans originated anaerobically. A holobenthic, interstitial and anaerobic origin for metazoans is advocated by Boaden, following analyses by Fenchel & Riedl and others of the thiobenthic organisms living today. These anaerobic meiofauna in the sulphide-rich deposits of most marine substrates are principally small bilateral organisms from many phyla, but exclude segmented forms, coelomates and cnidarians. Their distribution in relation to depth and hence the anaerobicity of conditions possibly reflects an evolutionary sequence (Fig. 1.10), those forms found deeper being the earlier in evolutionary terms, and having evolved at low levels of atmospheric oxygen (Fig. 1.10 B). It is true that all aerobic animals today also have anaerobic metabolic pathways but that the reverse does not hold, so there is a possibility that anaerobicity did come first; the geological information available would also support this view, as there is a preponderance of strongly reduced minerals in the earlier Pre-Cambrian rock. This story would also provide reasonable backing for an acoeloid type of ancestor called a **'thiozoon'** by Boaden, as it places acoel and catenulid turbellarians, gnathostomulids, and some aschelminths at the base of the metazoan radiation. However, the theory requires metazoan origins to be set very much earlier than is usually allowed – as long as 2000-2500 MYA – and there is no particular evidence for such a view. In fact the evidence amassed in favour of this thiobenthic sequence is all extrapolations from modern animals and their distribution, and could equally reflect a later radiation from an acoeloid form derived by any of the described

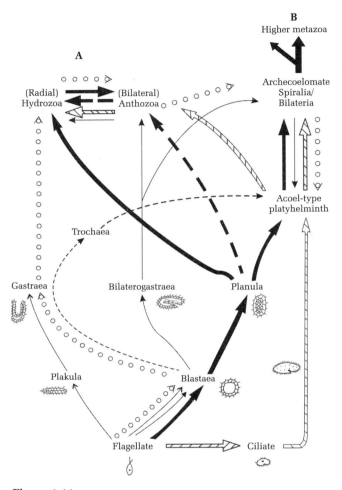

Figure 1.11

A summary of the different theories of *metazoan origins*. On the left, varying pathways between hypothetical ancestral forms and modern groups are shown: *Solid black arrows,* the planula theory of Hyman, Salvini-Plawen and others, with Hyman's variation for cnidarian evolution shown dashed; *circles*, gastraea theories derived directly from Haeckel, with archecoelomate ancestors (Remane, Siewing, etc); *continuous narrow lines*, Jägersten's bilaterogastraea version; *dashed narrow lines*, Nielsen's trochaea theory; *dotted line*, Grell's plakula theory; *diagonallybarred arrows*, synctial theories such as Hadzi's and Hanson's. On the right, the major questions raised by all these opposing theories are clarified.

routes from protists. It is very likely that some of the modern thiobenthic representatives are probably the result of secondary invasion of the habitat.

Boaden's theory has little to say on the actual mechanisms of metazoan origins. He in fact postulated a separate origin for non-bilateral groups that are absent from the thiobenthos (poriferans, cnidarians and ctenophores), though this would probably not be an essential part of a thiobiotic ancestry for metazoans. Further discussions of the thiobenthic theory are given by Plat and by Glaessner, whilst a severe critique is presented by Reise & Ax.

There are, then, almost a surfeit of possible theories as to the origin of the animal kingdom from the Protista, summarised in (Fig. 1.11). The process may have been mono-, di-, or polyphyletic, from ciliate or flagellate stocks. The first metazoan may have been benthic or pelagic, radial or bilateral; and it may have been a blastaea, gastraea, plakula, planula or acoeloid. The Figure 1.12 B indicates that there are certain additional key controversies, centred on the relationships amongst the cnidarians, and between 'higher Bilateria and the flatworms. Since the aim here is to make sense of the various theories and relate extant groups of lower metazoans to them, some of these controversial issues, and the relevant evidence such as it is, are examined in more detail below.

1. Whether the first metazoans were radial or bilateral

In some older classifications of the animal kingdom, phyla have actually been segregated according to their primary symmetry, three of the groups of lower metazoans – sponges, cnidarians and ctenophores – being united as the Radiata,. In this way they are distinguished from all the remaining bilateral groups (wormlike forms, and others which though apparently radial are clearly derived secondarily from bilateral ancestors). Often this view carried with it the suggestion that radiality in the three 'lower phyla preceded the evolution of bilaterality, implying that the first metazoans were radially symmetrical (Fig. 1.12).

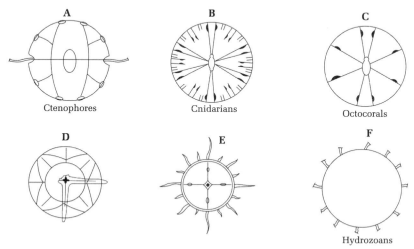

Figure 1.12
Various patterns of symmetry in radial groups of lower metazons. The ctenophores **A.** Show clear biradiality, whilst many cnidarians, **B-F.** Also have biradial axes, particulary in the disposition of mesenteries in actinarian anemones **B.** and octocorals **C**, only becoming truly radial in the advanced hydrozoans **F.** Where mesenteries are absent.

Many texts retain the idea of fundamental and primitive radiality in sponges (though many species have become irregular) and in diploblastic cnidarians and ctenophores (though a degree of biradiality is allowed as a secondary development). And in effect, all the theories that were outlined above, with the exception of the syncytial ciliate/acoeloid theory of Hadzi, begin with a radial ancestor. Hence for most authors, only the timing of the appearance of bilaterality is in dispute: it arises somewhat earlier in Jägersten's bilaterogastraea theory, and in the planula theories, than in Remane's and Marcus' versions of the gastraea theory.

Clearly there is no direct evidence as to the symmetry of the earliest metazoans, virtually all extant metazoans do show some element of bilaterality, or at least biradiality, in their organisation. The supposed Radiata are certainly a somewhat suspect group. Sponges are rarely symmetrical in any sense at all, but have elements of biradiality in their cleavage and in the amphiblastula larva; and ctenophores are very often clearly biradial. Even the cnidarians show clear longitudinal axes of symmetry in respect of the disposition of their mesenteries and muscles in all cases except some hydrozoan polyps, and the latter are among the most sophisticated of their kind, unlikely candidates as ancestral types. For these reasons, the planula/acoel type of metazoan precursor, of essentially biradial form, seems the most logical starting point; this is very much the conclusion reached by Salvini-Plawen, in his review supporting a blastaea/planula origin.

2. Whether the first metazoans were radial or pelagic or benthic?

This question relates very closely to the preceding one, in that it is widely taken for granted that pelagic forms are commonly radially symmetrical whilst epibenthic forms – living on the surface – tend to be bilateral (unless they are actually sessile or become interstitial in which case radiality may be resumed).

Pelagic ancestry is therefore commonly assumed, because in most theories radiality has already been taken for granted. Such an assumption was inherent in the views of Jägersten and Siewing reported earlier, where a pelagic radial blastaea form gave rise to a bilateral phase on becoming benthic. Indeed Jägersten's whole theory of life-history evolution revolved around this principle, with the newly benthic adult retaining a pelagic larval stage. So it might be assumed that if the ancestor was a radial blastaea/gastraea it was probably pelagic, and only if it was a bilateral acoel or planula by the direct routes envisaged by Hadzi or Reutterer was it likely to be benthic. Boaden of course also adopted a benthic ancestor, but for quite different reasons.

The basic assumption underlying all this is clearly an over-simplification, and will not bear close scrutiny. The ciliate forms from which a syncytial metazoan might have grown are not noticeably benthic, whilst the many small pelagic larvae now extant are rarely fully radially organised themselves, even in groups which may perhaps be primitively radial. Either type of ancestry seems equally possible: the phytoplankton may have formed a good food source favouring pelagic forms, as so many early theories required, but the surface deposits and associated decomposing bacteria and protists available to benthic animals are frequently more concentrated and easier to collect. Benthic animals would perhaps achieve higher growth rates for this reason, and since size increase may be an important factor in favouring multicellularity a benthic origin could be strongly advocated. But it is never going to be possible to decide on this issue; it must be left in abeyance, merely noting that whether the ancestral forms were benthic or pelagic is not so simply allied with their symmetry as some would like to believe.

3. Whether the first metazoans were flagellate or ciliate

The concern here is with the issue of monociliate (commonly called flagellate) cells, with a second centriolar structure in the basal body of the cilium giving a diplosomal condition; and multiciliate cells, with monosomal basal bodies. It is central to the issue of metazoan origins, for within the classic groups of lower metazoans both types of cell occur. Monociliated cells are characteristic of sponges, placozoans, cnidarians and gnathostomulids, and of course of flagellate protists; multiciliated cells occur in ciliates, and in all the spiralian groups including platyhelminths. Both types of cell occur in various kinds of gastrotrichs. Hence it must either be assumed that the ancestor was of one type and that in some derived lineage a switch to the other type occurred or that there were at least two separate origins for the Metazoa, some phyla deriving from a flagellate/monociliate ancestor and some from a ciliate form.

Since most protozoologists agree that flagellate/amoeboid protists are ancestral to all others, the arguments in fact centre around whether multiciliation was established before (Hadzi) or after (most others) the evolution of multicellularity. In the latter case, it must have arisen independently in the ciliates and in the early Metazoa, some of the pertinent arguments; for example, there can be little doubt that transformations between the two conditions have occurred at various points in evolutionary history (e.g. amongst the gastrotrichs and bryozoans, and en route from early chordate forms to the vertebrates). So it is not unlikely that a rather fundamental divergence into the different cell types should occur after a unitary origin of the metazoans. The most likely hypothesis would of course be that this unitary origin was from a flagellate, as the radiate groups and sponges are normally taken to be most primitive. The planuloid ancestor would probably have to be the point at which multiciliation then arose; some planulas could lead to the acoel flatworms, perhaps via gnathostomulid and gastrotrich-like groups Rieger and others to the ctenophores. This view resembles that advocated in Salvini-Plawen's review, where he maintains that uniflagellar (monociliate) diplosomal cells are a fundamental character of all true metazoans (making no mention of the rather obvious exceptions).

The alternative view would require a multiciliate monosomal ancestor derived from a ciliate protist, as Hadzi proposed. In fact modern ciliates have unusual and specialised basal bodies and seem implausible as direct ancestors for the multiciliate/monosomal condition. Instead an opalinid form might be invoked, or some other group of now extinct multiciliate protists derived from the flagellates – which would also avoid the criticism that modern ciliates are implausible ancestors because of their two types of nuclei and curiously complex conjugation-based sexual systems. This 'ciliate' ancestor would directly give rise to acoels and higher spiralians. But it would have to convert subsequently to the monociliate state and gain the second centriolar structure at least **three times**:

1. in cnidarians (and hence perhaps in the deuterostome groups if these were descended from the same stem)
2. in gnathostomulids, close to and probably primitive to the platyhelminths
3. in sponges (though Hadzi believed these to have been of independent origin reason given that they differ from all other diplosomal metazoans in the disposition of their second basal apparatus centriole.

Hadzi did not particularly concern himself with the ciliate or flagellate nature of cells in the lower metazoans, and other authors have not taken up seriously the implication that the first metazoans were multiciliate forms. If they were, the problem of explaining how all metazoans appear to have primitively uniflagellate spermatozoa is at first sight inseparable. But Hanson has reconsidered this issue, and points out that the widespread assumption of a single 'primitive' sperm type is itself largely dependent on underlying assumptions and preconditions about early metazoan phylogeny, and may not be tenable.

There remains the possibility that the two types of cell occur in the Metazoa because this is a polyphyletic assemblage, and that different ancestors were either mono- or multiciliated independently. This would at first sight be no more unlikely than either of the views so far considered; in fact Hadzi's phylogeny is already effectively diphyletic, with just sponges arising from flagellates, and many other theorists have in the past proposed the sponges to be a separate lineage from protists. It may only be necessary to extend these views somewhat, to ally cnidarians, sponges, gnathostomulids and placozoans with flagellate ancestors (perhaps the first pair from a gastraea type, and the second pair from planulas), and platyhelminths and ctenophores with ciliates, to be able to side-step many of the awkwardnesses imposed by a monophyletic phylogeny. The constant occurrence of a 'ciliary necklace' at the base of the axoneme in metazoans, a feature shared with flagellates but absent in ciliates, may seem damaging to such a polyphyletic view; but again an ancestral multiciliate protist derived from flagellates but not part of the highly modified Ciliophora could provide a way around the problem.

On this issue then, a reasonable conclusion would be that ancestral metazoans may have been just flagellate, are rather unlikely to have been just ciliate, but could quite possibly have been of both types (though not actual ciliophorans) independently

4. Whether the first metazoans were diploblastic or triploblastic

The theories considered have very different implications with respect to the nature of the first metazoans. Either they were simple 'diploblastic' gastraeas with ecto- and endoderm, readily transformable into classically 'diploblastic' cnidarians, or they were solid planula/acoeloid forms, perhaps with little internal differentiation but clearly with the' immediate possibility of producing a three-layered state. One of the main criticisms of Hadzi's views has always been the implausibility of commencing with a triploblastic group (the platyhelminths) and trying to derive diploblasts from it by an apparently regressive step.

Certainly the progression from unicell to diploblast and on to triploblast has a more immediate logic and the appeal of orderliness.

There are several important issues to be dealt before concluding whether the ancestors were diploblastic or triploblastic.

Firstly, the actual 'diploblast' nature of the cnidarians and ctenophores has long been a matter for dispute anyway. Both groups have a layer of noncellular mesogloea between ectoderm and endoderm, but in both this is invaded by cellular elements from the two epithelia, particularly in the anthozoan cnidarians. Nerve cell processes traverse the mesogloea, and the contractile tails of the cnidarian myoepithelial cells extend into it. Wandering amoeboid cells transport nutrients through it. In some respects it forms a perfectly good third layer of tissue; and the distinction between diploblasts and triploblasts becomes blurred.

Secondly, the old idea that cnidarians and ctenophores formed a close-knit group, within the Radiata or even as a phylum together (Coelenterata) has been largely abandoned recently (Harbison). They have nothing more in common than a very general plan, and no true synapomorphies exist (Salvini-Plawen). Cnidoblasts and colloblasts may perform similar functions but are clearly no more than analogous. Both groups have tentacles, but in cnidarians these are often hollow and are formed from both ectoderm and endoderm, whilst in ctenophores they are solid and purely ectodermal growths. Ctenophores have true muscle, not myoepithelial cells. And cnidarians are clearly monociliate/flagellate, whilst ctenophores have complex multiciliation.

Thirdly, it is perhaps not so implausible that triploblasts could appear to 'regress' into an almost diploblastic state. The mesogloea of cnidarians and ctenophores could readily be derived from a true mesoderm rather than from an acellular blastocoelic matrix. Neither the nerve nor muscle systems of these two groups are now seen as being so different from other metazoans as was once the case. Apparent radiality is itself a suspect character and of doubtful primitiveness. However, one issue cannot be readily explained away: flatworms have orderly spiral cleavage, whereas cnidarians show little or no trace of this, and it is difficult to see why this process should have been abandoned if cnidarians are indeed derived from platyhelminths.

Nevertheless, the cnidarian/ctenophoran designs suit particular ways of life, and the unique cnidoblasts and colloblasts respectively give benefits that may render more sophisticated architectures, locomotory systems and nervous systems (and perhaps even embryological patterns) superfluous. Thus regression from an apparently 'superior' triploblastic flatworm-like ancestor does not appear as incongruous in the light of some detailed modern knowledge as most earlier theorists maintained; and the whole question of whether ancestors were diplo- or triploblastic may be a rather unimportant matter of semantics.

5. Whether the first metazoans were syncytial or colonial

Either system of producing the first multicellular animals would be feasible in theory; perhaps the best approach here is to examine the processes as they actually occur in the protists, and in lower metazoans.

Colonial aggregations occur quite commonly in flagellate protists (Mastigophora), particularly in the plant-like phytoflagellates. Choanoflagellates also aggregate, often as stalked colonies, though Sphaeroeca is pelagic and the genus Proterospongia (Fig. 1.14) includes forms where the cells are embedded in a central gelatinous mass. Amoeboid protozoans, (Sarcodina), though closely related to the flagellates, do not have living colonial representatives; and examples are rare and usually rather specialised in the ciliates. Hence if a colonial origin is to be proposed, a flagellate ancestor certainly seems the most likely starting point; though it is a pity that extant protist colonies are all either plants or fresh-water animals.

A multinucleate condition that might precede multicellularity by subsequent cellularisation is quite common in protists. Several groups of zooflagellates contain two or more nuclei (notably diplomonads and oxymonads, though both are primarily symbionts or parasites). Many amoebas and heliozoans are multinucleate; and in foraminiferans the multinucleate protoplasm often gets divided up into a form of cell by membranes, though there may be no correspondence between the nuclei and these 'cells'. Ciliates have both macro and micronuclei, only the latter being concerned with reproduction. The micronucleus is often single, but up to 80 have been reported in some species. A few, such as Stephanopogon, possess only micronuclei, and in considerable numbers. And opalinids are particularly clear examples of multinucleation, having many identical nuclei within a large flattened multiciliated body. They would appear good candidates as plasmodial precursors for a syncytial metazoan stock, via Hadzi's route, but for their curious habits as commensals in the guts of frogs. However, since flagellate protists may have been the source of amoeboid and ciliate groups, as well as opalinids, we could return to the suggestion that another multiciliate and multinucleate group may have been derived from flagellates that in turn gave rise to metazoans. Thus either a colonial flagellate or a multinucleate 'derived flagellate/protociliate' seems plausible as the rootstock of metazoans.

However the latter proposal relies heavily on Hadzi's assumption that the most primitive living metazoans also exhibit ancestral syncytial organisation. This was long held to be true of the gut area of acoels, though their sexual, glandular and nervous cells appeared distinct . However, electron microscopic studies have blurred the whole issue of syncytial/cellular status, and some authors have denied that the Acoela are any more syncytial than other invertebrates. In any case, it is now clear that their syncytial tissues are secondary phenomena, as the gut region is clearly cellular during its early ontogeny. Secondary syncytialisation is a condition also found in other small interstitial metazoans, where it seems to be a derived condition relating to size and lifestyle. If acoel guts are not primitively syncytial, then Hadzi's views certainly lose a good deal of their force, though we cannot expect modern acoel turbellarians necessarily to reflect the condition of the first metazoans.

On balance, there are clearly problems with either theory given the present occurrence of colonial and multinucleate conditions in protists and simple metazoans. But at least one of these processes must have occurred to produce multicellular animal life; the only conclusion can be that both are possible and neither is inherently more unlikely than the other.

6. Whether the first metazoans were coelomate or acoelomate

A number of the theories outlined require the archemetazoan to be a small coelomate form with enterocoelic pouches derived at the gastraea stage from a cnidarian precursor. Acoelomates, and particularly flatworms, must then be secondary.

Archecoelomate theories do not bear up well in terms of functional considerations, as Clark clearly demonstrated. The earliest metazoan must surely have been very small, and would have moved efficiently with ciliary action; the acquisition of one or several fluid-filled cavities would not give mechanical benefits at this size range. Thus if the main benefit of a coelom is mechanical, as a hydrostatic skeleton, these ancestral forms should not have required or benefitted from their enclosed pouches. Alternatively, if the coelom serves mainly as a means of permitting size increase, then its acquisition at such an early stage is less implausible. But in that case its subsequent loss, with a return to ciliary-based locomotion and a loss of circulatory system and anus, by ctenophores, platyhelminths and some other groups not of notably minute size or interstitial habit, seems illogical and inconsistent with any conceivable selective pressures. And whatever the primary function of the coelom, it would have had to be a powerful necessity that overcame the disadvantages of actually sealing off gastric pouches whose initial function must have been to increase the absorptive surfaces available to the animal – particularly in an enlarging animal still lacking an organised through-gut! Enterocoely seems functionally most implausible in such early animal forms. The number of pouches proposed in the various theories also seems to raise rather than solve difficulties; whether there were four, five, six or eight initially, it is hard to explain why extant forms display such varieties of single paired, oligomeric or metameric coeloms nowadays. By contrast, if the coelom is indeed polyphyletic as seems only logical, then this diversity of modern coelomic organisation is only to be expected. Given polyphyletic coelom origins, a theory for metazoan origins that proposes an early and unitary development of coelomic pouches cannot be upheld. The platyhelminths probably preceded other worm-like groups and were the ancestral stock in which coelom formation was initiated; enterocoelic gastraea ancestors are not acceptable postulates.

The ancestral metazoan, then, remains enigmatic. Biradiality, and a relatively solid acoelomate form, seem most likely, and either a pelagic or benthic lifestyle would be perfectly plausible though the latter might offer greater food resources for growth. A colonial, flagellate/monociliate condition is most likely if all groups share descent from one ancestor, but multiple origins by different routes, including multinucleate ciliates, may be at least equally plausible. Now we need to pursue these possibilities in terms of their implications for evolution amongst the lower metazoan groups, and test them against available evidence from fossils and living animals.

The hypothetical first metazoan, the protometazoan (Fig. 1.14 A), may have differed from the premetazoan (Fig. 1.14 C) in several ways.

1. First, the surface cells probably closely adjoined or were in contact with each other (Fig. 1.14), thus facilitating intercellular communication and providing a regulatory barrier between the external environment and the ECM in the interior of the body.

2. Second, the close association of adjacent cells largely excluded the ECM from between the cells, thus separating the ECM into external and internal layers, each of which could then adopt independent functions (Fig. 1.14 D).

3. Third, the body was polarized along the anterior-posterior axis

4. Fourth, the separation of layers and body polarity promoted cell specialization (Fig. 1.14).

More recently, Greenberg and preston held that probably all Metazoa have not originated from any single ancestor. They have suggested a polyphyletic origin for the metazoans. It is fairly certain that

a) sponges developed by way of colonial flagellates e.g. Proterospongia,

b) Whereas other multicellular groups originated from the cellularization of syncytial protociliates or perhaps the mesozoans.

Evolution and Origin of Polarity in Metazoans

Most motile protozoans are polarized cells that have leading (anterior) and trailing (posterior) ends or if sessile and attached, they have oral free ends and aboral attached ends. Metazoans are similarly polarized, depending on whether they are motile or sessile, but how did the polarity of the multicellular metazoan body evolve from the unicellular polarity of protozoans? A clue to the answer is found in the eggs of several groups of metazoans. During oogenesis in these groups, the eggs express a rudimentary flagellum and a collar of microvilli at a site on the cell surface that corresponds to the animal pole of the zygote. In certain invertebrate taxa, the animal pole corresponds to the anterior end of the larva (although it is the posterior end in others). The current evidence suggests a line of descent from the polarity of a choanocyte to the primary anterior-posterior polarity of the metazon body (Fig. 1.13).

The protometazon probably was polarized along an anterior-posterior axis, but what environmental conditions might have selected for the evolution of such polarity? For an aquatic metazoan, the environment presents itself in gradients of light, temperature, oxygen, and food availability. If, for example, the availability of food is related to light (through photosynthesis), natural selection would favor any variant individual capable of detecting and following a light gradient. This is best accomplished with a direction-sensitive sensory capability coupled with a directional (polarized) locomotory system. Thus, selection may have favored any organism capable of tracking a resource concentraton gradient.

Wolpert suggests that body polarity evolved from attachment to a substratum. Attachment to a rock in water, for example, places an organism at an interface, a very steep gradient. Once attached, variants would be favored that adhere well at the attached end and perform other tasks, such as feeding at the opposite end. This again leads to a polarized body.

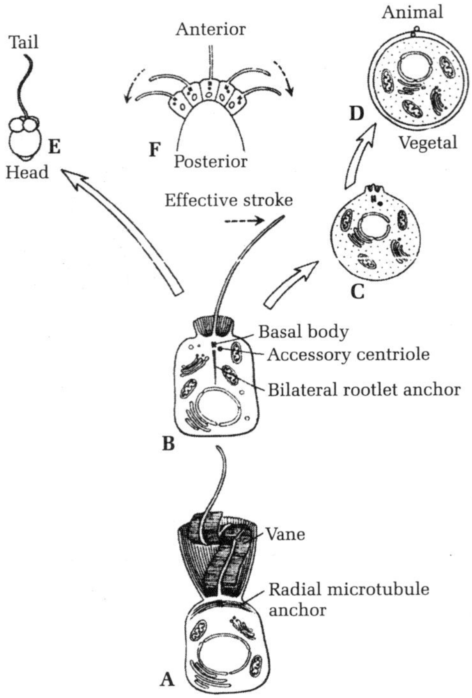

Once polarity was established, movement would create an environmental gradient along the locomotors axis that would favor differential expression of traits. For example, enhanced membrane sensitivity to environmental stimuli might be favored in cells at the anterior end of the body because they are the first to encounter changes in environmental conditions. Similarly, enhanced flagellar growth, density, or activity might be favored in cells at the equator, or widest part of the body, since those locations best contribute to locomotion. Cells with a capacity for division, leading to growth, might be favored at the posterior end, because in that position they contribute to and interfere least with locomotion. Thus, motility along a polar axis may itself promote cellular specialization because the cells occupy different fixed positions in an environmental gradient.

Figure 1.13

Origin of *metazoan polarity*. **A.** *Primitiye* collar cell with a flagellar vane, as occurs in protozoan choanoflagellates and metazoan sponges. **B.** Choanocyte-like monociliated cell from the surface epithelium (epidermis) of an anemone larva (Cnidaria), showing its power stroke in the direction of the accessory centriole. **C.** Early oocyte of a sea cucumber (Echinodermata) showing a low collar and rudimentary cilium; **D.** Sea cucumber zygote in the same orientation as the oocyte in C. Note that the animal and vegetal poles correspond to the apex and base, respectively, of the epithelial cell in **B. E.** A typical metazoan sperm showing its polarity in relation to an epithelial spermatogonium from which it arose. **F.** Anterior end of a simple larva (Cnidaria). Note the position of the accessory centrioles, which indicate posteriorly directed ciliary effective strokes and thus forward movement in an anterior direction.

According to Leo Buss; the origin of metazoan cell specialization may be related to a conflict between the demands for growth and locomotion. *Volvox, Proterospongia,* and planktonic blastula stages of metazoans require flagellated surface cells for locomotion, but most flagellated cells cannot divide by mitosis because the centrioles needed to form the mitotic spindles are already in use as the flagellar basal bodies. A metazoan flagellated cell can divide only *after* the flagellum regresses and its basal bodies are freed to form the mitotic apparatus. Thus, the options for **growth in a premetazoan** composed solely of flagellated cells (for example, the species of *Proterospongia* in (Fig. 1.14 A)) are:

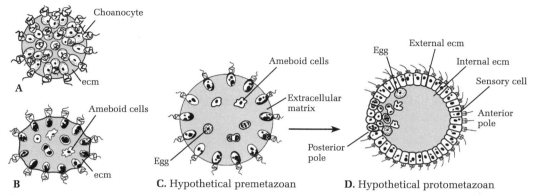

Figure 1.14 A
Origin of Metazoa: Choanoflagellate colonies metazoan evolution. **A** and **B** different forms of the choanoflagellate colony *Proterospongia haeckeli*. **C** A hypothetical premetazoan based on the choanoflagellates in **A** and **B**, **D**. A hypothetical protometazoan showing cellular specialization along an anterior-posterior axis. Surface cells are bound together and in mutual contact, thus allowing physiological regulation of the interior of the body.

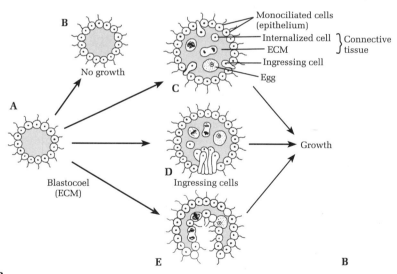

Figure 1.14 B
Origin of cell and tissue specialization: in *hypothetical protometazons*. **A.** Blastula-like developmental stage of a hypothetical organism; **B.** Hypothetical blastula-like organism that does not grow because its cells retain locomotory cilia and thus cannot divide by mitosis. **C-E.** Growth by internalizing nonflagellated mitotically active cells in three simple hypothetical metazoans. The patterns in **C-E** also represent three basic forms of gastrulation in actual animals. **C**, multipolar ingression; **D**, unipolar ingression, **E**, invagination (or emboly). Note how the embryonic blastocoel (ECM) is equivalent to adult connective tissue.

1. enlargement of existing cells;
2. disassembly of flagella, cell division, and then flagellar reassembly;
3. division of a few cell scattered throughout the embroyo while others retain flagella; or
4. division of a few localized cells set aside for growth (Fig. 1.14)

Option I, because of this restricted number of cells, limits ultimate body size, although *Volvox* daughter colonies and a few postembryonic micrometazoans grow by cell enlargement only.

Option 2, by requiring the regression of flagella, compromises locomotion.

Options 3 and **4** both permit growth and locomotion, but most invertebrate metazoans have adopted option 4.

The setting aside of mitotically active cells enables metazoan growth without compromising locomotion (Fig. 1.14 D) (A related evolutionary event was the setting aside of germ cells capable of meiosis.)

Surface flagella are a functional necessity for locomotion in the protometazoan, but what is the optimal position for the set-aside growth cells?

Buss suggests that placement at the surface might have resulted in overgrowth of locomotory cells or growth of a tumor like appendage, either of which would negatively affect motility. Internalizing the growth cells, however, would neither distort the body surface nor interfere with the locomotory cells. Perhaps for these reasons the protometazoan did internalize its growth cells and the process was preserved, as gastrulation, in its descendants ;. An internalization of cells capable of mitosis also evolved in *Volvox* and *Proterospongia haeckeli* (Fig. 1.14 B). A colony of *P. haeckeli* grows by the addition of new surface choanocyes that originate by mitosis of cells in the subsurface gel. Similarly colonies of *Volvox* consist of a thousand or more flagellated surface cells and invaginated subsurface. Pockets of nonflagellated dividing cells (gonidia). The gonidia can give rise to either new colonies or germ cells.

The simultaneous requirement for locomotion and growth in early metazoans favored a body with two layers of differentiated cells (Fig. 1.14 D). The layering of cells established tissues and an exterior-interior gradient. The surface cells had direct access to gases and other raw materials, but were constrained by the requirements for locomotion and for interfacing with the external environment. The interior cells, liberated from direct interaction with the environment, became specialized for reproduction, nutrient storage, and, with the evolution of a gut, the digestion of food.

There are three basic type of Body Plan

1. **The cell aggregate plan**—Body has large number of cells without clear distinction and the division of labour amongst them, There is no coordination amongst the cells as the tissues and organs are not formed *e.g.,* sponges (Fig. 1.15 A).
2. **The blind-sac plan**—Animals have an alimentary canal with one opening only to take in food and also to remove the undigested food *e.g.,* cnidaria and flatworms, the body cells are more specialized into tissues and organs and show the phenomenon of division of labour (1.15 B).

 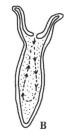 

Figures 1.15 A, B, C
Proposed types of body plan (amongst animals): **A.** A aggregate body plan, **B.** Blind-sac plan, **C.** Tube-within-a-tube plan.

3. **Tube-within-a-tube plan**—The alimentary canal is like a tube with one opening to take in the food (mouth) and a separate opening to remove the undigested food (anus), the food is digested and absorbed within the alimentary canal. This alimentary canal lies within the coelom or body cavity, hence the name tube-within-the tube.

   Animals are divided into two categories (1.15 C).

   a) **Protostomes** in which mouth opening develops first and then the anal opening embryonically. Mouth arises from the blastopore. It is seen in annelids, molluscs and arthropods.

   b) **Deuterostomes** in which mouth develops later on but anus develops first embryonically, from the balstopore. It is seen in echinoderms and chordates.

Metazoan Organization

The gross external morphology of metazoan animals show various patterns. These include:

1. (***symmetry***) (Fig. 1.16)

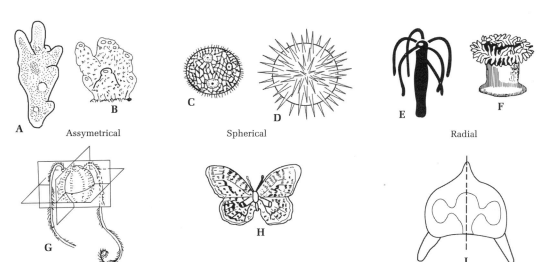

Figure 1.16
Types of symmetry in animals.

2. (***cephalization***), and polarity.
3. (***metamerism***), (Fig. 1.17)
4. grades of organization.

Similarly, the criteria of internal morphology are

5. differences in formation of body cavity or coelom

1. Symmetry and its Significance

Symmetry means an arrangement of body parts into geometrical designs. It refers to the division of body into equal parts by lines or planes (Fig. 1.16). Current zoological ideas about symmetry derive chiefly from Hackel. An animal is called *symmetrical* when a plane passing through its center will divide it into similar halves. When an animal cannot be divided into like parts by a plane, it is called asymmetrical. All animals are either asymmetrical or symmetrical. Examples of asymmetrical animals are most sponges, some Protozoa *(Amoeba)* and few others.

When explaining symmetry, An **axis** is an imaginary line passing through the center of body, such as **longitudinal axis** and **oral-aboral axis**. Either end of an axis is termed a *pole*. Thus, each axis has two poles. A *plane* of symmetry is a straight line that divides organisms into corresponding halves.

Metazoa display only two types of symmetry, **radial** and **bilateral** (Figs. 1.16 E-I). Two other types of symmetry are also recognized, **spherical** and **biradial**. Protozoa are not only asymmetrical but display all four types of symmetry in their diverse body forms.

1. **Spherical symmetry.** It is in animals whose body has the shape of a sphere (Fig. 1.16 D). All planes that pass through the center will cut it into similar halves. Some protozoans (e.g. Volvox, Heliozoa, Radiolaria) have spherical symmetry, and it is adapted for free-floating or rolling movements.

2. **Radial symmetry.** The body is in the form of a flat or tall cylinder (Fig. 1.16). Many similar body parts, called **antimeres**, are arranged around one main, central or longitudinal axis in a circular or radiating manner like the spokes of a wheel. All the lines passing through this longitudinal axis, in any plane, will divide the body into equal halves or **antimeres**. The surface having mouth is the oral surface, and the opposite surface is the abroal *surface*. Examples are echinoderms and most coelenterates (e.g. hydra)

Radial symmetry is best suited for a sessile existence. Most of them are attached by the aboral surface. Some are free-swimming but remain at the mercy of water currents. Due to similarity of antimeres, their sensory receptors are equally distributed all around the periphery. This enables them to receive stimuli and to meet the challenges of the environment equally from all directions. They can obtain food or repel enemies from all sides.

In the animal kingdom, radially symmetrical phyla are Porifera, Coelenterata, Ctenophora and Echinodermata. Out of these, only Coelenterata and Ctenophora display a fundamental radial symmetry. Both the phyla were grouped together by Hatschek (1888-91) under the Division **Radiata**. Adult Porifera are mostly asymmetrical, but they start life from a radially symmetrical larva. On the other hand, larval stage of Echinodermata has bilateral symmetry, but the adults become radially or pentaradially symmetrical.

3. **Biradial symmetry.** It is a variant form of radial symmetry found in Ctenophora and most. Anthozoa (Fig. 1.16 G) (e.g. anemones), and is best fitted for a floating life. Such symmetry has only 2 pairs of symmetrical sides. There are only 2 planes of symmetry, one through the longitudinal and sagittal axis, and the other through the longitudinal and transverse axis, which will divide the animal into equal halves.

4. **Bilateral symmetry.** In most higher animals, the longitudinal axis of body runs from the anterior end (head) to the posterior end (tail). There is a single plane, the *median longitudinal* or sagittal plane, through which the body can be divided into two similar right and left halves. This is called *bilateral symmetry* (Fig. 1.16). Besides right and left sides, an upper or *dorsal* surface and a lower or *ventral* surface are also recognizable, which are unlike because they are exposed to different conditions.

Bilateral symmetry is characteristic of the successful and higher animals, including the remaining invertebrates and all vertebrates. In most of them, the anterior end is differentiated into **a *head.***

First phylum of animal kingdom to exhibit bilateral symmetry is the phylum Platyhelminthes. All bilaterally symmetrical metazoans were grouped together by Hatschek under the **Division Bilateria**. As mentioned earlier, some Bilateria, such as echinoderms, display a radial symmetry which has been secondarily derived from bilateral ancestors due to assumption of an attached mode of life by adults.

Cephalization and Polarity

Bilateral symmetry is correlated with the locomotor movements brought about by these animals. One end of their body, usually containing the mouth, always moves forward in a particular direction (Fig. 1.17). It is the first to come in contact with the environment, so that there is great concentration of nervous tissue and sense organs at this anterior end called **head**. The posterior or rear end is usually equipped with some locomotory organ. This modification of anterior or oral end of the animal into a definite head is called ***cephalization*** which is characteristic of most bilateral animals.

Cephalization is always accompanied by a differentiation along an antero-posterior or oral aboral axis. This condition is known as ***polarity***, and it usually involves gradients which refers to ascending or descending activities between anterior and posterior ends.

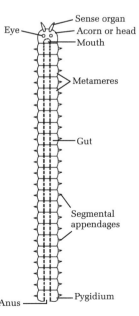

Figure 1.17
A metameric animal, in (annelid)

Metamerism

When the segmentation in bilateral animals, such as annelids, involves a longitudinal division of the body into a linear series of similar sections or parts, it is termed **metameric segmentation** or **metamerism** (Fig. 1.17). Each section or part is called a *segment, somite* or *metamere* typically repeats some or all of the various organ units. The term **metamerism** is applied only when organs of mesodermal origin are so arranged. The primary segmental divisions are the body wall musculature and sometimes the coelom. This in turn imposed a corresponding metamerism on the associated supply systems (nerves, blood vessels and excretory organs). Longitudinal structures such as gut, principal blood vessels and nerves extend the entire length of body, passing through successive segments. While other structures, such as gonads, are repeated in each or few segments only.

Metamerism is always limited to the trunk region of the body. The *head is* represented by the prostomium and bearing the brain and sense organs, and the *pygidium,* (Fig. 1.17) represented by the terminal part of the body which carries the anus, are not metameres. New segments arise just in front of the pygidium; thus the oldest segments lie just behind the head.

Metameric animals

Metameric segmentation of the body, encountered for the first time in Annelida, is of considerable interest because the most successful groups of animal kingdom, i.e., Arthrophoda and Vertebrata, also have their parts metamerically repeated. At least one group of Mollusca (Monoplacophora) also exhibits metamerism.

Metameric segmentation seems to have evolved three times independently in animal kingdom

    (i) in the annelid-arthropods,

    (ii) in the chordates and in

    (iii) cestodes.

External and internal metamerism

Metamerism is conspicuously visible in most annelids, both **externally** as well as **internally**. The common earthworm is a good illustration of both external and internal metamerism. Its body consists of a great number of similar segments and all the body organs, such as musculature, setae, blood vessels, nerves, ganglia excretory organs and gonads, etc., are repeated segmentally. Even the coelom is divided into segmental compartments by the intersegmental transverse mesenteries, called *septa.* Only the digestive tract remains unaffected, but it also extends through every segment. In arthropods, metamerism is chiefly *external,* while man and other vertebrates show an *internal metamerism* of body muscles, nerves, certain blood vessels, vertebrae and ribs.

Complete and incomplete metamerism

In annelid worms, metamerism is *complete,* affecting practically all the systems. The metameres are essentially alike or **homonomous,** each having segmental blood vessels, nerves, nephridia and coelomoducts. This condition is called ***homonomous metamerism***. On the other hand, higher animals such as arthropods and vertebrates, show ***incomplete metamerism***. Because of division of labour, the segments or metameres of different regions of their body become greatly dissimilar. This is called ***heteronomous metamerism***. However, incomplete metamerism should not be confused with the repetition of single organs such as shell plates or gills in certain unsegmented animals like molluscs.

In arthropods and vertebrates, metamerism is more complete and metameres are uniform and clear in the larval and embryonic stages. But, metamerism becomes obscure in the adult due to subsequent specialization or modification, so that the segments are no longer similar. It may result from simplification, by loss of metameres, by fusion of segments (cephalization), by differentiation between segments, by disappearance of organs, or by development of other structures, such as limbs. Heteronomous condition always appears first at the anterior end and progresses posteriorly. In segmented animals, varying degree of specialization are met with some of which are extreme.

Origin and evolution of metamerism

Various hypothesis have been proposed to explain the origin of metamerism, but none is acceptable in the absence of convincing evidence. The main theories concerning the origin of metamerism emphasize primarily either repetition of organs or mesodermal segmentation and correlate it with the origin of coelom.

Pseudometamerism theory

According to this theory metamerism developed secondarily as a result of repetition of body parts, such as muscles, nerves, nephridia, coelom, blood vessels etc. in a single individual.

Such serial repetition of organs, such as testes, yolk glands and transverse connectives of two nerve cords, is seen in some elongated turbellarians and nemerteans. Later, a segmented condition was obtained by the formation of cross-partitions in between them, so that each segment received a repeated part of each system. This process is witnessed even today in the formation of somites in larval and adult stages of some Annelida, in which cross-partitions develop after the basis segmentation is already laid down. Such segmentation was probably an adaptation for an undulatory mode of swimming. However, all ribbon-like animals swim in this way, whether segmented or not. This theory is supported by Hyman and Goodrich.

Cyclomerism theory

Proposed by Sedgwick and supported by Remane , the cyclomerism theory (Fig. 1.18) is a corollary of the enterocoelous theory for the origin of mesoderm and coelom. This theory assumes that coelom originated in some ancestral radiate actinozoan coelenterate, through the separation of four gastric or enterocoelic pouches from the central digestive cavity or gut. Division of two pouches resulted into three pairs of coelomic cavities–the protocoel, mesocoel and metacoel, in the protocoelomate or ancestral coelomate. Loss of protocoel and mesocoel led to the unsegmented coelomates, such as molluscs and sipunculids (Figs. 1.18 DE). Later subdivision of metacoel produced primary segments, leading to the segmented annelids. The phylogenetic implication of this theory is that all bilateral

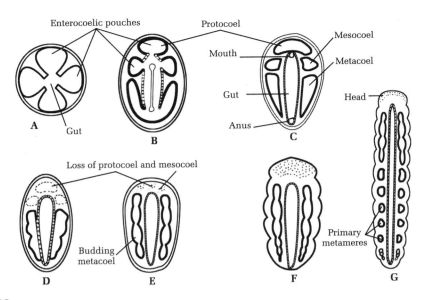

Figure 1.18
The *Enterocoel-Cyclomerism* theory illustrating of metamerism. **A-C** Transformation of enterocoelic pouches of a radiate ancestor into coelomic pouches of a bilateral ancestor. **D-G** Loss of protocoels and mesocoels to give rise to an unsegmented coelomate such as the mollusks; (D-F) Elongation of the metacoel and its division into paired compartments yields the annelid arthropod line according to the cyclomerism theory.

metazoans were originally segmented and coelomate, and that the acoelomate unsegmented groups (flatworms, nemerteans) have lost these characters secondarily.

Corm or fission theory

According to this theory, metameric segmentation resulted when some non-segmented ancestor divided by transverse fission repeatedly or by asexual budding producing a chain of subindividuals or zooids (Fig. 1.19), united end to end due to their incomplete separation. This occurs in some Platyhelminthes and Annelida even today. Later, with the passage of time, these subindividuals or segments gradually became integrated morphologically as well as physiologically into one complex individual. A segmented animal, according to this view, is a chain of completely coordinate subindividulas was supported by Perrier. The chief objection to this theory is the lack of gradations of age in such a chain of zooids, which is true of segments in a metameric animal, such as an annelid. In platyhelminthes and scyphozoan strobilae the sequence of zooid formation is never serial, fission occurs always

Figure 1.19
Formation of a chain of zooids in rhabdocoel flatworm *Stenostomum*, according to *Corm Theory*. Numbers indicate sequences of fission planes.

somewhere in the middle of the chain. In cestodes the proglottids are serially arranged but in a reversed order. Another objection is that reproduction by fission is usually confined to sessile animals, whereas the ancestors were probably free-swimming. Moreover fission is a more extensive and disruptive division of body than is metamerism.

Embryological theory

It explains original metameric segmentation mainly as an embryological accident. It suggests that mechanical stresses in the mesoderm of the elongating embryo or larva resulted in its fragmentation leading to segmental repetition of mesodermal derivatives in the adult.

Locomotory theory

It postulates that metamerism evolved as an adaptation to locomotion of different kinds. Annelid metamerism probably evolved as an adaptation for undulatory, serpentine swimming movements. Clark suggests that **coelom** evolved initially as a hydraulic skeleton to facilitate locomotion in response to the increasing body size. According to Clark, metamerism also evolved as an adaptation to burrowing in annelids. The ancestors of Annelida were in all probability elongate coelomate animals which burrowed in marine sand and mud. The evolution of compartmented coelom, due to development of septa and metameric segmentation, localized the function of the hydraulic skeleton. This allowed only part of the body to contract while other parts in the longitudinal axis relaxed. The locomotory movements can be more continuous and better controlled if the action of the body wall muscles is localized, i.e., restricted to sections of the body. Metamerism permits such localization and accounts for the evolution of this condition in annelids. The localization enabled a strong peristaltic wave to propagate down the body, an efficient type of locomotion for worm-like burrowers. Thus annelid metamerism is basically a modification of the coelom and the body wall muscles. The initial segmentation the coelom and body wall muscles led subsequent segmental organization for the nervous, circulatory and excretory systems.

In **chordates** metamerism evolved as an adaptation for undulating swimming movements. According to Berrill , notochord evolved in the early chordates to provide support for the body, followed by metameric segmentation of body wall musculature, so that alternate waves of contraction sweeping down the body could enable strong swimming. The initial muscle segmentation caused segmentation of the nervous and circulatory systems.

In cestodes, metamerism evolved as a response to reproduction. A reproductive package was formed which, once it had performed its function, was expendable proglottids are not additional to the body but are the body and so have to carry a complete series of organs. Metameric segmentation met in one phylum is not necessarily similar to that in another. It probably arose independently in more than one line of evolution, each time in adaptation to a major advantage for the group in question. It probably evolved in response to burrowing in annelids, to swimming in chordates and to reproduction in cestodes. Therefore, we cannot device one explanation for all the cases of segmentation or think of a common ancestor of all sorts of segmented animals.

Significance of metamerism

It is not clear just how and why metamerism evolved or how the primitive ancestors were benefited by it. Probably specialization of metameres for particular functions showed an advancement. It is possible that segmentation was initiated in the musculature of an elongated swimming worm. This breaking up of the body into metameres would facilitate swimming movements. Metamerism helps in locomotion in several ways.

1. The coordination of muscular action and fluid-filled coelomic compartments cause efficient swimming and creeping which is an advancement over the simple ciliary and creeping movement of lower invertebrates. Fluid-filled coelomic compartments also provide hydrostatic skeletons for burrowing. Precise movements can take place by differential turgor pressure effected by flow of coelomic fluid from one part of the body to the other.

2. Another advantage of segmentation or metamerism is the opportunity for different segments to specialize for different functions, thus leading to a rapid evolution of high grade of organization. It is not clearly marked in annelids, but is well developed in arthropods. Metamerism has, therefore, contributed towards the greater complexity of animal bodies and rapid evolution of high organization in animals. Thus, some indication of primitiveness of an animal can be determined by the degree of segmentation it displays.

The fact that cestodes, annelids, arthropods and chordates have metamerism does not necessarily indicate a close relationship among them, for the metameric condition may have arisen independently by convergent evolution.

Pseudometamerism

Pseudometamerism or **strobilization** of the tapeworms refers to superficial segmentation and could be termed as body annulation (Fig. 1.20). True segments of annelids are laid down in the embryonic stage. Whereas *proglottids* of tapeworm are not true metameres but rather complete reproductive individuals produced by strobilization, a type of budding, with the buds remaining attached. Table 1.1 shows the important differences between the two. However, the modern view now gaining favour is that cestodes are indeed metamerically segmented, although their metamerism is of a different type

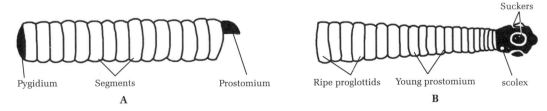

Figure 1.20
Showing differences between **A.** True metamerism of a generalized annelid. **B.** Strobilization of a generalized tapeworm.

Table 1.1: Differences between True Matamerism and Pseudometamerism.

True metamerism	Pseudometamerism
1. Number of segments is generally constant for each species; i.e., new segments are not added to the body after maturation except in asexual reproduction.	1. Number of segments or proglottids forming the body is not fixed as new segments are continually added throughout life.
2. Growth occurs due to simple elongation of pre-existing segments. The segments and end of body have a fixed relationship to one another throughout life.	2. Growth occurs due to addition of new segments from a region of proliferation. Just behind the scolex.
3. All segments are of the same age and at the same stage of development.	3. Proglottids differ from one another in age and in the degree of development.
4. Segments are functionally interdependent and integrated. Working in co-ordination, they preserve the individuality of body. For example, in a worm, during locomotion, muscles of each segment contract in a regular sequence so that rhythmical waves of contraction pass over the whole body which moves forward in an orderly manner.	4. Segments or *proglottids* are independent and self contained units, each having a full set of sex organs and a portion of excretory and nervous systems. They are productive units developed for detachment. A tapeworm represents a sort of colony or strobila made of a linear row of incomplete individuals.

Coelom

A true coelom may be defined as a secondary body cavity, formed by the splitting of mesoderm during embryonic development and bounded on all sides by a definite coelomic epithelium or peritoneum. It contains a colourless coelomic fluid, the excretory organs open into it and the reproductive organs arise from its walls.

However, all animals do not possess a coelom, such as sponges, coelenterates, ctenophores, flatworms and proboscis worms. They are said to be *acoelomate* (Fig. 1.21).

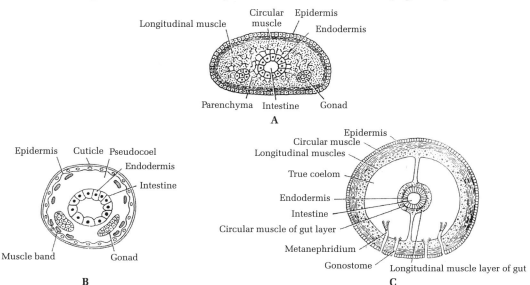

Figure 1.21
Metazoans on the basis of coelom. **A.** Acoelomate, with no body cavity. **B.** Pseudocoelomate. With body cavity not bounded by mesoderm. **C.** Coelomate or Eucoelomate with body cavity enclosed by mesoderm.

Body Cavity or Coelom

Body cavity can mean any internal space, or a series of spaces present inside body. Whereas coelom or true body cavity generally refers to a large fluid-filled space (cavity) lying between the outer body wall and the inner digestive tube. It arises as a secondary cavity between two layers of embryonic mesoderm and contains most of the visceral organs.

Types of coelom

Two types of body cavities or coelom occur in **Bilateria**, *Primary* and *secondary*:

1. **Primary coelom.** It is also called a false coelom or *pseudocoelom* (Fig. 1.21 B). It is derived from the blastocoel of the embryo, rather it represents a persistent blastocoel. Internal organs remain free in it since it is not bound by peritoneum or mesoderm. It is a space enclosed by ectoderm on the outside and endoderm (digestive tract) on the inside, and not by mesoderm on both sides. Such a pseudocoelom occurs in rotifers and roundworms, etc.

2. **Secondary coelom.** In highly developed Bilateria, the blastocoel is gradually obliterated by the embryonic archenterons, without forming a primary coelom, Instead, a *secondary or true coelom* or *eucoelom* develops within the embryonic mesoderm and lined by a characteristic layer of flattened mesodermal epithelial cells, known as *peritoneum,* which also surrounds the internal organs of the body. A true coelom probably appeared for the first time in annelids (Fig. 1.21 C).

On the basis of presence or absence of coelom, the Metazoa are divided into three major groups as follows:

1. **Acoelomata.** No body cavity or coelom is absent. Embryonic mesoderm remains as a solid layer, space between endoderm (gutwall) and ectoderm (bodywall) is filled with mesenchyme and muscle fibres, e.g.,: Porifera, Coelenterata, Ctenophora, Platyhelminthes and Nemertinea. (Fig. 1.21 A)

2. **Pseudocoelomata.** Body space is a pseudocoelom or false coelom. It is a persistent blastocoel enclosed between outer ectoderm and inner endoderm, and not lined by mesoderm. e.g.,: Acanthocephala, Ectoprocta and Aschelminthes (Rotifera, Gastrotricha, Kinorhyncha, Nematoda, Nematomorpha). (Fig. 1.21 B)

3. **Coelomata or eucoelomata.** Body space is a true coelom, enclosed by mesoderm on both sides. Remaining phyla of Bilateria, from Annelida to Arthorpoda, belong to Coelomata.

There are three different ways in which entomesoderm and coelom can arise during embryonic development. Accordingly, Hyman further divides coelomate Bilateria in three groups as follows:

   a) **Schizocoelomata.** Coelom arises by a splitting of endomesodermal bands which originate from blastoporal region of larva and extend between ectoderm and mesoderm (Fig. 1.22 A). It is a true coelom called a *schizocoel.* e.g.,: Most of the Protostomia (Annelida Arthropoda, Mollusca, etc.)

   b) *Mesenchymal coelomata.* Mesenchymal cells rearrange to enclose a space or coelom, which is regarded an aberrant schizocoel. It is seen only in Phoroida (Fig. 1.22 B)

c) ***Enterocoelomata.*** Coelom arises in the form of mesodermal pouches from larval archenteron (Fig. 1.22 C). After separation from endoderm, the pouches fuse and expand until they touch the gut and bodywall. Since the coelom arises from larval enteron, it is called an *enterocoel*. e.g.,: Deuterstomia (Chaetognatha, Echinodermata, Hemichordata, and Chordata) and Brachiopoda.

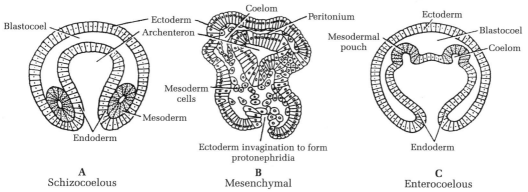

Figure 1.22
Three types of coelom formation. **A.** Schizocoelous, **B.** Mesenchymal, **C.** Enterocoelous.

Modifications of coelom

There are modifications of coelom in different animals. As mentioned, in aschelminths, it forms a *pseudocoelom not lined with mesoderm. In arthropods, it is formed by scattered spaces, collectively known as* haemocoel, in which blood circulates. In annelids, coelom is divided by *septa* into chambers corresponding to the somites. In mammals, coelom is divided by a muscular *diaphragm* into thoracic and abdominal cavities, with a separate *pericardial cavity around heart*.

Significance of coelom

Evolution of coelom is of great significance in animals. It plays an important role in the progressive development of complexity of structure. It permits greater size and contributes directly to the development of excretory, reproductive and muscular systems of body. In a triploblastic animal (e.g. earthworm), the appearance of a perivisceral coelom between gut and bodywall leads to several advantages.

1. It surrounds the internal organs like a water jacket and protects them from external shocks.
2. It provides flexibility to the body. It provides space and gives the digestive tract and other internal organs freedom of movement and opportunity for enlargement further differentiation and greater activity.
3. The coelomic fluid often functions as a hydraulic skeleton and also serves as a circulatory medium for the transport and distribution to nutritive substances and gases.
4. The excretory matter is collected into coelomic fluid and then passed out of the body through nephridia.

5. The gonads arise from its coelomic epithelium and project into coelom. Ova and sperms are extruded through special gonoducts connecting the coelom with the exterior.

No doubt, the evolution of coelom made a major advance in the evolution of Metazoa. It is evident from the great increase in size and diversity in structure and ways of life met within coelomate phyla in comparison with acoelomate and pseudocoelomate phyla.

Evolution of coelom

Origin of coelom in Metazoa is of great evolutionary significance. There is no direct evidence of evolution of coelom available from palaeontology. There is only indirect evidence which is based mainly on the embryology of present day metazoans. This has been differently interpreted by different workers resulting in many conflicting theories about the origin and evolution of coelom. Out of these, there are 4 principal theories as follows:

1. **Enterocoel theory.** The theory holds that the primitive mode of coelom formation was enterocoelous. According to this theory, the bilateral metazoan ancestor of coelenterates had gastric pockets which become separated from the central digestive cavity to form coelomic pouches (Figs. 1.23 A, B). However, the theory is mainly objected on the following grounds:

    a) Gastric pouches occur in highly organized coelenterates, such as Scyphozoa and Anhozoa, which are not suitable for ancestral types.

    b) Sealing of gastric pockets in the ancestor would defeat the purpose for which they were formed, that is, for increasing the surface area for digestion and absorption.

    c) Gastric pockets of coelenterates are not evaginations of gut endoderm, but formed differently by ingrowth of bodywall-septa.

2. **Gonocoel theory.** This is the most popular theory of origin of coelom. According to this theory, coelom represents a persistent expanded gonadial cavity or gonocoel (Figs. 1.24 A, B). The theory was proposed by Bergh, based on a idea earlier expressed by Hatschek.

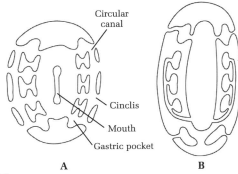

Figure 1.23
Coelom formation according to *Enterocoel theory*. **A.** Formation of gastric pockets in an anthozoan. **B.** Gastric pockets become coelomic pouches.

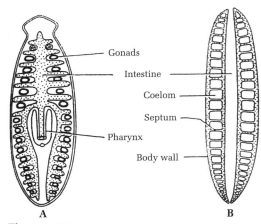

Figure 1.24
Gonocoel theory of coelom formation. **A.** Gonads alternating with intestinal branches in a triclad flatworm. **B.** Coelomic sacs or compartments in an annelid, formed by fusion of adjacent expanded gonads.

Meyer (1890) considered coelom initially unsegmented, arising from a single large pair of gonads. However Bergh and Lang believed that coelom initially arose in a segmented condition. In nemerteans and flatworms, especially in the triclad turbellarian *Procerodes lobata* (Fig. 1.24), they observed intestinal branches alternating with a linear series of gonads. If the intestinal branches withdraw and gonads enlarge with their walls meeting to become septa, they would resemble the row of coelomic compartments of an annelid. However, the chief draw backs of the gonocoel theory are:

a) It links the origin of coelom with that of metameric segmentation which is not acceptable phylogenetically.

b) It is not supported by embryology because gonads do not arise before coelom.

c) It does not account for unsegmented coelomate phyla, which are in majority, and there is no evidence that they have originated from segmented ancestors. Goodrich and Hyman have rejected the Gonocoel theory. They consider that metameric segmentation was preceded by a pseudometameric distribution of organs (such as gonads) in an elongated worm-like body.

3. **Nephrocoel theory.** According to this theory, proposed by Lankester in 1874, coelom originated as the expanded inner end of a nephridium. The theory is not seriously held because, (i) Protonephridia have been described in coelomates, and (ii) some coelomate groups, such as Echinodermata, do not have excretory organs.

4. **Schizocoel theory.** According to this theory, coelom has mesenchymal origin and has no relation with nephridia and gonads of Lower Bilateria and endodermal pouches of Radiata.

Conclusion

None of the above theories satisfactorily explains the origin of coelom. Firstly, they do not explain the advantages of intermediate stages passed through during the course of evolution. Secondly, the connection between evolution of coelom and metameric segmentation together, has not been made clear. Thirdly, no exact nature of a coelom has been defined. Which cavities should be regarded coelomic and which not has not been explained. Moreover, there is no evidence showing that secondary body cavity is homologous (with same origin) throughout the animal kingdom. Probably it is polyphyletic in origin. As Clark postulates, coelom might have arisen independently a number of times and in various ways in different animal groups- as *a persistent blastocoel* **in pseudocoelomates**, as an *enterocoel* **in deuterostomes**, as *a schizocoel* in **protostomes**, and as a gonocoel, in Nemerteans and flatworms.

Grades of Organization

Depending on the number and degree of complexity or specialization of cells present in individuals, we can divide animals into the following patterns or levels or grades of organizaion:

(i) protoplasmic,

(ii) cellular,

(iii) cell-tissue,

(iv) tissue-organ,

(v) and organ-system.

These are approximately in the order in which they have evolved. Most animals fit neatly into one of these groups or categories although some animals appear to be intermediate between certain categories (Fig. 1.25). For example, sponges only show a hint of tissue (cellular level), but not true tissue as shown by coelenterates (cell-tissue level).

1. **Protoplasmic or acellular level.** This type of organization occurs in Protozoa or acellular protists. All the activities are performed by one-cell body. However, specialized cytoplasmic structures or the organelles, carry on specific functions, thus illustrating division of labour.

   Colonial protozoans, such as *Volvox,* are differentiated into somatic and sex cells. But, the somatic cells do not show specialization and division of labour. This distinguishes colonial Protozoa from Metazoa.

2. **Cellular level.** Multicellular animals such as mesozoans and sponges, are made of loose association of cells. All the cells are in direct contact with the environment, that is water, so that respiration, excretion, etc. take place by the general body surface. However, some division of labour occurs in certain cells concerned with reproduction germ cells) and nutrition (choanocytes). But, such specialized cells do not form definite tissues. All cells act more or less independently and show little coordination. Sponges and mesozoans do not attain a higher level than this and are considered to be the simplest metazoans. Certain colonial protozoans are considered to have reached the cellular level.

3. **Cell-tissue level.** Multicellular animals, such as hydra and jelly fish, are supposed to make the beginning of the tissue plan. Most of their cells remain scattered, at the cellular level, and not organized into tissues.

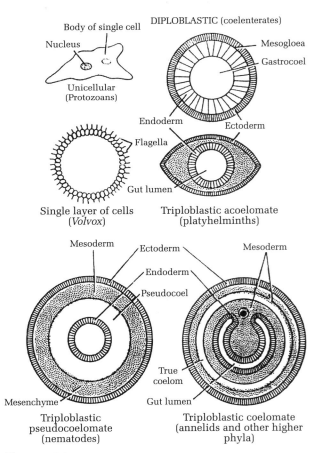

Figure 1.25
Grades of body organization in animal kingdom.

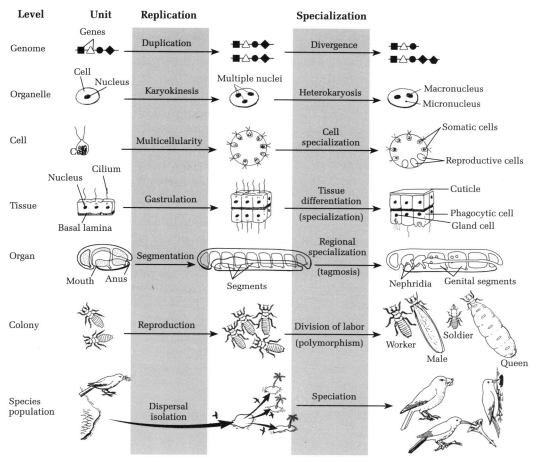

Figure 1.26
Origin of complexity at different levels of organization by replication and specialization of standard units.

Nerve cells appear for the first time but nerve ganglia are absent. But the nerve cells and their processes remain interconnected forming a *nerve net* which is a good example of a tissue with the function of transmission of impulses and coordination. Thus, first truly metazoan phylum of tissue grade organization is Cnidaria or Coelenterata. Their body is generally radially symmetrical and made of only two germ layers, ectoderm and endoderm. Such animals are called *diploblastic*.

4. **Tissue-organ level.** Further step of advancement is the aggregation of tissues into organs which have a more specialized function than tissues. Platyhelminths are the first bilaterally symmetrical, acoelomate metazoans to have reached the tissue-organ level of organization. They have a number of well-defined organs such as eyespots, proboscis, gonads, etc. They are also *triploblastic* because they have a third or middle cellular germ layer, called *mesoderm,* which lies between outer ectoderm and inner endoderm.

5. **Organ-system level.** In higher animals, several organs are associated to form a distinct system concerned with a specific function, such as digestion, respiration, circulation, excretion and reproduction. This is the highest level of organization. This is first seen in a group of marine worms known as nemerteans. They have a complete digestive system separate and distinct from circulatory system.

Members of superphylum Aschelminthes, such as nematodes, are bilaterally symmetrical, triploblastic and pseudocoelomate metazoans. They have a pseudocoelom but lack a true secondary coelom. On the other hand, annelids, arthropods, mollusks, echinoderms and chordates are triploblastic and eucoelomate because they possess a true coelom. The metazoans evolved from protozoan colonies in which initially similar cells became specialized for different functions. If so, the evolution of Metazoa can be described as a replication of similar units (cells) followed by unit specialization and integration into an organism at a new, higher level of complexity (Fig. 1.26). This **replication-specialization-integration of units** sequence is a general pattern in the evolution of large body size and complexity. Among metazoans, body is composed of a series of similar segments. Later, these segments became specialized and integrated into regions, such as the head, thorax, and abdomen. All levels of biological complexity are illustrated and exhibited in Figure 1.26.

Chapter 2

Recent Trends in Animal Classification

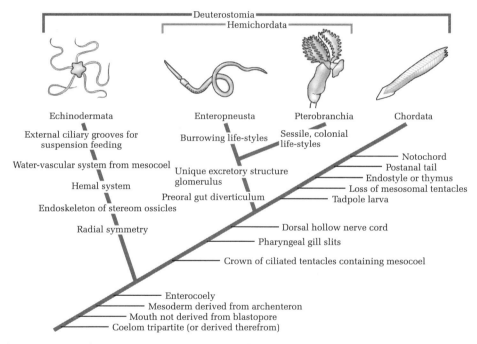

Inferred Relationship Among Deuterostome and interested in phylogeny of Deuterostome is perhaps became we are especially curious to know about our own deuterostome ancestry

Figure 2.0 Cladogram showing relationship among deuterostome the crown of ciliated tentacles a character borne by ancestors of lophophorates, hemichordates and chordates. The tentacular crown would have become the lophophore in lophophorate phyla and retained as a primitive character in pterobranchs. Because molecular evidence indicates that lophophores are protostomes, we removed them from this cladogram; the ciliated tentacular crown in pterobranchs and lophorates can be considered a convergent character. If recent molecular evidence is correct, this cladogram will require further revision to place the hemichordates sister taxon to the echinoderms rather than the chordates.

Considerable use of the terms "**primitive**," "**lower**, "**advanced**," and "**higher**" is inescapable in describing the animal classification. These are convenient but unfortunate words, which can mislead easily. If we assume that life arose during a fixed (Fig. 2.0) geological period. It is true that all organisms today share an equally long evolutionary history. None stopped evolving, thus remaining at primitive levels, while others continued evolving and so attained advanced status. Rather, animal evolution witnessed the retention of ancestral characters by some organisms and the development of new characters by others. The terms lower and primitive are reserved simply for those organisms most reminiscent of the ancestral type; advanced and higher species represent more divergent and specialized forms.

Related to this distinction is the concept of biological success. Commonly we speak of insects and mollusks as very successful groups, while horseshoe crabs and brachiopods are described as relative failures. In this context, biological success refers to species abundance and ecological diversity. Yet, in a very real sense, all living groups are profoundly successful, for they have survived the rigors of natural selection for millions of years. Members of smaller and/or more primitive groups may occupy only a few of the many possible ecological niches, but their very existence testifies to their success.

All those species alive today–the current ***spatial diversity*** *of* animals are the result of some 450 million years of the pressures of natural selection, which they may have survived by retaining generalized characteristics or by developing many and extreme specializations. How they changed through time–comprises their phylogenies, which we can call the ***temporal diversity of vertebrates***. Although the fossil record is limited (usually) to hard parts, it may be possible by various methods to reconstruct the soft parts and their temporal diversity.

One way is by *studying the spatial diversity*. Let us say, for instance, that we want to know about the temporal diversity of the mammalian heart. If we study the heart in living reptiles, birds, and mammals, we learn that at least one heavy-walled ventricle and thin walled atrium are found in all these forms. We could thus rather confidently infer that at least this much must have been present in the cotylosaur, which was their common ancestor.

Another way of *determining temporal diversity* of soft parts is by using Cuvier's principle of the correlation of parts, In the shoulder of all living forms, for instance, the supraspinatus muscle is found only in conjunction with the supraspinous fossa of a shoulder bone, the scapula. Therefore the presence or absence of this muscle in any fossil can be determined by the presence or absence of the supraspinous fossa.

Another useful correlation is the well-understood *relationship between structure and habitat*. The very rock in which a fossil is found reveals something about the habitat and therefore about the structure of the extinct animal.

Some insight on temporal diversity can be gained from the *ontogeny,* or life history, of the individual. Early development is similar in all vertebrates. Up to the pharyngula stage except that development occurs on a flat, broad plane of yolk in large-yolked eggs (e.g., chondrichthyeans, birds, reptiles), while in small-yolked eggs (e.g., most fish, amphibia, mammals) the entire egg is incorporated in development. As each group develops past the pharyngula stage, it becomes gradually more specialized until it has the precise adult form of its species. The changes which occur during this ontogeny can tell us something about the phylogenetic changes which comprise temporal diversity, but they must be interpreted carefully, since Haeckel was over generalizing when he declared that ontogeny recapitulates phylogeny.

No widely accepted evolutionary sequences can be established, as we can see from the many differences that exist in the proposals of different biologists. The disagreements arise largely from lack of data rather than from any doubt as to whether evolution did not occur; we can conclude that can be found in fossil records that evolution both spurred biologists to construct phylogenies and also made it very difficult for biologist to create totally satisfactory phylogeny. The study of the kinds and diversity of organisms and the evolutionary relationships among them is called systematics or taxonomy. These studies result in the description of new species and the organization of animals into groups (taxa) based on degree of evolutionary relatedness.

Table 2.1: The Six Kingdoms of Life

The Prokaryotes (the "domains" Eubacteria and Archaea)

Domain

Kingdom Eubacteria (Bacteria)

The "true" bacteria, including Cyanobacteria (or blue-green algae) and spirochetes. Never with membrane-enclosed organelles or nuclei, or a cytoskeleton; none are methanogens; some use chlorophyll-based photosynthesis; with peptidoglycan in cell wall; with a single known RNA polymerase.

Domain

Kingdom Archaea (Archaebacteria)

Anaerobic or aerobic, largely methane-producing microorganisms. Never with membrane-enclosed organelles or nuclei, or a cytoskeleton; none use chlorophyll-based photosynthesis; without peptidoglycan in cell wall; with several RNA polymerases.

The Eukaryotes (the "domain" Eukaryota, or Eukarya)

Domain

Cells with a variety of membrane-enclosed organelles (e.g., mitochondria, lysosomes, peroxisomes) and with a membrane-enclosed nucleus. Cells gain structural support from an internal network of fibrous proteins called a cytoskeleton.

Kingdom Fungi

The fungi. Probably a monophyletic group that includes molds, mushrooms, yeasts, and others. Saprobic, heterotrophic, multicellular organisms. The earliest fossil records of fungi are from the Middle Ordovician, about 460 mya. about 72,000 described species are thought to represent only 5-10 percent of the actual diversity.

Kingdom Plantae (Metaphyta)

The multicellular plants. Photosynthetic, autotrophic, multicellular organisms that develop through embryonic tissue layering. Includes some groups of alges, the bryophytes and their kin, and the vascular plants (about 240,000 of which are flowering plants). The described species are thought to represent about half of Earth's actual plant diversity.

Kingdom Protista

Eukaryotic single-celled microorganisms and certain algae. A polyphyletic grouping of perhaps 18 phyla, including euglenids, green algae, diatoms and some other brown algae, ciliates, dinoflagellates, foraminiferans, amoebae, and others. Many workers feel that this group should be split into several separate kingdoms to better reflect the phylogenetic lineages of its members. The 80,000 described species probably represent about 10 percent of the actual protist diversity on Earth today.

Kingdom Animalia (Metazoa)

The multicellular animals. A monophyletic taxon, containing 34 phyla of ingestive, heterotrophic, multicellular organisms. About 1.3 million living species have been described; estimates of the number of undescribed species range from lows of 10-30 million to highs of 100-200 million.

Even though von Linné did not accept evolution, many of his groupings reflect evolutionary relationships. Morphological similarities between two animals have a genetic basis and are the results of a common evolutionary history. Thus, in grouping animals according to shared characteristics, von Linné grouped them according to their evolutionary relationships. Ideally, members of the same taxonomic group are more closely related to each other than to members of different taxa.

Von Linné recognized five taxa. Modern taxonomists use those five and have added three others. The taxa are arranged hierarchically: **domain, kingdom, phylum, class, order, family, genus, and species** (Table 2.1). Domain, the broadest taxonomic category, was added recently, is not yet universally accepted.

In recent years, **molecular techniques** have provided important information for taxonomic studies. The relatedness of animals is reflected in the gene products animals produce and in the genes themselves (the sequence of nitrogenous bases in DNA). Related animals have DNA derived from a common ancestor. Genes and proteins of related animals therefore, are more similar than genes and proteins from distantly related animals. By comparing the sequence of amino acid in proteins, or the sequence of nitrogenous bases in DNA or RNA, and assuming a relatively constant mutation rate (referred to as a molecular clock), taxonomists can estimate the time elapsed since divergence from a common ancestor Sequencing the nuclear DNA and the mitochondrial DNA of animals has become common place. Mitochondrial DNA is useful in taxonomic studies because mitochondria have their own genetic systems and are inherited cytoplasmically. That is, mitochondria are transmitted from parent to offspring through the egg cytoplasm and can be used to trace maternal lineages. Using mitochondrial DNA involves relatively small quantities of DNA that change at a relatively constant rate.

Molecular techniques have provided a wealth of new information useful to animal taxonomists. These techniques, however, will not replace traditional taxonomic methods. The molecular clocks used to determine rates of evolutionary change have provided important information that helps fill in time gaps in the fossil record. Molecular clocks, however, apparently run at different rates, depending on whether one is looking at the sequence of amino acids in proteins, the sequence of bases in DNA from organelles like mitochondria, the sequence of bases in nuclear DNA, or data from different evolutionary lineages. Molecular and traditional methods of investigation will probably always be used to complement each other in taxonomic studies.

Whittaker described a system of classification that distinguished between kingdoms according to cellular organization and mode of nutrition (Fig. 2.1 A) According to this system, members of the kingdom **Monera** are the bacteria and the cyanobacteria. They are distinguished from all other organisms by being prokaryotic. Members of the kingdom **Protista** are eukaryotic and consist of single cells or colonies of cells. This kingdom includes *Amoeba, Paramecium,* and many others. Members of the kingdom **Plantae** are eukaryotic, multicellular, and photosynthetic. Plants have walled cells and are usually nonmotile. Members of the kingdom **Fungi** are also eukaryotic and multicellular. They also have walled cells and are usully nonmotile. Mode of nutrition distinguishes fungi from plants. Fungi digest organic matter extra-cellularly and absorb the breakdown products. Members of the kingdom Animalia are eukaryotic and muticellular, and they usually feed by ingesting other organisms or parts of other organisms. Their cells lack walls and they are usually motile.

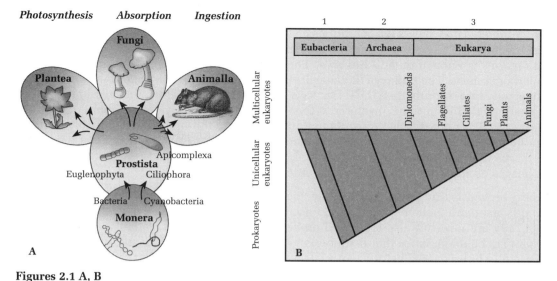

Figures 2.1 A, B

Classification of Organisms. **A.** Whittaker described a five-kingdom classification system based on cellular organization and mode of nutrition. **B.** Recent studies of ribosomal RNA indicate that a grouping into three domains more accurately portrays early evolutionary relationships.

In recent years, new information has challenged the five kingdom classification system. For the first two billion years the life on the earth, the only living forms were prokaryotic microbes. Fossil evidence from this early period is scanty; however, molecular studies of variations in base sequences of **ribosomal RNA** from more than two thousand organisms are providing evidence of relationships rooted within this two billion year period. The emerging picture is that the five previously described kingdoms do not represent distinct evolutionary lineages.

Ribosomal RNA is excellent for studying the evolution of early life on earth. It is an ancient molecule, and it is present and retains its function in virtually all organisms. In addition, ribosomal RNA changes very slowly. This slowness of change, called **evolutionary conservation**, indicates that the protein-producing machinery of a cell can tolerate little change and still retain its vital function. Evolutionary conservation of this molecule means that closely related organisms are likely to have similar ribosomal RNAs. Distantly related organisms are expected to have ribosomal RNAs that are less similar, but the differences are small enough that the relationships to some ancestral molecule are still apparent.

Molecular systematists compare the base sequences in ribosomal RNA of different organisms to find the number of positions in the RNAs where bases are different. They enter these data into computer programs and examine all possible relationships among the different organisms. The systematists then decide which arrangement of the organisms best explains the data.

Studies of ribosomal RNA have led systematists to the conclusions that all life shares a **common ancestor** and that there are three major evolutionary lineages (Fig. 2.1 B). Each of these lineages is called a **domain.** The domain thus supersedes the kingdom as the broadest taxonomic grouping. The **Archaea** are prokaryotic microbes that live in extreme environments, such as high-temperature rift valleys on the ocean floor, or high-salt or acidic environments. All members of the Archaea inhabit anaerobic environments. These environments may reflect the conditions on the earth at the time of life's origin. The Archaea are the most primitive life-forms known. Ancient archeans gave rise to two other domains of organisms Eubacteria and Eukarya. The **Eubacteria** are bacteria and are prokaryotic microorganisms. The **Eukarya** include all eukaryotic organisms. Interestingly, the Eukarya diverged more recently than the Eubacteria from the Archaea. Thus, the Eukarya are more closely related than the Eubacteria to the Archaea. The Eukarya arose about 1.5 billion years ago. The photosynthetic accumulation of oxygen in the atmosphere probably resulted in the production of ozone, which shielded the planet from deadly ultraviolet light. The Eukarya then underwent a late, but rapid, evolutionary diversification into the modern Lineages of protists, fungi, plants, and animals.

Many biologists find the relatively close relationships of all eukaryotic organisms depicted in this taxonomic scheme strange and unsettling. The relationships depicted in (Fig. 2.1 B) not only are new but also are based on cellular characteristics, rather than the whole-organism characteristics that taxonomists traditionally use. As a result, many questions remain to be answered before zoologists accept this system of higher taxonomy.

One of these questions involves how to deal with the eukaryotic kingdom lineages." Should the Animalia, Fungi, and Plantae still be considered kingdoms" Most zoologists would answer, "Yes." If so, molecular studies indicate that the separate protist lineages shown in (Fig. 2.1 B) should also be elevated to kingdom status. Questions such as this make systematics a lively and challenging field of study.

Animal Phylogeny

The goal of animal systematics is to arrange animals into groups that reflect evolutionary relationships. Ideally, these groups should include a single ancestral species and all of its descendants. Such a group is called a **monophyletic group** (Fig. 2.2). In searching out monophyletic groups, taxonomists look for animal characters that indicate relatedness. A **character** is virtually anything that has a genetic basis and can be measured—from an anatomical features to a sequences of nitrogenous bases in DNA or RNA. **Polyphyletic groups** have members that can be traced to separate ancestors. Since each group should have a single ancestor, a polyphyletic group reflects insufficient knowledge of the group. A **paraphyletic group** includes some, but not all, members of a lineage. Paraphyletic groups also result when knowledge of the group is insufficient (Fig. 2.2 A).

As in any human endeavor, disagreements have arisen in animal systematics. These disagreements revolve around methods of investigation and whether or not data may be used in describing distant evolutionary relationships. Three contemporary schools of systematics exist:

1. evolutionary systematics,
2. numerical taxonomy
3. phylogenetic systematics or cladistics.
1. **Evolutionary systematics** is the oldest of the three approaches. It is sometimes called the **"traditional approach,"** although it has certainly changed since the beginnings of animal systematics. A basic assumption of evolutionary systematists is that organisms

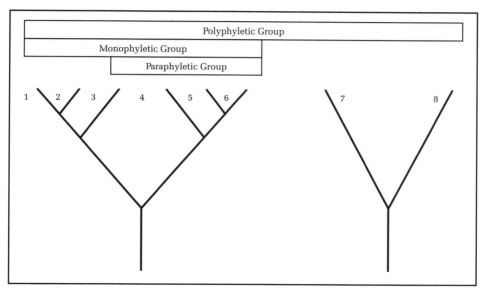

Figure 2.2 A
Evolutionary Groups. An assemblage of species 1-8 is a polyphyletic group because species 1-6 have a different ancestor than species 7 and 8. An assemblage of species 3-6 is a paraphyletic group because species 1 and 2 share the same ancestor as 3-6, but they have been left out of the group. An assemblage of species 1-6 is a monophyletic group because it includes all of the descendants of a single ancestor.

closely related to an ancestors will resemble that ancestor more closely than they resemble distantly related organisms.

Two kinds of similarities between organisms are recognized:

a) **Homologies** Homologies are resemblances that result from common ancestry and are useful in classifying animals. An example is the similarity in the arrangement of bones in the wings of *a bird, Pterodactyl, bat* and the *arm of human* (Fig. 2.2 B).

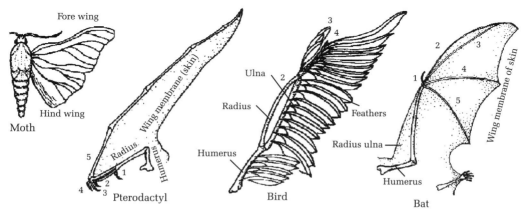

Figure 2.2 B
Homologous and analogous structures.

b) **Analogies** Analogies are resemblances that result from organisms adapting under similar evolutionary pressures. The latter process is sometimes called convergent evolution. Analogies do not reflect common ancestry and are not used in animal taxonomy. The similarity between the wings of birds and insects Moth (Fig. 2.2 B) is an analogy.

Evolutionary systematists often portray the results of their work on phylogenetic trees, where organisms are grouped according to their evolutionary relationships. Figure 2.3 is a phylogenetic tree showing vertebrate evolutionary relationships, as well as time scales and relative abundance of animal groups. These diagrams reflect judgment made about rates of evolution and the relative importance of certain key characters' (e.g., feathers in birds).

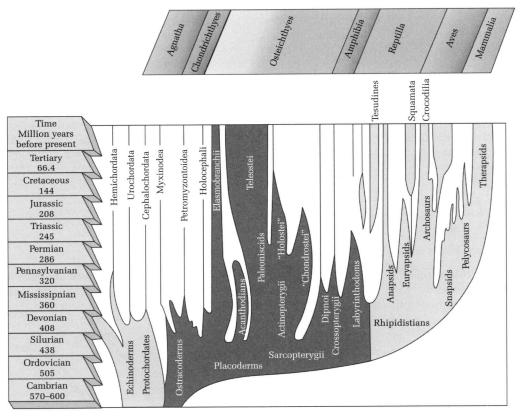

Figure 2.3
Phylogenetic Tree Showing Vertebrate Phylogeny. A phylogenetic tree derived from evolutionary systematics depicts the degree of divergence since branching from a common ancestor, which is indicated by the time periods on the vertical axis. The width of the branches indicates the number of recognized genera for a given time period. Note that this diagram shows the birds (Aves) as being closely related to the reptiles (Reptilia), and both groups as having class-level status.

2. **Numerical taxonomy** emerged during the 1950s and 1960s and represents the opposite end of spectrum from evolutionary systematics. The founders of numerical taxonomy believed that the criteria for grouping taxa had become too arbitrary and

vague. They tried to make taxonomy more objective. Numerical taxonomists use mathematical models and computer-aided techniques to group samples organisms according to overall similarity. They do not attempt to distinguish between homologies and analogies. Numerical taxonomists admit that analogies exist. They contend, however, that telling one from the other is sometimes impossible and that the numerous homologies used in data analysis overshadow the analogies. A second major difference between evolutionary systematics and numerical taxonomy is that numerical taxonomists limit discussion of evolutionary relationships to closely related taxa. Numerical taxonomy is the least popular of the three taxonomic schools; however, all taxonomists use the computer programs that numerical taxonomists developed.

3. **Phylogenetic systematics or cladistics** is a third approach to animal systematics. The goal of cladistics is similar to that described for evolutionary systematics–the generation of hypotheses of genealogical relationships among monophyletic groups of orgainisms. Cladists contend, however, that their methods are more open to analysis and testing, and thus are more scientific, than those of evolutionary systematists.

As do evolutionary systematists, cladists differentiate between homologies and analogies. They believe, however, that homologies of recent origin are most useful in phylogenetic studies. Characters that all members of a group share are referred to as symplesiomorphies. These characters are homologies that may indicate a shared ancestry, but they are useless in describing relationships within the group.

To decide what character is ancestral for a group of organisms, cladists look for a related group of organisms, called an **out-group**, that is not included in the study group. Figure 2.4 shows a hypothetical lineage for five taxa. Notice that taxon 5 is the outgroup for taxa 1-4. Character A is symplesiomorphic (ancestral) for the outgroup and the study group.

Characters that have arisen since common ancestry with the out-group are called derived characters or synapomorphies. Taxa 1-4 in figure 2.4 share derived character B. This character separates taxa 1-4 from the outgroup. Similarly, derived characters C and D arose more recently than character B. Closely related taxa 1 and 2 share character C. Closely related taxa 3 and 4 share character D. Taxa that share a certain synapomophy form a subset called a clade. Taxa 1 and 2 form a clade characterized by C.

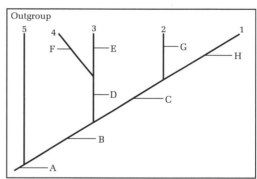

Figure 2.4
Interpreting Cladograms. This hypothetical cladogram shows five taxa (1-5) and the characters (A-H) used in deriving the taxonomic relationships. Charater A is symplesiomorphlc for the entire group. Taxon 5 is the outgroup because it shares only that ancestral character with taxa 1-4. All other characters are more recently derived. What single character is a synapomorphy for taxa 1 and 2, separating them from all other taxa?

The hypothetical lineage shown in figure 2.4 is called a cladogram. Cladograms depict a sequence in the origin of derived characters. A cladogram is interpreted as a family tree depicting a hypothesis regarding monophyletic lineages. New data in the form of newly investigated characters or reinterpretations of old data are used to test the hypothesis the cladogram describes.

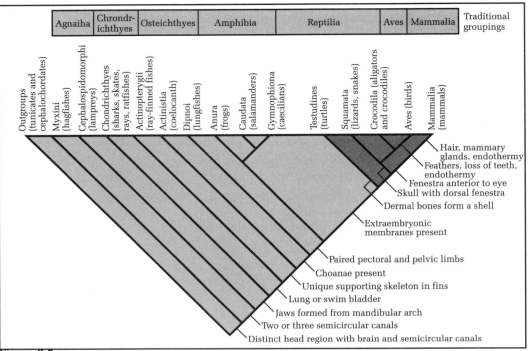

Figure 2.5
Chadogram Showing Vertebrate Phylogeny. A cladogram is constructed by identifying points at which two groups diverged. Animals that share a branching point are included in the same taxon. Notice that timescales are not given or implied. The relative abundance of taxa is also not shown. Notice that this diagram shows the birds and crocodilians sharing common branch, and that these two groups are more closely related to each other than either is to any other group of animals.

Figure 2.5 is a cladogram depicting the evolutionary relationships among the vertebrates. The tunicates and cephalochordates are an out-group for the entire vertebrate lineage. Derived characters are listed on the right side of the cladogram. Notice that extra embryonic membrane is a synapomorphy used to define the clade containing the reptiles, bird, and mammals. These extra embryonic membranes are a shared character for these groups and are not present in any of the fish taxa or the amphibians. Distinguishing between reptiles, birds, and mammals requires looking at characters that are even more recently derived than extra-embryonic membranes. A derived character, the shell, distinguishes turtles from all other members of the clade; skull characters distinguish the lizard and crocodile bird lineage from the mammal lineage; and hair, mammary gland, and endothermy is a unique mammalian character combination. Note that a synapomorphy at one level of taxonomy may be a symplesiomorphy at a different level of taxonomy. Extra embryonic membranes is a synapomorphic character within the vertebrates that distinguishes the reptile/bird/mammal clade. It is symplesiomorphic for reptiles, birds, and mammals because it is ancestral for the clade and cannot be used to distinguish among members of these three groups.

Zoologists widely accept cladistics, this has resulted in some nontraditional interpretations of animal phylogeny. A comparison of figures 2.3 and 2.5 shows one example of different interpretations derived through evolutionary systematics and cladistics. Generations of taxonomists have assigned class-level status (Aves) to birds. Reptiles also have had class-level status (Reptilia). Cladistic analysis has shown, however, that birds are more closely tied by common ancestry to the alligators and crocodiles than do any other group. According to the cladistic interpretation, birds and crocodiles should be assigned to a group that reflects this close common ancestry. Birds would become a subgroup within a larger group that included both birds and reptiles. Crocodiles would be depicted more closely related to the birds than they would be to snakes and lizards. Traditional evolutionary systematists maintain that the traditional interpretation is still correct because it takes into account the greater importance of key characters of birds (e.g., feathers and endothermy) that make the group unique. Cladists support their position by pointing out that the designation of "key characters" involves value judgments that cannot be tested.

As debates between cladists and evolutionary systematists continues, our knowledge of evolutionary relationships among animals will become more complete. Animal systematics is certainly going to be a lively and exciting field in future.

Role of the Blastopore

	PROTOSTOMES		DEUTEROSTOMES	
 Sprial cleavage	Cleavage mostly spinal	Cleavage mostly radial	Radial cleavage 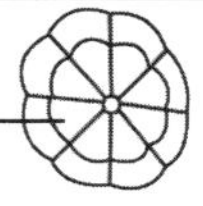
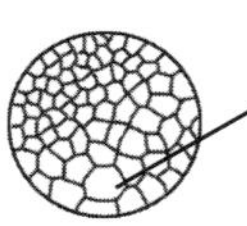 Cell from which mesoderm will derive	Endomesoderm usually from a particular blastomere designated 4d	Endomesoderm from enterocoelous pouching (except vertebrates)	Endomesoderm from pouches from primitive gut 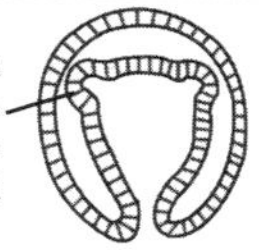
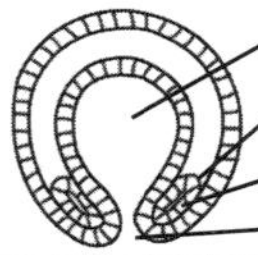 Primitive gut, mesoderm, coelem, Blastopore	In coelomate protostomes the coelom forms as a split in mesodermal bands (schizocoelous)	All coelomate, coelom from fusion of enterocoelous pouches (except vertebrates, which are schizocoelous)	Coelom, Mesoderm, Primitive gut, Blastopore 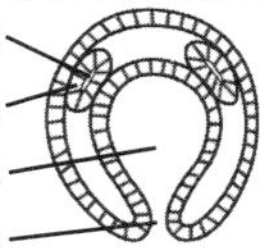
Anus, Annelid (earthworm), Mouth	Mouth forms from or near blastopore; anus a new formation. Embryology mostly determinate (mosaic). Includes phyla Platyhelminthes, Nemertea, Annelida, Mollusca, Arthropoda, Chaetognatha, Phorohida, Ectoprocta, Brachiopoda, minor phyla	Anus forms from or near blastopore, mouth a new formation. Embryology usually indeterminate (regulative). Includes phyla Echinodermata, Hemichordata, Chordata	Mouth, Anus

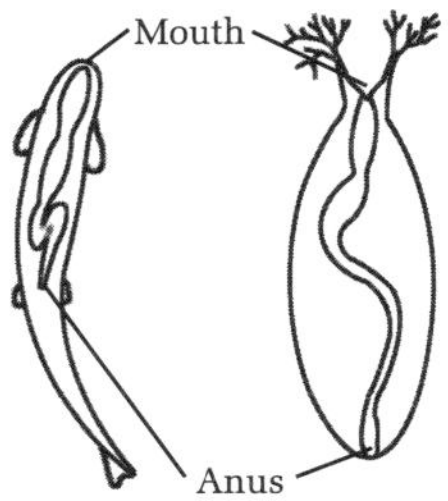

Figure 2.6
Basis for the distinction between Protostomes and Deuterostomes. Traditional classifications often place phyla Brachiopoda, Chaetognatha, Ectoprocta and Phoronida with deuterostomes, but recent molecular phylogenatic analyses place them with protostomes as shown here.

In the deuterostomes the blastopore of radially cleaving embryos becomes the adult anus and the mouth forms as a new structure. In most protostomes, the blastopore either becomes the mouth (and the anus form anew; (Fig. 2.6)) or gives rise to both mouth *and* anus (as in some mollusks, polychaetes, and onychophorans). When the latter occurs, the blastopore first forms a long furrow. Later, the furrow margins converge and fuse between the two extremes, one of which remains open as the mouth and the other as the anus. The blastopore of several taxa closes and disappears during early development and a new mouth and anus form later. In such cases, the assignment of a taxon to either Protostomia or Deuterostomia, based on this character, depends on where the new mouth or anus opens in relation to the old blastopore.

Morphologists have long speculated that the developmental origin of both mouth and anus from the blastopore was primitive for Bilateria. This, hypothesis is now supported by expression of the T-book gene, *Brachyury*. In cnidarians, the gene is expressed in the region of a sea anemone blastopore and around the mouth of *Hydra*. In bilaterians, its expression occurs in the embryonic mouth *and* anus of a polychaete (protostome) and a hemichoradate (deuterostome).

Morphological and molecular systematist agree on two broad divisions of Bilateria. These two major taxa are

Protostomia whereas protostomes include the bilaterians, such as molluscs, arthropods, annelids, and flatworms. Protostome cleavage is mostly spiral, the embryonic blastopore usually persists to become the adult mouth (or mouth and anus), and the mesothelium, when present, is typically schizocoelous in origin

Deuterostomia (Fig. 2.7). Deuterostomes include hemichordates, echinoderms, chordates, and sometimes chaetognaths. In deuterostomes, cleavage is usually radial, the blastopore typically becomes the anus, the mouth forms anew from the surface ectoderm, and the mesothelium arises by enterocoely.

Traditional morphology and contemporary molecular systematics are in agreement regarding deuterostomes, but disagree over the membership and major subdivision of protostomes (Fig. 2.7), and these disagreements are part of a lively debate and ongoing research. Morphology usually recognizes two protostome taxa

a) Spiralia – Spiralians are united by the synapomorphy "spiral cleavage" and include the flatworms, molluscs, annelids, and arthropods, among other Annelida, Onychophora, and Arthropoda together constitute the **Articulata,** so-called because the segmented worms velvet worms, and arthropods are all articulated, meaning that they share a segmented body and similar pattern of growth

b) Cycloneuralia includes Aschelminthes, Pseudocoelomata, or Nemathelminthes; (Fig. 2.7) Cycloneuralians are mostly small-bodied animal, such as gastrotrichs and nematodes, whose cleavage pattern differs from spiral and whose adult traits tend to isolate them from the remaining protostomes.

A third assemblage, Lophoporate (brachiopods, phoronids, bryozoans), is considered to be intermediate between protostomes and deuterostomes, but have closer ties to the deuterostomes (Fig. 2.7 A).

Molecular systemists also divide Protostomia into two groups, but taxon membership in them differs from the traditional scheme just described (Fig. 2.7 B).

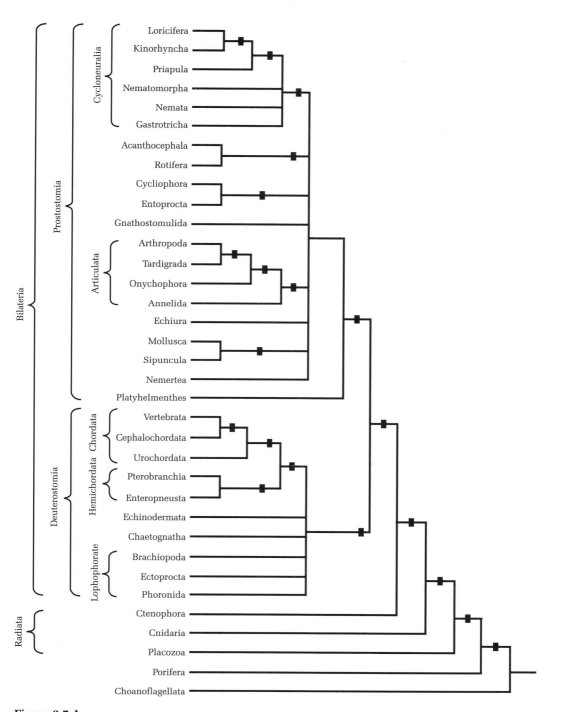

Figure 2.7 A
A phylogeny of the Metazoa. The cladogram hypothesizes the most primitive metazoans to be Porifera, followed by the placozoa, and the two radiate phyla (Cnidaria and Ctenophora). The Bilateria emerge as two large clades, the Deuterostomia and the Protostomia.

One taxon, called **Ecdysozoa**, unites all animals that periodically molt an exoskeleton. It includes arthropods and their allies (together Panarthropoda) and cycloneuralians, but not annelids, which do not molt.

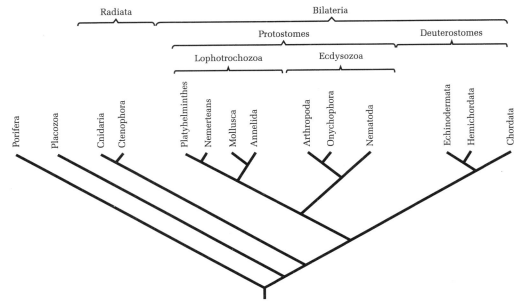

Figure 2.7 B
Phylogenetic relationships within major animal groups. Shows that chordates are deuterostomes along with hemichordates and echinoderms. The protostomes are a separate lineage,

The second taxon, **Lophotrochozoa**, encompasses all other protostomes, including the lophophorates. The Ecdysozoa grouping supports the monophyly of the panarthropods and provides a home for the cycloneuralians, but divorces the Annelida, casting it into the Lophotrochozoa, thus rejecting the longstanding annelid-arthropod alliance (Articulata). Because annelids and arthropods share a segmental body composed of many segments, their placement in separate evolutionary lines means that each evolved segmentation independently of the other, a conclusion vigorously opposed by most morphologist. Molecular systematics also views lophophorates as protostomes with no indication of a deuterostome alliance. Current evidence is insufficient to choose between these two alternative phylogenies.

The phylogenetic position of the lophophorates has been the subject of much controversy and debate. Sometimes they have been considered protostome with some deuterostome characters and at other times deuterostomes with some protostome characters. Brusca and Brusca contended that there is overwhelming evidence that they are a monophyletic clade and are deuterostomes. Their common possession of a lophophore is a unique synapomorphy. Other features, such as the U-shaped digestive tract, metanephridia (except in ectoprocts), and tendency to secrete outer casings may be homologous within the clade, but they are convergent with many other taxa.

The presence of some protostome characters suggests that the deuterostomes may have branched from the protostome line by way of a lophophorate type of ancestor. Brusca and

Brusca (1990) concluded that, although the lophophorates were derived from a common stock, their relationships were too uncertain to provide a cladogram.

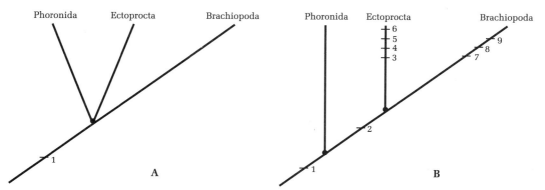

Figures 2.8 A, B
Lophophorate phylogeny. A. Without assumptions about the ancestral form, other than the origin of the lophophore (1), a trichotomy results where each group arises independently from the ancestor. **B.** Assuming a phoronid-like ancestor, the ectoprocts and brachiopods form a clade defined in part by the reduction of the circularoty system (2). The ectoprocts and brachioppds are subsequently defined by their unique synapomorphies. For ectoprocts these are: (3) colonial lifestyles, (4) retractable lophophore, (5) loss of nephridia, and (6) production of brown bodies. For brachiopods: (7) unique mantle and shell, (8) lophophoral skeleton, and (9) pedicle.

All **lophophorates** (Fig. 2.8) are suspension feeders, and most of their evolutionary diversification has been guided by this function. The tubes of phoronids vary according to their habitat. **Phoronids** (Fig. 2.10) are the least abundants of the lophophorates, living in tubes mostly in shallow coastal waters. **Ectoprocts** (Fig. 2.11) are abundants in marine habitat, living on a variety of submerged substrata,and a number of species are common in fresh water. Various ectoproct tend to build their protective exoskeletons of chitin or gelatin, which may or may not be impregnated with calcium and sand. **Brachiopods** (Fig. 2.12, Fig. 2.13) were a widely prevalent phylum in the paleozoic era but have been declining since the early Mesozoic era. They have a dorsal and a ventral shell and usually attached to the substrate directly or by means of a pedicel. Brachiopod variations occur largely in their shells and lophophores. The Phoronida, Ectoprocta, and Brachiopoda all bear a lophophore,which is a crown of ciliated tentacles surrounding the mouth but not the anus and containing the extention of mesocoel. The lophophore functions both as respiratory and a feeding structure. Because of their possession of unique synapomorphy, the lophophore, the lophophorates appear to form a monophyletic clade.

Despite continuing arguments, we remain convinced that the lophophorates are deuterostomes. They may also be monophyletic clade. In addition to the lophophore, these three phyla (Phoronida, Ectoprocta, and Brachiopoda) (Fig. 2.8) are united by their possession of U-shaped digestive tracts, peritoneal gonads, metanephridia (absent in ectoprocts), a diffuse nervous system, epistemial flaps, and a tendency to secrete outer casings. It is possible that these features are homologous within this clade but plesiomorphic or convergent with similar conditions in some other phyla. As discussed later in this section, each of the three lophophorate phyla displays enough derived character states to merit separate taxon status, but the idea of a "superphylum" (perhaps Lophophorata) may be warranted.

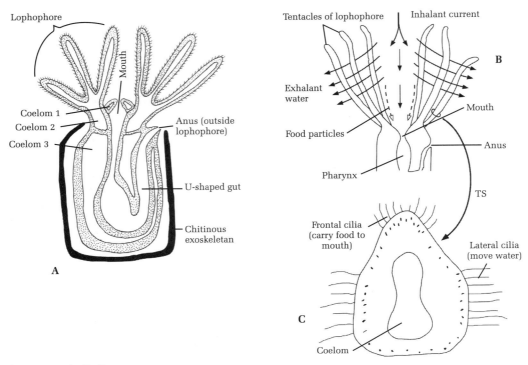

Figures 2.9 A, B, C
A. Hypothetical archetype for the lophophorate groups. **B.** Structure of the lophophore, showing the water currents and feeding tracts; **C.** The relevant ciliation on an individual tentacle (TS).

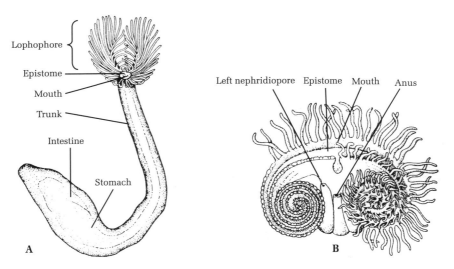

Figures 2.10 A, B
A. *External features* of a *phoronid* worm, *Phoronis architecta*; **B.** Dorsal view of the anterior end of *Phoronis australis*. The tentacles of the lophophore are cut away on the left side, and those of the innerridge are shortened on the right side to show their arrangement—actually all tentacles are the same length.

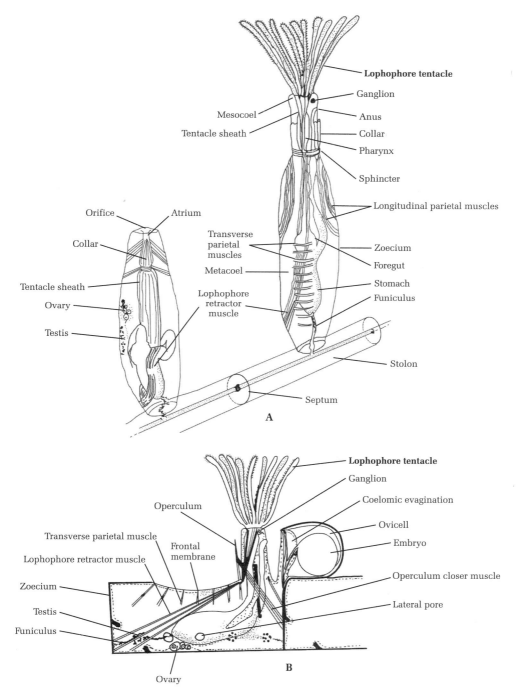

Figures 2.11 A, B

A. A colony of the bryozoan *Bowerbankia* showing two tubular zooids attached to a stolon. The left-hand zooid is fully retracted; the tentacles of the right-hand zooid are expanded; **B.** Diagrammatic section through a box-like cuboidal zooid of an encrusting bryozoan.

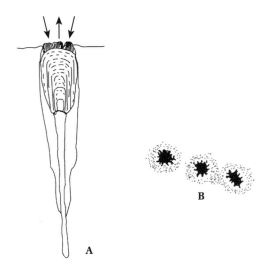

Figures 2.12 A, B
A. The brachiopod *Lingula* in feeding position in its burrow. Arrows show direction of water currents.
B. A *Lingula* burrow with its typical tripartite opening as seen from the surface of the substratum.

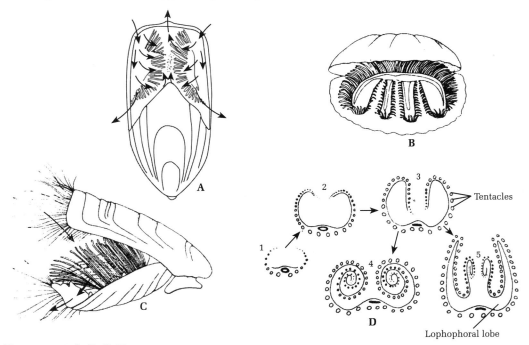

Figures 2.13 A, B, C, D
A. Dorsal view of *Lingula*; **B.** Dorso-frontal view of the quadrilobed lophophore of *Megathiris*; **C.** Lateral view of *Pumilus antiquatus* in feeding position. Arrows show the direction of the water current; **D.** The pattern of lophophore evolution in brachiopods shown schematically. Tentacles are sown as circles. Smaller circles represent younger tentacles. The primitive type 1. gains additional tentacles by becoming bilobed. 2. Further modification in this direction produces type 3. from which have evolved both the highly spiraled lophophore 4. and the horseshoe-shaped and spiraled lophophore 5.

Zimmer (1973) has critically made the case for the deuterostome nature of the lophophorates. They all show radial cleavage, enterocoely, and (except for the phoronids) a mouth that is not derived from the blastopore. In addition, the body plan and coelomic arrangement is clearly trimerous or obviously derived therefrom.

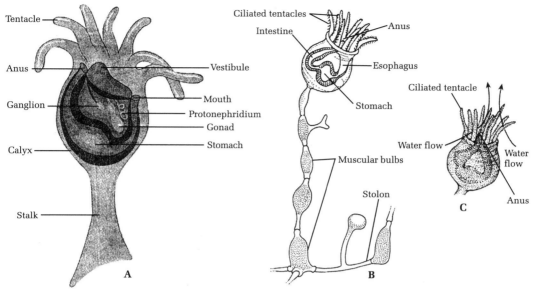

Figures 2.14 A, B, C
Phylum Entoprocta. The phyla Entoprocta lack lophophores but share other similarities with the lophophorates. Some zoologists think that all five phyla—Brachiopoda, Ectoprocta, Phoronida, Endoprocta, and Cycliophora—have evolutionary ties.

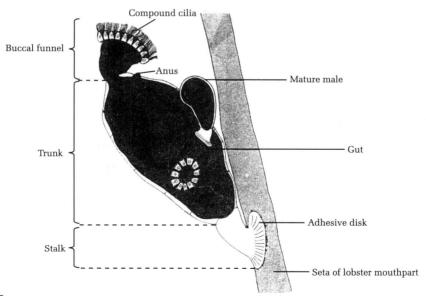

Figure 2.15
Phylum Cycliophora. The feeding stage of *Symbion pandora* attaches to the seta of the mouthpart of a lobster. A mature male is shown attached to the feeding stage, which is about 300 μm (0.3 mm) long.

An alliance between the groups Ectoprocta and Entoprocta (Fig. 2.14, Fig. 2.15). Proposed by Nielsen, is rejected on the basis of incompatibility with the idea of lophophorate unity and on direct comparative grounds. Entoprocts and cycliophora (Fig. 2.15) do not possess a lophophore as we have defined it. Furthermore, they lack any vestiges of a coelom and trimeric body plan. The feeding currents are virtually opposite in the two groups, and the methods of food capture and transport are entirely different. Entoprocts possess ducted gonads (Fig. 2.14), ectoprocts do not. Cleavage in entoprocts is spiral, whereas it is radial in ectoprocts. Larval forms and particularly metamorphosis are clearly different in the two groups. More important, if the two groups are related, then they must share common (homologous) characteristics—that is, synapomorphies. The similarities pointed out by Nielsen are superficial and common to many colonial sessile animals (e.g., budding, metamorphosis, and life cycles). The U-shaped guts are convergent adaptations to zooid life in "boxes"—no other condition would function. Thus, in the absence of unifying synapomorphies and the presence of multiple and significant differences, we can only consider one conclusion: the two groups are unrelated.

The origin of the lophophorates is puzzling, largely because it is tied, in part, to the origin of the entire deuterostome lineage, which is itself very uncertain. Most workers agree that the phoronids show the least amount of change from the presumed ancestral form.

That ancestor may have been a trimeric, coelomate, infaunal burrower. In any case, the ancestor probably evolved during the Precambrian as one evolutionary experiment with a coelomate bauplan. The first lophophorate was probably phoronid-like and became adapted to tube dwelling and feeding above the substratum. Modern phoronids may have changed little from this tube-dwelling protolophophorate.

The origin of the Ectoprocta clearly involved a reduction in body size and the development of colonial habits. The epidermal secretions became compartmentalized, with the exploitation of asexual budding as a means of colony formation. The acquisition of a retractable lophophore allowed protection of the soft tentacles. The absence of nephridia and circulatory structures provides space for the retraction of the polypide; short diffusion distances are associated with small size and the disappearance of these systems. Without nephridia as a means of gamete release, other avenues of egg and sperm escape arose in the form of coelomopores from the mesocoel and communication between the metacoel and the mesocoel.

The origin of the brachiopods (Fig. 2.12) is marked by the appearance of several novel features largely associated with the evolution of mantle folds, their secretion of valves, and the enclosure of the lophophore and body proper within the mantle cavity. The lophophore lost most of its hydraulic qualities and became more or less stationary, held by various structural support mechanisms. The circulatory system was reduced. The origin of a pedicle allowed a means of attachment in these solitary animals, supporting the body off the substratum. The first brachiopods may have been lingulid types that used the pedicle for anchorage in soft substrata.

Valentine has attempted to support a polyphyletic origin for the brachiopods, but he does recognize a monophyletic lophophorate clade, somewhat as we have described here. However, Rowell's cladistic treatment of the brachiopods, living and extinct, presents a convincing case for monophyly, although his subgroups do not correspond exactly with the Articulata-Inarticulata division.

The origin of the lophophore allowed various avenues of escape from the infaunal life of their Precambrian ancestor and the exploitation of three different lifestyles, all involving suspension feeding. However, in spite of the differences among the three phyla, and the unique qualities of each, evolution within the lophophorate clade remains obscure. Without making assumptions about the first lophophorate, the three taxa appear to have emerged separately from a common, lophophore-bearing ancestor (Fig. 2.8 A). Only by designating additional ancestral features can we eliminate the trichotomy. For example, if we assume that the ancestral lophophorate was phoronid-like, with a complete circulatory system, then the ectoprocts and brachiopods may form a distinct clade defined by the reduction and loss of the circulatory system (Fig. 2.8 B). Such assumptions carry implications about the deuterostome lineage in general and are explored.

There remain, of course, many questions and alternative hypotheses on the matter of lophophorate evolution. Not all zoologists are convinced that these animals are deuterostomes. For example, recent 18S rDNA molecular studies suggest that the lophophorates are more closely aligned with the protostomes than the deuterostomes, despite embryological and anatomical evidence to the contrary (Halanych et al. 1995; Mackey et al. 1996). Some other workers (Jeffries 1986) view the lophophorates as somewhat intermediate between the protostomes and deuterostomes.

Kingdom Animalia

This is the largest group of the animal classification. It includes the entire fauna (animal population) of the world. A brief description of the kingdom and its two divison i.e., Protostomia and Deuterostomia is given in this chapter.

It is divided into two subkingdoms: *Protozoa* and *Metazoa*.

Subkingdom I. Protozoa

Unicellular, microscopic animals. No tissues.

Phylum 1. **Protozoa** (first animals). Protozoans. About 50,000 species. Solitary or colonial. Cell organelles specialized. Single to many nuclei. Nutrition holozoic, holophytic or saprozoic. Free living commensal, symbiotic or parasitic. Freshwater, marine or moist terrestial. e.g.,: Euglena, Trypanosoma, Amoeba, Paramecium, Monocystis, etc.

Subkingdom II. Metazoa

Multicellular animals. Cells arranged in layers or tissues. Metazoans are subdivided into three branches: **Mesozoa, Parazoa** and **Eumetazoa.**

Branch 1. Mesozoa Digestive cells few, external, ciliated. No organs or tissues.

Phylum 2. **Mesozoa** (Middle animals). About 50 species. Worm-like, small, endoparasites of marine invertebrates. Body with an outer single layer of ciliated digestive cells enclosing one or several reproductive cells. e.g., s: *Dicyema, Rhopalura, etc.*

Branch 2. Parazoa Digestive cells many, internal, flagellated. No digestive cavity and mouth. tissues poorly defined, organs absent.

Phylum 3. Porifera (Pore bearers). Sponges. About 5,000 species. Body usually irregular with numerous pores and water canals, some lined by flagellated collar cells or choanocytes. Skeleton of minute calcareous spicules or of sponging fibres. Solitary or colonial. Sessile. Marine, a few freshwater. e.g.,: *Leucosolenia, Scypha, Sycon, Euspongia,* etc.

Branch 3. Eumetazoa Many-celled animals with organs, mouth and digestive cavity. Subdivided into two grades: *Radiata* and *Bilateria*.

Grade A. Radiata

Radially symmetrical, tentaculate, diploblastic animals with few organs. Digestive cavity opens externally through mouth.

Phylum 4. Coelenterata (With hollow intestine) or **Cnidaria** (Nettle-bearing) About 10,000 species. Symmetry radial or approaching bilateral. Two or three layers of cells. Mouth encircled by tentacles bearing nematocysts. Sac-like gastrovascular cavity. Sessile or free swimming. Solitary or colonial. Marine or freshwater. e.g.,: *Hydra, Obelia, Aurelia,* Corals, etc.

Phylum 5. Ctenophora (Comb-bearers). Comb-jellies. About 90 species. Symmetry biradial. 2 tentacles and 8 longitudinal rows of ciliated comb plates for locomotion. No nematocysts. No anus. Free swimming, marine. e.g.,: *Pleurobrachia, Cestum, Beroe, Ctenoplana, Coeloplana,* etc.

Grade B. Bilateria

Bilateraly symmetrical triploblastic animals with organ-systems. Digestive tract complete with anus. Mesoderm present. It is subdivided into two divisions: *Protostomia and Deuterostomia*.

Division 1. PROTOSTOMIA

Cleavage spiral and determinate. Mouth arises from or near blastopore. Protostomes are divided into 4 subdivisions:

(i) *Acoelomata,*

(ii) *Pseudocoelomata,*

(iii) *Lophophorate coelomata* and

(iv) *Schizocoelous coelomata.*

Subdivision (i) Acoelomata

No body cavity or coelom. Space between body wall and digestive cavity is occupied by mesenchym parenchyma.

Phylum 6. Platyhelminthes (Flatworms). About 15,000 species. Body dorso-ventrally flattened. Digestive tract branched or absent. No anus and circulatory system. Free-living or parasitic. Marine freshwater, a few terrestrial. e.g., *Planaria, Fasciola, Taenia,* etc.

Phylum 7. Rhynchocoela or Nemertinea (Ribbon worms). About 750 species. Body dorso-ventrally flattened and ciliated. With mouth and anus. Proboscis eversible. Mostly marine, few terrestrial and freshwater. e.g., *Cerebratulus, Lineus,* etc.

Subdivision (ii). Pseudocoelomata

Body cavity a pseudocoelom which is a persistent blastocoels, not lined by mesoderm. Anus present

Phylum 8. Acanthocephala (Spiny-headed worms). About 500 species. Minute worm-like endoparasites No digestive cavity. Protrusible proboscis with recurved spines. e.g., *Acanthocephalus, Gigantorhynchus*, etc.

Phylum 9. Entoprocta About 60 species. Sessile. Body of calyx and slender stalk. Digestive tube U-shaped. Mouth and anus close together and surrounded by a tentacular crown. Mostly marine. Solitary or colonial. e.g., Pedicellina, Loxosoma, Umatella, etc.

Superphylum Aschelminthes

Sac worms. An assemblage of pseudocoelomates with an anterior mouth, posterior anus and straight digestive tube. Predominantly aquatic. Free living, epizoic or parasitic. Incudes 5 classes

1. Class Rotifera,
2. Class Gastrotricha,
3. Class Kinorhyncha,
4. Class Nematoda and
5. Class Nematomorpha

These are also recognized as phyla.

Phylum 10 Rotifera (Wheel animalcules). About 1,500 species. Microscopic. Anterior end with a ciliated crown. Pharynx with internal jaws. Mostly freshwater, some marine. e.g.,: Philodina, Rotatoria, Hydatina, etc.

Phylum 11. Gastrotricha (Hairy stomach worms). About 175 species. Microscopic. Ventral surface flattened and ciliated. Cuticle with spines, plates or scales. Freshwater and marine. e.g., *Chaetonotus, Macrodasys*, etc.

Phylum 12. Kinorhyncha (Jaw-moving worms). About 100 species. Small. Cuticle segmented and with recurved spines. Spiny anterior end or proboscis retractile. Marine. e.g., **Echinoderms**, *Pycnophyes*, etc.

Phylum 13. Nematoda (Round worms). About 12,000 species. Body slender and cylindrical. Cuticle tough, often ornamented. Radial or biradial arrangement of structures around mouth. Free-living or parasitic. Freshwater, marine or in soil. e.g.,: *Ascaris, Trichinella, Wuchereria, Ancylostoma, Enterobius*, etc.

Phylum 14. Nematomorpha or **Gordiacea** (Horsehair worms). About 100 species. Body long, thread-like. Larval stage parasitic in insects. Adult free-living in water or damp soil. e.g., *Nectonema, Gordius*, etc.

Subdivision (iii) Lophophorate Coelomata

Coelom develops as schizocoel or enterocoel. With a crown of hollow tentacles (lophophore) surrounding mouth. Head indistinct. Digestive tract U-shaped.

Phylum 15. Phoronida. About 15 species. Worm-like unsegmented body enclosed in a chitinous tube. Lophophore horseshoe-shaped. Marine. e.g.,: *Phoronis, Phoronopsis*.

Phylum 16. Bryozoa or **Ectoprocta** (Moss animals). About 4,000 species. Sessile moss-like colonies. Body enclosed in a gelatinous, chitinous or calcareous covering. Lophophore V-shaped or circular. Mostly marine, few freshwater. e.g.,ples: *Plumatella Bugula, Pecinatella*, etc.

Phylum 17. Brachiopoda (Lamp shells). About 260 species. Body enclosd in two unequal calcareous shell valves. Lophophore W-shaped. Marine. e.g., *Lingula, Crania, Terebratulina* etc.

Subdivision (iv) Schizocoelous Coelomata

Coelom is a schizocoel which originates as a space by the splitting of the embryonic mesoderm.

Phylum 18 Priapulida. 8 species. Sausage or cucumber-shaped marine animals with a swollen anterior introvert or proboscis. Body surface covered with spines and tubercles. Peritoneum of coelom greatly reduced. e.g., *Pripulus, Halicryptus*.

Phylum 19. Sipunculida (Peanut worms). About 275 species. Body elongated and cylindrical with retractile anterior end (introvert). Lobes or tentacles around mouth. Anus dorsal. Marine. e.g.,: *Sipunculus, Aspidosiphon, Phascolosoma*, etc.

Phylum 20. Mollusca (Soft-bodies animals). About 80,000 species. Body soft unsegmented, wih ventral muscular foot. Mantle with shell glands. External limy shell of 1, 2 or 8 parts. Terrestrial, freshwater and marine. e.g.,: Chitons, Snails *(Pila)*, Mussels *(Unio)*, Squids *(Loligo)*, etc.

Phylum 21. Echiurida (Adder-tailed worms). About 60 species. Body cylindrical, unsegmented, with anterior retractile proboscis. One pair of large ventral setae below mouth. Marine. e.g., *Echiurus, Urechis*, etc.

Phylum 22. Annelida (Ringed worms). About 8,700 species. Body elongate, metamerically segmented. Setae for locomotion. Terestrial, freshwater and marine.e.g., Earthworms *(Pheretima), Nereis*, Leech *(Hirudinaria)*, etc.

Phylum 23. Tardigrada (Water bearers). About 180 species. Minute. Body segmented with 4 pairs of unsegmented legs terminating in claws. Freshwater, terrestrial and marine. e.g., *Echiniscus, Hypsibius*, etc.

Phylum 24. Onychophora (Claw-bearers). About 73 species. Worm-like unsegmented body covered by thin cuticle. A pair of anterior antennae. Many pair of short stumpy legs ending in claws. Moist soil. e.g., *Peripatus, Peripatopsis*.

Phylum 25. Arthropoda (Joint-footed animals). About one million species. Body segmented with jointed appendages. Exoskeleton chitinous. Coelom vestigial. Body cavity haemocoel. Terrestrial, freshwater and marine. e.g., Prawns, Scorpions, Flies, Centipedes, etc.

Phylum 26. Pentastomida (Tongue worms). About 70 species. Worm-like unsegmented body with two anterior appendages terminating in claws. Blood-sucking endoparasites of vertebrates. e.g., *Cephalobaena, Porocephalus*.

Division 2. DEUTEROSTOMIA

Cleavage radial and indeterminate. Mouth arises some distance away from blastopore.

Subdivision. Enterocoelous Coelomate

Coelom is an enterocoelic which originates as pouches of embryonic gut (archenterons).

Phylum 27. Chaetognatha (Arrow worms). About 50 species. Small elongated transparent body bearing postanal tail and lateral fins. Anterior end with grasping spines Planktonic and marine. e.g., *Sagitta, Spadella,* etc.

Phylum 28. Echinodermata (Spiny-skinned animals). About 6,000 species. Secondarily pentamerous radial symmetry. Calcareous endoskeleton of plates bearing external spines. A part of coelom as water vascular canals. Locomotion by tube feet. e.g., Starfish, Brittle stars, Sea urchins, Sea lilies, etc.

Phylum 29. Pogonophora (Beard worms). About 80 species. Body long, enclosed in a chitinous tube. Anterior end with one to many tentacles. No digestive tract. Deep water and marine. e.g., Siboglinum, Spirobrachina, Polybrachia, etc.

Phylum 30. Hemichordata (Acorn worms). About 80 species. Body worm-like divided into proboscis, collar and trunk. With gill slits. Embryo lacking a typical notochord. Marine. e.g.,: Acorn worms *(Balanoglossus), Cephalodiscus,* etc.

Phylum 31. Chordata. About 49,000 species. Dorsal tubular nerve cord, notochord and pharyngeal gill slits at some stage in life history. Tail postanal. Terrestrial, freshwater and marine. e.g.,s: Ascidians, Amphioxus, Fishes, Frogs, Snakes, Birds, Man, etc.

Since the time when the theory of organic evolution became the focal point for finding out relationships among groups of living organisms, zoologists have debated the question of chordate origin. It has been very difficult to reconstruct lines of descent because the earliest protochordates were probably soft-bodied creatures that stood little chance of being preserved as fossils even under the most ideal conditions. It was speculated that chordates evolved within the protostome lineage but discarded such ideas when they realized that supposed morphological similarities had no developmental basis. In the twentieth century when further theorizing became rooted in developmental patterns of animals, it became apparent that the chordates must have originated within the deuterostome branch of the animal kingdom. Deuterostomia, a grouping that includes the echinoderms, hemichordates, and chordates, has several important embryological features that clearly separate it from the Protostomia and establish its monophyly.

With only three major taxa to consider –Chaetognatha, Hemichordata, Choradata—you might think that deuterostome phylogeny would have been settled long ago, but nothing could be further from the truth. Even today, deuterostome phylogeny remains unresolved and key questions center on the pattern of relationships among the three major taxa

1. The nature of the deuterostome ancestor,
2. The relationships of Chaetognatha and Lophophorata to deuterostomes.
3. The evolutionary origin of chordates, and

Thus the deuterostomes are almost certainly a natural grouping of inter-related animals that have their common origin in ancient Precambrian seas. Several lines of anatomical, developmental, and molecular evidence suggest that somewhat later, at the base of the Cambrian period some 570 million years ago, the first distinctive chordates arose from a lineage related to echinoderms and hemichordates. Some workers suggested that Hemichordata is the sister group to Chordata, citing pharyngeal slits as a shared derived character. Others suggested that the chordate ancestry lies with an extinct free-swimming echinoderm. Despite the uncertainty of the identity of the long-sought chordate ancestor, we do know two living protochordate groups that descended from it.

In addition to the lophophorates and echinoderms, the deuterostome lineage includes three other phyla:

Phylum **Chaetognatha** (Greek, "spine-jaws"), The chaetognaths are called **arrow worms** (Fig. 2.16) and comprise about 100 species of marine, mostly planktonic creatures.

Phylum **Hemichordata** (Greek, "half-chordates"), The hemichordates (Fig. 2.17) include 85 or so species, most of which are benthic burrowers known as **tongue worms** or acorn worms.

Phylum **Chordata** (the chordates). The phylum Chordata includes three subphyla: **Urochordata** (Tunicata; the ascidians, larvaceans, and thaliaceans), **Cephalochordata** (the lancelets, or amphioxus), and **Vertebrata** (fishes, amphibians, reptiles, birds, and mammals). There are about 3,000 species of urochordates, 23 species of cephalochordates, and 46,670 species of vertebrates.

These three phyla are discussed here, and concluded with an overview of ideas about deuterostome phylogeny and comments on the origin of the vertebrates.

The first record of a chaetognath (Fig. 2.16) (arrow worm) was made by the Dutch naturalist Martinus Slabber in 1775. For nearly 100 years, as more and more descriptive work was conducted, the systematic position of the group was hotly debated. The arrow worms were at times allied with molluscs, arthropods, and certain blastocoelomates (particularly nematodes), generally within the catch-all taxon Vermes. Some of these arguments continued well into the twentieth century. Although the question of chaetognath phylogenetic affinities is still unsettled, embryological studies strongly favor a deuterostome relationship, and several unique characteristics support its separate phylum status.

Hemichordates (Fig. 2.17) were discovered in 1825 by Eschscholtz, who thought his specimen was a holothurian. Other early records allied these animals with nemerteans. Bateson conducted developmental studies on hemichordates and coined the present phylum name after recognizing similarities with chordate embryogeny. The chordate nature of tunicates (Urochordata) had been recognized by this time, on the basis of developmental studies. For many years the hemichordates were ranked as a subphylum of Chordata; but although they are clearly related, the hemichordates lack a true notochord—a defining synapomorphy of the chordates. The notochord is a dorsal, elastic, rodlike structure, derived from a mid dorsal strip of embryonic (archenteric) mesoderm, that provides structural and locomotory support in the body of larval or adult chordates.

Phylum Chaetognatha: Arrow Worms

Arrow worms (Figs. 2.16) are wholly marine largely planktonic animals of moderate size, ranging from about 0.5 to 12 cm in length. With the exception of a few benthic species (e.g., Spadella) and some that live just off the deep ocean floor (e.g., certain species of Heterokrohnia), chaetognaths are adapted to life as pelagic predators. At least one species, *Caecosagitta macrocephala*, is luminescent (Haddock and Case 1994). Arrow worms are distributed throughout the world's oceans and in some estuarine habitats. They often occur in very high numbers and sometimes dominate the biomass in mid-water plankton tows. They are most abundant in neritic waters, but some occur at great depths

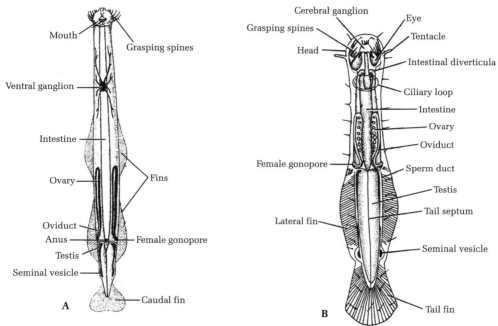

Figures 2.16 A, B
Arrow worms. A. Internal structure of *Sagitta*. **B.** The benthic chaetognath *Spadella* (dorsal view).

Characteristics of the Phylum Chaetognatha

1. Bilateral deuterostomes, with streamlined, elongate, trimeric body comprising head, trunk, and postanal tail divided from one another by transverse septa; head with single protocoel; trunk and tail with paired mesocoels and metacoels, respectively

2. Body with lateral and caudal fins, supported by "rays" (apparently derived from epidermal basement membrane)

3. Mouth surrounded by sets of grasping spines and teeth used in prey capture; mouth set in ventral vestibule; anterolateral fold of body wall forms retractable hood that can enclose vestibule

4. Longitudinal muscles of unusual type, arranged in quadrants; no circular muscle

5. No discrete gas exchange or excretory systems

6. With weakly developed hemal system between peritoneum and lined organs and tissues
7. Complete gut; anus ventral, at trunk-tail junction
8. Large dorsal (cerebral) and ventral (subenteric) ganglia connected by circumenteric connectives. Ciliary fans for detection of water-borne vibrations. Anterior ciliary loop (corona ciliata) of uncertain function. Inverted pigment-cup ocelli (of uncertain origin)
9. Hermaphroditic, with direct development. Cleavage radial, equal, and holoblastic. Mesoderm and body cavities form by enterocoely. Although blastopore denotes posterior end of body, both mouth and anus form secondarily, subsequent to closure of the blastopore
10. Strictly marine. Predatory carnivores. Largely planktonic, but some benthic species are known

The overall body form of chaetognaths, coupled with their high degree of transparency and locomotor abilities, have contributed to their great success as planktonic predators. They are frequently recovered in plankton hauls with their grasping spines firmly affixed to another animal.

Order phracmophora: With ventral transverse muscle bands (phragma) that appear whitish in living animals. Three families with about 30 species: Spadellidae (e.g., Gephyrospadella, Paraspadella, Spadella), Eukrohniidae (Aberrospadella, Bathyspadella, Eukrohnia, Heterokrohnia, Krohnittella, Kukrohnia, Zahonya), and Tokiokaspadellidae (Tokiokaspadella).

Order aphragmophora: Without ventral transverse muscle bands. Three families with about 70 species: Sagittidae (e.g., Bathybelos, Caecosagitta, Parasagitta, Sagitta), Pterosagittidae (monotypic: Pterosagitta draco), and Krohnittidae (Krohnitta).

The Chaetognath externally, are streamlined, with virtually perfect bilateral symmetry and transparent bodies, although some meso- and bathypelagic species have orange carotenoid pigmentation and some (phragmophorids) appear milky white because of the opaque ventral transverse musculature. The trimeric form includes a head, a trunk, and a tail—probably corresponding to the prosome, mesosome, and metasome of the general deuterostome bauplan. The trunk bears paired lateral fins and the tail bears a single tail fin. Chaetognath fins are simple epidermal folds enclosing a thick sheet of supportive extracellular matrix. The body surface bears various sensory structures, but the functions of most are not well understood. The head bears a ventrally placed mouth, set in a depression called the vestibule. Lateral to the mouth are heavy grasping spines, or "hooks," and in front of the mouth are smaller spines called teeth—both used in prey capture. Dorsally the head bears a pair of photoreceptors of unique structure. All chaetognaths possess an anterolateral folding of the body wall called the hood, which can be drawn over the front and sides of the head, enclosing the vestibule.

Other external features of note include a unique ciliary loop (or corona ciliata), of uncertain function, located on the dorsal surface at the head-trunk junction (Fig. 2.21). This organ consists of two rings of ciliated epithelial cells and may be involved in chemoreception, or perhaps sperm transfer. Male and female gonopores are located laterally and posteriorly in the tail and trunk, respectively. The anus is ventral at the trunk-tail junction.

Phylum Hemichordata

Hemichordates are usually described as, "**primitive chordates**" or "**invertebrate chordates**" (Fig. 2.17). Alliance with the chordates was based on the presence of gill slits and the so called notochord. It is now generally agreed that the hemichordate "notochord" is neither analogous nor homologous with the chordate notochord and that except the common possession of pharyngeal clefts the two groups are dissimilar.

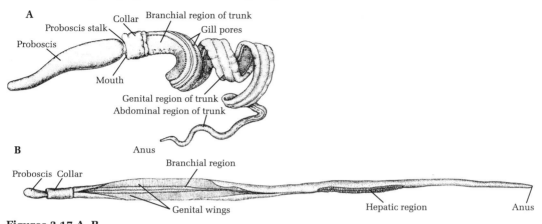

Figures 2.17 A, B
External anatomy of some representative hemichordates. **A.** *Saccoglossus* (class Enteropneusta). **B.** *Balanogossus* (Enteropneusta).

Characters

The phylum Hemichordata includes the acorn or tongue worms and their allies. They show the following characters:

1. Body is soft and unsegmented, and has wormlike or vase-like form and three distinct regions: proboscis, collar and trunk.
2. Symmetry is bilateral, i.e., right and left sides of the body are mirror images.
3. Hemichordates are triploblastic, i.e., develop from 3 germ layers: ectoderm, mesoderm and endoderm.
4. Hemichordates have organ-system level of organization.
5. There are no appendages. The collar may bear arms with tentacles.
6. The body wall consists of a single-layered epidermis with gland cells and nerve net, basement membrane, musculature of smooth mostly longitudinal fibres, and peritoneum.
7. Body cavity is a true coelom with 3 parts corresponding to the 3 body divisions: an unpaired proboscis coelom, a paired collar coelom and a paired trunk coelom. Proboscis and collar coeloms open out by dorsal pores, the coelomostomes.
8. Digestive tract is complete, either straight with terminal anus or U-shaped with anus near the mouth. Buccal diverticulum or stomochord and regarded as notochord in the past.

9. Respiration occurs by a pair to numerous pairs of gill slits.
10. Circulatory system includes a dorsal heart, two longitudinal vessels: a dorsal and a ventral, interconnected by small lateral vessels. Blood is colourless and without corpuscles.
11. Nervous system is diffuse, consisting of an epidermal plexus of nerve cells and nerve fibres. The plexus is thickened to form two longitudinal nerve cords: a dorsal and a ventral, connected together by a circumenteric nerve ring in the collar. The dorsal nerve cord may be transformed into a tubular and free collar cord in the collar. The collar cord foreshadows the future condition in chordates.
12. Excretory system comprises a proboscis gland or glomerulus situated in the proboscis and connected with blood vessels. There are no nephridia.
13. Sexes may be separate or united. The gonads may be in several pairs or only in one pair. Fertilization is external. Asexual reproduction by budding occurs in one group.
14. Development may include a free-swimming larval stage.
15. All hemichordates are marine. Some are solitary, naked and slow-moving, others sedentary, with investment and colonial. All feed on microorganisms and debris by filter or ciliary mechanism. There are about 70 species.

Classification

The phylum Hemichordata is divided into four classes:

a) Class Enteropneusta

b) Class Pterobranchia and

c) Class Planctosphaeroidea

d) Class Graptozoa

Class 1. Enteropneusta* The Enteropneusta are free, solitary, burrowing animals. The body is long (25-2500 mm.) and worm-like. Proboscis tapers anteriorly. Collar is without a lophophore. Trunk tongue bars. Alimentary canal is straight with terminal anus. Nerve cord is present in the collar. Sexes are separate. Gonads are numerous and sac-like. Development usually includes a free-swimming, ciliated tornaria larva. Asexual reproduction is lacking. The enteropneusts mainly live in shallow water. A few go deeper. One was dredged at 4500 m. off West Africa. Many live in burrows. Some occur under stones and shells. There are about 12 genera and 60 species, e.g., *Balanoglossus* (Fig. 2.22) *Saccoglossus* (*Dolichoglossus*) *Ptychodera*.

Class 2. Pterobranchia The Pterobranchia are sedentary, colonial, tube-dwelling animals. Body is quite short (1-7 mm.) and vase-like. Proboscis is shield-shaped. Collar has a lophophore of hollow, ciliated arms with tentacles. Trunk bears an aboral stalk. There is only a single pair of gill slits and even these may also be absent. The gill slits lack tongue bars. Alimentary canal is U-shaped with anus near the mouth. There is no nerve cord in the collar. Sexes may be separate or united. There is only a single pair or only a single gonad. Development may be direct or with a larval stage. Asexual reproduction occurs by budding. The pterobranchs live in both shallow and deep seas. There are only 3 genera and 10 species.

The class Pterobranchia is further divided into two orders: Rhabdopleurida and Cephalodiscida.

Order (i) **Rhabdopleurida**. The Rhabdopleurida have a two-armed lophophore, a single gonad and no gill slits. They are colonial. All the individuals in a colony are connected together by a common stolon. e.g., *Rhabdopleura* (Fig. 2.18 A).

Order (ii) **Cephalodiscida**. The Cephalodiscida have a many-armed lophophore, a pair of gonads and a pair of gill slits. They may be solitary or gregarious, several individual lying separately in a common case. e.g., *Cephalodiscus* (Fig. 2.18 B).

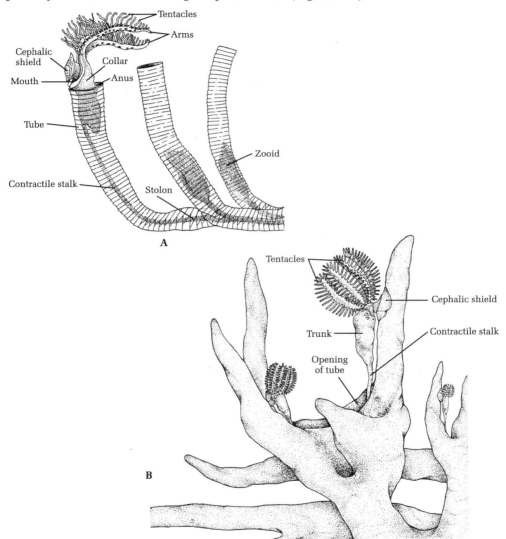

Figures 2.18 A, B
A. A colony of the pterobranch *Rhobdopleura* sp. **B.** Colony of *Cephalodiscus* sp., showing several individuals climbing about on their elaborate tubes Each zooid (excluding the stalk) is only about 2 mm long.

Class 3. Planctosphaeroidea This class is represented by a few transparent pelagic larvae with spherical form and branched ciliated bands on the surface. Alimentary canal is U-shaped. Coelomic sacs are present. e.g., *Planctosphaera pelagica*.

Class 4. Graptozoa The extinct graptolites, e.g., *Dictyonema*, included earlier in the phylum Coelenterata, are placed in the phylum Hemichordata under class Graptozoa. They were marine, colonial animals that lived in chitinous tubes in middle Cambrian to Carboniferous periods of Palaeozoic era. Their true affinity is with *Rhabdopleura*.

Affinities

Taxonomic position and phylogenetic relationship of the hemichordates was disputed for many years. They were earlier treated as a subphylum under the phylum Chordata. This classification was based on the presence in the hemichordates of a fore-gut diverticulum considered homologous to the notochord of the chordates. Recently, however, the hemichordates have been found to possess greater resemblance with the echinoderms than with the chordates. On this account, they have been given the status of an independent phylum in the non-chordates with a position close to the echinoderms.

Systematic Position of Hemichordata

The group *Enteropneusta*, to which *Balanoglossus* belongs, was established by Gegenbaur in 1870. Bateson (1885) proposed the name *Hemichordata* in place of Enteropneusta. Since then, due to their peculiar anatomical organization and embryology, the Hemichordata have been considered closer to the Chordata as well as most non-chordate phyla by different workers from time to time. Some of these views regarding the phylogenetic relationship and taxonomic position of the hemichordates are described below.

Affinities with Chordata

Some earlier workers, such as William Bateson (1885), proposed closer affinities between Hemi- chordata and Chordata. Their resemblance was based on the presence of the three fundamental chordate characteristics in Hemichordata, that is,

1. a notochord,
2. a dorsal hollow nerve cord and
3. the pharyngeal gill-slits (pharyngotremy).

Moreover, the structure and function of pharynx and branchial apparatus are similar to those of Urochordata and Cephalochordata. Also, the origin of coelom is enterocoelic in the form of five pouches from larval archenterons, as in *Branchiostoma*. Due to these similarities Hemichordata had been considered as a subphylum of the phylum choradata till recently, representing its lowest group, and probably having a common ancestry.

b) **Differences.** None of the chordate structures occurs in a typical condition in the hemichordates, for instance,

(i) The gill-slits of hemichordates are dorsal in position instead of lateral as in chordates.

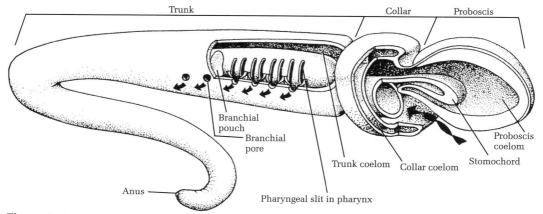

Figure 2.19
Hemichordate. Proboscis, collar, and trunk regions are shown in partial cutaway view, revealing the coelom in each region and the associated internal anatomy of the worm. Within the proboscis is the stomochord, an extension of the digestive tract. Excess water exits via the pharyngeal slits.

(ii) The so-called notochord of the hemichordates, unlike that of the chordates, is hollow, very short, ventral to the dorsal vessel, uncovered by sheaths and without a skeletal function. Moreover, it develops as a forward growth from the fore-gut and is not separated from the roof of the gut as happens in the chordates. Because of these differences, the zoologists now name it **stomochord** (Fig. 2.19). Komai has suggested that the stomochord may be homologised with the anterior pituitary gland.

(iii) The dorsal hollow nerve cord is confined to the collar region only. Elsewhere, it is replaced by an intraepidermal nerve strand without a cavity. Besides the dorsal nerve strand, there is a similar ventral nerve strand also and the two nerve strands are connected together by a circumenteric nerve-ring.

The hemichordates further differ from the chordates in—

(i) lacking cephalization, metamerism, paired appendages, tail, exoskeleton, living endoskeleton, cloaca, haemoglobin and red blood corpuscles,
(ii) having a single-layered ciliated epidermis,
(iii) containing a dorsal heart,
(iv) showing forward flow of blood in the dorsal and backward in the ventral vessel as in non-chordates,
(v) having open neurocoel, and
(vi) bearing numerous gonads.

Affinities with Echinoderms.

The hemichordates resemble the echinoderms in nervous system, coelom and larval form.

(i) **Nervous System** Both have a poorly developed nervous system consisting of an epidermal nerve plexus.

(ii) **Coelom** Both agree in the number and mode of formation of coelomic spaces. In both groups a portion of the ceolom opens out and is filled with sea-water to serve as a hydraulic mechanism. This device is not found in any other group of animals.

(iii) **Larva** Tornaria larva of Balanoglossus and bipinnaria larva of star-fish have an oval, transparent body with identical external tract with ventral mouth and posterior anus; and the same number of coelomic spaces, the anterior of which opens out by a dorsal pore.

The hemichordates and the echinoderms also have common habits and ecological niches, and possess a remarkable power of regeneration.

Resemblance of **hemichordates** with both the **chordate** and the **echinoderms** shows that all the three groups are related to one another. This relationship is strengthened by embryological, serological and biochemical studies.

a) Embryological studies reveal that in all these groups cleavage of the zygote is equipotential and radial, coelom develops opposite the blastopore, which forms the anus (deuterostomy).

b) Serological studies indicate that the proteins of chordates resembles those of hemichordates and echinoderms more closely than those of other non-chordates.

c) Biochemical studies show the presence of both phosphocreatine (a chordate energy compound) and phophoarginine in *Balanoglossus*.

The latter is, thus, a connecting link between chordates and non-chordates. From all this, it may be concluded that the three groups have a common evolutionary line possibly beginning with a dipleurula or tornaria lava. The hemichordates evolved from this ancestral line after the divergence of the ancient echinoderms, but before the rise of the true chordates. The early ancestral forms of these three groups were perhaps small and lacked hard parts. Hence, they left no fossil record. As the hemichordates arose from the common line prior to the chordates, they are often described as the **prechordates**.

Affinities with Rhynchocephalia (Nemertinea)

Feeding and burrowing habits are similar in *Balanoglossus* and Nemertinea. Body in both is elongated, vermiform, without external metamerism, with terminal anus, with smooth skin containing unicellular glands and ectodermal nerve plexes, and having metamerically arranged simple gonads. But Nemertinea differ in lacking a dorsal nerve cord and in having lateral nerve cords and a protrusible proboscis.

Affinities with Phoronida

Masterman (1897) advocated relationship of *Balanoglossus* with *Phoronis* on the following grounds:

1. Similar nature of epidermal nervous system.
2. The paired gastric diverticula of *Phoronis*, like the *buccal diverticulum* of *Balanoglossus*, forming so-called notochord.

3. Actinotroch larva of *Phoronis* has several enteropneust features of tornaria such as similar disposition of coelom, anus surrounded by a ciliary ring, presence of a proboscis pore and a sensory apical plate with cilia and eye spots.
4. Both have great power of regeneration.

Differences But, the chordate features of *Balanoglossus* like pharyngeal gills, are absent in Phoronis which also differs in having paired metanephridia. Moreover, Selys-Long Champ's (1940) account of development of phoronis does not corroborate Masterman's observations, so that relationships of these two groups are rejected.

Affinities with Pogonophora

Marcus tried to relate Hemichordata with Pogonophora due to following similarities:

1. Enterocoelous formation of coelom.
2. Body and coelom divided into three-regions.
3. Mesosome and metasome separated by a septum.
4. Nervous system intra-epidermal.
5. Pericardial sac in some pogonophores.
6. Gonads found in trunk.

Differences But pogonophores differ in having protocoelic nephridial coelomoducts and lacking an alimentary canal. Moreover, nervous system is concentrated in protosome in Pogonophora, but in mesosome in Hemichordata.

Affinities with Annelida

Spengel suggested affinities of Annelida and Hemichordata as follows:

1. Body vermiform and coelomate.
2. Burrowing habit, tubicolous life and ingesting mud which is passed out as castings through anus.
3. Collar of *Balanoglossus* (Fig. 2.20) similar to *clitellum* of earthworm.
4. Proboscis and prostomium similar and preoral.
5. Similar arrangement of blood vessels with blood flowing anteriorly in dorsal vessel and posteriorly in ventral vessel.
6. Dorsal position of heart.
7. Tornaria larva of *Balanoglossus* (Fig. 2.21) shows several structural resemblances with the trochophore larva of Annelida in being pelagic, ciliated, with apical plate, eye spots, sensory cilia and well developed alimentary canal with similar parts.

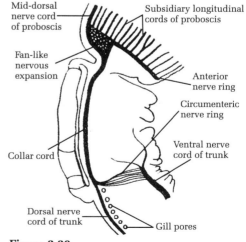

Figure 2.20

Balanoglossus Nerve cords in the anterior region of the body.

Differences However, the two groups show striking differences as follows:

1. Annelids do not have pharyngeal gill-slits, stomochord or buccal diverticulum and dorsal tubular never cord found in *Balanoglossus*.
2. *Balanoglossus* does not have double and solid ventral nerve cords and nephridia found in annelids.
3. In tornaria larva of *Balanoglossus*, preoral or proboscis coelom is present, nephridia are absent and blastopore becomes anus of the adult (Deuterostomia). In trochophore larva of annelids, preoral coelom is absent, nephridia present and the blastopore becomes the mouth (Proterostomia).

Thus, compared to their great fundamental differences, the similarities of the two groups are only superficial and quite insignificant indicating probably a convergent evolution due to similar habits and habitat.

Affinities with Echinodermata

Adult hemichordates and echinoderms and structurally quite different and it is difficult to suspect any phylogenetic relationship between them. They show few **resemblances** such as:

1. Enterocoelic origin of coelom and its division into three successive parts filled with sea water to serve a hydraulic mechanism.
2. Heart vesicle and glomerulus of enteropneusts are considered homologous to the dorsal sac and axial gland of echinoderms. Both the structures are related and combine vascular and excretory functions
3. Nervous system is poorly developed and forms epidermal nerve plexus
4. Proteins and phosphagens present in hemichordates closely resemble those of echinoderms
5. Common habits and ecological niches and remarkable power of regeneration.

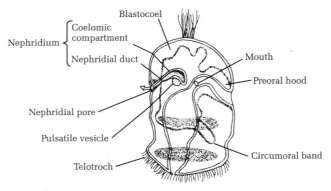

Figure 2.21

Hemichordate, tornaria larva. The simple gut begins at the mouth under a preoral hood and passes through the body of the larva. On the surface, a circumoral band of cilia runs along each side of the larva. A tuft of cilia projects from the anterior end, and the telotroch, an apron of cilia, runs along the posterior end, The excretory organ is a nephridium consisting of a coelomic compartment lined by podocytes that extends toward the exterior via a ciliated nephridial duct and opens through a nephridial pore.

Larval resemblances The two groups show a strong affinity on embryological ground as the tornaria larva of Balanoglossus has a striking structural similarity with an echinoderm larva, in particular the bipinnaria larva of asteroids. In fact, the tornaria was regarded an echinoderm larva for a long time by Johannes Muller, Krohn, Agassiz, etc., till Metschnikoff proved it to be an enteropneust larva.

The larvae of the two groups possess the following common features

1. Small, pelagic, transparent and oval
2. Identical ciliated bands taking up a similar twisted course
3. Enterocoelic origin and similar development of coelom.
4. Proboscis coelom opening to outside by proboscis pore of tornaria comparable to hydrocoel of echinoderm dipleurula.
5. Blastopore becomes the anus) and digestive tract is complete with mouth, anus and same parts.

Differences

1. However, the tornaria larva shows presence of apical plate with sensory hairs and eye spots and telotroch which are absent in echinoderm larvae.
2. The protocoel is single in tornaria but paired in echinoderm larva. This raises doubts about the echinoderm affinities of hemichordates. Fell (1993) and others believe that their larval similarities are only because of convergent evolution due to same mode of habits and habitat.

The peculiar anatomical organisation of *Balanoglossus* or hemichordates makes their systematic position uncertain and controversial. The earlier workers (Bateson, placed them as a *subphylum* under the phylum Chordata representing its lowest group. But the only chordate feature shown by them is the presence of pharyngeal gill-slits. Therefore, some recent workers like Van der Host (1939), Dawydoff, Marcus and Hyman (have chosen to remove hemichordates from the phylum Chordata and treat them as an independent invertebrate phylum. Since the group comprises only about 80 species, it is included in the category of a minor phylum.The name Hemichordata means they are "half" or "part" chordates, a fact that is undisputed.

Regarding phylogeny, the close affinities of Echinodermata, Phoronida, Pogonophora, Hemichordata and Chordata have led to the conclusion that they have arisen from a common ancestral stock, probably the dipleurula larva (Bather,), But Berril, Whitear (1957), Carter, Marcus, Hyman, Bone and many others do not contribute to this view.

Barrington's views based on the deuterostome line of chordate evolution, explain that Echinodermata deviated greatly from the ancestral stock and formed a blind branch. **Hemichordata also did not stand on the direct line of ancestry but formed a divergent offshoot from the main line of chordata evolution. Since the hemichordates arose from the ancestral line but before the divergence of the ancient Echinodermata and the rise of the true chordates, they are often called the prechordates.**

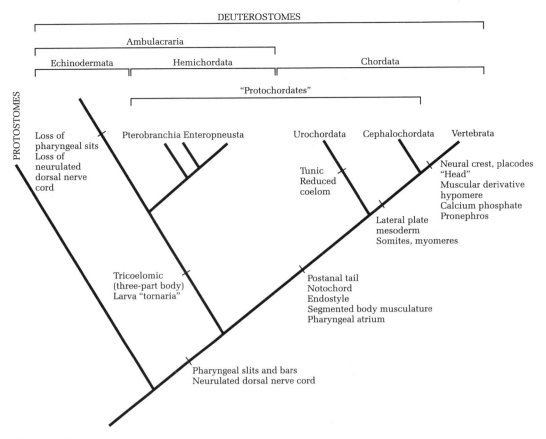

Figure 2.22
Phylogenetic relationships within the "protochordates." Protochordates are compared to echinoderms and more distantly, to protostomes.

Subphylum II. Urochordata or Tunicata

The tunicates adult body form varies, but usually lack obvious trimeric organization; body is covered by thick or thin tunic (test) of a cellulose-like polysaccharide; without bony tissue; notochord restricted to tail and usually found only in larval stage (and in adult appendicularians); gut U-shaped, pharynx (branchial chamber) typically with numerous gill slits (stigmata); coelom not developed; dorsal nerve cord present in larva stages; all marine.

Diagonastic Characters

1. Exclusively marine and cosmopolitan found in all seas and at all depths.
2. Mostly sedentary, some pelagic or free swimming.
3. Simple (solitary), aggregated in groups or composite (colonial).
4. Size 0.25 to 250 mm, shape and colour viariable.

5. Adult body degenerate, sac-like, unsegmented, without paired appendages and usually without tail.
6. Body covered by a protective tunic or test composed largely of tunicine, $(C_6H_{10}O_5)n$, similar to cellulose, hence the name *Tunicata*.
7. A terminal branchial aperture and a dorsal atrial cavity present which opens to outside through atrial aperture.
8. Coelom absent. Instead, an ectoderm-lined atrial cavity present which opens to outside through atrial aperture.
9. Notochord present only in larval tail, hence the name *Urochordata*.
10. Alimentary canal complete. Pharynx (branchial sac) large, with endostyle and two to several pairs of gill-slits. Ciliary feeders.
11. Respiration through test and gill-slits.
12. Blood-vascular system open. Heart simple, tubular and ventral. Flow of blood periodically reversed. Special vanadocytes in blood extract vanadium from sea water.
13. Excretion by neural gland, pyloric gland and nephrocytes
14. Dorsal tubular nerve cord only in larval stage, reduced to a single dorsal nerve ganglion in adult.
15. Mostly hermaphrodite. Fertilization cross and external.
16. Development indirect including a free-swimming tailed larva with basic chordate characters.
17. Metamorphosis retrogressive.
18. Asexual reproduction by budding common.

Classification

Subphylum Urochordata or Tunicata includes about 2,000 fixed and nearly 100 pelagic species. These have been variously classified by Herdman), Lahille, Garstang (1895), Perrier (1898), Hartmeyer (1909-11) and S.M. Dass (1957). The classification given has been adopted from Storer and Usinger. As usual, the subphylum Urochordata is divided into 3 classes.

Class 1. Ascidiacea

1. Solitary, colonial or compound. Bottom living.
2. Body form and size variable.
3. Test permanent, well developed and thick.
4. Atrium opens dorsally by atriopore.
5. Pharynx large with many persistent gill-slits.
6. Sexes united. Larva free-swimming and highly developed.
7. Adults usually sessile after retrogressive metamorphosis when larval notochord, nerve cord and tail are lost and brain reduced to a solid dorsal ganglion.
8. Stolon simple or none.

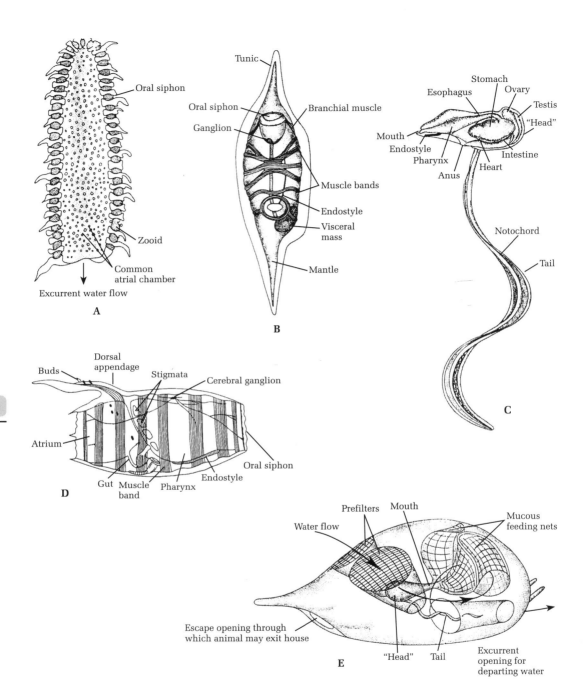

Figures 2.23 A, B, C, D, E

Thaliaceans and **appendicularians**. **A.** Sectional views of the colonial thaliacean *Pyrosoma*. **B.** A solitary thaliacean *Salpa*. **C.** The appendicularian *Oikopleura* removed from house. **D.** *Doliolum solitary,* and **E.** *Oikopleura* in its house. The arrows represent water currents.

Order 1. Enterogona

1. Body sometimes divided into thorax and abdomen.
2. Neural gland usually ventral to ganglion.
3. Gonad 1, lying in or behind intestinal loop.
4. Larva with 2 sense organs (ocelli and otolith).

Suborder 1. Phlebobranchia

1. Pharynx with internal longitudinal vessels.
2. Budding rare.
   e.g.,: *Ascidia, Ciona, Phallusia*.

Suborder 2. Aplousobranchia

1. Pharynx without longitudinal vessels.
2. Budding common.
   e.g., *Clavelina*.

Order 2. Pleurogona

1. Body compact, undivided.
2. Neural gland dorsal or lateral to ganglion.
3. Gonads 2 or more embedded in mantle wall.
4. Larva with otolith. Separate eye absent.
   e.g., Herdmania, *Botryllus, Molgula, Styela*.

Class 2. Thaliacea

1. Adults free-living, pelagic in warm and temperate seas. Solitary or colonial.
2. Body shape and size variable.
3. Tunic permanent, thin and transparent, with circular muscle bands.
4. Atriopore located posteriorly.
5. Pharynx with 2 large or many small gill-slits.
6. Sexes united, Larva formed or absent.
7. Adult without notochord, nerve cord and tail.
8. Asexual budding from a complex stolon.
9. Life history with an alternation of generations.

Order 1. Pyrosomida

1. Colony compact, tubular, closed at one end and phosphorescent in life.
2. Zooids embedded in a common test.

3. Muscle bands confined to body ends.
4. Gill-slits tall, numerous upto 50.
5. No free-swimming larval stage.
    e.g., *Pyrosoma*. (Fig. 2.23 A).

Order 2. Doliolida

1. Body characteristically barrel-shaped.
2. Muscle bands form 8 complete rings.
3. Gill-slits small, few to many.
    A tailed larva with notochord present.
    e.g., *Doliolum* (Fig. 2.23 B) *Doliopsis*.

Order 3. Salpida

1. Body cylindrical or prism-shaped.
2. Muscle bands incomplete ventrally.
3. Pharynx communicates freely with atrium through a large gill-slit.
4. Tailed larva absent.
    e.g., *Salpa, Scyclosalpa*.

Class 3. Larvacea (Appendicularia)

1. Small, solitary, free-swimming, pelagic, neotenic, larva-like forms with persistent tail, notochord, nerve cord and brain.
2. Test forming a temporary house, renewed periodically.
3. Atrium and atrial aperture absent.
4. Gill-slits 2, opening directly to outside.
5. Sexes united. No metamorphosis.

Order 1. Endostylophora

1. House bilaterally symmetrical, with separate inhalent and exhalent apertures.
2. Pharynx with endostyle.
    e.g., Oikopleura (Fig. 2.23 D).

Order 2. Polystylophora

1. Body biradially symmetrical, with single aperture
2. Pharynx without endostyle.
    e.g., *Kowalevskia*.

Systematic Position of Herdmania (Urochordata)

The true systematic status of Urochordata had long remained controversial (Fig. 2.24). Lamarck (1816) first called them *Tunicata*. Cuvier (1817) classified them along with *Mollusca*, but Lamarck put them between *Radiata* and *Vermes*. Milne Edwards(1843) created the class Molluscoidea to include Bryozoa with Tunicata, to which Huxley (1853) also added Brachiopoda. However, Kowalevsky (1886) put an end to the controversy when he described the development of a simple ascidian and established their chordate nature. Since then, the urochordates have been placed as a class or subphylum under the pylum Chordata with certainty.

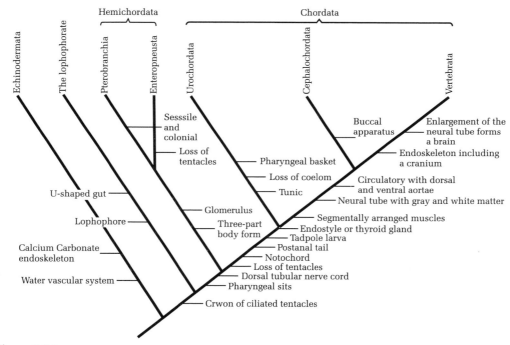

Figure 2.24
Interpretation of Deuterostomate Phylogeny. The dorsal tubular nerve cord and pharyngeal slits are possible synapomorphies that link the Hemichordata and the Chordata. The notochord, postanal tail, and endostyle or thyroid gland are important characteristics that distinguish the Hemichordata and Chordata. Some of the synapomorphies that distinguish the chordate subphyla are shown.

Affinities with Nonchordates

In some features, urochordates show some similarities with certain nonchordate groups.

1. Porifera, Coelenterata and urochordates are sessile in nature.
2. Mechanism of filter feeding and respiration through a water current is parallel with that of sponges, molluscs (oysters) and lophophorates.
3. Budding chain of new zooids is common with coelenterates and annulets.
4. Larval eyes and otocysts are found in many invertebrates.

5. Colonial mode of life of simply and composite fixed ascidians is observed in a number invertebrates.
6. Presence of typhlosole in intestine.

None of the above features establishes any non chordate relationship. These similarities are only due to similar mode of life and parallel evolution.

Affinities with Chordates

The **chordate affinities** of urochordates are beyond any doubt. The ascidian tadpole larva possesses all the basic chordate characters such as:

1. rod-like *notochord* forming axial skeleton of tail,
2. dorsal tubular nerve cord, and
3. gill-slits in the pharyngeal wall.

This is probably because both urochordates and the other chordates have originated from a common ancestor.

Affinities with Hemichordata

The assumption that nearest relatives of urochordates are the existing hemichordates, is based on the following similarities:

1. Same structural plan of pharynx perforated by gill-slits and having similar accessories.
2. Similar development of central part of nervous system.
3. Occurrence of restricted notochord.

Differences However, the two groups have important differences

1. Balanoglossus lives in burrows, but urochordates may be fixed, or pelagic.
2. Hemichordate body is divisible into proboscis, collar and trunk which is not found in urochordates.
3. A true notochard is present in the tail of ascidian larva, whereas the buccal diverticulum of *Balanoglossus* is no longer considered a notochord Thus even inclusion of hemichordates as true chordates has become doubtful so that they are considered nowadays as an independent non chordate phylum.

Affinities with Cephalochordata

The adult urochordate bears following structural similarities with *Branchiostoma*:

1. Similar ciliary filter feeding or food concentration mechanism and respiratory mechanism.
2. Large pharynx with similar accessories.
3. Branchial tentacles are similar to velar tentacles.
4. Similar endostyle and associated parts.
5. Similar atrial complex.

Besides the fundamental three chordate characteristics (notochord, dorsal tubular nerve cord and pharyngeal gill-slits the ascidian tadpole larva and Branchiostoma also share the following similarities:

1. Identical early stages of development.
2. Tail with median vertical fins without fin rays.
3. Median sensory organs (otocyst, ocellus, statocyst).
4. Pharynx with endostyle.
5. Atrial complex similar.

Affinities with Vertebrata

Ascidian tadople larva can be compared with a larval fish. It resembles higher chordates in having:

1. dorsal tubular nerve cord,
2. axial skeletal notochord,
3. pharyngeal gill-slits,
4. postanal tail covered by vertical caudal fin and
5. type of cleavage and gastrulation.

Beside, in the **adult ascidian,**

1. neural gland is homologous with vertebrate pituitary,
2. endostyle with vertebrate thyroid, and
3. typhlosole comparable to intestinal spiral valve of elasmobranch fishes.

From above account it is obvious that urochordate are primitive and degenerate descendants of ancestral chordates. The tadpole larva represents the relic of a free-swimming ancestral chordate. Degeneration is a secondary modification due to sedentary and sessile mode adult life.

Subphylum. Cephalochordata or Acrania

Lancelets (amphioxus) (Fig. 2.25) is a. Small fish like chordates with notochord, gill slits, dorsal nerve cord, and postanal tail present in adults, but without vertebral column or cranial skeleton structure; gonads numerous (25-38) and serially arranged. Marine and brackish water, usually associated with clean sand or gravel sediments in which they burrow, (e.g., Asymmetron, Branchiostoma (Fig. 2.25))

Diagonastic Characters

1. Marine, widely distributed in shallow waters.
2. Mostly sedentary and buried with only anterior body end projecting above bottom sand.

3. Body small, 5 to 8 cm long, slender, fish-like, metameric, and transparent.
4. Head lacking. Body has trunk and tail.
5. Paired appendages lacking. Median fins present.
6. Exoskeleton absent. Epidermis single-layered.
7. Muscles dorso-lateral, segmented into myotomes.
8. Coelom enterocoelous, reduced in the pharyngeal region by development of atrial cavity.
9. Notochord rod-like, persistent extending from rostrum to tail, hence the name Cephalochordata.
10. Digestive tract complete. Pharynx large perforated by numerous persistent gill-slits opening into atrium. Filter feeders.
11. Respiration through general surface. No special organs for respiration present.
12. Circulatory system well developed, closed and without heart and respiratory pigment. Hepatic portal system developed.
13. Excretion by protonephridia with solenocytes.
14. Nerve cord dorsal, tubular, without ganglia and brain. Dorsal and ventral nerve roots separate.
15. Sexes separate. Gonads numerous and metamerically repeated. Gonoducts lacking. No asexual reproduction.
16. Fertilization external in sea water.
17. Development indirect including a free-swimming larva.
18. The Cephalochordata comprise about 30 species mostly of the genus Branchiostoma and all put in the class Leptocardii.

Classification

Subphylum Cephalochordata

Class Leptocardii,

family Branchiostomidae, and only two genera

genera, Branchiostoma with 8 species and

genera **Asymmetron** with 7 species.

Asymmetron differs from *Branchiostoma* in having unpaired gonads on the right side of body and asymmetrical metapleural folds.

Branchiostoma has primitive, Degenerate & Specialized Characters of (Cephalochordta)

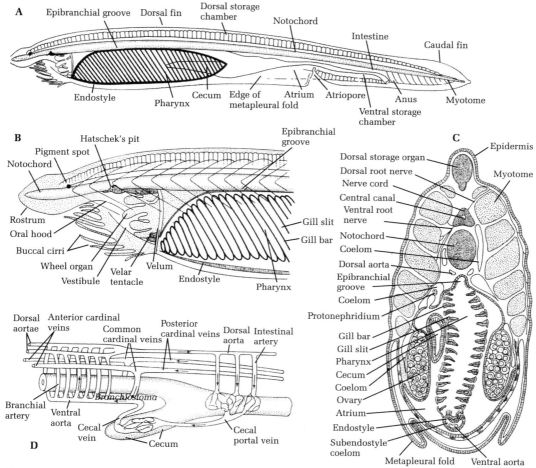

Figures 2.25 A to D

Anatomy of a cephalochordate (*Branchiostoma*). **A.** External and internal anatomy. **B.** The anterior end. **C.** The region of the pharynx (cross section). **D.** The major blood vessels in the area of the gut cecum.

1. **Primitive characters** Branchiostoma is regarded to be a primitive chordate. It differs from vertebrates because it retains several primitive chordate characters in most typical or unmodified form, as relics from its ancestors. The most prominent of these primitive characters are as enumerated below:

    a) Asymmetrical body as in echinoderms which are regarded to have common ancestry with chordates.

    b) Absence of a specialized head or cephalization.

    c) Absence of paired limbs or fins.

    d) Epidermis one-cell thick. Dermis absent.

    e) Coelom enterocoelous, arising as lateral pouches of larval archenterons.

f) Metamerically arranged muscles or myotomes.

g) Notochord persistent throughout life. Vertebral column or any other enoskeleton not developed.

h) Jaws absent. Alimentary canal straight.

i) Pharynx large, perforated by persistent gill slits and specialized for ciliary mode of feeding by drawing a water food current. Endostyle present without modification.

j) Liver represented by a midgut diverticulum.

k) Blood vascular system is simple, without a heart and any distinction between arteries and veins. Hepatic portal system is primitive.

l) No special respiratory organs and respiratory pigment.

m) Excretion by segmentally arranged protonephridia which are not coelomoducts.

n) Neural tube hollow and lying dorsally above notochord. Specialized brain lacking. Dorsal and ventral roots of spinal nerves separate. Dorsal roots without ganglia so that impulses pass directly from skin to neural tube.

o) Sensory organs simple and not paired.

p) Gonads several pairs, alike, segmentally arranged and without gonoducts.

q) Eggs are small, almost yolkless. Blastula is spherical, hollow and one-layered. Gastrulation embolic.

2. **Degenerate characters** Gregory consider that *Branchiostoma* was once a highly developed organism but due to semi sedentary mode of life underwent specialization and degeneration during later course of evolution. The degenerated characters include:

a) Poorly developed brain (cerebral vesicle) and simple sensory organs.

b) Lack of any cartilaginous or bony endoskeleton.

c) Lack of gonoducts.

3. **Specialized, peculiar or secondary characters** These changes developed perhaps due to their special mode of life and also probably prevented their further evolution. Because of their degenerate and special character, they are not considered in direct line of evolution of chordates but as a side offshoot.

a) Peculiar asymmetry of adult and early stages of development.

b) Anterior projection of notochord into rostrum making it stronger for burrowing. Over-development of notochord may be responsible for the lack of brain.

c) Mouth surrounded by oral hood with sensory oral cirri mean for filtering and concentrating food particles from water.

d) Elaborate velum with sensory tentacles to permit only small food particles to enter pharynx.

e) Large, spacious and elaborate pharynx with ciliated gill clefts which are more numerous than actual body segments, to enable sufficient food collection.

f) Wheel organ and Hatschek's groove and pit developed to help in ciliary feeding.

g) Delicate pharynx surrounded by a protective atrial cavity opening to outside through atriopore.

h) Coelom displaced and reduced due to devlopment of atrium

Systematic Position of Cephalochordata

Cephalochordates (*Branchiostoma*) are unique in showing affinities with chordates as well as nonchordates.

Non-chordate affinities

Cephalochordates have been regarded to be phylogenetically related to several non-chordate groups at one time or other. Only those with more important groups are being summarized below.

Affinities with Annelida

Some of the common features which **show resemblances** are:

1. Body bilaterally symmetrical and metamerically segmented,
2. metamerically arranged protonephridia with solenocytes (as in some polychaetes),
3. well developed coelom,
4. closed and similarly disposed blood vascular system, and
5. filter feeding method as in some polychaetes.

Differences

1. In cephalochrdates, unlike annelids, metamerism is restricted only to myotomes and gonads.
2. Coelom is enterocoelic and not schizocoelic as in annelids.
3. The flow of blood in main blood vessels is in opposite directions in the two groups.
4. Above all, the three basic chordate characters of Cephalochordata are not present in Annelida.

Affinities with Mollusca

There are some **resemblances** between these two groups.

1. It was Pallas (1778) who first described and named amphioxus as *Limax lanceolatus* considering it to be a slug
2. But the ciliary mode of feeding and respiratory machanism through water current which are common features of the two groups may be due to similar mode of life.

Differences

1. Their anatomy is completely different.
2. Moreover, molluscs are unsegmented
3. locomotory podium is absent in cephalochordates.

Affinities with Echinodermata.

1. Echinoderms have enterocoelic coelom and similarly formed mesoderm.
2. Perforations in the calyx of some fossil echinoderm,. look similar to gill-slits of amphioxus.
3. As in Branchiostoma, ophiuroids have similar features may be because of a very remote common ancestry of the two groups.

Chordate Affinities

Cephalochordata (*Branchiostoma*) shows the three basic chordate features, viz

1. The notochord,
2. dorsal tubular nerve cord and
3. pharyngeal gill-slits, in the most typical manner and there is no doubt about its chordate nature. It shows relationships with all the major groups of phylum Chordata

Affinities with Hemichordata

Hemichordata and Cephalochordata **resemble** in having similar

1. pharyngeal apparatus with numerous gill slits and gill bars,
2. filter feeding mechanism,
3. respiratory mechanism,
4. enterocoelic coelom, and
5. numerous gonads without gonoducts.

Differences

1. Body muscles in Hemichordata are unsegmented
2. nervous system distinctly of nonchordate type
3. gill-slits dorsal in position instead of lateral, and
4. postanal tail is lacking.

Moreover, inclusion of Hemichordata under Choradata is also uncertain because of doubtful nature of notochord. As such, Hemichordata are without question more primitive than Cephalochordata.

Affinities with Urochordata.

Branchiostoma (Cephalochordata) and *Herdmania* (Urochordata) are regarded to be very closely related because **of resemblances**.

1. primitive ciliary feeding and respiratory mechanisms,
2. large pharynx bearing numerous lateral gill slits, epipharyngeal groove, endostyle and peripharyngeal bands,
3. an ectoderm-lines atrial cavity opening to outside through atriopore (atrial siphon),
4. identical early stages, (holoblastic cleavage, gastrulation by invagination) of development, and
5. the ascidian larva having a continuous notochord, above it a dorsal hollow nerve cord, and a post-anal tail with median caudal fin without fin rays.

Differences But the adult urochordates are extremely degenerate and sedentary animals having several features unrepresented in cephalochordates, such as

1. body unsegmented,
2. covered by a test made of cellulose,
3. with enterocoelic coelom,
4. without notochord, and hollow nerve cord,
5. with a liver,
6. a well-developed muscular heart covered by peritoneum,
7. without nephridia,
8. sexes united with hermaphrodite gonads and
9. larva undergoing retrogressive metamorphosis to become the adult.

These differences show that inspite of close similarities reflecting upon a probable common ancestry, the cephalochordates are better evolved than the urochordates.

Affinities with Cyclostomata

The *Ammocoete* larva of lamprey (Fig. 3.8) (Cyclostomata) and Branchiostoma show a striking similarity in many characters, such as:

1. elongated, slender fish-like body,
2. continuous dorsal median fin,
3. mouth surrounded by an oral hood and
4. guarded by a velum, and
5. pharynx having endostyle and gill slits.

Besides these fundamental chordates characters, their adults show metameric myotomes, persistent gill slits, velum and a postanal tail.

Affinities with Other Vertebrates

Besides cyclostomes, *Branchiostoma* also resembles other vertebrates in several ways, such as

1. metamerically arranged myotomes,
2. true coelom lined by mesodermal epithelium,
3. postanal tail,
4. midgut diverticulum comparable with liver,
5. well-formed hepatic portal system and
6. similar arrangement of main longitudinal vessels with forward flow of main longitudinal vessels with forward flow of blood in ventral and backward flow in dorsal blood vessel.

Differences Celphalochordates differs from cyclostomes and other vertebrates in having primitive features such as" lack of head, paired limbs, skull, vertebral column, muscular heart, red blood corpuscles, brain, specialized nephridia, atrium, numerous gonads, asymmetry, etc"

Fossil **invertebrate chordates** are rare and known primarily from two fossil beds—the Cambrian Burgess Shale of Canada and the recently discovered early Cambrian fossil beds of Chengjiang and Haikou, China. An ascidian tunicate and **Yunnanozoon,** a probable **cephalochordate**, are known Chengjiant. Slightly better known is **Pikaia** (Fig. 2.26), a ribbon-shaped, somewhat fishlike creature about 5 cm in length discovered in the Burgess Shale. The presence of myomeres and a notochord clearly identifies Pikaia as a chordate. The superficial resemblance of Pikaia to living amphioxus suggests it may be an early **cephalochordate.**

Cephalochordates (*Branchiostoma*) possess all the important chordate so that their inclusion in the phylum Chordata is beyond doubt and conclusive. However, their true systematic place in the phylum remains controversial. They are definitely more evolved than the hemichordates.Wiley) stated that Amphioxus is a prototype chordate evolution. According to Garstang (1928) and Berrill (1958), cephalochordates and vertebrates both evolved from a neotenic ascidian larva which failed to metamorphose. But this view is no longer upheld. The specialized characters of Branchiostoma indicate that it is not on the direct line of evolution of chordates.

Costa (1834) and Yarrel (1936) and more recently Gregory (1936) consider Branchiostoma to be a modified and degenerate form, some workers even considered Branchiostoma to be a permanent paedogenetic larval form of some species of cyclostomes. But the absence of craniate or vertebrate characters (head, cranium, skull, vertebral column, special sense organs, etc.) and the presence of protonephridia unknown in chordates, show that it has a very primitive character.

Branchiostoma possesses a peculiar admixture of primitive, degenerate as well as specialized or secondary characters. As such, it is regarded to be a generalized and primitive type of chordate very close to the ancestral vertebrates. But, because of its many differences, it is not included in the vertebrates. Instead it is placed in an independent subphylum **Cephalochordata.**

Although phylogenetic analysis of a group of organisms as large and diverse as the Metazoa involves such complex issues and such a magnitude of information as to be beyond the comprehension of any one individual. Many of the accomplished practitioners have a cartholic knowledge of structure and function, a talent for synthesis, and sound logical ability. Such individuals, many of whose names are mentioned in this book, have been largely responsible for our current knowledge of animal relationships and for new idea about the evolution of organs and other structures.

In this respect, phylogenetic speculations are hypotheses similar to others in science, that is, they are provisional ideas that are subject to scrutiny and reevaluation. The progress may not be rapid, but with the growing interest in evolutionary studies, future advances are assured..we can only treat the problem as a continuing challenge and constantly seek to refine our knowledge and render our discrimination more acute."

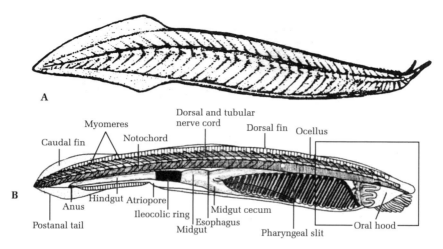

Figures 2.26 A, B
Cephalochordate. A. *Pikaia*, a possible fossil cephalochordate. B. *Bronchiostoma lanceolatum*, a living cephalochordate known as amphioxus.

Chapter 3
Agnathostomes

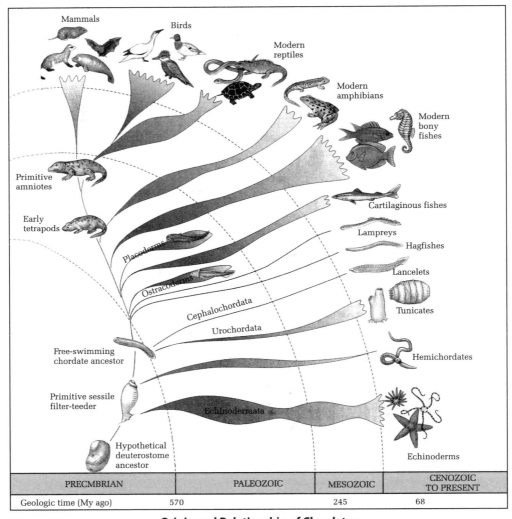

Origin and Relationship of Chordates

Figure 3.0 Phylogenetic tree of the chordates, suggesting probable origin of vertebrate and relationships. The relative abundance in numbers of species of each group through geological time, is suggested by the bulging and thinning of that group's line of descent.

Zoologists do not know what animals were the first vertebrates. Closer to the origin of vertebrates is Haikouella (Fig. 3.1 A) and Haikouichthys, (Fig 3.1 B) recently discovered in 530-million-year old sediments near Haikou. Haikouella is known from over 300 specimens, including 32 nearly complete fossils. This amazing amount of material provided considerable insight into the anatomy of these animals. It possessed several characters that clearly identify it as a chordate, including notochord, pharynx, and dorsal nerve cord, but also had several characters that are more typical of vertebrates. Haikouella seems to have had dorsal and ventral aortas, a heart, gill filaments, and a tripartite brain, although there is no evidence of a cranium. The mix of vertebrates and protochordates characters suggests that evolution of "vertebrates" soft characters may have preceded evolution of an endoskeleton.

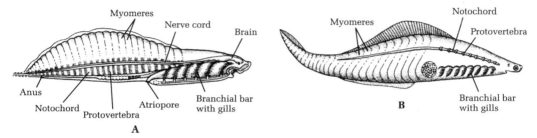

Figures 3.1 A, B
Early vertebrate fossils A. *Haikouella* from Cambrian and also. **B.** *Haikouichthys* from early Cambrian, period are early vertebrates.

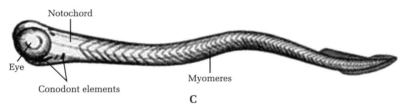

Figure 3.1 C
Conodont, whole restored animal.

Though for almost a century and a half, Conodont a tooth like microfossil known as conodont elements was an important index fossil in many geological survey. The mystery was solved in 1980 with the discovery of fossil impressions of a laterally compressed slender, soft bodied animal bearing a complete set of conodont elements in its mouth. These fossils bore evidence that conodonts were infact vertebrates. The trunk had a series of v-shaped myomeres, a notochord and caudal fin rays on what could be described as post-anal tail. Above the notochord was a streak representing a dorsal nerve cord. The unusual feeding apparatus, locomotor system and relatively large eyes suggests that conodont selected and fed on large food, not suspended material, within the marine water where they lived. In many ways conodonts differed significantly from vertebrates. The conodonts feeding apparatus was a very specialized structure. It carried on a tongue like structure, it would function similar to the lingual feeding mechanism of hagfish. The pharyngeal slits are present within at least one specimen. A lower jaw is absent. Dentin and dentin plus enamel were present in the conodont calcified feeding apparatus. Most likely bone must have evolved after conodonts.

Recent cladistic analysis of vertebrate evolution indicates that a group of fishes, called hagfishes, are the most primitive vertebrates known. Fossilized bony fragments indicate that bone was present at least 510 million years ago. These fossils are from bony armor that covered animals called ostracoderms (Fig. 3.2, Fig. 3.3). These Ostracoderms were relatively inactive filter feeders that lived on the bottom of prehistoric lakes and seas. They possessed neither jaws nor paired appendages; however, the evolution of them resulted in both jaws and paired appendages as well as many other structures. The first vertebrates were probably marine because ancient stocks of other deuterostome phyla were all marine. Vertebrates, however, adapted to freshwater very early, and much of the evolution of fishes occurred there.

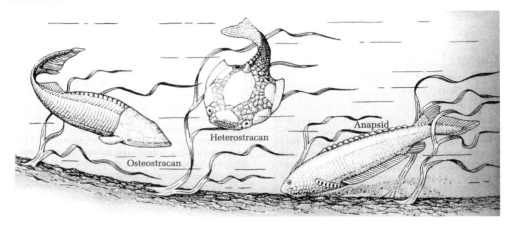

Figure 3.2 A
Three ostracoderms, jawless fishes of Silurian and Devonian times. All were probably filter feeders, but employed a strong pharyngeal pump to circulate water rather than the much more limiting mode of ciliary feeding used by their protovertebrate ancestors.

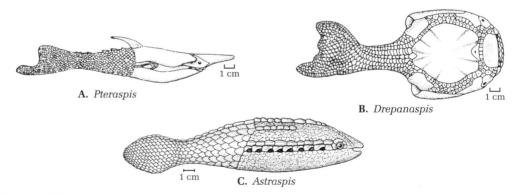

Figure 3.2 B
Pteraspidomorphs. All are extinct fishes of the early Paleozoic, with plates of bony armor that developed in the head. **A.** The heterostracan *Pteraspis*. **B.** The heterostracan *Drepanaspis*. **C.** *Astraspis* from North America.

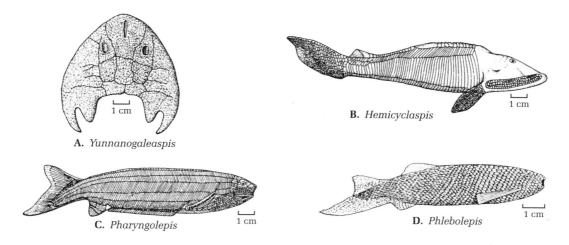

Figure 3.3
Other ostracoderms. **A.** The galeaspid *Yunnanogaleaspis*, for which only the head shield is known. **B.** The osteostracean *Hemicyclaspis*. **C.** The anaspid *Pharyngolepis*. **D.** The thelodontid *Phlebolepis*.

The taxonomy of fishes has been the subject of debate for many years. The classification used in this text divides fishes into two superclasses based on

    a) **superclass Agnatha** lack jaws and paired appendages

    b) **superclass Gnathostomata** possess jaws and paired appeddages (Fig. 3.4).

Diagonastic Characters of Cyclostomata

1. Body elongated, eel-like.
2. Median fins with cartilaginous fin rays, but no paired appendages, Tail diphycercal.
3. Skin soft, smooth, containing unicellular mucous glands but no scales.
4. Trunk and tail muscles segmented into myotomes separated by myocommata.
5. Endoskeleton fibrous and cartilaginous, Notochord persists throughout life. Imperfect neural arches (arcualia) over notochord represent rudimentary vertebrae.
6. Jaws absent in group Agnatha.
7. Mouth ventral, suctorial and circular, hence the class is named Cyclostomata.
8. Digestive system lacks a stomach. Intestine with a fold like typhlosole.
9. Gills 5 to 16 pairs in lateral sac-like pouches off pharynx, hence another name of class, Marsipobranchii. Gill-slits 1 to 16 pairs.
10. Heart 2-chambered with 1 auricle and 1 ventricle. Many aortic arches in gill region. No conus and renal portal system. Blood with leucocytes and nucleated circular erythrocytes. Body temperature variable.
11. Two mesonephric kidneys with ducts to urinogenital papilla.
12. Dorsal nerve cord with differentiated brain 8 to 10 pairs of cranial nerves.
13. Single median olfactory pit, 1 to 3 semicircular canals.

14. Sexes separate or united. Gonad single, large, without gonoduct.
15. Fertilization external.
16. Development direct or with a prolonged larval stage.

Superclass Agnatha

Members of the superclass Agnatha lack jaws and paired appendages, but possess a cartilaginous skeleton and a notochord that persists in the adult stage. Zoologists believe that ancient agnathans are ancestral to all other fishes (Fig. 3.4).

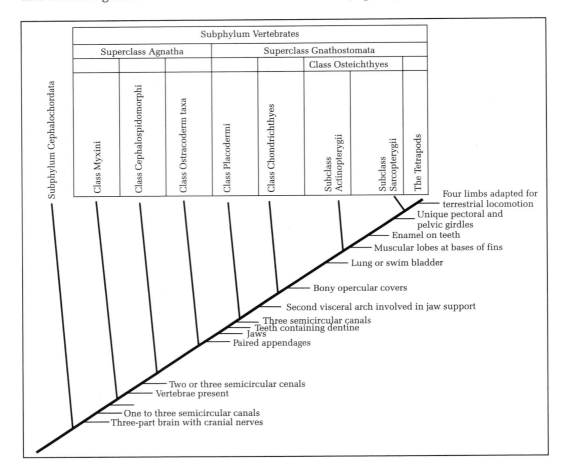

Figure 3.4
Interpretation of the Phylogeny of Fishes. This cladogram shows few selected ancestral and derived characters. Each lower taxon has numerous synapomorphies that are not shown. The position of the lampreys in fish phylogeny is debated. Recent evidence indicates that both lampreys and ostracoderms are more closely related to jawed vertebrates than to hagfishes. Most zoologists consider the ostracoderms a paraphyletic group (having multiple lineages). Their representation as a monophyletic group is an attempt to simplify this presentation. Daggers (†) indicate group whose members are extinct.

Ostracoderms are extinct agnathans that belonged to several classes. The sluggish ostracoderms had, bony armor and they used it as defense organs. Ostracoderms were bottom dwellers, often about 15 cm long. Most were probably filter feeders, either filtering suspended organic matter from the water or extracting annelids and other animals from muddy sediments. Some ostracoderms may have used bony plates around the mouth in a jaw like fashion to crack gastropod shells or the exoskeletons of arthropods.

Class Myxini

Hagfishes are members of the class Myxini. Hagfishes live buried in the sand and mud of marine environments, where they feed on soft-bodied invertebrates and scavenge dead and dying fish. When hagfishes find a suitable fish, they enter the fish through the mouth and eat the contents of the body, leaving only a sack of skin and bones. The hagfishes are the most primitive group of vertebrates.

Class Cephalaspidomorphi

Lampreys are agnathans and belong to the class Cephalaspidomorphi. They are common inhabitants of marine and freshwater environments in temperate regions. Most adult lampreys (Fig. 3.5) prey on other fishes, and their larvae are filter feeders. The mouth of an adult is sucker like and surrounded by lips that are sensory and help in attachment. Numerous epidermal teeth line the mouth and cover a movable tongue like structure. Adults attach to prey with their lips and teeth, and use their tongue to rasp away scales. Lampreys have salivary glands with anticoagulant secretions and feed mainly on the blood of their prey. Some lampreys, however, are not predatory. Members of the genus Lampetra are called brook lampreys.

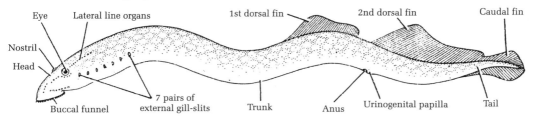

Figure 3.5
Petromyzon marinus (sea lamprey).

Agnathans constitute the oldest group of vertebrates and includes the extinct **ostracoderms** and the living **cyclostomes**. The living cyclostomes are regarded as highly specialized relics of the ostracoderm line. Stensio (1927) proposed that cyclostomes had descended from some group of ostracoderms by the evolution of a sucking mouth, loss of bony exoskeleton and paired limbs, and development of cartilage. Their structural organization has been considered higher than that of Cephalochordata but lower than that of Gnathostomata.

There were two major groups of ostracoderms:

1. The **Cephalaspides** having several pairs of gill slits, paired fin-like appendages, heterocercal tail and single median nostril.
2. The **Pteraspids** having a single pair of gill-slits, no paired appendages, and two separate nostrils.

According to Sensio the present day agnathans are derived from these lines and the lamprey and hag fishes are their representatives.

The **lampreys** of order Petromyzontiformes represent **cephalaspid line**.

The **hagfishes** and **slime eels** of the order Myxiniformes are probably derived from **pteraspid line**.

This view holds that existing cyclostomes are diphyletic in origin. The Jarvick support the idea that cyclostomes and gnathostomes betray several features and are even more primitive than the earliest recorded cyclostomes. The status or systematic position of cyclostomes remains doubtful. They represent a dead end now. There are researchers who believe that the gnathostomes probably had their origin much earlier than cyclostomes in the ostracoderm line.

About 50 species of the living jawless fishes are recognized. They belong to two major divisions as orders.

a) **Petromyzontiformes** and

b) **Myxiniformes**

Because they possess a jawless, round mouth, they are put in the class **Cyclostomata**. The similarity of these two groups is probably the result of convergent evolution. However they show important and basic morphological differences which can be attributed to their long phylogenetic separation and different habits and habitats

Order 1. Petromyzontiformes

1. Mouth ventral, within a suctorial buccal funnel beset with many horny teeth.
2. Nostril dorsal Naso-hypophyseal sac closed behind not connected to pharynx.
3. Gill pouches and gill slits 7 pairs each, opening a separate respiratory pharynx.
4. Dorsal fin well developed.
5. Branchial basket complete.
6. Dorsal and ventral roots of spinal nerves remain separate.

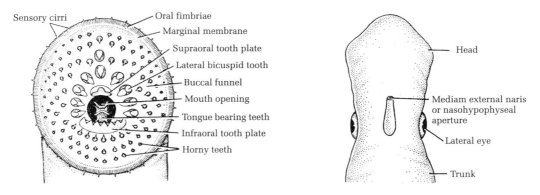

Figure 3.6
Petromyzon, Buccal funnel in ventral view. *Petromyzon*, Head in dorsal view.

7. Ear with two semicircular canals.
8. Eggs numerous, small. Development indirect with a long larval stage.
9. Both marine and freshwater forms.
   e.g.,: Lampreys, (Fig. 3.6) Petromyzon, Lampetra, Entospherus, Ichthyomyzon.

Order 2. Myxiniformes

1. Mouth terminal with 4 pairs of tentacles and few teeth. No buccal funnel.
2. Nostril terminal. Nasohyophyseal duct opens behind into pharynx.
3. Gill pouches 6 to 15 pairs. Gill slits 1 to 15 pairs.
4. Dorsal fin feeble or absent.
5. Branchial basket poorly developed.
6. Dorsal and ventral roots of spinal nerves united.
7. Ear with only 1 semicircular duct.
8. Eggs few, large, development direct.
9. Hagfishes are all marine 15 species.
   e.g., Myxine (Fig. 3.7), Eptatretus (Bdellostoma) (Fig. 3.7).

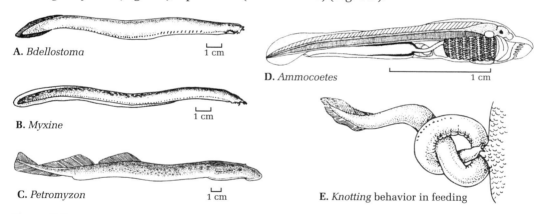

Figure 3.7
Living agnathans. Myxinoidea, hagfishes, lamprey larva, Ammocoetes and Petromyzontida, lampreys. **A.** The slime hag *Bdellostoma*. **B.** The hagfish *Myxine*. **C.** Lamprey, *Petromyzon*. **D.** Ammocoetes, lamprey larva, **E.** "Knotting" behavior. (Hagfishes are scavengers, When pulling pieces of food off dead prey, they can twist their bodies into a "knot" that slips forward to help tear pieces free).

Affinities of Cyclostomata

The chordate characteristics of cyclostomes show that they represent the most primitive members among living vertebrates. The similarity of several features of larva of lamprey with Cephalochordata indicates primitive relationship. But the adult cyclostomes have certain specialized as well as degenerate features as adaptations to a parasitic mode of life. Thus, the affinities of cyclostomes may be best explained by discussing their characters. i.e.

1. primitive, characters
2. advanced, characters
3. specialized characters, and
4. degenerate characters,

Primitive Characters

Resemblances with Cephalochordata

The adult cyclostomes and Amphioxus have many characters in common such as:

1. Lack of jaws, exoskeleton, paired fins and gonoducts.
2. Segmental muscle blocks or myotomes.
4. Numerous gill slits
5. Straight and simple alimentary canal.
6. Dorsal and ventral roots of spinal nerves separate in lamprey.

Besides these, the ammocoete larva Figs. 3.8 A, B of lampreys further resemble amphioxus in:

1. Fish-shaped body.
2. Vesitibule or oral hood anterior to mouth
3. Continuous median dorsal and caudal fins
4. Ciliated gut
5. Microphagus filter feeder
6. Endostyle functions in feeding.

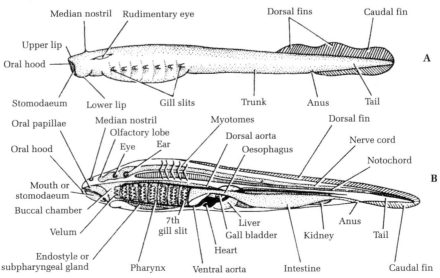

Figures 3.8 A, B
A. Ammocoete larva. External features in lateral view. **B.** Ammocoete larva showing general internal structure.

Dissimilarities from fishes

Both cyclostomes and fishes are aquatic vertebrates, but cyclostomes present many primitive characters in which they differ from fishes, such as:

1. Absence of biting jaws, scales, true teeth, paired appendages, true fin rays, girdles, ribs, stomach, spleen and gonoducts
2. Diphycercal caudal fin
3. Continuous median dorsal fin.
4. Single median nostril rather than paired
5. Incomplete or poorly developed cranium, vertebra, intestinal spiral valve, pancreas, brain, sympathetic nervous system and lateral line organs.
6. Heart S-shaped tube without conus arteriosus.
7. 9th and 10th cranial nerves not enclosed by cranium.
8. Non-medullated nerves,
9. One or two semicircular canals in ear instead of 3 of advanced vertebrates.

Affinities with ostracoderms

The oldest known fossils of vertebrates are fragments of ostracoderms belonging to Ordovician. The Ostracoderms became abundant in Silurian but died out in Devonian. Palaeontologists believe that they were the fore runners of higher fish. The fossil ostracoderms and living cyclostomes are grouped together under Agnatha due to the following structural similarities:

1. Absence of biting jaws.
2. Single nasal opening.
3. Pineal eye.
4. No paired limbs,
5. Pouch-like branchial sacs,
6. Internal ear with two semicircular canals.
7. Lateral line system

Advanced characters

Cyclostomes are undoubtedly vertebrates as they have many advanced though simple features similar to those of fishes and higher vertebrates. These are:

1. Distinct head bearing paired eyes and internal ears.
2. Brain like embryonic vertebrates with several pairs of cranial nerves.
3. Cranium for housing brain.
4. Segmental vertebrae.
5. Stratified or multilayered epidermis.

6. Dorsal root ganglia on spinal nerves.
7. Sympathetic nervous system.
8. Lateral line organs.
9. Gills primarily used for respiration and not for food collection as in Branchiostoma.
10. Water enters in pharynx by muscular activity and not by ciliary activity as in Branchiostoma.
11. E-shaped myotomes as in fishes.
12. Presence of liver, gall bladder, bile duct, pineal and parietal eyes, pancreatic cells in midgut wall and thyroid and pituitary glands.
13. Well developed circulatory system and heart.
14. Blood with erythrocytes and leucocytes.
15. Hepatic portal system.
16. Lymphatic system, and
17. Mesonephric kidneys.

Specialized characters

Adult cyclostomes may be considered either too specialized or too degenerative in many respect . It is true that many adult features are adaptations for parasitic mode of feeding.

Some of these **specialized features** are:

1. Suctorial mouth and buccal funnel with armature of horny spikes in lampreys for attachment to host body.
2. Powerful, muscular tongue heavily armed with sharp horny teeth serves as a rasping organ while feeding.
3. Production of anticoagulants in saliva to feed on blood and body fluids of prey.
4. Peculiar sac-like gill pouches located far behind head.
5. Posterior position of gill openings, probably an adaptation to burrowing.
6. Complete separation of ventral sac-like respiratory pharynx from dorsal oesophagus.
7. Respiratory water entering gill pouches as well as leaving them through external gill openings and not through mouth which mostly remains attached to rocks or fishes for feeding.
8. Prominent mucous glands secreting enormous quantities of mucus in hagfishes.
9. Single dorsal nostril high on head in lampreys.

Degenerate characters

The degenerate characters of cyclostomes include:

1. Simple, cylindrical eel-like body shape as compared to broad fish-like shape of ostracoderms.

2. Lack of bony armour or exoskeleton.
3. Lack of bony endoskeleton which is cartilaginous.
4. Absence of paired fins and girdles.
5. Vestigial eyes covered by thick skin and muscle in hagfishes.
6. Reduced liver and disappearance of gall bladder and bile duct in adult lamprey.

The Phylogenetic status of cyclostomes remains doubtful. They represent a dead end. It is quite clear that the gnathostomes had their origin much earlier than cyclostomes in the ostracoderm line.

Chapter 4

Gnathostomes

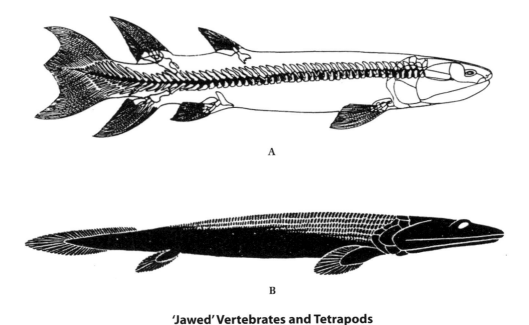

'Jawed' Vertebrates and Tetrapods
Sarcopterygians—"Rhipidistians."
These Devonian fishes are closely related to tetrapods.

Figure 4.0 **A.** The osteolepiform *Eusthenopteron* has paired pectoral and pelvic fins with internal bony supports. **B.** The panderichthyid *Panderichthys*, also equipped with paired pectoral and pelvic fins, has a flattened body, eyes on top of the head, and lacks dorsal and anal fins.

All jawed vertebrates, whether extinct or living, are collectively called gnathostomes ("jaw mouth") in contrast to the jawless vertebrates, the agnathostomes ("without jaw"). Gnathostomes are a monophyletic group since presence of jaws is a derived character state shared by all jawed fishes and tetrapods (Fig. 4.1 A). Agnathans, however, are defined principally by the absence of jaws, a character that is not unique to jawless fishes since jaws are lacking in vertebrate ancestors. Thus group, Agnatha is paraphyletic.

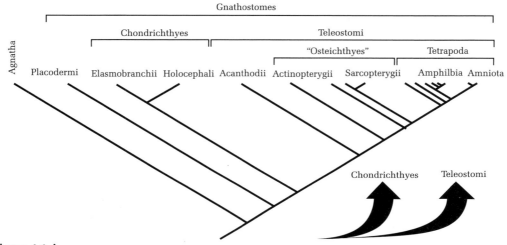

Figure 4.1 A
Phylogenetic relationships. gnathostomes, above the placoderms, evolved along two major lines—the Chondrichthyes and the Teleostomi.

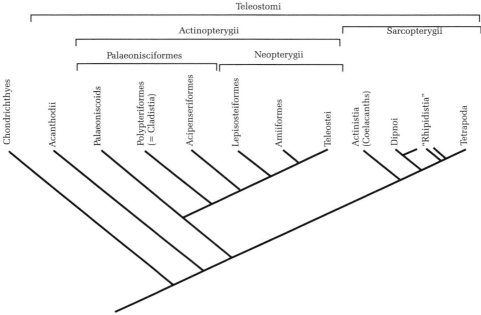

Figure 4.1 B
Teleostomi, phylogenetic relationships.

The origin of jaws was one of the most important events in vertebrate evolution. The utility of jaws is obvious since with jaws they could feed on large animals and catch active living animals in forms of food not available to jawless vertebrates. Ample evidence suggests that jaws arose through modifications of the first or second cartilaginous gill arches. The first mandibular arch may have become enlarged to assist gill ventilation, perhaps to meet the increasing metabolic demands of early vertebrates. Later, the anterior gill arches became hinged and bent forward into the characteristic position of vertebrate jaws. Evidence for this remarkable transformation is threefold.

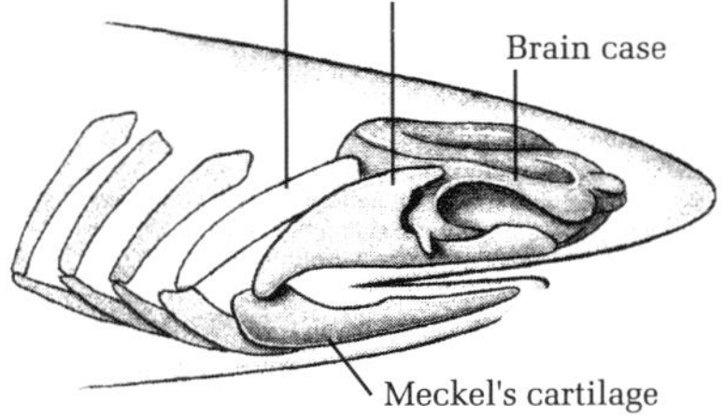

First, both gill arches and jaws are formed from upper and lower bars that bend forward and are hinged in the middle (Fig. 4.2).

Figure 4.2
Resemblance between jaws and gill supports of primitive fishes, such as Carboniferous shark, suggests that the upper jaw (palatoquadrate) and lower jaw (Meckel's cartilage) evolved from structures that originally functioned as gill supports. The gill supports immediately behind the jaws are hinged like jaws and served to link the jaws to the braincase. Relics of the transformation are seen during the development of modern sharks.

Second, both gill arches and jaws are derived from neural crest cells.

Third, the jaw musculature is homologous to the original gill support musculature as evidenced by cranial nerve distribution. This drastic morphological remodeling became the subsequent evolutionary fate of jaw bone elements—their transformation into ear ossicles of the mammalian middle ear.

An additional feature characteristic of all gnathostomes is the presence of paired pectoral and pelvic appendanges in the form of fins or limbs. These appendages originated as stabilizers to check yaw, pitch, and roll generated during active swimming. Pectoral fins of fishes are appendages usually just behind the head, and pelvic fins are usually located ventrally and posteriorly. Both sets of paired fins give fishes a more precise steering mechanism and increase fish agility. The **fin-fold hypothesis** has been proposed to explain the origin of paired fins. According to this hypothesis, paired fins arose from paired continuous, ventro-lateral folds from fin-forming zones. The addition of skeletal supports in the fins enhanced their properties of providing stability during swimming. Evidence for this hypothesis is found in the paired flaps of Myllokunmingia, Haikouichthys, and anaspids and in the multiple paired fins of acanthodians. However, pectoral fins appear in the fossil record before pelvic fins, suggesting a more complex evolutionary scenario. In one fish lineage the muscle and skeletal supports in the paired fins became strengthened, allowing them to become adapted for locomotion on land as limbs.

Recent studies of the distribution of homeobox-containing genes that control the body plan of chordate embryos (homeobox genes) suggest that Hox genes were duplicated at about the time of the origin of vertebrates. One copy of Hox genes is found in amphioxus and other invertebrates whereas living gnathostomes have four copies. Perhaps these additional copies of body-plan-controlling genes provided genetic material free to evolve a more complex kind of animal. The origin of jaws and paired appendages may be lined to a second Hox duplication, near the origin of the gnathostomes. The appearance of both jaws and paired fins were major innovations in vertebrate evolution and were, among the most important reasons for the subsequent major radiations of vertebrates that produced the modern fishes and tetrapods.

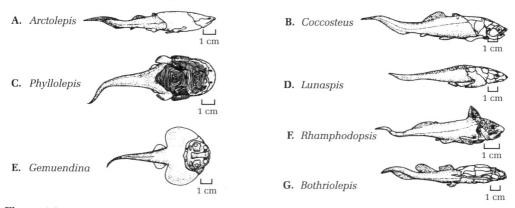

Figure 4.3
Placoderms. Most placoderms possessed a dermal armor composed of bony plates that were broken up into small scales on the midbody and tail. Many placoderms were large and most were active predators. **A.** The arthrodire *Arctoplepis*. **B.** The arthrodire *Coccosteus*. **C.** The phyllolepid *Phyllolepis*. **D.** The petalichthyid *Lunaspis*. **E.** The rhenanid *Gemuendian*. **F.** The ptyctodontid *Rhamphodopsis*. **G.** The antiarch *Bothriolepis*.

Among **the first jawed** vertebrates were the heavily armoured placoderms (Fig. 4.3). They first appeared in the fossil record in the early Silurian period (Fig. 4.4). Placoderms evolved a great variety of forms, some very large (upto 10 m in length) and grotesque in appearance. They were armored fishes covered with diamond-shaped scales or with large plates of bone. All became extinct by the end of the Devonian period and appear to have left no descendants. However, contemporary with placoderms were the acanthodians (Fig. 4.4, Fig. 4.5), a group of early jawed fishes characterized by fins with large spines, and may have given rise to the great radiation of bony fishes that dominates the waters of the world today.

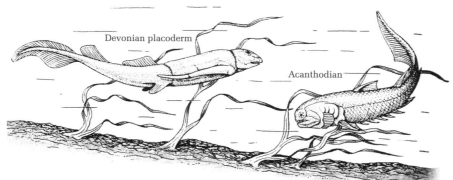

Figure 4.4
Early jawed fishes a placoderm (*left*) and an acanthodian (*right*). Jaws and gill supports from which the jaws evolved develop from neural crest cells, a diagnostic character of vertebrates. Most placoderms were bottom dwellers that on benthic animals although some were active predators. Acanthodians carried less armor than placoderms and had a bony endoskeleton and prominent spines on paired fins. Most were marine but several species entered fresh water.

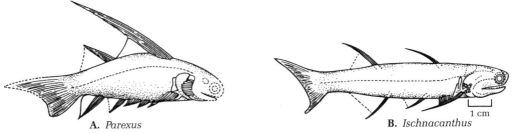

Figure 4.5
Acanthodians. The spines along the body of each that in life supported a web of skin. **A.** *Parexus*, Lower Devonian. **B.** *Ischnacanthus*, Lower Middle Devonian.

"Did ancestral fishes live in freshwater or in the sea"?

The answer to this question is not simple. The first vertebrates were probably marine because ancient stocks of other deuterostome phyla were all marine. Vertebrates, however, adapted to freshwater very early, and much of the evolution of fishes occurred there. Apparently, early vertebrate evolution involved the movement of fishes back and forth between marine and freshwater environments. The evolutionary history of some fishes took place in ancient seas, and most of the evolutionary history of others occurred in freshwater. The importance of freshwater in the evolution of fishes is evidenced by the fact that over 41% of all fish species are found in freshwater, even though freshwater habitats represent only a small percentage of the earth's water resources.

Two classes of gnathostomes still have living members:

1. The class, **placoderms** (Fig. 4.3, Fig. 4.4) contained the earliest armored jawed fishes they are now extinct and apparently left no descendants.
2. The class **acanthodians** (Fig. 4.5) another groups of ancient fishes **(extinct)**, may be more closely related to the bony fishes.
3. The **Chondrichthyes** cartilaginous fishes (Figs. 4.6 A, B).
4. The **Osteichthyes** bony fishes (Fig. 4.10 A, Fig. 4.11).

The First Jawed Vertebrates

In the mid-Silurian, as the ostracoderms were disappearing, a host of more efficient and jawed fishes appeared. Earlier grouped in a single category, they are now placed into **two** separate classes;

    a) classes **Placodermi**
    b) classes **Acanthodii**

Placoderms (Fig. 4.3) were earliest jawed vertebrates of fossil record. They appeared in Silurian, flourished in Devonian and Carboniferous and became extinct in Permian. They probably lived both in fresh water as well as seas. Some primitive agnathan ostracoderms were probably the ancestors of placoderms. But their fossil record does not show any connecting link between the jawless and the jawed fishes. Fossil evidence for the ancestry of Placodermi simply does not exist.

Acanthodians (Fig. 4.5) were little shark-like fishes covered by diamond-shaped scales. A common fossil genus, *Climatius,* was about 8 cm long. Between larger pectoral and pelvic fins, they had extra smaller paired fins some of which in the form of stout spines, hence the name "spiny sharks". Some of them featured true teeth, especially on the lower jaw. The acanthodians survived into early Permian before they became extinct.

The Important Features of placoderms show that they combined the heavy external bony armour of the ostracoderms with powerful jaws and efficient fins.

1. **External Bony armour.** In their times the placoderms were highly diversified, ranging in length from a few centimeters to 3 metres or more. The giant predator, *Dunkleosteus*, grew to 10 meters. Despite the differences among them, all were characterized by the presence of a bony skeleton. Some in particular (Dunkleosteus) featured a heavy armour of bony plates over the head and anterior part of trunk, while the rest of the body was nearly naked. The name Placodermi means "armoured fish" or "plate-skinned". Their dermal armour links them genetically with their predecessors, the ostracoderms.

2. **Powerful Jaws.** All placoderms possessed jaws which are supposed to have originated from the first pair of gill bars (mandibular arch) in front of the first gill slit (Fig. 4.2). With paired fins and bony jaws they became more active, predaceous and specialized than the jawless ostracoderms. Jaws enabled them to spread into new habitats, make use of a wider variety of food and evolve extensively. Jaws also provided protection or defense and the ability to manipulate objects. With a biting mouth, heavy external armour probably became unnecessary and tended to disappear with the result that descendants could grow still larger and swim faster. No wonder the success of jawed fishes probably contributed to the extinction of the ostracoderms and to the limitation of cyclostomes.

3. **Efficient Paired fins.** Most placoderms possess paired fins. In an aquatic environment, development of strong mobile fins was coincident with the evolution of jaws, for swimming faster. The lateral fins served very much like the elevators and ailerons of an aircraft to produce turning movements in any direction and to prevent roll, pitch, and yaw when swimming in a straight path.

Classification

Two orders of class Placodermii are recognized as follows:

Order 1. Arthrodiriformes

1. Earliest placoderms Resembled ostracoderms in appearance and habitat.
2. Heavy bony armour shields of head and trunk meeting in a movable joint.
3. Powerful gaping jaws with sharp shearing blades. Violently predaceous in Devonian times.
   e.g., Coccosteus (Fig. 4.3), Dunkleosteus, Dinichthyes.

Order 2. Antiarchiformes

1. Bottom-dwellers and mud-feeders in fresh water.
2. Ecologically similar to and competing with flat agnaths.

3. Head and anterior part of trunk covered by a depressed bony shield
4. Pectrol fins long in Devonian.
   e.g., Bothriolepis (Fig. 4.3), Pterichthyodes

Biological Significance

Placoderms were the earliest of the known fossil vertebrate creatures of the Devonian period with jaws. Their bony armour links them genetically with their predecessors. All the extinct earliest jawed fishes were once believed to be the ostracoderms., with bony jaws and paired fins, and were more specialized than ostracoderms. A new era in the history of fishes opened with the advent of jaws. "Perhaps the greatest of all advances in vertebrate history was the development of jaws and the consequent revolution in the mode of life of early fishes". There is no fossil evidence for the ancestry of placoderms. Some primitive ostracoderms were probably their ancestors. According to one view, the placoderms all became extinct by the end of the Palaeozoic era, without giving rise to any surviving forms. According to another view, the cartilaginous fishes and the bony fishes, both arose in early Devonian from some primitive group of placoderms. Thus bone, rather than cartilage, is considered a primitive characteristic.

Chondrichthyes

The Cartilaginous fishes

Members of the class Chondrichthyes include the sharks, skates, rays, and rat fishes Figs. 4.6 A, B. Most chondrichthians are carnivores or scavengers, and are marine. In addition to their biting mouth parts and paired appendages, chondrichthians possess epidermal placoid scales and a cartilaginous endoskeleton.

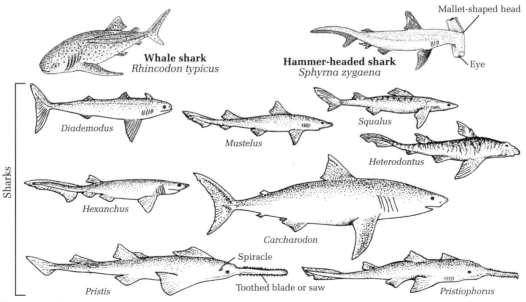

Figure 4.6 A
Class chondrichthyes. Some cartilaginous fish.

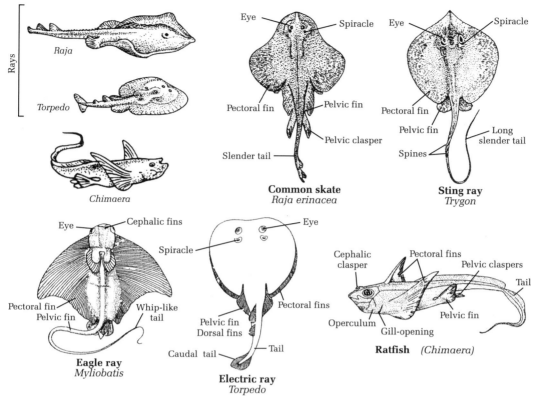

Figure 4.6 B
Chondrichthyans. B. Elasmobranchs, including various sharks and rays.

General Characters

1. Mostly marine and predaceous.
2. Body fusiform or spindle shaped.
3. Fins both median and paired, all supported by fin rays. Pelvic fins bear claspers in male. Tail heterocercal.
4. Skin tough containing minute placoid scales and mucous glands.
5. Endoskeleton entirely cartilaginous, without true bones. Notochord persistent. Vertebrae complete and separate. Pectoral and pelvic girdles present.
6. Mouth ventral. Jaws present. Teeth are modified placoid scales. Stomach J-shaped. Intestine with spiral valve.
7. Respiration by 5 to 7 pairs of gills. Gill-slits separate and uncovered. Operculum absent. No air bladder and lungs.
8. Heart 2-chambered with one auricle and one ventricle. Sinus venosus and conus arteriosus present. Temperature variable.
9. Kidneys opistho-nephric. Excretion ureotelic. Cloaca present.

10. Brain with large olfactory lobes and cerebellum. Cranial nerves 10 pairs.
11. Olfactory sacs do not open into pharynx. Membranous labyrinth with 3 semicircular canals. Lateral line system present.
12. Sexes separate. Gonads paired. Gonoducts open into cloaca. Fertilization internal. Oviparous or ovoviviparous. Eggs large, yolky. Cleavage meroblastic. Development direct, without metamorphosis.

Classification given by Storer and Usinger is followed and given below.

The **class Chondrichthyes Fig 4.6 A, B** is also called Elasmobranchii, include the sharks, rays, skates and chimaeras, comprises about 600 living species.

Subclass 1. Selachii

1. Multiple gill slits on either side protected by individual skin flaps.
2. A spiracle behind each eye.
3. Cloaca present.

Order 1. Squaliformes or Pleurotremata

1. Body typically spindle-shaped.
2. Gill slits lateral, 5 to 7 pairs. Spiracles small.
3. Pectoral fins moderate, constricted at base.
4. Tail heterocercal.
   e.g,. **True sharks**. about 250 living species. **Dogfishes** (Scoliodon, Chiloscyllium, Mustelus, Carcharinus), **spiny dogfish** (Squalus), seven gilled shark (Heptanchus), **zebra shark** (Stegostoma), **hammer-headed** (Sphyrna), **whale shark** (Rhineodon).

Order 2. Rajiformes or Hypotremata

1. Body depressed, flattened dorso-ventrally.
2. Gill slits ventral, 5 pairs.
3. Pectoral fins enlarged, fused to sides of head and body.
4. Spiracles large, highly functional.
   e.g., **Skates** and **rays**. About 300 species. Skate (Raja), stingray (Trygon), electric ray (Torpedo), eagle ray (Mylobatis), guitar fish (Rhinobatus), sawfish (Pristis) (Fig. 4.6 B).

Order 3. Holocephali

1. Single gill opening on either side covered by a fleshy operculum.
2. No spiracle, cloaca and scales.
3. Jaws with tooth plates.
4. Single nasal opening.
5. Lateral line system with open groove.
   e.g., Rat fishes or chimaeras (Fig. 4.6 B). About 25 species. Hydrolagus (Chimaera).

Osteichthyes

The Bony fishes

General Characters

1. Inhabit all sort of water—fresh, brackish or salt; warm or cold.
2. Body spindle-shaped and streamlined.
3. Fins both median and paired, supported by fin rays of cartilage or bone. Tail usually homocercal.
4. Skin with many mucous glands, usually with embedded dermal scales of three types; ganoid, cycloid or ctenoid. Some without scales. No placoid scales.
5. Endoskeleton chiefly of bone. Cartilage in sturgeons and some others. Notochord replaced by distinct vertebrae. Pelvic girdle usually small and simple or absent. Claspers absent.
6. Mouth terminal or sub-terminal. Jaws usually with teeth. Cloaca lacking, anus present.
7. Respiration by 4 pairs of gills on bony gill arches, covered by a common operculum on either side.
8. An air (swim) bladder often present with or without duct connected to pharynx. Lung-like in some Dipnoi.
9. Ventral heart 2-chambered (1 auricle and 1 ventricle). Sinus venosus and conus arteriosus present. Aortic arches 4 pairs. Erythrocytes oval, nucleated. Temperature variable.
10. Adult kidneys mesonephric. Excretion ureotelic.
11. Brain with very small olfactory lobes, small cerebrum and well developed optic lobes and cerebellum. Cranial nerves 10 pairs.
12. Well developed lateral line system. Internal ear with 3 semicircular canals.
13. Sexes separate. Gonads paired. Fertilization usually external. Mostly oviparous, rarely ovoviviparous or viviparous. Eggs minute to 12 mm. Cleavage meroblastic. Development direct, rarely with metamorphosis.

Classification

Class Osteichthyes includes a large assemblage of true bony fishes. There are well over 20,000 living species, both freshwater and marine. Some of the freshwater forms are the carp, perch, bass, trout, catfish, sucker, etc. Representatives of marine fishes are the tarpon, mackerel, tuna, sailfish, barracuda, flying fish, etc.

The classification of class Osteichthyes given below is most accepted. Only important orders of each subclass have been included

Two subclasses are recognized:

1. Subclass   Actinopterygii
2. Subclass   Sarcopterygii

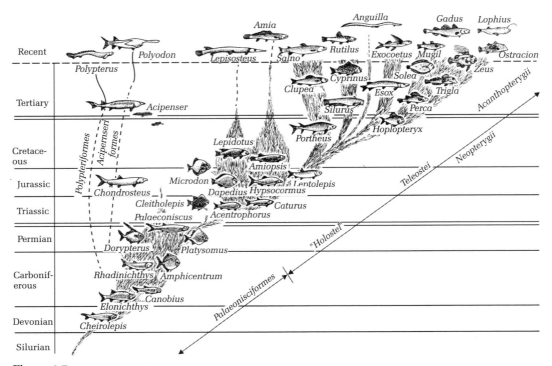

Figure 4.7
Actinopterygian phylogeny.

Subclass I. Actinopterygii

1. Paired fins thin, broad, without fleshy basal lobes, and supported by dermal fin rays.
2. One dorsal fin, may be divided.
3. Caudal fin without epichordal lobe.
4. Olfactory sacs not connected to mouth cavity.
5. Popularly called ray-finned fishes (Fig. 4.7, Fig. 4.8).

Divided into 3 **superorders:**

   a) Superorder   Chondrostei
   b) Superorder   Holostei
   c) Superorder   Teleostei

Superorder A. Chondrostei

1. Mouth opening large.
2. Scales usually ganoid.
3. Tail fin heterocercal.
4. Primitive ray-finned fish or cartilaginous ganoids.

Order 1. Polypteriformes

1. Rhomboid ganoid scales and lobed pectoral fins.
2. Dorsal fin of 8 or more finlets.
3. Ossified skeleton.
   e.g., Polypterus (bichir*) (Fig. 4.7).

Order 2. Acipenseriformes

1. Scaleless except for bony (ganoid) scutes.
2. Skeleton largely cartilaginous.
   e.g., Acipenser (sturgeon*) (Fig. 4.7), Polyodon (Fig, 4.7) (paddlefish).

Superorder B. Holostei

1. Mouth opening small.
2. Ganoid or cycloid scales.
3. Tail fin heterocercal.
4. Intermediate ray-finned fish, transitional between Chondrostei and Teleostei.

Order 1. Amiiformes

1. Thin, overlapping cycloid scales.
2. Snout normal, rounded.
3. Long dorsal fin.
   e.g., Amia (bowfin) (Fig. 4.7).

Order 2. Semionotiformes

1. Scales rhomboidal ganoid in oblique rows.
2. Snout and body elongated.
   e.g., Lepidosteus or Lepisosteus (Fig. 4.7) (garpike) (Fig. 4.8 A).

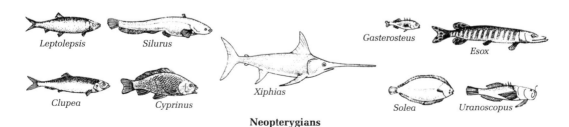

Neopterygians

Figure 4.8 A
Representative actinopterygians

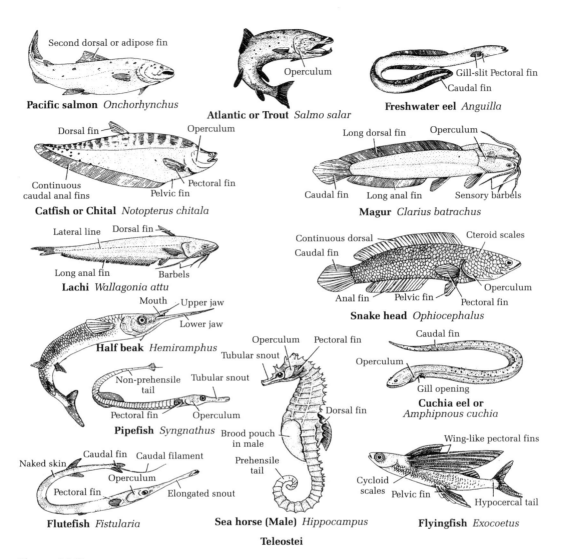

Figure 4.8 B
Representative Actinopterygians. **A.** Neopteragians, **B.** Teleostei

Superorder C. Teleostei

1. Mouth opening terminal, small.
2. Scales cycloid, ctenoid or absent.
3. Tail fin mostly homocercal.
4. A hydrostatic swim bladder usually present.
5. Advanced or modern ray-finned fishes.

Order 1. Semionotiformes

1. Scales cycloid . Head and operculum no scale
2. Fins without spines.
3. Tail fin homocercal
4. Pelvic fins abdominal
5. Air bladder with open duct to pharynx .

    e.g., Salmon (Fig. 4.8), Sardinops (Pacific sardine), Esox (pike) (Fig. 4.8 A), Notopterus (Fig. 4.8) (chital fish).

Order 2. Scopeliformes

1. Deep sea forms with phosphorescent organs
2. Mouth wide with numerous minute teeth.
3. Swim bladder absent
4. Dorsal and anal fins without spines

    e.g., Harpodon (Bombay duck).

Order 3. Cypriniformes or Ostariophysi

1. Air bladder with duct to pharynx.
2. Weberian ossicles between air bladder and internal ear

    e.g., Cyprinus (carp), Labeo rohita (rohu), Catla, Botia, Carassius (goldfish) (Fig. 4.8 B), Clarius (magur), Heteropneustes or Saccobranchus (singhi), Wallago (lachi), Mystus (tengra), Electrophorus (electric eel).

Order 4. Anguiliiformes

1. Body long and slender, snake-like.
2. Scales vestigial or absent.
3. Dorsal and anal fins long and confluent.
4. Pelvic fins, if present, abdominal.
5. Air bladder with duct.

    e.g., *Anguilla* (freshwater eel) (Fig. 4.8 B), *Muraena (moray)*.

Order 5. Beloniformes or Synentognathi

1. Scales cycloid.
2. Pectoral fins large and high on body.

    e.g., *Exocoetus and Cypselurus* (flying fishes) (Fig. 4.8 B), *Hemirhamphus* (half beak), *Belone* (garfish).

Order 6. Syngnathiformes or Solenichthyes

1. Protective scales or bony rings on body
2. Snout tubular with suctorial mouth.
3. Swim bladder closed..

   e.g., *Hippocampus* (Fig. 4.8 B) (*sea horse*), *Syngnathus* (*pipe fish*), *Fistularia* (*flute fish*) (Fig. 4.8 B).

Order 7. Ophiocephaliformes or Channiformes

1. Head depressed with plate-like scales
2. Air bladder long and without duct.
3. Accessory respiratory organs present

   e.g., (Fig. 4.8 B), *Ophiocephalus* or Channa (snake head).

Order 8. Symbranchiformes

1. Body elongated, eel or snake-like
2. Gill slits join to form a transverse ventral slit.
3. Paired fins, fin rays and air bladder lacking

   e.g., (Fig. 4.8 B), *Amphipnous, Symbranchus* (eels).

Order 9. Mastacembeliformes

1. Body eel-like.
2. Free spines in front of dorsal fin.
3. Nostrils on tubular tentacles at end of snout

   e.g., *Mastacembelus, Macrognathus*

Order 10. Perciformes

1. Fin spines present
2. Dorsal fins two
3. Weberian apparatus absent
4. Air bladder without duct.

   e.g., *Anabas* (climbing perch), *Perca* (Fig. 4.8 A) (yellow perch), *Lates* (bhetki).

Order 11. Scorpaeniformes

1. Enlarged heads and pectoral fins
2. Projecting spines from gill covering.

   e.g., Pterios (Scorpion Fish)

Order 12. Pleuronectiformes

1. Lying on one side.
2. Head asymmetrical, both eyes present on upper or dorsal side
3. Dorsal and anal fins fringing body
4. Bottom dweller, body flat.
   e.g., Flatfishes: *Pleuronectes, Synaptura*, (Fig. 4.8 B) *Solea*

Order 13. Echeneiform

1. Dorsal fin forms a flat oval adhesive disc or sucker on head.
2. Scales cycloid.
3. No Air bladder
   e.g., *Echeneis* or Remora (sucker fish).

Order 14. Tetraodontiformes or Plectognathi

1. Strong jaws with a sharp beak.
2. Scales often spiny.
3. Some inflate by swallowing water.
   e.g., *Diodon* (porcupine fish), *Tetrodon* (globe fish), *Ostracion* (Fig. 4.8 A) (trunk fish).

Order 15. Lophiiformes

1. A few dorsal flexible spines with a bulblike tip over head to lure prey into wide mouth.
2. Luminescent organs present
   e.g.; *Lophius* and *Antennarius*, (Fig. 4.8 A) (angler fishes).

Ganoid Fish

The Actinopterygians fishes belonging to the superorders Chondrostei and Holostei are known as ganoid fish (Figs. 4.9 A, B). They are all characterized by having scales covered with an enamel-like ganoin and heterocercal tail fin.

Polypterus. The bichirs, represented by species of polypterus and Calamoichthys, are restricted to the rivers of Africa. These small (less than 1 meter) fishes probably are the most primitive living actinopterygians. Their pectoral fins with fleshy bases are used as supporting limbs. The caudal fin has secondarily become homocercal. Skeleton is well ossified and intestine has a spiral valve. Dorsal fin is divided into a series of separate finlets, each with a single supporting bony ray. Body surface is well armoured with thick, interlocking, multilayered scales. The paired ventral air bladders serve as lungs. The larva resembles the tadpole of Amphibia.

Acipenser. The chondrosteans or cartilaginous ganoids are represented by sturgeons and paddle fishes which are skeletally degenerate or specaialized forms. The tail is elongate and heterocercal. The skeleton is largely cartilaginous and the intestine has a spiral fold.

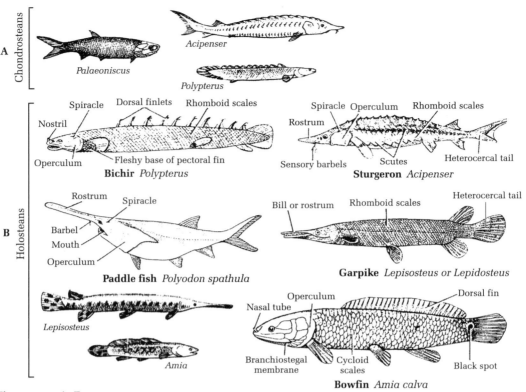

Figures 4.9 A, B
Ganoid fish **A.** Chondrostei and **B.** Holostei

Subclass 2. Sarcopterygii

1. Paired fins leg-like or lobed, with a fleshy, bony central axis covered by scales.
2. Dorsal fins two. Caudal fin heterocercal with an epichordal lobe.
3. Olfactory sacs usually connected to mouth cavity by internal nostrils or choanae, hence the previous name of subclass, Choanichthyes (Gr., choana, funnel + ichthyes, fish).
4. Popularly called fleshy or lobe-finned, or air breathing fish (Fig. 4.10).

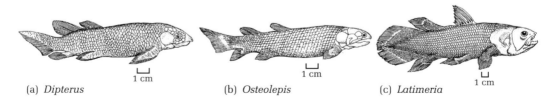

(a) *Dipterus*   (b) *Osteolepis*   (c) *Latimeria*

Figure 4.10 A
Sarcopterygians. A. *Dipterus*, fossil lungfish of the Devonian. Note the heterocercal tail. **B.** *Osteolepis*, **C.** *Latimeria* is a living sacropterygian (coelacanthiformes) exhibiting a diphycercal tail.

5. Divided into two superorders or orders:

Order 1. Crossoptergii

1. The fins of crossopterygians differ from those of all other fishes. Each is borne on a fleshy, lobe like, scaly stalk extending from the body.
2. Pectoral and pelvic fins have similarities with tetrapod limbs.

Suborder Rhipidistia

1. The fins of Eusthenopteron have been clearly homologized with those in the limbs of tetrapods (Fig. 4.10).
2. Fossils of primitive amphibians have been found together with those of rhipidistians in the same geological deposits (Fig. 4.11). Now there are enough evidences that it was through members of this suborder that evolutionary progress occurred.

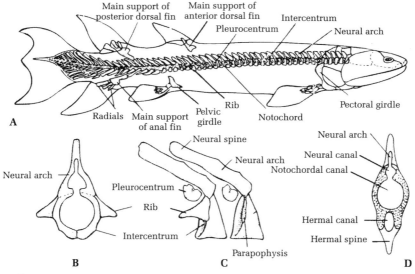

Figure 4.10 B
Axial skeleton of the fossil rhipidistian *Eusthenopteron*. A. Restored axial skeleton. Cross section. **B.** lateral views. **C.** of trunk vertebrae. **D.** Cross section of a caudal vertebra.

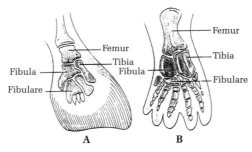

Figure 4.10 C
Origin of Tetrapod Appendages. A comparison of **A.** The fin bones of a *rhipidistian* and **B.** The limb bones of a *tetrapod* suggests that the basic bone arrangements in tetrapod limbs were already present in primitive fishes.

Suborder Coelocanthini

1. The coelacanth crossopterygians showed little change for millions of years.
2. There is no doubt that *Latimeria* (Fig. 4.11 A) is a survivor of the oldest stock represented among living vertebrates. It would seem that the ancient rhipidistians did not persist long as fish, but gave rise to coelacanths on the one hand and land vertebrates on the other.

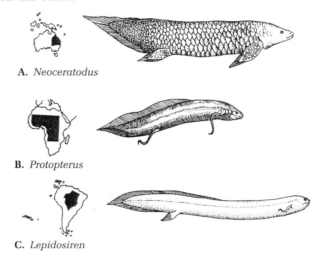

Figure 4.11 A
Sacropterygians—Living lungfishes. A. Australian lungfish, *Neoceratodus*. **B.** African lungfish, *Protopterus*. **C.** South American lungfish, *Lepidosiren*.

Order 2. Dipnoi

Three genera of lungfishs (Fig. 4.11 B) are included in this order.

1. The Australian lung fish-*Epiceratodus*
2. The south American *Lepidosiren*
3. The south African *Protopterus* –they all posses a pair of internal nares or nostrils.

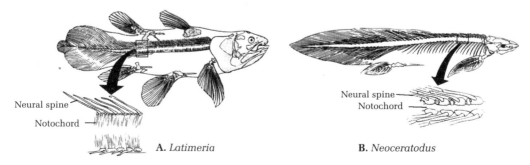

Figure 4.11 B
Axial skeletons of living sarcopterygians. A. Enlarged lateral view of posterior axial skeleton of the coelacanth *Latimeria*. **B.** Enlarged lateral view of trunk vertebrae and notochord of the lungfish *Neoceratodus*.

Diagonastic characters of Dipnoi

1. Median fins continuous to form diphycercal tail.
2. Premaxillae and maxillae absent.
3. Internal nares present and spiracles absent.
4. Air bladder single or paired, lung-like.

   e.g., Lung fishes. Only 3 living genera: Epiceratodus, Protopterus and Lepidosiren.

Significance of Crossopterygii

The crossopterygians were dominant fishes of fresh water in Devonian period. By the end of the Palaeozoic most were extinct. One branch, the coelacanth, persisted through Mesozoic as large predaceous marine fishes, to become finally extinct in Cretaceous. With the exception of the sole living coelacanth, Latimeria, in the coast of South Africa, crossopterygians are all extinct now,

They are of special interest because of their resemblance *to **amphibians***:

1. The skeletal elements of paired fins lobes resembled the proximal skeletal elements of tetrapod limbs.
2. Skull was similar to that of earliest amphibians.
3. Freshwater forms migrated from one body of water to another because they could use their air bladders like lungs.
4. Many had internal nares piercing the roof of mouth cavity, although not meant for breathing. Because of these and other morphological traits, one group of extinct Crossopterygii, the freshwater carnivorous rhipidistians, are believed to have given rise to the earliest amphibians. There is evidence to show that this group was ancestral not only to Amphibia and all bony fishes, but also to all higher terrestrial vertebrates.

Dipnoi (Lung-fish)

The Dipnoi, commonly known as lung fish, are considered to be specialized or degenerate descendants of the more primitive lobe fins fishes which they closely resemble. They were once supposed to be the ancestors of the amphibians, a view no longer held. They are large, bizarre fishes represented only by three living genera and furnishing an example of discontinous distribution. In all three, the air bladder serves functionally as a lung by which they can breathe air when necessity arises.

The dipnoans, belonging to the order Dipnoi of the subclass Sorcopterygii of Class Osteichthyes, are generally called "**lung fishes**". The name Dipnoi means double breathers", as they respire through gills as well as lungs.

Classification

Order Dipnoi is divided into 2 suborders.

Suborder. 1. Monopneumona

1. Lung single

2. Lateral jointed rays of archipterygium (paired fins) well developed.e.g., Living Australian neoceratodus forsteri and extinct Triassic Deratodus.

3. The endangered monotypic Australian lungfish, **Neoceratodus** (Epiceratodus forsteri), restricted to fresh waters, attains a length of 1.5 meters or more. It has a single lung (**Monophneumona**) and cannot survive out of water.

Suborder 2. Dipneumona

1. Lung double.
2. Lateral rays of archipterygium (paired fins) vestigial or absent.

   e.g., African Protopterus and Sough American lepidorsiren paradoxa.

While the African lungfish, **Protopterus,** has four or so related species. These two genera have weekly developed gills so that they will drown if prevented from using their paired lungs **(Dipneumona)**. They live on muddy swamps or marshes. During drought, when the marshes dry up, they aestivate by secreting mucous cocoons in mud, in which they become dormant and breathe atmospheric air through burrow openings. These 1 to 2 meters long, elongated, eel-like or snake-like forms have unique filamentous but highly mobile paired appendages. Lungfishes are mostly cartilaginous. They have specialized crushing tooth plates to feed effectively upon shell fish. Their embryogeny is more primitive than that of other living bony fishes. Cleavage of eggs is holoblastic and gastrulation occurs by invagination. Their tadpole like larva has external gills as accessory respiratory structures.

Dipnoi

Affinities of Dipnoi

With special features of their own, the Dipnoi combine characteristics in which they resemble different groups of fishes as well as Amphibia.

1. **Affinities with fishes.** Lung fishes are true fishes beyond doubt as they resemble them in general in the following features:

   a) Body spindle-shaped and streamlined.
   b) Locomotory appendages fin .
   c) Diphycercal caudal fin.
   d) Largely ossified, slender dermal fin rays.
   e) Body covered by overlapping cycloid scales.
   f) Notochord persistent.
   g) Vertebrae without centra.
   h) Skull with little ossification and with several investing bones.
   i) Branchial arches 4 to 6 pairs present.
   j) Aquatic respiration by gills.
   k) Lateral line sensory system.

2. **Affinities with Elasmobranchii.** Lung **fishes resemble** cartilaginous fishes in the following primitive characters:
   a) Intestine with a spiral valve.
   b) Nephrostomes lacking in kidney tubules.
   c) Similar diencephalon.
   d) Similar female reproductive system.
   e) Each gill arch with 2 efferent arteries.
   f) Similar conus arteriosus.

   The main **differences** from **elasmobranches** are presence of diphycercal tail, operculum, absence of claspers and external fertilization.

3. **Affinities with Holocephali.** According to Jarvick, dipnoans and holocephalians resemble remarkably with each other as follows:
   a) Teeth fused into dental plates on jaws.
   b) Gills covered by operculum.
   c) Operculo-gular membranes of both sides fused.
   d) Intestine with a spiral valve.
   e) Jaw suspension autostylic.
   f) Excurrent nostrils opening into mouth cavity.
   g) Similar kidneys, gonads and ducts.
   h) Similar cranial muscles.
   i) Each gill arch with 2 efferent arteries.

   Main **differences** from **holocephalians** are presence of lungs, absence of claspers and external fertilization.

4. **Affinities with Actinopterygii.** Lung fishes resemble subclass Actinopterygii in the following characters:
   a) Blunt snout.
   b) Body covered with cycloid scales.
   c) Paired inferior jugular veins.
   d) Powerful palatine and splenial teeth.
   e) Presence of swim bladder.
   f) Presence of operculum over gills.

   However, the Actinopterygii belong to a separate evolutionary line. They have thin, broad fins modified for swimming, and the external nostrils never penetrate into the mouth. Most of them are small in size, have reduced snout, large eyes, single separate dorsal fin and a homocercal caudal fin.

5. **Affinities with Crossopterygii.** The two orders, Dipnoi and Crossopterygii were included under the subclass Sarcopterygii by Romer. Instead of specializing for aquatic habitats, they have adapted for a semiaquatic existence. They show many close similarities as follows:

   a) Powerful leg-like lobate fins.
   b) Caudal fin diphycercal.
   c) Internal nostrils piercing roof of mouth cavity in some cases.
   d) Similar skull bones.
   e) Vertebral column reaching upto the end of caudal fin.
   f) Air bladder modified as a lung with good pulmonary circulation.
   g) Contractile conus arteriosus.
   h) Larval forms in some cases with external gills as accessory respiratory organs.

   Besides, the fossil crossopterygians rhipidistians such as Devonian Osteolepis and fossil dipnoans such as Devonian Dipterus they show even closer affinity in:

   (i) Similar body shapes and sizes.
   (ii) Separate 2 dorsal, 1 anal and 1 heterocercal caudal fin, supported by dermal bony rays.
   (iii) Paired fins somewhat lobate with a fleshy scaly central axis. Pectoral fins placed high.
   (iv) Presence of internal nares in some.
   (v) Cycloid scales modified from cosmoid
   (vi) Similar number and disposition of dermal bones on skull and pectoral girdle.
   (vii) No vertebral centra.
   (viii) Similar opercular and gular bones.
   (ix) Similar lateral line sensory system.
   (x) Comparable lower jaw.

   This led to the belief that the dipnoans are degenerate descendants of the crossopterygians, which the early dipnoans closely resembled. But according to jarvick and others, certain structures are more specialized in dipnoans than in crossopterygians. The basic differences are in structural organization of food crushing apparatus, fin skeleton, vertebral column, visceral skeleton, neural endocranium, snout, division of heart, atrium, blood supply of swim bladder, etc.

6. **Affinities with Amphibia.** Dipnoans resemble amphibians in several features, such as:

   a) Semiaquatic or marshy habitat.
   b) Internal nostrils piercing roof of mouth cavity.

c) Multicellular skin glands.
d) Dermal scales present in Gymnophiona.
e) Lungs capable of pulmonary respiration.
f) Spiracles lacking.
g) Vomerine teeth present.
h) Auricle and sinus venosus partially divided into right and left halves.
i) Conus arteriosus spirally twisted and longitudinally partitioned.
j) Ventral aorta short.
k) Presence of anterior abdominal vein, posterior vena cava and pulmonary artery and vein.
k) Pericardium is thin-walled.
m) Jaw suspension autostylic.
n) Brain similar in structure of cerebrum and cerebellum.
o) Sperms carried through excretory part of mesonephric kidney.
p) Similar structure of egg and development.
q) Larval stages having suckers and external gills.

These close similarity led early workers to conclude that lung fishes gave rise to amphibians, a view no longer held nowadays. According to Dolo, these similarities probably were due to convergent evolution on account of similar habits and habitat.

On the other hand, the lung fishes have the **following special features** by which they **differ** from **amphibians.**

(i) Lobate fins instead of limbs for locomotion.
(ii) Fin skeleton not like that of primitive tetrapods.
(iii) Peculiar crushing plates instead of teeth.
(iv) Skull mainly cartilaginous with little ossification.
(v) Few anterior vertebrae fused with skull.
(vi) Lungs located dorsal to gut.
(vii) Urinary bladder develops from dorsal wall of cloaca in Dipnoi but from ventral wall in Amphibia.

Phylogeny and Significance

The origin and evolution of Dipnoi remains problematic due to diverse opinions. They have characteristics in which they resemble almost all the other groups of fishes as well as Amphibia. Fossil primitive Dipnoi (e.g. Dipterus) show geater similarity with fossil crossopterygians (e.g. Osteolepis), than do their living members. During the course of their evolution, the modern Dipnoi have undergone several changes or specializations such as:

1. Anterior dorsal fin was reduced and eventually lost.
2. Remaining median fins elongated and fused so that originally heterocercal tail became symmetrically diphycercal.
3. Reduction in number of dermal bones of skull and operculum.
4. Thick bony cosmoid scales modified into thin cycloid scales.
5. Extensive sheet of cosmine covering head and body was lost.
6. Fusion of conical teeth into crushing tooth plates to feed effectively upon shell fish.
7. Air bladders became functional lungs not unlike those of higher vertebrates.

Evidence indicates that Dipnoi are degenerate descendants of Crossopterygii which early dipnoans closely resembled. Romar (1945) thought that Dipnoi and rhipidistian (Crossopterygii)s had a common ancestor. On the other hand close similarity between Dipnoi and Amphibia led early workers to conclude that dipnoans gave rise to amphibians, a view no longer held now. However, it is accepted that Amphibia may have originated directly from rhipidistian Crossopterygian or from their .common ancestor.

Chapter 5

Amphibia

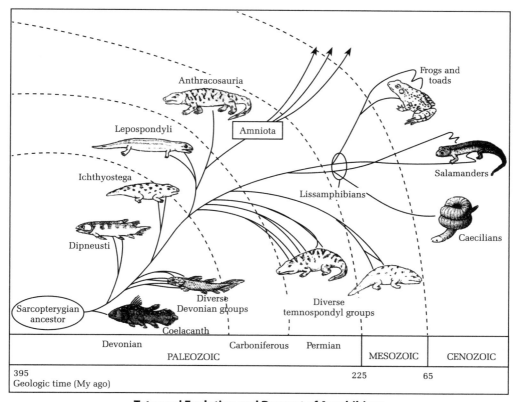

Tetrapod Evolution and Descent of Amphibians

Figure 5.0 Tetrapods share ancestry with diverse Devonian groups. Amphibians share most recent common ancestry with diverse temnospondyls of the Carboniferous and Permian periods of the Paleozoic, and Triassic period of the Mesozoic.

During the first 250 million years of vertebrate history, adaptive radiation resulted in vertebrates filling most aquatic habitats. Prehistoric waters contained many active, powerful predators. Land, however, was free of vertebrates predators and, except for some arthropods, was free of enemies. Animals that moved around the edges of water bodies were not having fear to be preyed upon by the animals living on the land. With lungs for breathing air and muscular fins for scurrying across mud, these animals probably found major component of their diet such as the insects and other arthropods around the edges of these water bodies.

Origin of amphibians from ancient sarcopterygians and adaptive radiation of amphibians resulted in a much greater variety of forms than exists today. Later convergent and parallel evolution and widespread extinction clouded evolutionary pathways. The evolutionary transition from water to land occurred not in weeks but over millions of years. A lengthy series of alterations cumulatively fitted the vertebrate body plan for life on land. The origin of land vertebrates is no less a remarkable feat for this fact—a feat that would have a poor chance of succeeding today because well-established terrestrial competitors would eliminate poorly adapted transitional forms. Amphibians are the only living vertebrates that have a transition from water to land in both their ontogeny and phylogeny. Even after 350 million years of evolution, only few amphibians are completely land adapted; most are quasi-terrestrial, hovering between aquatic and land environments. This double life is expressed in their name. Even the amphibians best adapted for a terrestrial existence cannot stay far from moist conditions. Many, however, have developed ways to keep their eggs out of open water where their larvae would be exposed to enemies.

No one knows, therefore, what animal was the first amphibian, but the structure of limbs, skulls, and teeth suggests that Ichthyostega is probably similar to the earliest amphibians (Figure 5.1). Two lineages of early amphibians from the late Devonian and early Carboniferous periods can be distinguished , on the basis of attachment of the roof and the posterior portion of the skull to each other.

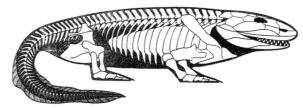

Figure 5.1

Ichthyostega: **An Early Amphibian.** Terrestrial adaptations are heavy pectoral and pelvic girdles and sturdy limbs. Strong jaws suggest that it was a predator in shallow water. Other features include a skull that is similar in structure to ancient sarcopterygian fishes and a finlike tail. Bony rays dorsal to the spines of the vertebrae support the tail fin. This pattern is similar to the structure of the dorsal fins of fishes and is unknown in any other tetrapod.

a) **One lineage** flourished into the Jurassic period. Most of this lineage became extinct, but not before giving rise to the three orders of living amphibians. This lineage is called the **nonamniote** lineage.

b) **Second lineage** of amphibians became extinct late in the Carboniferous period. An amniotic egg that resisted drying, evolved in this group. This lineage, called the **amniote** lineage, left its descendants in the form of the reptiles, birds, and mammals (Fig. 5.2).

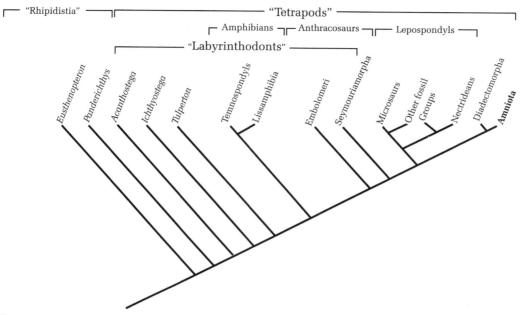

Figure 5.2
Tetrapod phylogenetic relationships.

Position in the Animal Kingdom

Amphibians are ectothermic, primitively quadruped vertebrates, with glandular skin. Many depend on water for their reproduction. They are one of two major groups of living descendants of early Devonian tetrapods

1. The first vertebrates to evolve adaptations to breathe, support themselves, move, and detect airborne sounds and odors on land, while minimizing water loss.
2. The other group is amniotes: reptiles, birds, and mammals whose evolving adaptations freed them from dependence on aquatic habitats for reproduction.

Amphibians developed following Characters

1. Strong skeletal framework to support body weight on land, and a tetrapod limb with associated shoulder/hip girdle for walking on land.
2. A respiratory system with lungs (some modern amphibians are gilled, and some lack both lungs and gills) and paired internal nostrils (choanae), which enable breathing through the nose.
3. Double circulation with functionally separated pulmonary and systemic circuits. Pulmonary arteries and veins supply the lungs and return oxygenated blood to the heart. Heart chambers include a sinus venosus, two atria, one ventricle, and a conus arteriosus.
4. Ancestral aquatic sensory receptors were modified for life on land. The ear with tympanic membrane (eardrum) and stapes (columella) for transmitting vibrations to

the inner ear is designed to detect airborne sounds. For vision in air, the cornea rather than the lens became the principal refractive surface for bending light; eyelids and Lachrymal glands evolved to protect and wash the eyes. A well-developed olfactory epithelium lining the nasal cavity evolved to detect airborne odors.

5. Integument includes modifications for cutaneous respiration and granular glands associated with secretion of defensive compounds.

Classification of Class Amphibia

Order **Gymnophiona (Apoda)**: Caecilians. Body elongate; limbs and limb girdle absent; mesodermal scales present in skin of some; tail short or absent; 95 to 285 vertebrae; pan tropical, 5 families, 33 genera, approximately 160 species.

Order **Urodela (Caudata)**: Salamanders. Body with head, trunk, and tail; no scales; usually two pairs of equal limbs; 10 to 60 vertebrate; predominantly holarctic; 10 living families, 61 genera, approximately 500 species.

Order **Anura (Salientia)**: Frogs, toads. Head and trunk fused; no tail; no scales; two pairs of limbs; large mouth; lungs; 6 to 10 vertebrae including urostyle (coccyx); cosmopolitan, predominantly tropical; 29 families; 352 genera; approximately 4840 species.

Adaptation for life on land was a major theme of the remaining vertebrate groups (Fig. 5.3). These animals form a monophyletic unit known as tetrapods. Amphibians and amniotes (including reptiles, birds, and mammals) represent the two major extant branches of tetrapod phylogeny.

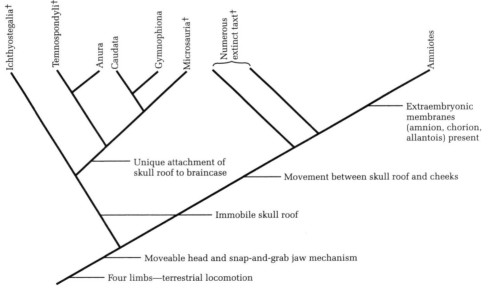

Figure 5.3
Evolutionary *Relationships* among the Amphibians. A nonamniotic lineage gave rise to three classes of modern amphibians and numerous extinct taxa. Some zoologists think that the class Amphibia is a paraphyletic group. If this is true, the three modern classes should be represented an momophyletic taxa. The amniotic lineage of early tetrapods gave rise to reptiles, birds, mammals, and other extinct taxa. Daggers (†) indicate extinct taxa.

Movement on to Land

Movement from water to land is perhaps the most dramatic event in animal evolution because it involves invasion of a physically hazardous habitat. Life originated in water. Animals are mostly water in composition, and all cellular activities occur in water. Nevertheless, organisms eventually invaded land, carrying their watery composition with them. Pulmonate snails, and tracheate arthropods made this transition much earlier than vertebrates, and winged insects were diversifying approximately the same time that the earliest terrestrial vertebrates evolved. Although invasion of land required modifications of almost every system in the vertebrate body, aquatic and terrestrial vertebrates retain many basic structural and functional similarities. We see a transition between aquatic and terrestrial vertebrates most clearly today in the many living amphibians that make this transition during their own life histories.

Beyond the obvious difference in water content, there are several important physical differences that animals must accomodate when moving from water to land. These include

1. oxygen content,
2. density,
3. temperature regulation, and
4. habitat diversity.

Oxygen is at least 20 times more abundant in air and it diffuses much more rapidly through air than through water. Consequently, terrestrial animals can obtain oxygen far more easily than aquatic ones once they possess appropriately adapted lungs and other respiratory structures. Air, however, is approximately 50 times less viscous. It therefore provides relatively less support against gravity, requiring terrestrial animals to develop strong limbs and to remodel their skeleton to achieve adequate structural support. Air fluctuates in temperature more readily then water does, and terrestrial environments therefore experience harsh and unpredictable cycles of freezing, thawing, drying, and flooding. Terrestrial animals require behavioral and physiological strategies to protect themselves from thermal extremes; one such strategy is homeothermy (regulated constant body temperature) of birds and mammals.

Despite its hazards, the terrestrial environment offers a great variety of habitats including coniferous, temperate, and tropical forests, grasslands, deserts, mountains, oceanic islands, and other regions. Safe shelter for protection of vulnerable eggs and young may be found much more readily in many of these terrestrial habitats than in aquatic ones.

Evolution of Terrestrial Vertebrates

The Devonian period, beginning some 400 million years ago was a period of mild temperatures and alternating droughts and floods. During this period, some primarily aquatic vertebrates evolved two features that would be important for permitting subsequent evolution for life on land:

   a) lungs
   b) limbs

Devonian freshwater environments were unstable. During dry periods, many pools and streams evaporated, water became foul, and dissolved oxygen disappeared. Only those fishes which were able to acquire atmospheric oxygen survived such conditions. Gills were unsuitable because in air the filaments collapsed, dried and quickly lost utility. Virtually all freshwater fishes surviving this period, including lobe-finned fishes and lungfishes, had a kind of **lung** that developed as an outgrowth of the pharynx. The efficiency of the

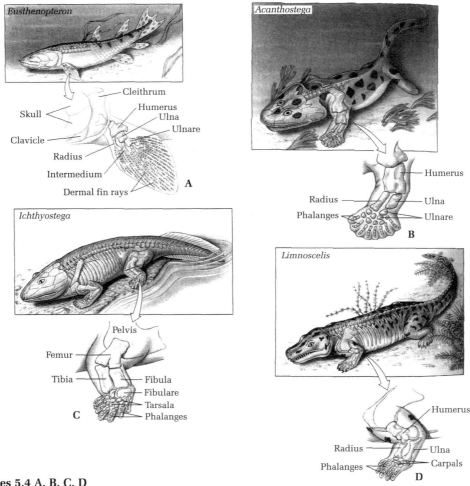

Figures 5.4 A, B, C, D
Evolution of tetrapod limbs. *Eusthenopteron*, **A.** A late Devonian lobe-finned fish (rhipidistian) had paired muscular fins supported by bony elements that foreshadowed the bones of tetrapod limbs. The anterior fin contained an upper arm bone (humerus), two forearm bones (radius and ulna), and smaller elements homologous to wrist bones of tetrapods. As typical of fishes, the pectoral girdle, consisting of the cleithrum, clavicle, and other bones, was firmly attached to the skull. In *Acanthostega*. **B.** One of the earliest known dermal fin rays of the anterior appendage were replaced by eight fully evolved fingers. *Acanthostega* was probably exclusively aquatic because its limbs were too weak for travel on land. *Ichthyostega*. **C.** A contemporary of *Acanthostega*, had fully formed tetrapod limbs and must have been able to walk on land. The hindlimb bore seven toes (the number of front limb digits is unknown). **D.** *Limnoscelis* as anthracosaur of the Carboniferous had five digits on both front and hindlimbs, the basic pentadactyl, model that became the tetrapod standard.

air-filled cavity was enhanced by improving the vascularity with a rich capillary network, and by supplying with arterial blood from the last (sixth) pair of aortic arches. Oxygenated blood returned directly to the heart by a pulmonary vein to form a complete pulmonary circuit. Thus the double circulation characteristic of all tetrapods originated: a systemic circulation serving the body and a pulmonary circulation supplying the lungs.

Vertebrate limbs also arose during the Devonian period. Although fish fins at first appear very different from the jointed limbs of tetrapods, an examination of bony elements of the paired fins of lobe-finned fishes shows that they broadly resemble homologous structures of amphibian limbs. In Eusthenopteron (Fig. 5.4), a Devonian lobe-fin, we can recognize an upper arm bone (humerus) and two forearm bones (radius and ulna) as well as other elements that we can homologize with wrist bones of tetrapods (Fig. 5.4). *Eusthenopteron* could walk—more accurately flop—along the mud of pool bottom with its fins, since backward and forward movement of the fins was limited to about 20 to 25 degrees. *Acanthostega* (Fig. 5.4) one of the earliest known Devonian tetrapods, had well-formed tetrapod legs with clearly formed digits on both fore-and hind limbs, but these limbs were too weakly constructed to support the animals' bodies off the surface for proper walking on land. *Ichthyostega*, however, with a fully developed shoulder girdle, bulky limb bones, well-developed muscles, and other adaptations for terrestrial life, must have been able to pull itself onto land, although it probably did not walk very well.

Until recently zoologists thought that the early tetrapods had five fingers and five toes on their hands and feet, the basic pentadactyl plan of most living tetrapods. However, newly discovered fossils of Devonian tetrapods have more than five digits, indicating that the five-digit pattern became stabilized later in tetrapod evolution.Movement onto land was clearly a revolution in vertebrate history.

"How did it occur?"

Alfred Romer proposed that when freshwater pools of the Devonian evaporated during seasonal droughts, aquatic vertebrates were forced to move to other pools. Fleshy fins of sarcopterygians (the living coelacanth and lung-fishes, and extinct Devonian groups, could be adapted to paddle across land in search of water. Those with strong fins lived to reproduce. According to this hypothesis, land travel and gradual development of limbs originated as a means for finding new sources of water.

Recent discovery of more complete fossils of the earliest known tetrapods changes this view. Although Acanthostega had tetrapod limbs (Fig. 5.4), in all other respect it was a fully aquatic animal. A consensus emerging now is that tetrapods evolved their limbs underwater and only later, for reasons unknown, moved onto land.

Evidence points to lobe-finned fishes as the closest relatives of tetrapods; in cladistic terms they contain the sister group of tetrapods (Fig. 5.5). Both lobe-finned fishes and early tetrapods such as *Acanthostega* and *Ichthyostega* shared several characteristics of their skull, teeth, and pectoral girdle. *Ichthyostega* represents an early offshoot of tetrapod phylogeny that possessed several adaptations, in addition to jointed limbs, that equipped it for life on land. The lepospondyl condition appeared early. It may have given rise to modern amphibians, but more likely it was restricted to the lepospondyls, the taxonomic group named for this vertebral type. The solid vertebrae of modern amphibians probably arose independently. The rhachitomous vertebra, inherited from rhipidistian fishes, evolved along two major lines: temnospondyl and anthracosaur. In the temnospondyl line, the intercentrum enlarged at the expense of the pleurocentrum, in the anthracosaur line, however, the pleurocentrum came to predominate. Thus stronger vertebrae and associated muscles to support their body in air, new muscles to elevate their head, strengthened shoulder and hip girdles, protective rib cage, modified ear structure for detecting airborne sounds, fore shortening of the skull, and lengthening of the snout that improved olfactory powers for detecting dilute airborne odors. Yet *Ichthyostega* still resembled aquatic forms in retaining a tail complete with fin rays and in having opercular (gill-covering) bones.

Bones of Ichthyostega (Fig. 5.4 C), the most thoroughly studied of all early tetrapods, were first discovered on an East Greenland mountain side by Gunnar Save-Soderberg who also uncovered skulls of Ichthyostega. On the return of the expedition to Greenland site and exploration again at this site they found the remainder of Ichthyostega's skeleton ,but it was Erik Jarvik, who, examined the skeleton in detail. This became his life's work, producing a description of Ichthyostega that remains the most detailed of any Paleozoic tetrapod.

The Devonian period was followed by the Carboniferous period, characterized by a warm, wet climate during which mosses and large ferns grew in profusion on a swampy landscape. Tetrapods radiated quickly in this environment to produce a great variety of forms, feeding on abundant insects, insect larvae, and aquatic invertebrates. Evolutionary relationships of early tetrapod groups remain controversial. A tentative cladogram (Fig. 5.5),depicting,

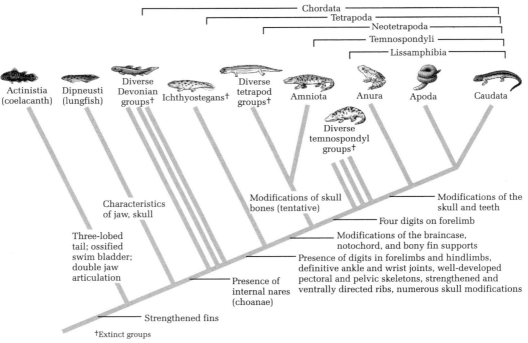

Figure 5.5
Cladogram of the Tetrapoda with emphasis on descent of amphibians. Especially controversial are relationships of major tetrapod groups (Amniota, Anthracosauria, Lepospondyli, Temnospondyli) and outgroups (Actinistia, Dipneusti, extinct Devonian groups). All aspects of this cladogram are controversial, however, including monophyly of Lissamphibia. Relationships shown for the three groups of Lissamphibia are based on recent molecular evidence.

several extinct lineages plus Lissamphibia, which contains modern amphibians, are placed in a group called **temnospondyls** (Fig. 5.5). This group generally has only four digits on the forelimb rather than the five characteristic of most tetrapods.

Lissamphibians diversified during the Carboniferous to produce ancestors of the three major groups of extant amphibians

1. **Anura** or Salientia- frogs
2. **Caudata** or Urodela- salamanders
3. **Apoda** or Gymnophiona- caecilians.

Amphibians improved their adaptations for living in water during this period. Their bodies became flatter for moving through shallow water. Early salamanders developed weak limbs, and their tail became better developed as a swimming organ. Even anurans (frogs and toads), which are now largely terrestrial as adults, developed specialized hind limbs with webbed feet better suited for swimming than for movement on land. All amphibians use their porous skin as a primary or accessory breathing organ. This specialization was encouraged by swampy environments of the Carboniferous period but presented serious desiccation problems for life on land.

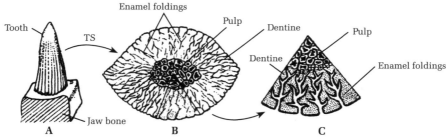

Figures 5.6 A, B, C
Structure of a labyrinthine tooth. **A.** Entire tooth. **B.** T.S. tooth. **C.** A portion of T.S. magnified.

Amphibian phylogeny is controversial as far as the relationship among the three orders of modern amphibians are concerned. Some zoologists place **anurans, urodeles**, and **caecilians** into a single subclass, **Lissamphibia.** This placement implies a common ancestry for modern amphibians and suggests that they are more closely related to each other than to any other group. Supporters of this classification point to common characteristics, such as the stapes/operculum complex, the importance of the skin in gas exchange, and aspects of the structure of the skull and teeth, as evidence of this close relationship. (Fig. 5.6) depicts this interpretation.

Other zoologists think that modern amphibians were derived from at least two non-amniotic lineages. They note that fine details of other structures, such as the vertebral column, are different enough in the three orders to suggest separate origins. If this is true, the class Amphibia is a paraphyletic group and should be divided into multiple monophyletic taxa. This controversy is not likely to be settled soon.

The focus is on descendants of the **amniote lineage** (Fig. 5.7). Anthracosaurs are often cited as amphibian ancestors of these animals, but support for this conclusion is weak. Three sets of evolutionary changes in amphibian lineages allowed movement onto land. Two of these occurred early enough that they are found in all amphibians.

a) **One was the set of changes** in the skeleton and muscles that allowed greater mobility on land.

b) **A second change** involved a jaw mechanism and movable head that permitted effective exploitation of insect resources on land. A jaw-muscle arrangement that permitted rhipidistian fishes to snap, grab, and hold prey was adaptive when early tetrapods began feeding on insects in terrestrial environments.

c) **The third set of changes** occurred in the amniote lineage—the development of an egg that was resistant to drying. Although the amniotic egg is not completely independent of water, the extraembryonic membranes that form during development protect the embryo from desiccation, store wastes, and promote gas exchange. In addition, this egg has a leathery or calcified shell that is protective, yet porous enough to allow gas exchange with the environment. These evolutionary events eventually resulted in the remaining three vertebrate groups: reptiles, birds, and mammals.

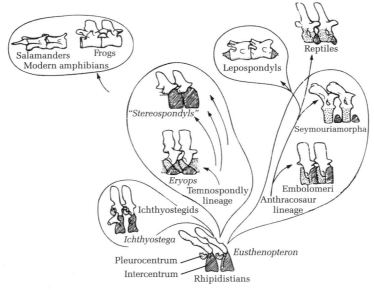

Figure 5.7
Traditional view of the evolution of tetrapod vertebrae.

Two additional generally recognized but nonetheless controversial groupings of Carboniferous and Permian tetrapods, **lepospondyls** and **anthracosaurs,** are judged on the basis of skull structure to be closer to amniotes than to temnospondyls (Fig. 5.7). Together they form a second major branch of tetrapod phylogeny.

Modern Amphibians

The three living amphibian orders comprise more than 5400 species. Most share general adaptations for life on land, including skeletal strengthening and a shifting of special sense priorities from the ancestral lateral-line system to the senses of smell and hearing. The olfactory epithelium and ear are redesigned to improve sensitivities to airborne odors and sounds, respectively.

Nonetheless, most amphibians meet problems of independent life on land only halfway. In the ancestral life history of amphibians, eggs are aquatic and hatch to produce an aquatic larval form that uses gills for respiration. A metamorphosis follows in which gills are lost and lungs, which are present throughout larval life, are then activated for breathing air. Many amphibians retain this general pattern, but important exception include some salamanders that lack a complete metamorphosis and retain a permanently aquatic, larval morphology throughout life. Some caecilians and other salamanders live entirely on land and have no aquatic larval phase. Both alternatives are evolutionarily derived conditions. Some frogs also maintain a steady terrestrial existence by eliminating an aquatic larval stage. Some frogs, salamanders, and caecilians that undergo a complete metamorphic life cycle nonetheless remain in water as adults rather than moving onto land during their metamorphosis.

Even the most terrestrial amphibians remain dependent on very moist if not aquatic environments. Their skin is thin, and it requires moisture for protection against desiccation in air. An intact frog loses water nearly as rapidly as a skinless frog. Amphibians also require moderately cool environments. Being exothermic, their body temperature is determined by and varies with the environment, greatly restricting where they can live. Cool and wet environments are especially important for reproduction. Eggs are not well protected from desiccation, and they must be shed directly into water or onto moist terrestrial surfaces. Completely terrestrial amphibians may lay eggs under logs or rocks, in the moist forest floor, in flooded tree holes, in pockets on the mother's back , or in folds of the body wall. One species of Australian frog even broods its young in the stomach while discontinuing normal digestion, and males of one South American species brood the young in their vocal pouches.

We now highlight special characteristics of the three major groups of amphibians

Order Gymnophiona (Apoda)

The order Gymnophiona contains approximately 160 species of elongate, limbless, burrowing creatures commonly called caecilians (Fig. 5.9). They occur in tropical forests of South America (their principal home), Africa, and Southeast Asia. Caecilians possess a long, slender body, small scales in the skin of some, many vertebrae, long ribs, no limbs, and a terminal anus. Eyes are small, and most species are totally blind as adults. Special sensory tentacles occur on the snout. Because they are almost entirely burrowing or aquatic, caecilians seldom are seen by humans. Their food consists mostly of worms and small invertebrates, which they find underground. Fertilization is internal, and males have a protrusible copulatory organ. Eggs usually are deposited in moist ground near water. Larvae may be aquatic, or complete larval development may occur in the egg. In some species eggs are carefully guarded during their development in folds of the body. Viviparity also is common in some caecilians, with embryos obtaining nourishment by eating the wall of the oviduct.

Amphibian are between the fish on one hand, and the reptiles on the other.

Diagonastic Characters of Amphibians

1. Aquatic or semiaquatic (freshwater), air and water breathing, carnivorous, cold-blooded, oviparous, tetrapod vertebrates.

2. Head distinct, trunk elongated. Neck and tail may be present or absent.
3. Limbs usually 2 pairs (tetrapod), some limbless. Toes 4-5 (pentadactyl) or less. Paired fins absent. Median fins, if present, without fin rays.
4. Skin soft, moist and glandular. Pigment cells (chromatophores) present.
5. Exoskeleton absent. Digits clawless. Some with concealed dermal scales.
6. Endoskeleton mostly in body. Notochord does not persist. Skull with 2 occipital condyles.
7. Mouth large. Upper or both jaws with small homodont teeth. Tongue often protrusible. Alimentary canal terminates into cloaca.
8. Respiration by lungs, skin and mouth lining. Larvae with external gills which may persist in some aquatic adults.
9. Heart 3-chambered (2 auricles 1 ventricle). Sinus venosus present. Aortic arches 1-3 pairs. Renal and hepatic portal systems well developed. Erythrocytes large, oval and nucleated Body temperature variable (poikilothermous).
10. Kidneys mesonephric. Urinary bladder large. Urinary ducts open into cloaca. Excretion ureotelic.
11. Brain poorly developed. Cranial nerves 10 pairs.
12. Nostrils connected to buccal cavity. Middle ear with a single rod-like ossicle, columella. Larval forms and some aquatic adults with lateral line system.
13. Sexes separate. Male without copulatory organ. Gonoducts open into cloaca. Fertilization mostly external. Females mostly oviparous.
14. Development indirect. Cleavage holoblastic but uequal. No extra-embryonic membranes. Larva a tadpole which metamorphoses into adult.

The living amphibians belong to only 2,500 species, a very much smaller number than that of other principal classes of vertebrates. However, these represent a mere shadow of the great amphibian radiations of the past, ranging from mid-Palaeozoic (Devonian) to early Mesozoic (Triassic). They dominated the World during Carboniferous, but most of them have become extinct since long. About 10 orders of extinct Amphibia are known only by fossil remains.

The classification by Kingsley Noble recognizes 3 orders of extinct and 3 orders of living amphibians.

1. **Subclass Stegocephalia** in the past, **all extinct** groups of Amphibia were placed under this single subclass stegocephalia
2. **Subclass Lissamphibia** all living groups in subclass Lissamphibia

Subclass 1. Stegocephalia (Extinct)

1. Limbs pentadactyl.
2. Skin with scales and bony plates.
3. Skull with a solid bony roof leaving openings for eye and nostrils. Permian to Triassic.

Order 1. Labyrinthodontia

1. Oldest known tetrapods called stem Amphibia.
2. Freshwater or land forms. Salamander or crocodile like.
3. Teeth large with characteristically much folded dentine similar to their crossopterygian ancestors. Carboniferous to Triassic. e.g.,: *Eryops* (Fig. 5.8).

Figure 5.8
Eryops, a fossil labyrinthodont.

Order 2. Phyllospondyli

1. Small salamander-like, Head large, flat.
2. Vertebrae tubular
3. Notochord and spinal cord housed in common cavity.
4. Believed to be ancestors of modern Salientia and Urodela. Carboniferous to permian. e.g.,: *Branchiosaurs* (Ichthyostega) (Fig. 5.4 C).

Order 3. Lepospondyli

1. Small salamander or eel-like.
2. Vertebrae cylindrical, each made of a single piece.
3. Neural arch and centrum continuous
4. Ribs articulating intervertebrally.
5. Regarded ancestral to modern caecilians (Gymnophiona). Carboniferous to Permian e.g.,: *Diplocaulus, Lysorophus*.

Subclass II. Lissamphibia (living)

1. Modern amphibia lacking dermal bony skeleton.
2. Teeth small, simple.

Order 1. Gymnophiona or Apoda

1. Limbless, blind, elongated worm like, burrowing tropical forms known as caecilians.

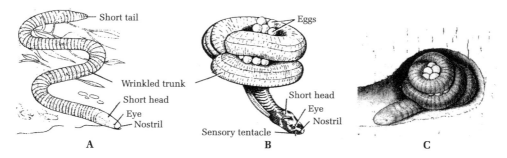

Figures 5.9 A, B, C
Ichthyophis glutinosa **A.** Male **B.** Female guarding eggs **C.** Female caecilian coiled around eggs in burrow.

2. Tail short or absent, cloaca terminal.
3. In some dermal scales embedded in skin which is transversely wrinkled.
4. Skull compact, roofed with bone.
5. Limb girdles absent.
6. Males have protrusible copulatory organs. e.g.: *Ichthyophis* (Fig. 5.9), *Uroaeotyphlus*.

Order 2. Urodela or Caudata

1. Lizard-like amphibians with a distinct tail.
2. Limbs 2 pairs, usually weak, almost equal.
3. Skin devoid of scales and tympanum.
4. Gills permanent or lost in adult.

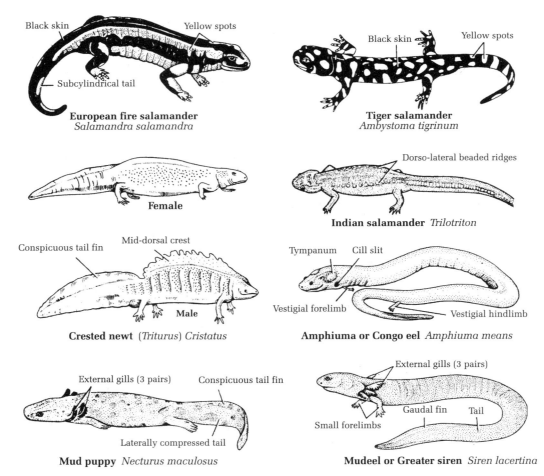

Figure 5.10
Urodele amphibians. Salamanders and newts

5. Males without copulatory organs.
6. Larvae aquatic, adult-like, with teeth, about 300 species in 5 suborder.

Suborder 1. Cryptobranchoidea

1. Most primitive. Permanently aquatic.
2. Adults without eyelids and gills.
3. Angular and pre-articular separate.
4. Premaxillary spine short.
5. Fertilization external.
   e.g.,: *Cryptobranchus, Megalobatrachus*.

Suborder 2. Ambystomatoidea

1. Adults terrestrial with eyelids.
2. Angular fused with prearticular.
3. Premaxillary spine large.
4. Vertebrae amphicoelous.
5. Fertilization internal.
   e.g.,: *Ambystoma*. (Fig. 5.10).

Suborder 3. Salamandroidea

1. Vertebrae opisthocoelous.
2. Teeth on palate and prevomers.
3. Three sets of cloacal glands.
4. Fertilization internal.
   e.g.,*Triton* and *Triturus* (newts), *Salamadra* (salamander), *Desmognathus, Amphiuma* (Fig. 5.10) (congo eel*), Plethodon*.

Suborder Proteida

1. Aquatic bottom dwellers representing permanent larval forms, without eyelids.
2. Adults with 3 pairs of external gills and 2 pairs of gill slits
3. Skull cartilaginous, without maxillae.
4. Jaws with teeth.
   e.g., *Proteus, Necturus* (Fig. 5.10).

Suborder 5. Meantes

1. Aquatic representing permanent larvae.
2. Forelimbs small, hindlimbs absent.
3. Three pairs of external gills.

4. No eyelids, no cloacal glands.
5. Jaws with horny covering.

   e.g.,. *Siren* (mud eel), *Pseudobranchus*.

Order 3. Salientia or Anura

Specialized Amphibia without tail in adults.

1. Hindlimbs usually adapted for leaping and swimming.
2. Adults without gills or gill openings.
3. Eyelids well-formed. Tympanum present.
4. Skin loosely-fitting, scaleless; Mandible toothless.
5. Pectoral girdle bony. Ribs absent or reduced. Vertebral column very small of 5-9 presacral vertebrae and a slender urostyle.

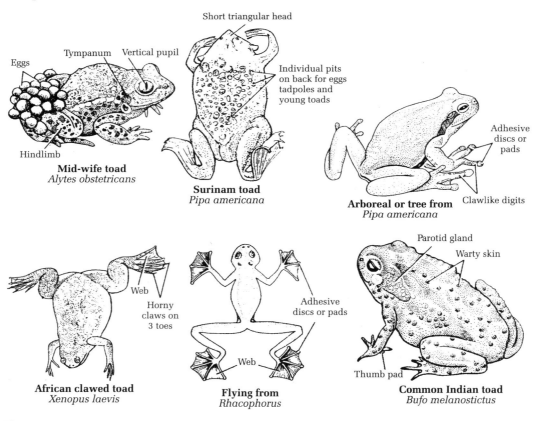

Figure 5.11
Anuran amphibians. Frog and toads

6. Fertilization always external.
7. Fully metamorphosed without neotenic forms.
8. About 2,200 species of frogs and toads in 5 suborders.

Suborder 1. Amphicoela

1. Vertebrae amphicoelous. presacral
2. Free ribs and 2 relict tail muscles.
3. Fertilization internal.

   e.g., *Leiopelma, Ascaphus*

Suborder 2 Opisthocoela

1. Vertebrae opisthocoelous. Scapula small.
2. Ribs free in adult or larva.

   e.g.,. *Alytes* (Fig. 5.11) (midwife toad), *Bombinator, Discoglossus, Pipa* (Fig. 5.11).

Suborder 3. Anomocoela

1. Vertebrae procoelous or amphicoelous.
2. Free ossified ribs absent.
3. Upper jaw with teeth.

   e.g., *Pelobates. Scaphiopus.*

Chapter 6

Reptilia

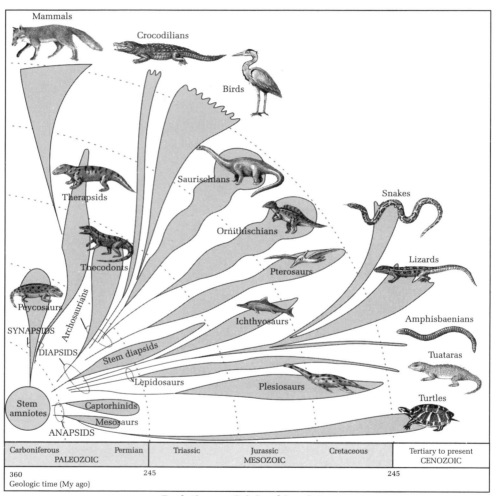

Evolutionary Origin of Amniotes

Figure 6.0 The evolutionary *origin of amniotes* occurred by evolution of an amniotic egg that made reproduction on land possible. The amniote assemblage, which includes reptiles birds, and mammals, evolved from a lineage of small, lizardlike forms that retained the anapsid skull pattern of early tetrapods. First to diverge from the primitive stock was a lineage that evolved a skull pattern termed the synapsid condition. All other amniotes, including birds and all living reptiles except turtles, have a skull pattern known as diapsid. Turtles retain the primitive, anapsid skull pattern. The great Mesozoic radiation of reptiles may have resulted partly from the increased variety of ecological habitats that amniotes could exploit.

The amphibians, with well-developed limbs, redesigned sensory and respiratory systems, and modifications of the postcranial skeleton for supporting the body in air, have made a notable conquest on land. But, with shell-less eggs and often gill-breathing larvae, their development remained hazardously tied to water. The lineage containing reptiles, birds, and mammals developed an egg that could be laid on land. This shelled egg (Fig. 6.1) perhaps more than any other adaptation, unshackled early reptiles from the aquatic environment by freeing the developmental process from dependence on aquatic or very moist terrestrial environments. In fact, the "pond-dwelling" stages were not eliminated but enclosed within a series of extra embryonic membranes that provided complete support for embryonic development. One membrane, the amnion, encloses a fluid-filled cavity, the "pond", within which the developing embryo floats. Another membranous sac, the allantois,

Figure 6.1
Seymouria with cleidoic eggs

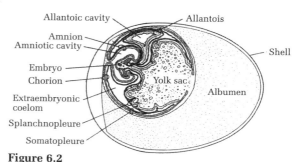

Figure 6.2
Cross section of a bird embryo within the shelled egg after about eight hours of incubation. Shows early formation of the allantois and the amnion.

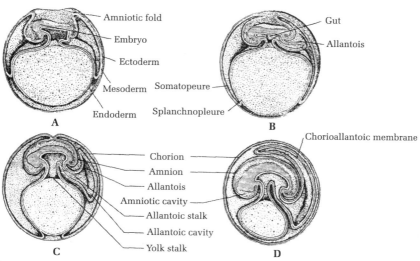

Figure 6.3
Extraembryonic membrane formation in a bird (sagittal sections). Somatopleure lifts upward **A.** Forming amniotic folds that join, **B.** And fuse, **C.** Above the embryo to produce the chorion and the aminion. The expanding allantois comes into association with the chorion to produce the chorioallantoic membrans, **D.** An extensive vascular network forms within the mesoderm and serves as a site of respiratory exchange for gases passing through the porous shell.

serves both as a respiratory surface and as a chamber for the storage of nitrogenous wastes (Fig. 6.2, Fig. 6.3). Enclosing these membranes is a third membrane, the chorion, through which oxygen and carbon dioxide freely pass. Finally, surrounding and protecting everything is a porous, parchment like or leathery shell.

With the last ties to aquatic reproduction severed, conquest of land by vertebrates was ensured. Paleozoic tetrapods that developed this reproductive pattern were ancestors of a single, monophyletic assemblage called Amniota, named after the innermost of the three extraembryonic membranes, the amnion. Before the end of the Paleozoic era amniotes had diverged into multiple lineages that gave rise to all reptilian groups, birds, and mammals.

Amniotes evolved in the late Paleozoic and most paleontologists agree that the amniotes arose from a group of amphibian-like tetrapods, the anthracosaurs, during the early Carboniferous period of the Paleozoic. By the late Carboniferous amniotes had separated into **three groups** (Figs. 6.4).

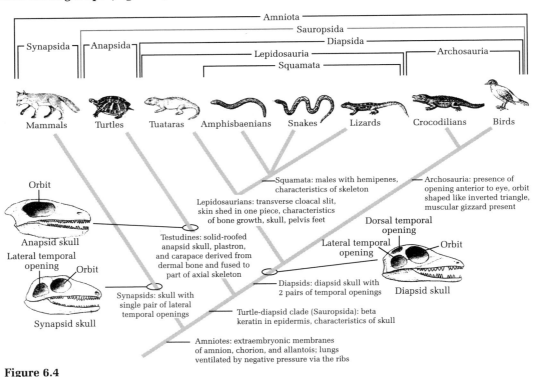

Figure 6.4
Cladograms of living Amniota showing monophyletic groups. Some shared derived characters (synapomorphies) of the groups are given. The skulls represent the ancestral condition of the three groups.

1. **The first** group, the **anapsids**, is characterized by a skull having no openings in the temple area behind the eye sockets, the skull behind the orbits being completely roofed with dermal bone (Fig. 6.4). This group is represented today only by turtles. However, the relationship of turtles to other amniotes is controversial. Although some view the anapsid turtle skull as secondarily derived and place turtles within the diapsid lineage, as per the traditional view that turtles are not diapsids (Fig. 6.4).

Turtle morphology is an odd mix of ancestral and derived characters that has scarcely changed at all since they first appeared in the fossil record in the Triassic some 200 million years ago.

2. The **second group**, the **diapsids**, includes all other reptilian groups and birds (Fig. 6.4). The diapsid skull is characterized by two temporal openings; one pair located low on the cheeks, and a second pair positioned above the lower pair and separated from them by a bony arch (Fig. 6.4). Four subgroups of diapsids appeared.

   a) The **lepidosaurs** include all modern reptiles except turtles and crocodilians.

   b) The **archosaurs** comprised dinosaurs and their relatives, and living crocodilians and birds.

   c) A third, smaller subgroup, **the sauropterygians** included several extinct aquatic groups, most conspicuous of which were the large, long-necked plesiosaurs.

   d) The fourth subgroup comprised **Ichthyosaurs,** represented by extinct, aquatic dolphin like forms (Fig. 6.4).

3. The **third group** is the **synapsids**, the mammals and extinct forms traditionally termed mammal-like reptiles. The synapsid skull had a single pair of temporal openings located low on the cheeks and bordered by a bony arch (Fig. 6.4). Synapsids were the first amniote group to diversify, giving rise first to pelycosaurs, later to therapsids, and finally to mammals (Fig. 6.4).

Peculiar adaptations attained by reptiles

1. The shelled, amniotic egg that evolved with the earliest Paleozoic amniotes is supplied with extra embryonic membranes that provide a complete life-support system for the enclosed embryo. This innovation allowed amniotes to lay larger eggs and in drier habitats. In some viviparous reptiles the extra embryonic membranes are restructured into a placenta suggestive of a somewhat paralleling the evolution in the synapsid lineage of the more complex mammalian placenta.

2. A tough, dry, heavily keratinized skin that provides protection against desiccation and injury. Scales in reptiles and feathers in birds arise as epidermal elevations overlying a nourishing dermal layer.

3. Larger and stronger jaw muscles permit powerful jaw closure. Temporal vacuities in the diapsid skull provide space for bulging temporal muscles.

4. Internal fertilization, with sperm introduced directly into the female reproductive tract with a copulatory organ.

5. Effective adaptations for water conservation include a metanephric kidney that excretes nitrogenous wastes as uric acid (diapsids and desert turtles) or urea (most turtles). Such adaptations allowed reptiles (and birds) to occupy many terrestrial habitats.

Diagonastic Characters of Class Reptilia

1. Body varied in shape, compact in some, elongated in others; body covered with keratinized epidermal scales with the addition sometimes of bony dermal plates; integument with few glands

2. Two paired limbs, usually with five toes, and adapted for climbing, running, or paddling; limbs vestigial or absent in snakes and some lizards and amphisbaenians
3. Skeleton well ossified; ribs with sternum (absent in snakes) forming a complete thoracic basket; skull with one occipital condyle
4. Respiration by lungs; cloaca, pharynx, or skin by some because of absence of gills.
5. Circulatory system functionally divided into pulmonary and systemic circulation; heart typically consisting of a sinus venosus, an atrium completely divided into two chambers, and a ventricle incompletely divided hence it is three chambered; crocodilians with a sinus venosus, two atria, and two ventricles
6. Ectothermic; many thermoregulate behaviorally
7. Metanephric kidney (paired); uric acid main nitrogenous waste
8. Nervous system with optic lobes on dorsal side of brain; 12 pairs of cranial nerves in addition to nervus terminalis; enlarged cerebrum
9. Sexes separate; fertilization internal
10. Eggs covered with calcareous or leathery shells; extra embryonic membranes are amnion, chorion, and allantois during embryonic life; no aquatic larval stages

Class Reptilia is no longer recognized by cladists as a valid taxon because it is not monophyletic hence important changes have been made in the traditional classification of reptiles. As customarily defined, class Reptilia excludes birds, which descended from the most recent common ancestor of the reptiles. Consequently, reptiles are a paraphyletic group because they do not include all descendants of their most recent common ancestor. Reptiles and birds share several derived characters, including several skull characteristics and largely aglandular skin with a special type of harder keratin called beta keratin, which unites them as a monophyletic group (Fig. 6.5). Although we recognize that Reptilia, as historically used, is not a monophyletic group, we can use it as a term of convenience to refer to all amniotes that have beta keratin is their epidermis that are not birds. Thus we are using the word "reptiles" to refer to the living turtles, snakes, lizards, amphisbaenians, tuataras, and crocodilians in addition to a number of extinct groups such as plesiosaurs, ichthyosaurs, pterosaurs, and dinosaurs.

Crocodilians and birds are sister groups; they are more recently descended from a common ancestor than either is from any other living reptilian lineage. In other words, birds and crocodilians belong to a monophyletic group apart from other reptiles and, according to the rules of cladistics, should be assigned to a clade that separates them from the remaining reptiles. This clade is in fact recognized; it is Archosauria (Fig. 6.4), a grouping that also includes the extinct dinosaurs. Therefore, according to cladistics, birds should be classified as reptiles. Archosaurs plus their sister group, the lepidosaurs (tuataras, lizards, snakes, and amphisbaenids), comprise a monophyletic group that cladists call Reptilia. The term "Reptilia" is thereby redefined to include birds in contrast to its traditional usage.

"However, evolutionary taxonomists argue that birds represent a novel adaptive zone and grade of organization whereas crocodilians remains within the traditionally recognized reptilian adaptive zone and grade".

In this view, the morphological and ecological novelty of birds has been recognized by maintaining the traditional classification that places crocodilians in class Reptilia and

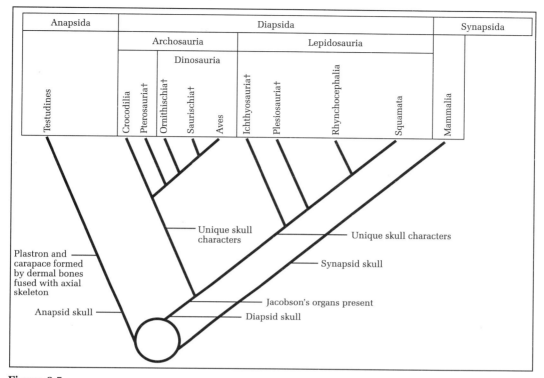

Figure 6.5

Amniote Phylogeny. This cladogram shows one interpretation of amniote phylogeny. The circle at the base of the cladogram indicates that the ancestral amniotes are not described. Some researchers believe that the Diapsida is nor a single lineage and that the Archosauria and Lepidosauria should be elevated to subclass status. Synapomorphies used to distinguish lower taxa are not shown. Daggers (†) indicate some extinct taxa.

birds in class Aves. Such conflicts of opinion between proponents of the two major competing schools of taxonomists i.e., cladistics and evolutionary taxonomists have had the healthy effect of forcing zoologists to reevaluate their views of amniote genealogy and how vertebrate classification should represent genealogy and degrees of divergence.

Distinctive Characters that differentiate Reptiles from Amphibians

Reptiles have tough, dry, scaly skin offering protection against desiccation and physical injury. The skin consists of a thin epidermis and a much thicker, well-developed dermis. The dermis is provided with chromatophores, color-bearing cells that give many lizards and snakes their colorful hues. Resistance to desiccation is provided by hydrophobic lipids in the epidermis. The epidermis also contains a hard form of keratin called beta keratin that is unique to reptiles. The characteristic scales of reptiles, formed largely of beta keratin, provide protection against wear and tear in terrestrial environments. They are derived mostly from the epidermis and thus are not homologous to fish scales, which are bony, dermal structures. In some reptiles, such as alligators, the scales remain throughout life, growing gradually to replace wear. In others, such as snakes and lizards, new scales grow beneath the old, which are then shed at intervals. Turtles add new layers of keratin under the old layers of the platelike scutes, which are modified scales. In snakes the old skin is turned inside out when discarded; lizards split out of the old skin leaving it mostly intact and right side out, or it may slough off in pieces. Crocodiles and many lizards possess bony plates called osteoderms located beneath the keratinized scales in the dermis.

Most tortoises are rather slow moving; an hour of determined trudging carries a large Galapagos tortoise approximately 300 m. Their low metabolism probably explains their longevity, for some are believed to live more than 150 years.

The shell, like a coat of armour, offers obvious advantages. The head and appendages can be drawn in for protection. The familiar box tortoise, Terrapene Carolina, has a plastron that is hinged, forming two movable parts that can be pulled up against the carapace so tightly that one can hardly force a knife blade between the shells. Some turtles, such as the large eastern snapping turtle, have reduced shells, making complete withdrawal for protection quite impossible. Snappers, however, have another formidable defense, as their name implies. They are entirely carnivorous, living on fishes, frogs, waterfowl, or almost anything that comes within reach of their powerful jaws. An alligator snapper lures unwary fish into its mouth with a pink, wormlike extension of its tongue that serves as a "bait". Alligator snappers are wholly aquatic and come ashore only to lay their eggs.

Diapsid Reptiles: Subclass Diapsida

Diapsid reptiles, those reptiles having a skull with two pairs of temporal openings (Fig. 6.3), are classified into *three* lineages.

Order Squamata: Lizards, Snakes, and Worm Lizards (Fig. 6.4).

Squamates are the most recent and diverse products of diapsid evolution, comprising approximately 95% of all known living reptiles.

a) Lizards appeared in the fossil record as early as the Jurassic, but they did not begin their radiation until the Cretaceous period of the Mesozoic era when dinosaurs were at the climax of their radiation.

b) Snakes appeared during the late Jurassic period, probably from a group of lizards whose descendants include the Gila monster and monitor lizards. Two specializations in particular characterize snakes: extreme elongation of their body and accompanying displacement and rearrangement of internal organs; and specializations for eating large prey.

c) Amphisbaenians (worm lizards), which first appear in the fossil record of the early Cenozoic era, have structural specializations associated with a burrowing habit.

Viviparity in living reptiles is limited to squamates, and has evolved at least 100 separate times. Evolution of viviparity is usually associated with cold climates and occurs by increasing the length of time eggs are kept within the oviduct. Developing young respire through extraembryonic membrances and obtain nutrition from **yolk sacs** (lecithotrophy) or via the **mother** (placentotrophy), or some combination of each.

Skulls of squamates are modified from the ancestral diapsid condition by loss of dermal bone ventral and posterior to the lower temporal opening. This modification has allowed evolution in most lizards and snakes of a mobile skull having movable joints. Such a skull is called a **kinetic skull.** The quadrate, which in other reptiles is fused to the skull, has a joint at its dorsal end, as well as its usual articulation with the lower jaw. In addition, joints in the palate and across the roof of the skull.

The earliest members of the class Reptilia were the first vertebrates to possess amniotic eggs (Fig. 6.1). Amniotic eggs have extraembryonic membranes that protect the embryo from desiccation, cushion the embryo, promote gas transfer, and store waste materials (Fig. 6.2, Fig. 6.1). The amniotic eggs of reptiles and birds also have hard or leathery shells that protect the developing embryo, albumen that cushions and provides moisture and nutrients for the embryo, and yolk that supplies food to the embryo. All of these features are adaptations for development on land. The amniotic egg is not, however, the only kind of land egg: some arthropods, amphibians, and even a few fishes have eggs that develop on land. The amniotic egg is the major synapomorphy that distinguishes the reptiles, birds, and mammals from vertebrates in the nonamniote lineage. Even though the amniotic egg has played an important role in vertebrates' successful invasion of terrestrial habitats, it is one of many reptilian adaptations that have allowed members of this class to flourish on land. Living representatives of the class Reptilia include the turtles, lizards, snakes, worm lizards, crocodilians, and the tuatara (Cladogram 6.4, 6.5).

Even though fossil records of many reptiles are abundant, much remains to be learnt of reptilian **origin.** The adaptive radiation of the early amniotes began in the late Carboniferous and early Permian periods. This time coincided with the adaptive radiation of terrestrial insects, the major prey of early amniotes. The adaptive radiation of the amniotes resulted in certain lineages described below. Skull structure, particularly the modifications in jaw muscle attachment, is one way these lineages are distinguished (Fig. 6.5).

Classification of Living Reptiles

Class Reptilia

Dry skin with epidermal scales; skull with one point of articulation with the vertebral column (occipital condyle); respiration via lungs; metanephric kidneys; internal fertilization; amniotic eggs.

Order Testudines Chelonia

Teeth absent in adults and replaced by a horny beak; short; broad body; shell consisting of a dorsal carapace and ventral plastron. e.g., Turtles. (Fig. 6.6).

Order Rhynchocephalia

Contains very primitive, lizardlike reptiles; well-developed parietal eye. A single species, e.g., Sphenodon punctatus, (Fig. 6.7) survives in New Zealand. Tuataras.

Order Squamata

Recognized by specific characteristics of the skull and jaws (temporal arch reduced or absent and quadrate movable or secondarily fixed); the most successful and diverse group of living reptiles. e.g., Snakes, lizards, worm lizards (Fig. 6.8).

Order Crocodilia

Elongate, muscular, and laterally compressed; tongue not protrusible; complete ventricular septum. e.g., Crocodiles, alligators, caimans, gavials. (Fig. 6.11).

General Characters

Reptiles represent the first class of vertebrates fully adapted for life in dry places on land.

1. Predominantly terrestrial, creeping or burrowing, mostly carnivorous, air-breathing, cold-blooded, oviparous and tetrapodal vertebrates.
2. Body bilaterally symmetrical and divisible into 4 regions—head, neck, trunk and tail.
3. Limbs 2 pairs, pentadactyl. Digits provided with horny claws. However, limbs absent in a few lizards and all snakes.
4. Endoskeleton bony. Skull with one occipital condyle (monocondylar). A characteristic T-shaped interclavicle present.
5. Heart usually 3-chambered, 4-chambered in crocodiles. Sinus venous reduced. 2 systemic arches present. Red blood corpuscles oval and nucleated. Cold-blooded.
6. Respiration by lungs throughout life.
7. Kidneys metanephric. Excretion uricotelic.
8. Brain with better development of cerebrum than in Amphibia. Cranial nerves 12 pairs.
9. Lateral line system absent. Jacobson's organs present in the roof of mouth.
10. Parental care usually absent.

Classification

According to Bogert, there are more than 7,000 living and several extinct species of reptiles, grouped into approximately 16 orders of which only 4 are living.

Class Reptilia is divided into 5 major groups or subclasses on the basis of presence or absence of temporal vacuity of the skull.

Subclass I Anapsida

Primitive reptiles with a solid skull roof. No temporal openings

Order 1. Chelonia or Testudinata

1. Body short, broad and oval.
2. Limbs clawed and/or webbed, paddle-like.
3. Body encased in firm shell of dorsal carapace and ventral plastron, made of dermal bony plates. Thoracic vertebrae and ribs usually fused to carapace.
4. Skull anapsid, with a single nasal opening and without a parietal foramen. Quadrate is immovable.
5. No sternum is found.

6. Teeth absent. Jaws with horny sheaths.
7. Cloacal aperture a longitudinal slit.
8. Heart incompletely 4-chambered with a partly divided ventricle.
9. Copulatory organ single and simple.
10. About 400 species of marine turtles, freshwater terrapins and terrestrial tortoises. e.g., *Chelone, Chrysemys, Testudo, Trionyx, Dermochelys*. (Fig. 6.6).

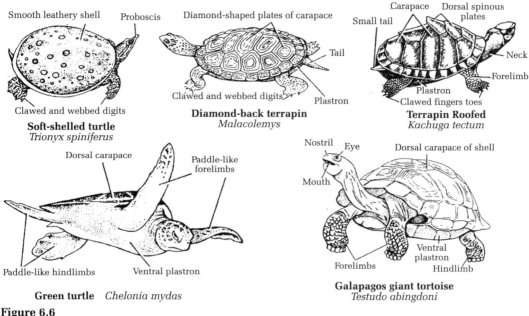

Figure 6.6
Turtles and tortoises (Order Chelonia)

Subclass II Euryapsida (extinct)

Skull with a single dorso-lateral temporal opening on either side bounded below by postorbital and squamosal bones.

Subclass III Parapsida (extinct)

Skull a single dorso-lateral temporal opening on either side bounded below by the supratemporal and postfrontal bones.

Subclass IV Synapsida (extinct)

Skull with a single lateral temporal opening on either side bounded above by the postorbital and squamosal bones.

Subclass V Diapsida

Skull with two temporal openings on either side separated by the bar of postorbital and squamosal bones.

Order 2. Rhynchocephalia

1. Body small, elongated, lizard-like.
2. Limbs pentadactyl, clawed and burrowing.
3. Skin covered by granular scales and a mid-dorsal row of spines.
4. Skull diapsid. Nasal openings separate. Parietal foramen with vestigial pineal eye present. Quadrate is fixed.
5. Vertebrae amphicoelous or biconcave. Numerous abdominal ribs present.
6. Teeth acrodont. Cloacal aperture transverse.
7. Heart incompletely 4-chambered.
8. No copulatory organ in male. e.g. Represented by a single living species, the "tuatara" or *Sphenodon punctatum*.

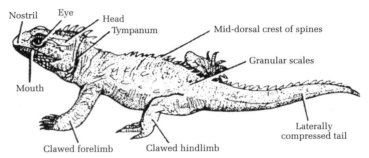

Figure 6.7
Sphenodon punctatum

Order 3. Squamata

Advanced, small to medium, elongated animals.

1. Limbs clawed, absent in snakes and few lizards.
2. Exoskeleton of horny epidermal scales, shields and spines.
3. Skull diapsid. Quadrate movable.

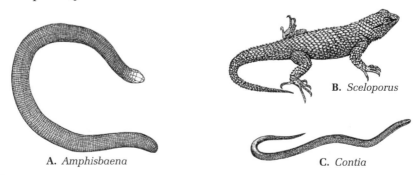

Figure 6.8
Lepidosaurs. A. Amphisbaenian, burrowing lepidosaur. **B.** Lizard (*Sceloporus*). **C.** Snake (Contia).

4. Vertebrae procoelous. Ribs single – headed.
5. Teeth acrodont or pleurodont.
6. Heart incompletely 4-chambered.
7. Cloacal aperture is transverse.
8. Male with eversible double copulatory organs (hemipenes).
9. About 6,800 species of lizard and snakes. These are divided into 2 distinct suborder- Lacertilia (Fig. 6.9) and Ophidia (Fig. 6.10) with contrasting characters,

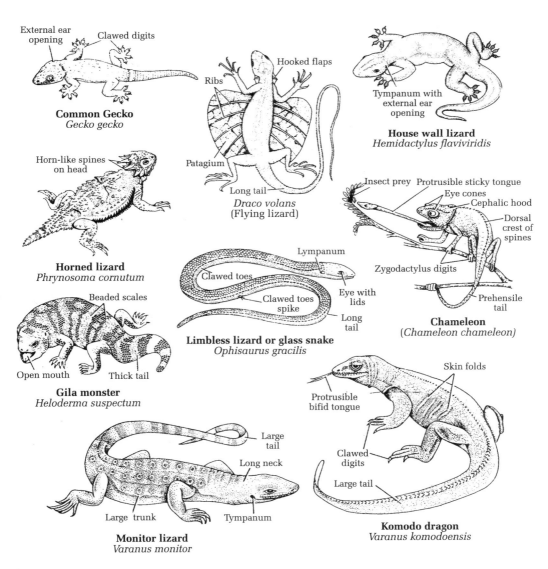

Figure 6.9
Common lizards

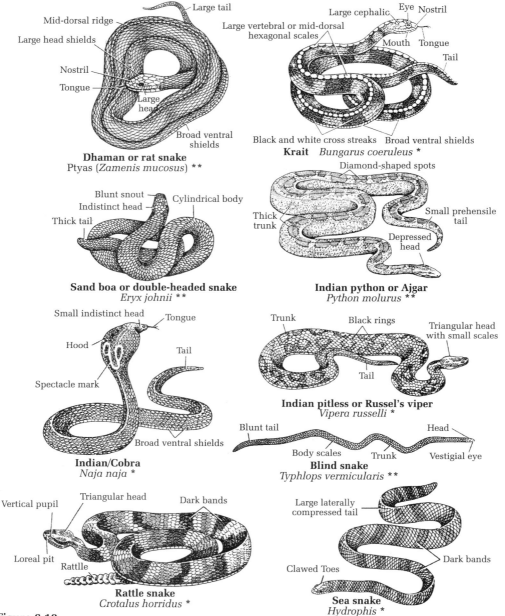

Figure 6.10
Non-poisonous Indian snakes (Suborder Ophidia or Serpentes) and Poisonous snakes of India
** Non-poisonous   * Poisonous

Order 4. Crocodilia

1. Large-sized, carnivorous and aquatic reptiles.
2. Tail long strong and laterally compressed.
3. Limbs short but powerful, clawed and webbed.

4. Skin thick with scales bony plates and scutes.
5. Skull diapsid. Quadrate immovable. No pariental foramen. A pseudopalate present.
6. Ribs bicephalous. Abdominal ribs present.
7. Teeth numerous, thecodont, lodged in sockets.
8. Heart completely 4-chambered.
9. Cloacal aperture is a longitudinal slit.
10. Male with a median, erectile, grooved penis. e.g.,. *Crocodylus. Gavialis, Alligator.* (Fig. 6.11).

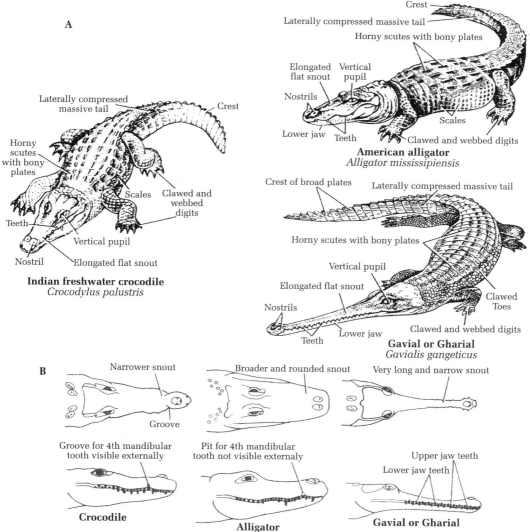

Figures 6.11 A, B

A. Common crocodilians. **B.** Heads of crocodile, alligator and gharial to illustrate their differences. Upper row shows shape of heads and width of snouts in dorsal view. Lower row shows arrangement of teeth in lateral view

Origin and Evolution of Reptiles

Evolution of reptiles is of special importance because it was cleidoic egg, which finally made the vertebrates independent of water. Thus, reptiles became the first vertebrates to wander at will on land without fear of dehydration and without needing water for laying eggs.

It is generally agreed that the reptiles originated from the labyrinthodont stegocephalian amphibians. The labyrinthodonts possed characteristically folded or labyrinthine teeth, similar to crossopterygian ancestors. They flourished through Carboniferous and Permian periods before extinction in the Triassic. We cannot however point to a single ancestor of reptiles. Probably they arose polyphyletically along a dozen or more independent lines. From the Permian through the Cretaceous periods, they were the most abundant of vertebrates. The early reptiles, were the cotylosaurs. The best known of these was Seymouria, (Fig. 6.1) a small aquatic tetrapod fossil from Taxas. It is a connecting link between the amphibians and the reptiles as it has characters of both these groups. Its limbs were directed laterally from the body and provided the same awkward type of locomotion as in amphibians. Later forms gradually pulled their legs under their bodies and evolved into birds and mammals. During Car-boniferous period of late Palaeozoic Era, about 250 million years ago, some labyrinthodont amphibians gradually took on reptilian characters. These earliest reptiles are called the stem reptiles. They belong to the order Cotylosauria of the subclass Anapsida. The transition was so gradual that often it is difficult to decide whether some fossil skeletons are those of advanced amphibians or primitive reptiles.

Seymouria, found in the Lower Permian of Texas (U.S.A.), was about 250 million years old. It was a lizard–like animal about 60 cm long with a comparatively thick body, relatively small pointed head with dorsally placed nostrils, and a short tail. Structure of Seymouria was intermediate between the amphibians of that time and the early reptiles.

Phylogenetic status of Seymouria

Amphibian affinities. Seymouria resembled early Amphibia or Labyrinthodontia in many features.

1. Skull is flat with reduced ossification.
2. An inter temporal bone is present.
3. Palate is primitive.
4. Teeth are labyrinthine and also found on vomers and palatines.
5. Position of fenestra ovalis below the basal level of brain is amphibian.
6. There are traces of lateral line canals in head region.
7. Vertebrae show little differentiation.
8. One pair of sacral ribs are present.
9. Neck is short so that pectoral girdle lies close behind skull.

Reptilian affinities. Seymouria had several characteristically reptilian features.

1. Limbs are muscular and rise mid-ventrally.

2. Skull is anapsid and monocondylic.
3. Pelvic girdle is attached to vertebral column by sacral vertebrae.
4. Number of phalanges 2:3:4:5:3 or 4 is more reptile like.

Seymouria is not directly ancestral to all reptiles. At the time it was living, the reptiles had already been present for some 50 million years. It is so perfectly intermediate between an amphibian and a reptile that its true position remains uncertain. Romar treats it as a reptile under the order Cotylosauria, whereas others classify it with primitive Amphibia. Perhaps Seymouria is a connecting ling between Labyrinthodontia and Cotylosauria. The oldest fossils, clearly judged to be the most primitive reptiles, comprise a group of so-called stem reptiles or Cotylosauria. They range from the upper Carboniferous to the upper Triassic when they became extinct. They closely resembled the labyrinthodont Amphibia from which they had evolved. These primitive or generalized reptiles had thick body, small pointed head with dorsal nostrils, short tail and short muscular pentadactyl limbs. Two characteristics were common; the complete roofing of skull (anapsid condition), and the flattened plate-like pelvic girdle. The size ranged from 30 cm to 3 meters. From this generalized condition in the succeeding periods, arose several lines of radiation or specializations, some leading to the great array of reptiles, both extinct as well as living, and others to birds and mammals.

Evolution of reptiles bears a special significance. They represent the first terrestrial vertebrates adapted for life in dry places on land. The dryness of skin to prevent loss of moisture from body, method of reproduction including cleidoic eggs capable of development on land, and the devices for economizing in the use of water, are some of the achievements in their terrestrial adaptations.

We shall not undertake to discuss the whole gamut of the extinct reptiles but only the more notable lines based on the morphology of the skull, such as anapsid, synapsid, euryapsid, parapsid and diapsid.

1. **Anapsid line.** The modern Chelonia (turtles and tortoises) represent a direct and an early offshoot of cotylosaurs retaining anapsid skull. They have remained virtually unchanged since Triassic, some 160 million years ago.
2. **Synapsid line.** The mammal – like reptiles or Synapsida had a single temporal cavity in skull ventral to postorbital and squamosal. Early Pelycosauria or Theromorpha were similar to cotylosaurs. Later Therapsida with differentiated dentition and improved locomotion were more mammal – like. Before disappearing in Jurassic they gave rise to ancestral mammals.
3. **Euryapsid line.** The euryapsids or plesiosaurs had a single temporal fossa in skull above the joint of postorbital and squamosal. They were large, marine, turtle-like, heavy-bodied and long-necked creatures. They were obviously fish-eaters. All became extinct towards the end of Cretaceous.
4. **Parapsid Line.** There was another marine blind alley, like Euryapsida, represented by fish-like or porpoise-like ichthyosaurs. They also became extinct near the close of Mesozoic Era.
5. **Diapsid line.** Most of the present-day reptiles are diapsid with two temporal openings on either side of skull separated by squamosal and postorbital bones. The earliest diapsids divided into two branches,

a) **Lepidosauria.** The Lepidosauria were probably ancestral to modern Squamata (snakes and lizards) and Rhynchocephalia (Sphenodon).

b) **Archosauria.** The Archosauria were the "ruling reptiles" dominating the Mesozoic Era. They represent the extinct ***Pterosauria,*** the extinct ***Dinosaurs*** and the modern **Crocodilia.** They also gave rise to the **modern birds.**

   (i) **Pterosauria.** The extinct flying reptiles called Pterosauria or Pterodactyla were of light built. Their forelimbs evolved into membranous wings or patagia. First 3 fingers were short, hooked and probably used for clinging to rocks. 4th finger was greatly elongated to support the edge of patagium. 5th finger was lost. Rhamphorhynchus (Fig. 6.12 A) of late Jurassic was a primitive pterosaur with 1 meter wingspan, a long balancing tail and toothed jaws. Pteranodon (Fig. 6.12 B) of Cretaceous had a 9 meter wingspan but no tail. Its huge horny and toothless beak was balanced by a backward bony projection of head.

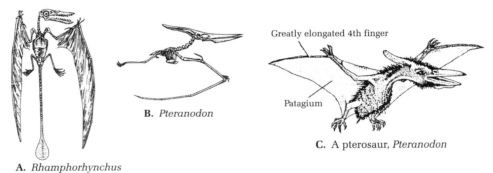

A. *Rhamphorhynchus*
B. *Pteranodon*
C. A pterosaur, *Pteranodon*

Figure 6.12
Pterosaurs. The lengthened forelimb of pterosaurs supported a membrane derived from skin to form the wing. **A.** *Rhamphorhynchus.* Wingspan was about 1.5 m. **B.** Skeleton of *Pteranodon.* Wingspan was about 8. m. **C.** *Pteranodon.* A pterosaur.

   (ii) **Dinosaurs.** At the end of Triassic, Thecodontia or Pseudosuchia, the early descendants of Archosauria, gave rise to the most fantastic Mesozoic reptiles, the dinosaurs, which means "terrible lizards". They subdivided early into two orders depending on the structure of their pelvis.

   a) Order **Saurischia**

   b) Order **Ornithischia**,

   a) **Saurischia** means *"reptile hips"* (Fig. 6.13 A). They possessed a triradiate pelvis with pubis entirely separate and anterior to ischium.

   Suborder **Theropoda-** included all flesh-eating and bipedal carnivores. Smaller Cretaceous ostrich – like forms, such as Struthiomimus and Ornithomimus, walked on 3 fingers of which one was opposable like a thumb and used for grasping. Jurassic Allosaurus, a monster carnivore, was 10 meters long. But the largest and the most fearful predator that ever walked the face of earth was Tyrannosaurus rex (Fig. 6.14 A) from

Cretaceous of North America. It was 15 meters long and stood 6 meters high. Its head was disproportionately great with large jaws armed with dagger-like teeth 15 cm long. The 3 toed massive hind legs were adapted for running, but extremely short forelimbs were almost useless.

Suborder **Sauropoda-** included huge herbivorous quadrupedal dinosaurs. Some of them were the largest and heaviest of all terrestrial and amphibious vertebrates that ever lived. Apatosaurus (Brontosaurus) (Fig. 6.14 B), Diplodocus, and Branchiosaurus were enormous Jurassic reptiles each more than 25 meters long and weighing over body would be supported partly by the buoyancy of water. They had long necks and tails, small head with exceptionally small brains and weak jaws.

b) **Ornithischia** means *"bird hips"* (Fig. 6.13 B). They had a typical tetraradiate bird like pelvic girdle with pubis directed backwards parallel to ischium. They were all herbivorous. Bipedal guanodon (Lower Cretaceous) grew to 10 meters. Its sharp dagger-like thumb was probably used for defence. Quadrupedal Stegosaurus (Fig. 6.14) (Jurassic) measured 8 meters and weighed 10 tons. Forelimbs were much shorter than hindlimbs. Skull was small and brain much smaller than the lumbar swelling of spinal cord. It had a parapet of heavy triangular plates on its neck, back and tail. The tail was also furnished with formidable sharp spikes or long and stood 3 meters high. Its enormous head carried 3 huge horns projecting from a large bony frill or collar protecting the vulnerable skull.

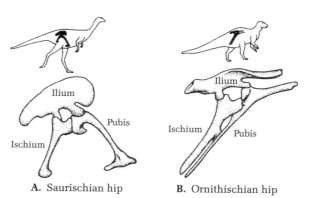

A. Saurischian hip B. Ornithischian hip

Figure 6.13
Dinosaur hips. Two types of hip structures define each group of dinosaurs. **A.** Saurischians all possessed a pelvic girdle with three radiating bones. **B.** *Ornithischians* had a hip with pubis and ischium bones lying parallel and next to each other.

Adaptive Radiation

Adaptive radiation. Because of the competition for food and living space, a single ancestral species evolves into different forms which occupy different habitats. This is called adaptive radiation or divergent evolution. Perhaps reptiles have shown the greatest groups. Their adaptive radiation took place twice, first in the Palaeozoic and secondly in the Mesozoic.

Palaezoic radiation. During Palaeozoic, there were no serious competitors on land, cotylosaurs multiplied rapidly occupying all ecological niches available to them. Their radiation involved adaptations to different methods of locomotion and feeding. During this period distinct anapsid and synapsid forms dominated.

Mesozoic radiation. The ancestral cotylosaurs had disappeared by the end of Palaeozoic. Their descendants produced a second and bigger radiation during Mesozoic. They dominated not only the land but also the sea and the air throughout the Mesozoic Era. They lasted over a great span of time, about 130 million years. The reptiles displayed an amazing pattern of adaptive radiation in the Triassic period. This was correlated with the new ecological niches provided by the climatic and geological changes which were taking place at that time. These changes included shift in climate from hot to cool, mountain building, terrain transformation and varying plant life. From rather small stem reptiles, they evolved into a great variety of forms that dominated the land and that is the reason Mesozoic Era is called the **Age of Reptiles**.

Outstanding among the ancient reptiles were the **dinosaurs** (Fig. 6.14). Many of them got adapted for land life and attained huge size. They developed **a two-legged gait.** In correlation with this mode of locomotion, they had reduced forelimbs, enlarged hindlimbs, and a heavy tail that could serve as a counterbalance for the trunk. Belonging to this group was **Tyrannosaurus** (Fig. 6.14 A) the largest terrestrial carnivore that ever lived. It was about 6 metres high and had 15 cm. long teeth in its large jaws. It was a ferocious animal and used its short forelimbs for seizing and tearing the prey.

Some dinosaurs became herbivorus swamp-dwellers and reverted to a **four-legged trait**. The bipedal gait of their ancestors was reflected in their long hind limbs. The buoyancy of water permitted some of them to grow to enormous size. **Brontosaurus** (Fig. 6.14 B), for instance, attained a length of 25 metres.

Certain dinosaurs became terrestrial herbivores. They also reverted to a four-legged gait. They formed bony plates on the body (Fig. 6.14 C) as in **Stegosaurus** (6 m. long), or horns on the head (Fig. 6.14) as in **Triceratops** (6 m. long).

Some ancient reptiles, called the **pterosaurs** (Fig. 6.14 D), got adapted for aerial life. The wings of these flying forms consisted of a membrane of skin, known as the patagium, supported by the forelimb and its greatly elongated outer (fourth) finger (Fig. 6.14). The 5th finger was lost, and the others were probably used for slinging to the cliffs. The hind limbs were very weak. This made them nearly helpless on land. Some pterosaurs attained a very large size, Pteranodon with a wingspread of over 6 meters was the largest.

Certain ancient reptiles returned to water. Marine forms became fully adapted to the aquatic mode of life. The **plesiosaurs** (Fig. 6.15 A) developed turtle-like body with a long neck. They propelled themselves through water with long paddle-shaped limbs. Elasmosaurus about 15 metres long, is an example of this group. The **ichthyosaurs** (Fig. 6.15 B) acquired fish-like body with fin-like limbs and even a structure looking like the dorsal fin of a fish. They moved by lateral undulations of the trunk. **Ophthalmosaurus**, about 21 metres long, is an example of this group. Two evolutionary lines of the ancient reptiles gave rise to birds and mammals directly before the age of reptiles ended.

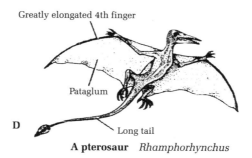

Figure 6.14
Varied types of Mesozoic *dinosaurs*.

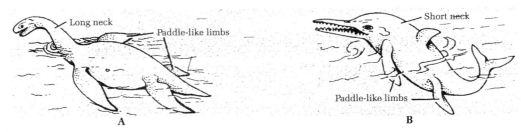

Figures 6.15 A, B
A. A plesiosaur *Euryapsid*, B. An ichthyosaur *Parapsid*

At the prime of their evolution in the Mesozoic era, the reptiles were represented by as many as 16 different orders. Most of these orders disappeared by the Late Cretaceous times. The present-day reptiles form 4 orders only.

The existing living reptile groups have also undergone an extensive adaptive radiations. This has led to the forms with various food habits, modes of locomotion, semiaquatic and aquatic life, and even development of flight.

Lizard are mostly diurnal, terrestrial quadrupeds. Many have become adapted for climbing trees and rocks, and have developed interesting devices for this purpose. Chamaeleons possess prehensile tail and grasping hands and feet for arboreal life. Geckos have dilated adhesive digit pads that cling to walls and rocks by vacuum. Some lizards have taken to a burrowing mode of life. They have assumed worm-like body form and have lost limbs. Anguis, the blind worm, and Ophisaurus, the glass snake, are examples of the burrowing lizards (Fig. 6.9). A few lizards have become Volant, e.g., Draco, the flying dragon (Fig. 6.9). It has patagia supported by ribs to sustain it in the air.

Snakes are mostly nocturnal, terrestrial animals. They are generally adapted to live in burrows, in the heaps of bricks, stones and logs, and in thick grass and bushes. They have developed a long, slender body devoid of limbs and have evolved an interesting feeding device (Fig. 6.10). Some snakes have reverted to the sea and have acquired laterally compressed tail to help in swimming. Hydrus, the seasnake, is an example. Certain snakes live on the trees, and a few can glide through the air by flattening the body.

Tortoises and **turtle**s have evolved directly from the stem reptiles (Fig. 6.6). They have developed a shell for protection. Tortoises are mostly adapted to live on land or in fresh water. They have a relatively high carapace and normal appendages. **Turtles** mostly inhabit the sea. They possess a low carapace and paddle-like swimming limbs.

Crocodiles and their allies have assumed an amphibious mode of life (Fig. 6.11). They swim in water with their powerful, laterally compressed tail and roam about on land with their well developed limbs.

Causes of Extinction

After thriving and dominating the earth for 130 million years, the great dinosaurs and their contemporaries became suddenly extinct by the end of Cretaceous period. Various factors have been suggested for their total extinction such as catastrophism, epidemic, food poisoning, racial senescence, climatic changes, over-specialization, interspecific competition. None of these has been accepted as being completely satisfactory. Probably a combination of several factors was responsible for their extinction.

The desire to explain this extraordinary exuberance of reptiles has attracted much attention. It has been argued that some of them, at least, may have been endothermic, mainly on the ground that the bone structure shows the well developed Haversian systems characteristic of high metabolic activity (Life of mammals). Tuna fish show the same structure. However, this bone structure may have been needed to support the retained ectothermic heat for long periods thus permitting maintenance of high metabolic rate as inertial 'homoiotherms'.

There is also evidence that large dinosaurs possessed special arrangements for cooling the brain and spinal cord. The nasal crests of the Hadrosaurs probably served to provide large nasal surfaces, either for special olfactory epithelia or for cooling surfaces (or both). The need to cool the central nervous system could of course arise in either an ectotherm or endothermous animal (Wheeler 1978).Their batteries of teeth show some wear and tear, so they must have been feeding for some time, presumably kept together by the parents. This gives some evidence of their social life.

Unique Terrestrial adaptations of Reptiles

1. adaptations for water conservation,
2. adaptation for burrowing life,
3. adaptation for defence,
4. survival during winter and land breeding.
5. adaptation for breeding on land

1. **Adaptations for Water Conservation.** Reptiles is able to conserve water in several ways–

    a) A complete covering of impermeable, horny scales prevents the loss of water by evaporation from the body surface.

    b) The kidney tubules are so modified that only small amount of water is removed from the blood in their initial parts, and much of the water that is removed is reabsorbed in the rest of the kidney tubules.

    c) Water is absorbed from the urine in the urodeaum of the cloaca and also in the urinary bladder, so much so that the urine, when passed out, is almost solid.

    d) Uric acid, which is the chief excretory product in Reptiles, is much less toxic and water-soluble than ammonia and urea. Therefore, it is not required to be flushed out of the tissues so rapidly nor with a large volume of water.

    e) Water is also absorbed from the feces, which is nearly dry when passed out.

2. **Adaptation for Burrowing Life.** Many adaptive features help than in fossorial life–

    a) Sharp, horny claws on the digits form efficient digging tools.

    b) Selection of sandy soil for habitation has facilitated burrowing. Long, narrow, low body is suitable for moving into holes or underneath vegetation, stones, etc.

c) During burrowing, the overlapping lips guard the mouth, thickenings of mucous membrane close the nares and movable lids protect the eyes.

   d) Delicate tympanum is protected by its position in a depression.

3. **Adaptations for Defence.** Reptiles have a number of protective devices.

   a) Life in the burrows protect Reptiles from many predators, like birds of prey and large carnivores, which cannot go into the burrows.

   b) Closing the burrow after getting into it at night affords an additional protection.

   c) Ability to run fast is another device to escape from the enemies.

   d) Colour of skin harmonized with the background, making it nearly invisible from a distance.

   e) Spiny tail is a nice defensive weapon as it can cause injury to the enemies.

   f) A warning posture involving raising of the front part of the body and opening of mouth wards off the predators.

   g) Laying eggs in the burrows protects the progeny.

4. **Adaptations for Survival During Winter.** Reptiles manages to survive in winter by locking themselves up in a burrow (hibernation). Total lack of food in a burrow is compensated by reduced metabolism due to inactive life and reserve food stored in the fat bodies.

5. **Adaptations for Breeding on Land.** There are a variety of good adaptations for breeding on land.

   a) External fertilization is not possible on land. Therefore, Reptiles resorts to internal fertilization and, for this, copulatory organs have been developed in the male.

   b) Shell around the egg protects the embryo from mechanical injury and desiccation at the same time permitting gaseous exchange through its pores.

   c) Abundance of yolk in the egg enables the embryo to complete its development into a young animal capable for looking after itself, excluding the necessity of a larval stage, which cannot survive on land.

   d) Embryonic membranes provide aquatic medium within the egg by having fluid in them and also help in feeding, respiration and storing waste matter.

   Loss of lateral-line sense organs is also an adaptation for land life. Careful review of the various systems will reveal many less obvious adaptations.

Chapter 7

Aves

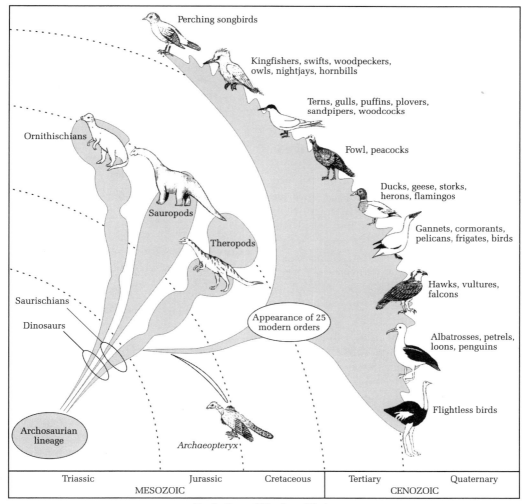

Evolution of Modern Birds

Figure 7.0 The earliest known bird, *Archaeopteryx,* lived in the Upper Jurassic. *Archaeopteryx* uniquely shares many specialized aspects of its skeleton with the smaller theropod dinosaurs and is considered to have evolved within the theropod lineage. Evolution of modern bird orders occurred rapidly during the cretaceous and early Tertiary periods.

Birds are the most fascinating, the most melodious, and the most beautiful animals. With more than 9900 species distributed over nearly the entire earth, birds out number any other vertebrate group except fishes. Some birds live in total blackness in caves, finding their way by echolocation, and others dive to depths greater than 45 m to prey on aquatic life. The "bee" hummingbird of Cuba, weighing only 1.8 g, is one of the smallest vertebrate endotherms. The single unique feature that distinguishes birds from other living animals is their feathers. If an animal has feathers, it is a bird; if it lacks feathers, it is not a bird. No other living vertebrate group bears such an easily recognizable and foolproof identification tag. Feathers also were present in some theropod dinosaurs, but these feathers were not capable of supporting flight.

There is great uniformity of structure among birds. Despite approximately 150 million years of evolution, during which they proliferated and adapted to specialized ways of life, we have little difficulty in recognizing a living bird as a bird. In addition to feathers, all birds have forelimbs modified into wings (although not always used for flight); all have hind limbs adapted for walking, swimming, or perching; all have keratinized beaks; and all lay eggs. The reason for this great structural and functional uniformity is because the birds evolved into flying machines and this greatly restricted morphological diversity.

Figure 7.1
Archaeopteryx, a 147-million-year-old ancestor of modern birds. Cast of the second and most nearly perfect fossil of *Archaeopteryx* which was discovered in a Bavarian stone quarry.

Approximately 147 million years ago, a flying animal drowned and settled to the bottom of a shallow marine lagoon in what is now Bavaria, Germany. It was rapidly covered with a fine silt and eventually fossilized. There it remained until discovered by a workman splitting slate in a limestone quarry. The fossil was approximately the size of a crow, with a skull not unlike that of modern birds except that the beaklike jaws bore small bony teeth set in sockets like those of reptiles (Fig. 7.1). The skeleton was decidedly reptilian with a long bony tail, clawed fingers, and abdominal ribs. It might have been classified as a theropod dinosaur except that it carried an unmistakable imprint of feathers, those marvels of biological engineering that only birds possess. *Archaeopteryx lithographica*, as the fossil was named, was an especially fortunate discovery because the fossil record of birds is disappointingly meager. The finding was also dramatic because it demonstrated beyond reasonable doubt the phylogenetic relatedness of birds and theropod dinosaurs.

A bird's entire anatomy is designed around flight. An airborne life for a large vertebrate is a highly demanding evolutionary challenge. A bird must, of course, have wings for support and propulsion. Bones must be light yet serve as a rigid airframe. The respiratory system must be highly efficient to meet the intense metabolic demands of flight and also serve as a thermoregulatory device to maintain a constant body temperature. A bird must have a rapid and efficient digestive system to process an energy-rich diet; it must have a high metabolic rate; and it must have a highly efficients circulatory system. Above all, birds must have a finely tuned nervous system and acute senses, especially superb vision, to handle the complex problems of head first, high-velocity flight.

Birds are a lineage of endothermic, diapsid amniotes that evolved flight in the Jurassic period of the Mesozoic. Phylogenetically, they are most closely related to theropod dinosaurs, a group of bipedal carnivores with birdlike skeletal characteristics. Their closest living relatives are crocodilians (shared ancestry of birds and crocodiles). The morphological characteristics and great uniformity of structure of birds relate almost entirely to the strict demands of flight, and the mobility that flight provides is responsible for many distinctive aspects of their behavior and ecology.

Unique characterisic adaptation acquired by birds

1. Feathers distinguish birds from all other living animals. The evolution of feathers was the single most important event leading to the capacity for flight in birds.

2. In addition to feathers, several other essential adaptations contribute to the two prime requirements for flight:

    a) increase in power and

    b) decrease in weight.

    These adaptations include forelimbs modified as strong wings, hollow bones, keratinized bill (rather than heavy jaws and teeth), endothermy, high metabolic rate (six to ten times as high as reptiles of similar weight and body temperature), large hearts and highly efficients circulation, highly efficient respiratory system, keen vision, and excellent neuromuscular coordination.

3. Birds occupy almost every available habitat on the earth's surface and, within the constraints imposed by requirements for flight, have radiated modestly in body form, especially in bill adaptations.

4. The unparalleled mobility of birds has enabled many to benefit from the advantages of making seasonal and long-distance migrations. Migration enables birds to secure seasonal habitats most beneficial for breeding, finding food, avoiding predators, and reducing interspecific competition.

Zoologists had long recognized the similarity of birds and reptiles. The skulls of birds and reptiles abut against the first neck vertebra by a single occipital condyle . Birds and reptiles have a single middle ear bone, the stapes . Birds and reptiles have a lower jaw composed of five or six bones, whereas the lower jaw of mammals has one bone, the dentary. Birds and reptiles excrete their nitrogenous wastes as uric acid whereas mammals excrete as urea. Birds and most reptiles lay similar yolked eggs with the early embryo developing on the surface by shallow cleavage divisions.

Thomas Henry Huxley called birds the "glorified reptiles" and classified them with a group of dinosaurs called theropods that displayed several birdlike characteristics (Fig. 7.2). Theropod dinosaurs share many derived characters with birds, the most obvious of which is the elongate, mobile, S-shaped neck.

Dromeosaurs, a group of theropods that includes Velociraptor, share many additional derived characters with birds, including a furcula (fused clavicles) and lunate wrist bones that permit swiveling motions used in flight. Additional evidence linking birds to dromeosaurs comes from recently described fossils from late Jurassic and early Cretaceous deposits in Liaoning Province, China. These spectacular fossils, including

Protarchaeopteryx and Caudiperyx, are dromeosaur-like theropods, but with feathers! It is unlikely these feathered dinosaurs could fly, however, as they had short forelimbs and symmetrical vaned feathers (the flight feathers of modern flying birds are asymmetrical). While these feathers could not have been used for powered flight, they may have been useful for controlling glides of jumps from trees. These primitive feathers, like modern feathers, may have been colorful and used in social displays. Theropod dinosaurs recently unearthed in China, such as *Sinosauropterex*, are covered with filaments that appear to be homologous with feathers. The filamentous covering of these dinosaurs likely served as insulation and was a precursor to vaned feathers. Other fossils from Spain and Argentina of birds document the development of the keeled sternum and alula, loss of teeth, and fusion of bones characteristic of modern birds. Clearly, a phylogenetic approach to classification would include birds with theropod dinosaurs. With this view, dinosaurs are not extinct—they are with us today as birds!

Living birds (Neornithes) are divided into two groups:

1. **Paleognathae**, the large *flightless* ostrichlike birds and the kiwis, often called ratite birds, which have a flat sternum with poorly developed pectoral muscles,
2. **Neognathae**, *flying* birds that have a keeled sternum on which powerful flight muscles insert.

This division assumes that flightless ratites form a monophyletic group. However, evidence supporting this grouping is weak, and relationships of ratites to other birds are controversial. Ostrich like paleognathids clearly have descended from flying ancestors. Furthermore, not all neognathous birds can fly and many of them even lack keels. Flightlessness has appeared independently among many groups of birds; the fossil record reveals flightless wrens, pigeons, parrots, cranes, ducks, auks, and even a flightless owl. Penguins are flightless although they use their wings to "fly" through water. Flightlessness almost always has evolved on islands where few terrestrial predators are found. Flightless birds living on continents today are the large paleognathids (ostrich, rhea, cassowary, emu), which can run fast enough to escape predators. An ostrich can run 70 km (42 miles) per hour, and claims of speeds of 96 km (60 miles) per hour have been made.

Classification of Class Aves

Class **Aves** contains more than 9900 species distributed among 25 orders of living birds and a few fossil birds. There is little consensus between the relationships among bird orders, and the monophyly of several bird orders is in question. Until recently classifications have relied on shared, derived morphological characters. An alternate taxonomy has been proposed that is based on DNA hybridization studies by Sibley and Ahlquist (1990). This study suggests a number of surprising relationships, including placement of penguins, loons, grebes, albatrosses, and birds of prey in order Ciconiiformes, which traditionally only contained the herons and relatives. Relationships implied by this biochemical study have not been tested with other phylogenetic studies using molecular techniques.

Questions persist about the validity of the "molecular clock" which is critical for DNA hybridization studies. Given this uncertainty, we choose to present a traditional classification, primarily based on morphological characters.

Class Aves

Subclass **Archaeornithes.** Birds of the late Jurassic and early Cretaceous bearing many primitive characteristics. e.g., Archaeopteryx.

Subclass **Neornithes.** Extinct and living birds with well-developed sternum and usually with keel; tail reduced; metacarpals and some carpals fused together. Cretaceous to Recent.

Superorder **Paleognathae** Modern birds with primitive archosaurian palate. Ratites (with unkeeled sternum) and tinamous (with keeled sternum (Fig. 7.3)).

Order **Struthioniformes** e.g., ostrich, rheas, cassowaries, emus, kiwis. Fifteen species of flightless birds of Africa, South America, Australia, New Guinea, and New Zealand. The ostrich, *Struthio camelus* (Fig. 7.3), of Africa, is the largest of living birds, with some specimens being 2.4 m tall and weighing 135 kg. The feet are provided with only two toes of unequal size covered with pads, which enable the birds to travel rapidly over sandy ground. The kiwis, about the size of domestic fowl, are unusual in having only the merest vestige of a wing.

Order **Tinamiformes**: tinamous. Ground-dwelling, grouse like birds of Central and South America. About 47 species.

Superorder **Neognathae.** Modern birds with flexible palate.

Order **Sphenisciformes**: penguins. Web-footed marine swimmers of southern seas from Antarctica north to the Galapagos Islands. Although penguins are carinate birds, they use their wings as paddles for swimming rather than for flight. About 17 species.

Order **Gaviiformes**; loons. The five species of loons are remarkable swimmers and divers with short legs and heavy bodies. They live exclusively on fish and small aquatic forms. The familiar great northern diver, *Gavia immer*, is found mainly in northern waters of North America and Eurasia.

Order **Podicipediformes**: grebes. These are short-legged divers with lobate-webbed toes. The pied billed grebe, *Podilymbus podiceps*, is a familiar example of this order. Grebes are most common in old ponds where they build their raft like floating nests. Twenty-one species, worldwide distribution.

Order **Procellariiformes**: albatrosses, petrels, fulmars, shearwaters. All are marine birds with hooked beak and tubular nostrils. In wingspan (more than 3.6 m in some), albatrosses are the largest of flying birds. About 115 species, worldwide distribution.

Order **Pelecaniformes**; pelicans, cormorants, gannets, boobies, and others. These are colonial fish-eaters with throat pouch and all four toes of each foot included within the web. About 65 species, worldwide distribution, especially in the tropics.

Classification of Living Birds of Class Aves

Order **Ciconiiformes**: herons, bitterns, storks, ibises, spoonbills, flamingos, vultures. These are long-necked, long-legged, mostly colonial waders and vultures. A familiar eastern North American representative is the great blue heron, *Ardea herodias*, which frequents marshes and ponds. About 120 species, worldwide distribution.

Order **Falconiformes** eagles, hawks, falcons, condors, buzzards. Diurnal birds of prey. All are strong fliers with keen vision and sharp, curved talons. About 310 species, worldwide distribution.

Order **Anseriformes:** swans, geese, ducks. The members of this order have broad bills with filtering ridges at their margins, a foot web restricted to the front toes, and a long breast bone with a low keel. About 160 species, worldwide distribution.

Order **Galliformes:** quail, grouse, pheasants, ptarmigan, turkeys, domestic fowl. Chicken like ground-nesting herbivores with strong beaks and heavy feet. The bobwhite quail, *Colinus virginianus*, is found all over the eastern half of the United States. The ruffed grouse, *Bonasa umbellus*, is found in about the same region, but in woods instead of the open pastures and grain fields, which the bobwhite frequents. About 290 species, worldwide distribution.

Order **Gruiformes:** cranes, rails, coots, gallinules. Mostly prairie and marsh breeders. About 215 species, worldwide distribution.

Order **Charadriiformes:** gulls, oyster catchers, plovers, sandpipers, terns, woodcocks, turnstones, lapwings, snipe, avocets, phalaropes, skuas, skimmers, auks, puffins. All are shorebirds. They are strong fliers and are usually colonial. About 330 species, worldwide distribution.

Order **Columbiformes:** pigeons, doves. All have short necks, short legs, and a short, slender bill. The flightless dodo, *Raphus cucullatus*, of the Mauritius Islands became extinct in 1681. About 320 species, worldwide distribution.

Order **Psittaciformes:** parrots parakeets. Birds with hinged and movable upper beak, fleshy tongue. About 370 species, pantropical distribution.

Order **Musophagiformes:** Turacos. Medium to large birds of dense forest or forest edge with a conspicuous patch of crimson on the spread wing. Bill brightly colored, wings short and rounded. Twenty-three species restricted to Africa.

Order **Cuculiformes:** cuckoos, roadrunners. European cuckoos, *Cuculus canorus*, lay their eggs in nests of smaller birds, which rear the young cuckoos. American cuckoos, black billed and yellow billed, usually rear their own young. About 150 species, worldwide distribution.

Order **Strigiformes:** owls. Nocturnal predators with large eyes, powerful beaks and feet, and silent flight. About 185 species, worldwide distribution.

Order **Caprimulgiformes:** goatsuckers, nighthawks, whippoorwills. Night and twilight feeders with small, weak legs and wide mouths fringed with bristles. Whippoorwills, *Antrostomus veciferus*, are common in the woods of the eastern states, and nighthawks, *Chordeiles minor*, are often seen and heard in the evening flying around city buildings. About 115 species, world-wide distribution.

Order **Apodiformes:** swifts, hummingbirds. These are small birds with short legs and rapid wing beat. The familiar chimney swift, *Chaetura pelagia*, fastens its nest in chimneys by means of saliva. Most species of hummingbirds are found in the tropics, but there are 14 species in the United States, of which only one, the ruby-throated hummingbird, is found in the eastern part of the country. About 435 species, worldwide distribution.

Order **Coliiformes:** mouse birds. Small crested birds of uncertain relationship. Six species restricted to southern Africa.

Order **Trogoniformes:** trogons. Richly colored, long-tailed birds. About 40 species, pantropical distribution.

Order **Coraciiformes:** kingfishers, hornbills, and others. Birds with strong, prominent bills that nest in cavities. In the eastern half of the United States, belted kingfishers, *Megaceryle alcyon*, are common along most waterways of any size. About 220 species, worldwide distribution.

Order **Piciformes:** woodpeckers, toucans, puff birds, honey-guides. Birds with highly specialized bills and having two toes extending forward and two backward. All nest in cavities. There are many species of woodpeckers in North America, most common of which are flickers and downy, hairy, red-bellied, redheaded, and yellow-bellied woodpeckers. Largest is the pileated woodpecker, which is usually found in deep and remote woods. About 410 species, worldwide distribution.

Order **Passeriformes:** perching songbirds. This is the largest order of birds, containing 56 families and 60% of all birds. Most have a highly developed syrinx. Their feet are adapted for perching on thin stems and twigs. The young are altricial. To this order belong many

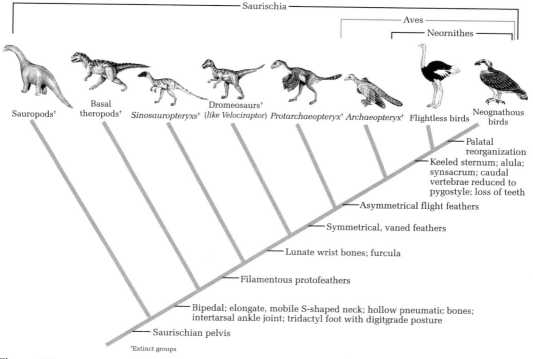

Figure 7.2
Cladogram of the Saurischia, showing the relationship of several taxa to modern birds. Shown are a few of the shared derived characters, mostly related to flight, that were used to construct the genealogy. The ornithischians are the sister group to the saurischians and all are members of the clade Archosaurla.

birds with beautiful songs such as thrushes, warblers, mockingbird, meadow lark, and hosts of others. Others of this order, such as swallows, magpie, starling, crows, raven, jays, nut hatches, and creepers, have no songs worthy of the name. More than 5900 species, worldwide distribution.

More than 9900 species of living birds are egg-laying, endothermic vertebrates with feathers and forelimbs modified as wings. Birds are closest phylogenetically to theropods, a group of Mesozoic dinosaurs with several birdlike characteristics. The oldest known fossil bird, Archaeopteryx from the Jurassic period of the Mesozoic era, had numerous reptilian characteristics and was almost identical to certain theropod dinosaurs except that it had feathers. It is probably the sister taxon of modern birds.

Feathers, the hallmark of birds, are complex derivatives of reptiles scales and combine lightness with strength, water repellency, and high insulative value. Body weight is further reduced by elimination of some bones and fusion of others (to provide rigidity for flight). The light, keratinized bill, replacing the heavy jaws and teeth of reptiles, serves as both hand and mouth for all birds and is variously adapted for different feeding habits.

Phylogenetic Relationships

Birds are traditionally classified as members of the class Aves. The major characteristics of this class are adaptations for flight, including appendages modified as wings, feathers, endothermy, a high metabolic rate, a vertebral column modified for flight, and bones lightened by numerous air spaces. In addition, modern birds possess a horny bill and lack teeth.

The similarities between birds and reptiles are so striking that birds are often referred to as "glorified reptiles." Like the crocodilians, birds have descended from ancient archosours. Other flying reptiles in this evolutionary lineage e.g., pterosaurs and pterodactyls are ruled out of bird ancestry because these reptiles lost an important avian feature, the clavicles, long before birds appeared. The fused clavicles, or "wishbone", is one of the attachment points for flight muscles. Thus, these reptiles could not have been as strong at flying as are modem birds. In addition, instead of feathered wings, the flight surfaces of the wings of these primitive reptiles were membranous folds of skin. Modern birds are derived from the Saurischian lineage of dinosaurs that also included bipedal carnivorous dinosaurs like Tyrannosaurus. According to cladistic interpretations (Fig. 7.3), the birds can be thought of as modern dinosaurs.

Diversity of Modern Birds

Archaeopteryx, Sinornis, and Eoalulavis provide the only evidence of the transition between reptiles and birds. Zoologists do not know, however, whether or not one of these birds is the direct ancestor of modern birds. A variety of fossil birds for the period between 100 million and 70 million years ago has been found. Some of these birds were large, flightless birds; others were adapted for swimming and diving; and some were fliers. Most, like Archaeopteryx, had reptile like teeth. Most of the lineages that these fossils represent became extinct, along with the dinosaurs, at the end of the Mesozoic era.

Birds show less diversification than any other group of vertebrate animals. The great homogeneity of birds, therefore, fails to present convenient external features, such as the teeth of mammals, for their classification. According to Wetmore (1960), there are 34 orders, 27 orders of living birds of which two have recently become extinct, and 7 orders of fossil birds.

Diagonastic Characters of Birds

1. Body usually spindle-shaped, with four divisions: head, neck, trunk, and tail; neck disproportionately long for balancing and food gathering
2. Paired limbs, with the forelimbs usually adapted for flying; posterior pair variously adapted for perching, walking, and swimming; foot with four toes (chiefly)
3. Epidermal covering of feathers and leg scales; thin integument of epidermis and dermis; no sweat glands
4. Skeleton ossified, with air cavities; skull bones fused with one occipital condyle; each jaw covered with a horny sheath, forming a beak; no teeth; ribs with strengthening processes; tail not elongate; sternum well developed with keel or reduced with no keel; single bone in middle ear
5. Nervous system well developed, with brain and 12 pairs of cranial nerves
6. Circulatory system of four-chambered heart, with the right aortic arch persisting; nucleated red blood cells
7. Endothermic
8. Respiration by slightly expandable lungs, with thin air sacs among the visceral organs and skeleton; syrinx (voice box) near junction of trachea and bronchi
9. Excretory system of metanephric kidney; ureters open into cloaca; no bladder; semisolid urine; uric acid main nitrogenous waste
10. Sexes separate; females with left ovary and oviduct only
11. Fertilization internal; amniotic eggs with much yolk and hard calcareous shells; embryonic membranes in egg during development; incubation external; young active at hatching (precocial) or helpless and naked (altricial)

Class Aves is divided into two subclasses as follows:

Subclass I Archaeornithes

1. Extinct, archaic, Jurassic birds of Mesozoic Age, about 155 million years ago.
2. Wings primitive, with little power of flight.
3. Tail long, tapering, lizard-like, bearing two lateral rows of rectrices.
4. Each hand bearing three unfused and clawed fingers.
5. Skull with teeth in both jaws, embedded in sockets.
6. Vertebrae amphicoelous.
7. Tail with 18-20 free caudal vertebrae, without pygostyle.
8. Sternum without a keel.

9. Carpals and metacarpals free.
10. Thoracic ribs slender, without uncinate processes.
11. Abdominal ribs present.
12. Cerebellum small.

This subclass includes a single order.

Order Archaeopterygiformes

e.g., *Archaeopteryx lithographica*, from Jurassic of Bavaria, Germany; one specimen lying in the British museum, London, the other lying in the Berlin Museum, Berlin,

Subclass II Neornithes

1. Modern as well as extinct post-Jurassic birds.
2. Wings usually well-developed and adapted for flight, with few exceptions.
3. Tail short and reduced, with rectrices arranged in a fanlike manner.
4. Wing composed of 3 partly fused fingers without claws.
5. Teeth absent except in some fossil birds.
6. Vertebrae heterocoelous in living forms.
7. Few caudal vertebrae free. Rest fused into a pygostyle.
8. Sternum usually with a keel.
9. Distal carpals fused with metacarpals to form carpometacarpus.
10. Thoracic ribs usually with uncinate processes.
11. Abdominal ribs absent.
12. Cerebellum large.

The subclass is divisible into 4 super-orders:

Super-order 1. Odontognathae

1. Extinct, Upper Cretaceous birds.
2. Jaws bear teeth, "so advantageous for catching fish."
3. Brain of the avian type.

Order 1. Hesperornithiformes

1. Large flightless marine birds.
2. Sharply pointed pleurodont teeth, present in grooves rather than in sockets.
3. Vertebrae amphicoelous.
4. Shoulder girdle reduced.
5. Sternum without a keel.
   e.g.,: Hesperornis, Enaliornis, Baptornis, etc.

Order 2. Ichthyornithiformes

1. Whether teeth were present is not definite.
2. Neck vertebrae amphicoelous.
3. Shoulder girdle well-developed.
4. Sternum with a well-developed keel.
   e.g.,: *Ichthyornis, Apatornis*.

Super-order 2. Palaeognathae or Ratitae

1. Modern big-sized, flightless, running birds, without teeth.
2. Wings vestigial or rudimentary; feathers devoid of interlocking mechanism.
3. Rectrices absent or irregularly arranged.
4. Pterylae are irregular.
5. Oil gland is absent, except in Tinamus and Kiwi.
6. Skull is dromaeognathous or palaeognathous that is, vomer is large and broad and interpolated between palatines.
7. Skull sutures remain distinct for a long time.
8. Quadrate articulates by a single head with skull.
9. Sternal keel vestigial, absent or flat, raft-like.
10. Uncinate processes are vestigial or absent.
11. Tail vertebrae free. Pygostyle small or absent.
12. Scapula and coracoid are comparatively small and fused at an abtuse angle (more than a right angle).
13. Clavicles are small or absent.
14. iIium and ischium not united posteriorly except in Rhea and Emu.
15. Pectoral muscles poorly developed.
16. Syrinx is absent.
17. Male has a large and erectile penis; female has a clitoris.
18. Young are precocious.
19. Distribution is restricted.

The flightless birds or ratites are not represented in India. They are grouped in 7 orders as follows;

Order 1. Struthioniformes

1. Legs strongly developed, each with two toes 3rd and 4th with stunted nails.
2. Pubis form a ventral symphysis.
   e.g.,: True ostriches (*Struthio camelus*) of Africa and western Asia (Arabia).

Order 2. Rheiformes

1. Each leg bears three clawed toes.
2. Ischia form a ventral symphysis.
   e.g.,: American ostriches or common rhea (*Rhea Americana*) represented by two species in South American pampas; Darwin's rhea (*Pteroncemia pennata*).

Order 3. Casuariformes

1. Forelimbs greatly reduced.
2. Head bears a comb-like structure.
   e.g.,: Cassowaries (Casuarius) of Australia, and New Guinea and Emus (*Dromaius navaehollandiae*) of New Zealand.

African ostrich
Struthio camelus

South American ostrich
Rhea americana

Australian cassowary
Casuarius casuarius

Australian Emu
Dromaius novaehollandiae

Tinamou
Eudromia elegans

Large grey kiwi
Apteryx haasti

Figure 7.3
Flightless birds (Ratitae)

Order 4. Apterygiformes

1. Feathers simple, hair-like or bristle-like.
2. Wings vestigial.

3. Long bill with nostrils near the tip.
   e.g.,: Kiwis (Apteryx) of New Zealand.

Order 5. Dinornithiformes

1. Giant birds, became extinct nearly 700 years ago.
2. Wings almost absent, beaks short and massive legs bearing four toes each.
   e.g.,: Moas (*Dinornis maximus*) of New Zealand.

Order 6. Aepyornithiformes

1. Recently exterminated, rather later than moas.
2. Wings tiny, but legs powerful and 4-toed.
   e.g.,: Giant Elephant-birds of Africa and Madagascar. *Aepyornis titan*, Mulleornis.

Order 7. Tinamiformes

1. Small terrestrial birds, not flightless but essentially great runners (cursorial).
2. Sternum is keeled.
   e.g.,: Tinamou (*tinamus*), Eudromia.

Super-order 3. Impennae

Order 1. Sphenisciformes

1. Modern, aquatic, flightless, with paddle-like wings of flippers.
2. Feet are webbed.
   e.g.,: Penguins (*Aptenodytes*) of Southern Hemisphere.

Super-order 4. Neognathae or Carinatae

1. Most modern, usually small-sized, flying birds.
2. Wings well-developed; feathers with interlocking mechanism.
3. Rectrices present and arranged regularly.
4. Pterylae are regular.
5. Oil gland is present.
6. Skull is neognathous, that is, vomer is short allowing palatines to meet.
7. Skull sutures disappear very early.
8. Quadrate is double-headed.
9. Sternum with a well-developed keel.
10. Uncinate processes are present.
11. Pygostly is present.
12. Scapula and coracoid meet at a right angle or acute angle.
13. Clavicles are always well developed.

Figure 7.4
Common Indian birds

14. Ilium and ischium are united posteriorly.
15. Pectoral muscles large.
16. Male had no copulatory organ.
17. Young are altricious.
18. Distributed all over the world.

The super-order Neognathae includes several orders. For the sake of convenience they are grouped into five homogeneous ecological groups, as follows:

1. Arboreal Bird
2. Terrestrial Birds
3. Swimming and diving birds
4. Shore or wading birds
5. Aerial birds

Group A. Arboreal Birds

Under this group are placed the majority of birds spending most of their lives in and around shrubs and trees.

Order 1. passeriformes

This is the largest of all the bird orders including half the known species. Feet are adapted for perching, while beaks are adapted for cutting (Fig. 7.4).

e.g., Common house sparrow (*Passer domesticus*), common house crow (*Corvus spledens*), Indian jungle crow (*Corvus macrorhynchos*), common myna (*Acridotheres tristis*), bank myna (*Acridotheres ginginianus*), Indian robin (*Saxicoloides fulicata*), black drongo or kingcrow (*Dicrurus macrocercus*), golden oriole (*Oriolus oriolus*), tailor bird (*Orthotomus sutorius*), weaver bird or baya (*Ploceus philippinus*), flycatchers (*Muscicapa*), swallows, bulbuls (*Molpestes*) magpie (*Pica*), crossbill (*Loxia*), starling (*Sturnus*), larks (*Alauda*).

Order 2. Piciformes

It includes woodpeckers, toucans, sap-suckers and their allies. e.g.,: Yellow fronted pied woodpecker (*Dendrocopos mahrattensis*), golden-backed woodpecker (*Dinopium benaghalensis* (Fig. 7.4))

Order 3. Columbiformes

It includes doves and pigeons. e.g.,: Blue rock pigeon (*Columba livia*), green pigeon (*Crocopus*), crowned pigeon (*Goura cristata*), passenger pigeon (*Ectopistes migratiorius*), ringed turtle dove (*Streptopelia risoria*), spotted dove (*Streptopelia chinensis*), extinct dodo (*Raphus*) and solitaire (*Pezophaps*).

Order 4. Psittaciformes

It includes parrots, parakeets, cockatoos, macaws, love-birds, etc., denizens of the equatorial jungles.

e.g.,: Large Indian parakeet (*Psittacula eupatria*), green parrot (*Psittacula krameri*), budgeriger (*Melopsittacus* (Fig. 7.4))

Group B. Terrestrial Birds

These birds are perfectly able to fly but spend most of their time walking or running on ground.

Order 5. Galliformes

It includes gamebirds notable for their palatability, massive scratching feet, short and powerful flight and largely graminivorous diet. e.g.,: Red jungle fowl (*Gallus*), peafowl (*Pavo cristatus*), quail (*Cotrunix cturnix*), grey partridge (*Francolinus pondicerianus*), chukor (*Alectoris grecca*), pheasants (*Phasiaus*).

Order 6. Cuculiformes

It includes cuckoos and their allies. e.g.,: Cuckoo (*Cuculus canorus*), Koel (*Eudynamis scolopaceous*), crow-pheasant (*Centropus sinensis*).

Group C. Swimming and Diving Birds

Order 7. Anesriformes

Aquatic birds such as geese, swans and ducks belong to this order. e.g.,: Wild duck or mallard (*Anas*), common teal (*Nettion crecca*), mergansers (*Mergus*).

Order 8. Coraciiformes

It includes kingfishers and their allies. e.g.,: White breasted kingfisher (*Halcyon smynensis*), pied kingfisher (*Ceryle rudis*), great horn bill (*Dichoceros bicornis*), grey hornbill (*Tochus birostris*), hoopoe (*Upupa epops*).

Order 9. Gaviiformes

It includes marine birds, called loons (*Gavia*) represented by only four species.

Order 10. Podicipediformes or Colymbiformes

It includes diving bird. It includes grebes (*Podicipes*), often called divers because or their habits.

Order 11. Procellariformes

It includes tube-nosed, long and oily winged seabirds such as albatrosses (*Diomedea*), Petrels (*Procellaria*), shearwaters

Order 12. Pelecaniformes

It includes pelicans, darters, gannets and cormorants. e.g.,: Pelicans (*Pelecanus*), little cormorant (*Phalacrocorax niger*), Indian darter (*Anhinga melanogaster*).

Group D. Shore Birds and Wading Birds

These aquatic birds seldom swim or live beneath the water to any great extent.

Order 13. Charadriiformes

These order includes a rather diverse group of water frequenting shore birds characterized by long wading legs, webbed toes and mud probing beaks. e.g.,: Red wattled lapwing (*Lobivanellus indicus*), pheasant-tailed jacana (*Hydrophasianus chirugus*), sand piper (*Tringa glariola*), snipe (*Capella*), gull (*Larus*), curlew (*Numenius*).

Order 14. Ciconiiformes

It includes long-legged, marshy wading birds with long snake-like neck and javelin or pincer-like beak for piercing their aquatic prey. e.g.,: Cattle egret (*Bubulcus ibis*), heron (*Ardea herodias*), night heron (*Nycticorax*), spoonbill (*Platalea leucorodia*), strok (*Ciconia*), flamingo (*Phonicopterus*).

Order 15. Gruiformes

It includes crane-like wading birds with long legs and partially webbed feet. e.g.,: Common coot (*Fulica atra*), sarus crane (*Antegone antegone*), bustard (*Choriotis*), rail.

Group 16. Falconiformes

The diurnal birds of prey with sharp hooked beaks and strong claws. e.g.,. Common pariah kite (*Milvus migrans*), brahminy kite (*Haliaster indus*), spar hawk (*Astur badius*), white backed vult (*Pseudogyps bengalensis*), king vulture (*Sarcogcalvus*), peregrine falcon (*Falco peregrinus*), eagle (*Aquila*) (Fig. 7.4).

Order 17. Strigiformes

It includes owls which are nocturnal birds of prey characterized by large heads, huge yellow frontal eyes and powerful grasping feet feathered up to toes. e.g.: Brown fish owl (*Ketupa zeylonensis*), great horned owl (*Bubo*), Tylopus.

Group E. Aerial Birds

These birds are mostly on wing, and have weak or vestigial perching feet.

Order 18. Micropodiformes or Apodiformes

It includes swifts and humming birds. e.g.,: Indian swift (*Micropodus*), palm swift (*Cypsiurus*).

Order 19. Caprimulgiformes

It includes shy, nocturnal, insectivorous bird such as night hawks (*Chordeiles*), whippoorwill (*Phaelanenoptilus*), goat suckers (*Caprimulgus*).

Origin and Ancestry of Birds

Although Archaeopteryx is supposed to be a connecting link between reptiles and birds, the gap between it and the actual reptilian ancestors of birds remains yet to be filled. As we know, the Mesozoic was the period of great adaptive radiation for various groups of reptiles, especially the diapsid archosaurians or ruling reptiles. One large ancestral group of ruling reptiles was called **Pseudosuchia** or **Thecodontia** (e.g., *Saltoposuchus, Euparkeria*). They were small, bipedal, carnivorous reptiles having hind limbs much longer than forelimbs, three toes forward and one (hallux) backward, large orbits with sclerotic ring and teeth set in sockets of elongated jaws. They were sufficiently generalized to be the probable distant ancestors of birds. Among pseudosuchians, only three groups—Pterosauria, Saurischia and Ormithischia—may be singled out as potential avian ancestors.

Proaves

But all these potential avian ancestors did not have a clavicle or wishbone. The fact that all flying birds, including Archaeopteryx, possess a V-shaped wishbone means that the immediate ancestors could hardly have been without it. Heilmann gives the name proavis to this hypothetical connecting link between the rather generalized pseudosuchians and the first birds

Diphyletic origin of birds

The earliest known fossil birds include both

    a) flying (*Archaeopteryx*, *ichthyornis*) types as well as

    b) flightless (*Hesperornis*, *Diatryma*) types.

The recently extinct Elephant birds were also flightless. The most primitive living birds or Ratitae (Ostrich, Rhea, Cassowary, etc.) and Penguins are also flightless. This led Lowe, to believe in the diphyletic (two-lines-of-descent) origin of birds. They maintain that the flightless and flying birds of today have descended from different flightless ancestors. According to Lowe, the present-day flightless birds were never capable of flight, and their wings are not degenerate now, but better developed than at any time in their past history.

Monophyletic origin of birds

In Ratitae, the legs are well-developed and powerful, the wings vestigial, and the feathers are fluffy. But a recently discovered fossil of *Eleuheronis*, a probably ancestor of the present-day ostrich from the Eocene of Switzerland, shows closer affinities to flying forms than does the present-day ostrich, and poses a serious blow to the concept of diphyletic origin of birds.

Today most palaeonotolgists believe that the Carinate are more primitive. Presumably the Ratitae evolved from flying ancestors but readapted to a terrestrial mode of life in areas with abundant food and few competitiors or enemies. The more usually accepted view today maintains that birds have a monophyletic (one-line-of-descent) origin, i.e., all birds have evolved from a single ancestor, perhaps close to Archaeopteryx. Accordingly, the flightless birds have evolved by loss of flight from flying ancestors. The weight of the known evidence also favours this view. Although Archaeopteryx is supposed to be a connecting link between reptiles and birds, the gap between it and the actual reptilian ancestors of birds remains yet to be filled.

Chapter 8

Mammalia

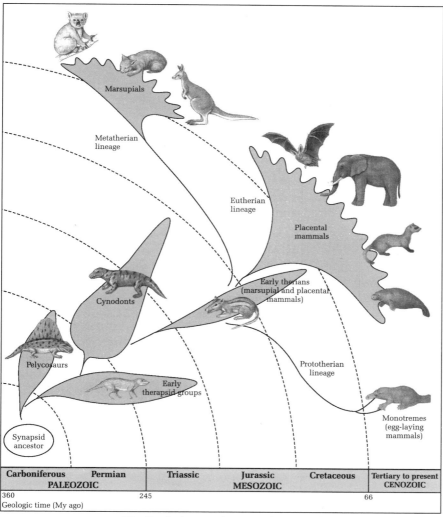

The Evolutionary descent of Mammals

Figure 8.0 Evolution of major groups of synapsids. The *synapsid lineage*, began with the pelycosaurs. Pelycosaurs radiated extensively and evolved changes in jaws, teeth, and body form that presaged several mammalian characteristics. These trends continued in their successors, the therapsids, especially in the cynodonts. One lineage of cynodonts gave rise in the Triassic to therians, the *placental* mammals. Fossil evidence, as indicates that all three groups of living mammals—monotremes, marsupials, and placentals—are derived from the same lineage.

The evolutionary descent of mammals from their earliest amniote ancestors is perhaps the most fully documented transition in vertebrate history. From the fossil record, we can trace the derivation over 150 million years of endothermic, furry mammals from their small, ectothermic, hairless ancestors. Skull structures and especially teeth are the most abundant fossils, and it is largely from these structures that we can identify the evolutionary descent of mammals.

The structure of the skull roof permits us to identify three major groups of amniotes that diverged in the Carboniferous period of the Paleozoic era, the synapsids, anapsids, and diapsids.

1. The synapsid group which includes the mammals and their ancestors, has a pair of temporal openings in the skull for attachment of jaw muscles. Synapsids were the first amniote group to radiate widely into terrestrial habitats.

2. The anapsid group is characterized by solid skulls and includes turtles and their ancestors.

3. The diapsids have two pairs of temporal openings in the skull (Figs. 8.1 C) and contain dinosaurs, lizards, snakes, crocodilians, birds, and their ancestors.

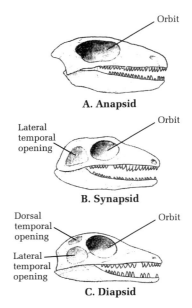

Figures 8.1 A to C

Skull of early amniotes, showing the pattern of temporal openings that distinguish the three groups.

The earliest synapsids radiated extensively into diverse herbivorous and carnivorous forms often collectively called pelycosaurs (Fig. 8.2). These early synapsids were the most common and largest amniotes in the early Permian. Pelycosaurs share a general outward resemblance to lizards, but this resemblance is misleading. Pelycosaurs are not closely related to lizards, which are diapsids, nor are they a monophyletic group. From one group of early carnivorous synapsids arose the therapsids (Fig. 8.2), the only synapsid group to survive beyond the Paleozoic. Therapsids had for the first time an efficient erect gait with upright limbs positioned beneath the body. Since stability was reduced by raising the animal from the ground, the cerebellum, muscular coordination center of the brain, assumed an expanded role. Changes in the morphology of the skull and mandibular adductor muscles associated with increased feeding efficiency began with the early therapsids. Therapsids radiated into numerous herbivorous and carnivorous forms; however most early forms disappeared during the great extinction event at the end of the Permian. Previously, pelycosaurs and therapsids have been referred to as "mammal-like reptiles," but in reality they are not a part of the reptilian lineage.

Peculiar Mammalian characteristics

Modern mammals are descendants of the synapsid lineage of amniotes that appeared in the Carboniferous period. The synapsid lineage is characterized in the primitive condition by having a skull with a single temporal opening (Figs. 8.1 B). Modern mammals are endothermic and homeothermic, have bodies partially or wholly covered with hair, and have mammary glands that secrete milk for the nourishment of the young. These derived characteristics, together with several distinctive skeletal characteristics, a highly developed nervous system, and complex individual and social behavior, distinguish mammals from all other living amniotes. Their genetic plasticity and numerous derived adaptations have enabled mammals to invade almost every environment that supports life on earth.

Mammals have been thoroughly studied and adequately classified. They include approximately 5,000 living species and numerous fossil forms. On the basis of the following characters they have been classified into different orders.

1. mode of caring for their young.
2. nature of dentition,
3. foot posture,
4. nails, claws and hoofs,
5. complexity of nervous system, and
6. systematics.

Simpson, recognized 18 living and 14 extinct orders of mammals. We shall refer to only living orders of mammals which are divided into two subclasses:

a) Subclass **Prototheria**
b) Subclass **Theria**.

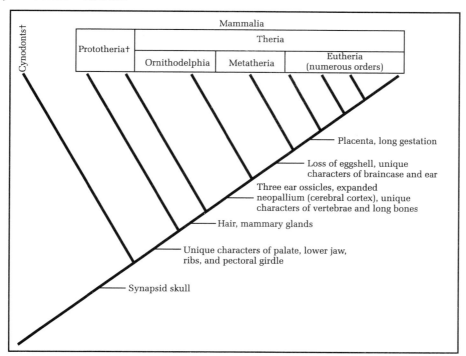

Figure 8.2
Mammalian Phylogeny, A cladogram showing the evolutionary relationships among mammals. Selected characters are shown. Daggers (†) indicate some extinct taxa.

From the point of view of convenience, we have discussed the subclasses separately under three sections.

1. **Section A** Subclass – Prototheria
2. **Section B** Subclass – Theria
   Infraclass Metatheria
3. **Section C** – Infraclass – Eutheri

Section A

Prototheria

The subclass Prototheria includes primitive egg-laying mammals and present many features in common with the Sauropsida.

Diagonastic Characters

1. Restricted mainly to the Australian region including Australia, Tasmania, New Guinea and neighboring island.
2. Aquatic or terrestrial, burrowing, incompletely warm-blooded, quadruped, oviparous or egg-laying mammals.
3. Body small and covered by hairs and spines. Snout produced into a beak. External ears inconspicuous or absent. Tail present or absent. Mammary glands are without teats or nipples. Male carries a hollow, horny, tarsal spur on each hind leg, connected internally to a crural poison gland. A temporary mammary pouch, equivalent to teats, develops during breeding season on the abdomen of female.
4. Includes epidermal horny hairs, spines, beaks and claws. Skin glandular.
5. Divided by a typical mammalian diaphragm into an anterior thoracic and a posterior abdominal cavity.
6. Skull is dicondylic. Skull sutures become obliterated. Orbits continuous with temporal fossae. Lacrimals and alisphenoids absent, but distinct pterygoids present. Tympanic bone ring-like and no tympanic auditory bulla formed. Ear ossicles three. Malleus, stapes and incus relatively large. Each half of mandible made by a single bone, the dentary. Mandibular symphysis weak and coronoid process small. Vertebral epiphyses indistinct or absent. Cervical vertebrae 7 and without zygapophyses. Ribs unicephalous, with only capitulum. Coracoids well developed. Spine of scapula present at its anterior border. A large T-shaped interclavicles similar to that of reptiles, present. Ischia and pubis fuse at a long ventral symphysis. A pair of small rod-like epipubic or marsupial bones present in ventral abdominal wall.
7. Teeth develop only in embryos, replaced by horny beak in adults. Rectum opens into a cloaca.
8. Respiration pulmonary,
9. Heart 4-chambered. Right auriculo-ventricular valve incomplete and fleshy. Right valve tricupid. Chordae tendineae absent. Single left aortic arch persists. Ventral abdominal vein or its mesentery present. R.B.C small, circular and non-nucleated. Body temperature variable (25°-28°).
10. Kidneys metanephric. Ureters open into a urinogenital sinus which terminates into common cloaca.

11. Brain relatively small, simple and without corpus callosum. Optic lobes four (corpora quadrigemina). Cochlea slightly bent and with a lagena.

12. In male, testes abdominal and penis retractile passing out sperms but not urine. In female, right ovary reduced and oviducts lead separately into cloaca. There are no uterus and vagina.

13. Females oviparous. Eggs large with much yolk and plastic shells. Cleavage meroblasitc. No uterine gestation. Newly hatched young very immature, fed on milk in abdominal pouch till fully developed.

Classification of Living Prototherian Mammals

Class **Mammalia**

Subclass **Prototheria**.

Infraclass **Ornithodelphia** – Monotreme mammals.

Order Monotremata: egg-laying (oviparous) mammals: duck-billed platypus, echidnas. Three species in this order are from Australia, Tasmania, and New Guinea. The most noted member of the order is the duck-billed platypus, *Ornithorhynchus anatinus*. Spiny anteaters, or echidnas, Tachyglossus, have a long, narrow snout adapted for feeding on ants, their chief food (Fig. 8.3).

Affinities

In most characters the **Prototheria** resemble the Sauropsida (reptiles and birds), though showing a little advancement over them, and thus coming nearer to their mammalian relatives.

1. **Reptilian affinity.** Although there is no definite connecting link known, there is enough evidence to show that mammals had a reptilian ancestry. This view is further supported by the following resemblances between the monotremes and the living reptiles:

    a) Presence of cloaca.

    b) Skull with large pterygoids, epipterygoids, dumb bell shaped prevomers and ring-like tympanic bones; but without alisphenoids and tympanic bulla.

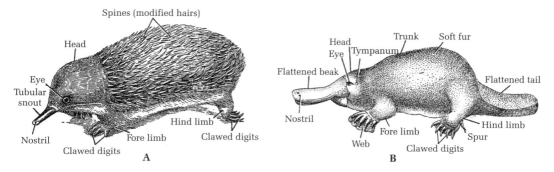

Figure 8.3
A. *Tachyglossus aculeata* (Spiny ant eater)   B. *Ornithorhynchus anatinus*
Egg-laying mammals (monotremes)

c) Vertebrae without epiphyses and with cervical ribs.
d) Thoracic ribs are single headed.
e) A median T-shaped inter-clavicle present.
f) Large coracoids and plate-like pre-coracoids.
g) Both ischium and pubis form ventral symphyses.
h) Acetabulum in echidna is perforated.
i) Can withstand starvation for a long period.
j) Anterior abdominal vein or its mesentery present.
k) A large anterior commissure but no corpus callosum joining the two cerebral hemispheres.
m) Body temperature not constant.
n) Cochlea of internal ear with lagena .
l) Ureters lead into a urogenital sinus.
o) Testes are abdominal.
p) Penis is simple with retractile groove conducting only sperms.
q) Oviducts separately open into cloaca and without uterus and vagina.
r) Females are oviparous. No uterine gestation.
s) Eggs large, cleidoic, with a leathery shell.
t) Newly hatched young born with a caruncle or an egg tooth.

2. **Mammalian affinity:** The prototherians are essentially mammals as they possess the following typical mammalian characters:

a) Body covered by hairs. Pinnae present.
b) Skin richly glandular, containing sweat and sebaceous glands.
c) A typical mammalian diaphragm divides body cavity.
d) Chondrocranium is typically mammalian.
e) Skull is dicondylic.
f) Middle ear cavity has 3 ear ossicles.
g) Each ramus of lower jaw made of a single bone, the dentary.
h) Cervial vertebrae typically seven.
i) Sternum is segmented.
j) A slender caecum demarcates two intestines.
k) Lobes of liver typically mammalian.
l) Heart 4-chambered.
m) Only left aortic arch present.
n) R.B.C small, circular and non-nucleated.

o) Presence of four optic lobes or corpora quadrigemina.

  p) Presence of milk glands secreting milk.

3. **Peculiar characters:** The following characters are peculiar to Prototheria:

   a) Presence of tarsal spurs in males:

   b) Milk glands derived from sweat glands (not sebaceous as in other mammals), and without teats.

   c) Temporary abdominal mammary pouch in female during breeding season.

   d) Teeth replaced in adults by horny plates.

   e) Jaws elongated forming a beak or rostrum.

   f) Ear ossicles relatively large showing transition.

   g) marsupial or epipubic bones present.

   h) Imperfectly warm blooded with body temperature varying from 25° to 28°C.

   i) Right ovary smaller and usually functionless.

Distinctive affinities of Prototheria lead us to some definite conclusions:

1. Peculiar blending of reptilian and mammalian characters suggests an intermediate stage between two groups.

2. Possession of primitive, degenerate and highly specialized characters indicates that they represent an early separate side line from main mammalian stock. This justifies their placement in a separate subclass Prototheria from the rest of mammals which are placed under the subclass Theria.

3. Monotremes show that reptiles, birds and mammals together constitute a natural group more homogeneous than the group Ichthyopsida (fishes and amphibians) or even the superclass Pisces.

Section B

Theria

Theria: This subclass includes the modern, young-producing mammals. Ear usually has a pinna. Brain is mostly with corpus callosum. Vertebrae generally bear epiphyses. Coracoid is reduced. Testis commonly descend into scrotal sacs. Cloaca is lacking. Consequently digestive and urino-genital system open out by separate apertures. Each oviduct is differentiated into anterior Fallopian tube and posterior uterus. Uteri join to form a vagina. Teeth usually occur in both the youngs and the adults. Mammary glands open on nipples or teats. Eggs are small with little yolk and no shell. They develop in the uterus and living young are produced.

The subclass Theria is divided into two

   A) Infraclasses: **Metatheria** (Marsupialia) and

   B) Infraclassess **Eutheria** (Placentalia).

Infraclass Metatheria

Infraclass Metatheria includes pouched mammals without a true placenta, and placed in a single order Marsupialia or Didelphia.

Distinctive Characters

1. Almost entirely confined to the Australian region with the exception of the American opossums.
2. Terrestrial, burrowing or arboreal, rarely aquatic, herbivorous, carnivorous or omnivorous, nocturnal or diurnal, air-breathing, warm-blooded, viviparous and pouched mammals.
3. Body furry, that is, covered with soft hairs, External ear lobes or pinnae are well-developed. Tail generally long, often prehensile Mammary glands are modified sebaceous glands and have teats. Female usually with a ventral abdominal pouch, called marsupium, or with marsupial folds enclosing mammary nipples.
4. Includes epidermal horny hairs and claws. Skin richly glandular.
5. Typical muscular diaphragm present.
6. Skull is dicondylic. Skull top flat. Cranium small with well developed sagittal and occipital crests. Paraoccipital process greatly Gognaed. Orbit and temporal fossa confluent due to absence of postorbital bar. Nasal bones large and expanded posteriorly. Jugal (malar bone) large and reaches up to glenoid articulation. Pterygoid is small. Tympanic bulla absent. In some cases, alisphenoid large forming the so-called alisphenoid bulla. Palate imperfectly ossified having large posterior vacuities. Middle ear cavity with three small ear ossicles. Angle of dentary inflected or inturned.
7. Vertebrae with epiphyses. Cervical vertebrae seven. Cervical ribs and odontoid process fuse early with their respective vertebrae. Thoracic ribs bicephalous. Interclavicle absent. Large clavicles present. Coracoids reduced. Scapula with an additional anterior spine. A pair of epipubic or marsupial bones present in front of pubic symphysis for the support of marsupium.
8. Teeth generally exceed typical eutherian number of 44. Only one set present (monophyodont). Anus and urinogenital aperture open into a shallow cloaca surrounded by a common sphincter.
9. Respiratory, Circulatory and excretory systems typically mammalian.
10. Brain relatively small and less convoluted. Olfactory lobes large. Cerebral hemispheres small. Cerebellum small and exposed. Anterior commissure large. Corpus callosum feebly developed or absent. Cochlea of internal ear spirally coiled.
11. In males, penis well-developed, often bifurcated at the tip. Scrotal sacs containing Testes lie in front of penis. In females, two oviducts open separately into urinogenital sinus, so that there are two uteri and two vaginae. Clitoris in female may also be double.
12. Females are viviparous. Gestation period for uterine development small 2 weeks in opossum to 5 weeks in kangaroo. A true allantoic placenta absent except in Perameles. Young are born exceedingly small, naked and blind. They are kept in marsupium and nourished on milk until fully formed. Thereafter, small young retreat to marsupium for shelter.

Classification

Marsupials are more widespread as fossils. 8-10 families include about 240 living genera. Previously, they were placed in two groups, based on their dentition or foot structure, as follows:

Order **Didelphimorphia**: American opossums. These mammals, like other marsupials, are characterized by an abdominal pouch, or marsupium, in which they rear their young. Most species are found in Central and South America, but one species, the Virginia opossum, *Didelphis virginiana*, (Fig. 8.5 A) is widespread in North America; 66 species.

Order **Paucituberculata**: shrew opossums. Tiny, shrew-sized marsupials found in western South America; seven species.

Order **Microbiotheria**: Monito del Monte. A South American mouse-sized marsupial that may be more closely related to Australian marsupials; one species only.

Order **Dasyuromorphia**: Australian carnivorous mammals. In addition to a number of larger carnivores, this order includes a number of marsupial "mice," all of which are carnivorous. Confined to Australia, Tasmania, and New Guinea; 64 species.

Order **Peramelemorphia**: bandicoots. Like placentals, members of this group have a chorio-allantoic placenta and a high rate of reproduction for marsupials. Confined to Australia, Tasmania, and New Guinea; 22 species.

Order **Notoryctemorphia**: marsupial moles. Bizarre, semi-fossorial marsupials present in Australia; two species.

Order **Diprotodontia**: koalas, wombats, possums, wallabies, kangaroos. Diverse marsupial group containing some of the largest and most familiar marsupials. Present in Australia, Tasmania, New Guinea, and many islands of the East Indies; 131 species .

Family 1. Didelphidae. American marsupials. Common or Virginian opossum (*Didelphis marsupialis*) (Fig. 8.4), water opossum (*Chironectes*).

Family 2. Dasyuridae. Carnivorous or insectivorous. Native cat (*Dasyurus*), marsupial wolf (*Thylacinus*), banded anteater (*Myrmecobius*).

Family 3. Notoryctidae. Insectivorous. Masupial moles (*Notoryctes*).

Family 4. Peramelidae. Bandicoots (*Perameles*).

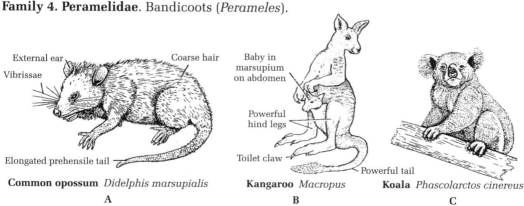

Common opossum *Didelphis marsupialis* **Kangaroo** *Macropus* **Koala** *Phascolarctos cinereus*
A B C

Figure 8.4
Pouched mammals (marsupials)

Family 5. Caenolestidae. Shrew-like (*Caenolestes, Rhyncholestes*).

Family 6. Phalangeriae. Phalanger, flying phalanger (*Petaurus*), koala (*Phascolarctos*). (Fig. 8.4 C).

Family 7. Phascolomidae. Wombat (Phascolomys).

Family 8. Macropodidae. Herbivorous, Kangaroo Macropus, (Fig. 8.4 B).

Distinguishing characterisitic

This infraclass includes the pouched mammals. Epipubic bone is present. Testes descend into scrotum, which is in front of the penis. Anal and urinogenital apertures are controlled by a common sphincter muscle. Cranium is small. Corpus callosum is absent. Uteri and vaginae of the sides remain separate. Female has a ventral pouch, the marsupium, of folds surrounding the nipples present on the abdomen. Placenta is small and simple. The young have a brief intrauterine development and are born in a very imperfect condition. After birth they are nursed by the mother in the marsupium. Here each young one attaches itself firmly by its mouth to a nipple and remains there as a "mammary foetus" till fully formed. After this the young return to marsupium for shelter. Body temperature ranges from 32-36°C.

As per the old classification ,the infraclass Metatheria includes a single order, viz.,

Infraclass Metatheria

Order-Marsupialia, which is divided into two suborder

 a) suborder **Didactyla**

 b) suborder **Syndactyla**.

Suborder. Didactyla. All the digits are free. e.g., *Didelphis marsupials* (Virginiana).

Didelphis marsupialis — The American Opossum (Fig. 8.5 A) Didelphis occurs in America. Head and trunk measure about 50 cm. and tail about 33 cm. It has a relatively long and bare muzzle and a scaly and prehensile tail. It is nocturnal and omnivorous. It feigns death, when disturbed. It is hunted for eating. Young ones, which are 11 mm. Long, spend about two months in the marsupium.

Suborder Syndactyla. Second and third toes are bound together. e.g., *Macropus major*.

Macropus major — The Great Grey Kangaroo (Fig. 8.5 B) The kangaroo inhabits Australia and Tasmania. It has a small head, a large body and a long muscular tail. Its hind-limbs are much longer and more powerful than the forelimbs. It walks in a normal way, using all the four limbs. During running, it takes long leaps

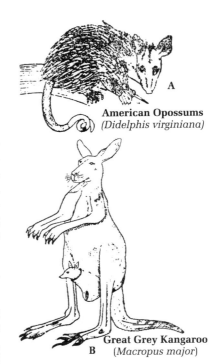

American Opossums
(*Didelphis virginiana*)

Great Grey Kangaroo
B (*Macropus major*)

Figures 8.5 A, B
Representation of infraclass Metatheria, order Marsupialia.

on the hind limbs alone, balancing the body on the tail. The animal is herbivorous and gregarious. The herd travels under the watchful eye of an old male, who also maintains discipline in the herd. The foetus has a short intrauterine development and is born in a very helpless condition. It is blind, naked and only 25 mm. Long. The mother rears it in the marsupium, where milk is forced into its mouth by the contraction of certain muscles of her belly, as it is incapable of sucking milk.

Distribution of Metatheria

The marsupials are interesting not only from the point of view of their structure, even their extant as well as past distribution is equally interesting .

1. During Cretaceous, period, marsupials were found over much of the earth. Today, they are almost entirely confined to the Australian region. The only exceptions are the opossums (*Didelphis*) and opossum rats (*Caenolestes*) of North and South America. Thus they furnish a good example of discontinuous distribution.

2. In Australia, where marsupials dominate at present have, no fossil remains prior to Pleistocene. However, their earliest known fossils belong to the Jurassic of Europe. This suggests that their earliest home or seat of origin was probably in Europe.

3. Australia has no native eutherian mammals. Except monotremes, nearly all the remaining Australian mammals are marsupials. We may assume that at the time when the marsupials came into existence, Australia was connected to the main land by a land bridge. It seems that Australia became isolated, from the rest of the world, soon after marsupials entered it but before the higher mammals could arrive .Similarly, the discovery of strictly dasyurine fossils from South America supports the widely accepted theory that even after separation from the Old World, South America was connected to Australia by an Antarctic land bridge. Some of the Australian marsupials may have wandered into S. America through this land connection.

4. The cause of total or universal extinction of marsupials outside Australia, except the American opossums, remains a mystery. Probably they disappeared due to stiff competition against the ancestral placental mammals which bear their young alive without need for protection in a pouch.

5. In Australia, in the absence of competition from higher mammals and isolated and protected by a wall of water or sea for countless ages from the rest of the world, the marsupials prospered successfully. They developed along several lines of parallel evolution similar to higher mammals in other parts of the world. In addition to opossums and kangaroos, there are marsupial mice, squirrels, rabbits, cats, dogs, wolves and the native bear, the koala. They furnish splendid illustration of the effect of isolation by a natural barrier (sea) on the perpetuation of hereditary variations.

Affinities

Metatherians show a mixture of primitive and advanced characters.

1. **Affinity with Prototheria**. They have certain primitive characters also found in Prothotheria such as:

    a) Presence of cloaca.

b) Presence of clavicles, epipublic bones and ring like tympanic membrane.
   c) Absence of tympanic bulla.
   d) Brain, relatively simple with large olfactory lobes and anterior commissure, but without corpus callosum.
   e) Absence of true allantoic placenta.

However, metatherians differ from prototherians mainly in being viviparous, having permanent marsupial pouch, teats in mammary glands, well developed external ears, vertebrae with epiphyses, ribs bicephalous, no interclavicle and separate coracoids, teeth in the adult, cochlea spirally coiled, penis bifid, testes in scrotal sacs, uterine gestation and viviparity.

2. **Affinity with Eutheria.** Metatheria possess many advanced characters similar to eutherians or higher placental mammals:
   a) Presence of hairs and external ears.
   b) Mammary glands sebaceous and with teats.
   c) Brain with four optic lobes. Cochlea spirally coiled.
   d) Coracoids reduced. Interclavicle absent. Ribs bicephalous.
   e) Teeth heterodont.
   f) Male with penis. Testes in scrotal sac.
   g) Presence of uterus and vagina.
   h) Female viviparous. Ova small, yolkless. Uterine gestation and placenta.

Systematic Position of Marsupials

However, metatherians differ form eutherians mainly in restricted distribution, having shallow cloaca, marsupial pouch, flat small cranium, no tympanic but alisphenoid bulla, epipubic bones, inflected mandibular angle, two vaginae and two uteri, bifid penis, scrotal sac in front of penis, gestation period small, no true allantoic placenta, etc.

It is obvious that the metatherians are more advanced than the primitive, reptile-like, oviparous Prototheria. They are more closely related with the eutherians, but do not belong to the same grade of evolution. Therefore, they are put under a separate infraclass Metatheria, while the rest of the higher and truly placental mammals are placed in the infraclass Eutheria, and both combined in the subclass Theria.

Section C

Infraclass Eutheria

Mammals form the "highest" group in the animal kingdom. The name of the group, refers to the milk-producing dermal mammary glands in the females for suckling the young. Parental care is highly developed in this group, and it reaches its climax in the human species. The reason of their success is that they adapt themselves readily to new conditions and have improved nervous system. The mammals possess a covering of hair and walk on all four limbs therefore, called as the hairy quadrupeds.

Characters

The mammals have the following characters:

1. Body is variously shaped and divisible into head, neck, trunk and tail.
2. There are two pairs of pentadactyl limbs. These are variously adapted for walking, running, burrowing, swimming or flying. Each foot bears 5 (or fewer) toes provided with horny claws, nails or hoofs.
3. Skin is glandular and mostly covered by a horny epidermal exoskeleton of hair, which conserve body heat.. The oil and sweat glands are abundant in both the sexes, and milk glands in the females only.
4. Endoskeleton is fully ossified. Skull has two occipital condyles, large cranium and fewer bones. Each half of the lower jaw consists of a single bone, the dentary, and articulates with the squamosal instead of quadrate. A tympanic bone supports the tympanic membrane. A secondary palate has been formed separating the respiratory passage from the food passage. Teeth are heterodont, diphyodont and thecodont and occur in both the jaws. Vertebrae have epiphyses and generally flat centra. Neck contains 7 vertebrae irrespective of its size. Pectoral girdle is typically without a coracoid.
5. Coelom is completely divided into anterior smaller thoracic cavity and posterior larger abdominal cavity by a muscular partition, the diaphragm. The thoracic cavity contains a pair of large pleural cavities that enclose the lungs and a small ventral pericardial cavity which has the heart. Other viscera, except the kidneys, lie in the abdominal cavity.
6. Buccal cavity contains true salivary glands. Tongue is usually mobile. Alimentary canal usually leads directly to the exterior by anus, there being no cloaca.
7. Respiration occurs only by lungs. The lungs are spongy organs enclosed in special pleural cavities. Larynx contains well-developed vocal cords for sound production. Glottis is guarded by epiglottis.
8. Heart is 4-chambered, having two auricles and two ventricles and keeping oxygenated and de-oxygenated blood completely separate. Right auriculo-ventricular valve consists of three flaps. There is neither sinus venosus nor truncus arteriosus. Only the left systemic arch is present. Renal portal system is lacking. Red blood corpuscles are circular, biconcave and denucleated in most cases.
9. Brain is highly developed. Cerebrum is large and cerebellum complex. Corpus callosum connects the two halves of the cerebrum. Optic lobes are divided into four corpora quadrigemina. There are 12 pairs of cranial nerves.
10. Jacobson's organs are developed only in lower forms. Olfactory sacs have turbinals and open by internal nares far back into the pharynx. Eyes have movable lids. Sclerotic consists of dense fibrous tissue and lacks ossicles. Ear has a large fleshy pinna, three ear ossicles -malleus, incus and stapes, and a large coiled cochlea with an organ of Corti.
11. A well developed endocrine system that conforms to the general vertebrate plan is present in mammals.

12. Kidneys are metanephric and bean shaped. Their concavity, the hilus, faces inward, and blood vessels and ureter enter or leave here. Ureters open into the urinary bladder. In monotremes, ureters and urinary bladders open into the cloaca by separate apertures.

13. Sexes are separate. Gonads are paired. Testes usually descend into scrotal sacs through inguinal canal, in the adult. Male has a copulatory organ (penis) containing sinusoidal erectile tissue. Gonoducts lead directly to the exterior. Fertilization is internal. Mammals are viviparous, except the Prototheria.

14. Egg are minute with little yolk and no shell. Development takes place in the uterus of the female. Embryo has an amnion, chorion and allantois. A placenta fixes the embryo to uterine wall for nourishment, respiration and excretion. Development is direct, there is no larval stage. Young are nourished on milk for some time after birth and brought up with a great care and love.

15. Body temperature is regulated (homoeothermous).

16. They occur in all sorts of habitats from the polar regions to the tropics, including the dense forests and driest deserts. Some live in the sea. Mammals show a great diversity in the habits and include fossorial borrowing), saltatorial (leaping), cursorial (running), scansorial (climbing), Volant (flying), natatorial (aquatic) and cave-dwelling forms.

Peculiar Characterisitics of Class Mammalia

1. Mammals share with birds both endothermy and homeothermy, which permit a high level of activity at night, and year-round penetration into low temperature habitats denied to ectothermic vertebrates.

2. The placenta allows developing young to obtain nourishment and grow in a protected environment during the most vulnerable period of their lives. After birth the young continue to feed by suckling from mammary glands. A long period of parental care and training allows the young to acquire skills necessary for survival.

3. Specialization of mammalian teeth for different functions permit utilization of a broad variety of foods. The secondary palate, which separates the air passage from the food passage, enables mammals to hold and partially break down food in their mouths without interrupting breathing.

4. Convoluted turbinate bones in the nasal cavity provide a high surface area for warming and moistening the inspired air and for reducing loss of moisture during exhalation.

5. The highly evolved brain, has bequeathed mammals with a well-developed memory and capacity to learn rapidly and to respond appropriately to situations not previously encountered. Highly elaborated sense organs, particularly those of hearing, smell, and touch, contribute an inflow of environmental information that, together with their processing brain centers(neo-cortex), provide mammals with a level of environmental awareness and responsiveness unequaled in the animal kingdom.

Mammals, with their highly developed nervous system and numerous ingenious adaptations, occupy almost every environment on earth that supports life. Mammals are exceedingly diverse in size, shape, form, and function. They range in size from the recently discovered Kitti's hognosed bat in Thailand, weighing only 1.5 g, to blue whales, exceeding 130 metric tons.

Identifying Characters of Class Mammalia

1. Body mostly covered with hair.
2. Integument with sweat, scent, sebaceous, and mammary glands
3. Skull with two occipital condyles and secondary palate, turbinate bones in nasal cavity; jaw has a joint between squamosal and dentary bones, middle ear with three ossicles (malleus, incus, stapes); seven cervical vertebrae (except in Xenarthrans, and manatees); pelvic bones fused
4. Mouth with diphyodont teeth (milk, or deciduous, teeth replaced by a permanent set); teeth heterodont in most (varying in structure and function); lower jaw a single enlarged bone (dentary)
5. Movable eyelids and fleshy external ears called pinna.
6. Circulatory system of a four-chambered heart (two atria and two ventricles), persistent left aorta, and nonnucleated, biconcave red blood corpuscles
7. Respiratory system comprises, lungs with alveoli, and larynx; secondary palate anterior portion is bony palate and posterior portion is of soft tissue, the soft palate separates air and food passages; muscular diaphragm separates thoracic and abdominal cavities
8. Excretory system of metanephric kidneys with ureters that usually open into a bladder
9. Brain has highly developed, cerebral cortex; 12 pairs of cranial nerves
10. Endothermic and homeothermic
11. Cloaca present only in monotremes (present but shallow in marsupials)
12. Separate sexes; Male reproductive organs are a penis, testes (usually in a scrotum), Female reproductive organs are ovaries, oviducts, and uterus.
13. Internal fertilization; embryos develop in a uterus with placental attachment (placenta absent in monotremes); fetal membranes (amnion, chorion, allantois)
14. Young nourished by milk from mammary glands

One therapsid group to survive into the Mesozoic era was the cynodonts. Cynodonts evolved several features that supported a high metabolic rate; increased and specialized jaw musculature, permitting a stronger bite; several skeletal changes, supporting greater agility; heterodont teeth, permitting better food processing; turbinate bones, in the nasal cavity, aiding retention of body heat; and a secondary bony palate, enabling an animal to breathe while holding prey or chewing food. The secondary palate would be important to subsequent mammalian evolution by permitting the young to breathe while suckling. The upright posture in cynodonts, led the long bones became more slender and developed bony processes at the joints for firmer muscle attachment. Loss of lumbar ribs in cynodonts is correlated with the evolution of a diaphragm and also may have provided greater dorsoventral flexibility of the spinal column. Within the diverse cynodont clade a small carnivorous group called trithelodontids most closely resembles the mammals, sharing with them several derived features of the skull and teeth

Infraclass **Eutheria** has placental mammals. The following important orders have been described in the present text,

Order 1. **Insectivora**: insect-eating mammals: shrews, hedgehogs, tenrecs, moles (Fig. 8.6). The principal food is insects. Insectivores, widely distributed over the world except Australia and New Zealand, are small, sharp-snouted animals with primitive characters that spend a great part of their lives underground. Shrews are among the smallest mammals known; 440 species.

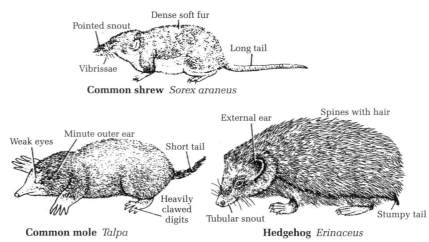

Figure 8.6
Insect-eating mammals (Insectivora)

Order 2. **Macroscelidea**: elephant shrews. These are secretive mammals with long legs, a snout like nose adapted for foraging for insects, large eyes. Widespread in Africa; 15 species.

Order 3. **Dermoptera**: flying lemurs. (Fig. 8.7) These are related to true bats and consist of a single genus Galeopithecus. They are not lemurs (which are primates) and cannot fly but glide like flying squirrels. They are found in the Malay peninsula in the East Indies; two species.

Order 4. **Chiroptera**: bats (Fig. 8.8). Wings of bats, the only true flying mammals, are modified forelimbs in which the second to fifth digits are elongated to support a thin integumental membrane for flying. The first digit (thumb) is short with a claw. Common North American forms are little brown bats, *Myotis*, free-tailed bats, *Tadarida*, and big brown bats, *Eptesicus*. In Old World tropics fruit bats, or "flying foxes," *Pteropus*, are largest of all bats, with a wingspread of 1.2 to 1.5 m; they live chiefly on fruits; 977 species.

Order 5. **Scandentia**: tree shrews. Tree shrews are small, squirrel-like mammals of the tropical rain forests of southern and southeastern Asia. Despite their name, many are not especially well-adapted for life in trees, and some are almost completely terrestrial; 16 species.

Order 6. **Primates**: prosimians, monkeys, apes, humans.

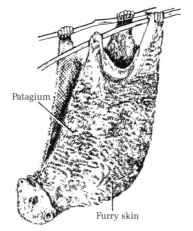

Figure 8.7
Flying lemur *Cynocephalus*

It includes the most intelligent animals, like the lemurs, loris, monkeys, apes and man. They have large, highly-convoluted cerebral. Hemispheres, which incompletely or almost completely cover the cerebellum. Correlated with the greater development of the brain, the cranium is large and rounded with the foramen magnum on the lower side. They are plantigrade and have long limbs, each bearing five digits protected by nails. The first digit of each limb (pollex and hallux) can be brought opposite the remaining digits to make the hands and feet grasping organs. This is associated with the arboreal life to which most of the primates are adapted. The eyes are directed forwards to give binocular vision, i.e., both eyes see the same object but from slightly different angles, which provides depth to the images and enables the animal to judge distances correctly. The females have a single pair of teats, which are located on the thorax. Usually a single young one is produced at a time and it gets care and affection from the mother. The primates are omnivorous and gregarious.

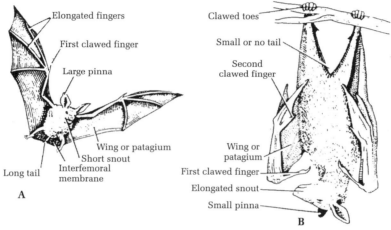

Figure 8.8
Flying mammals or bats (chiroptera) **A.** An insect-eating bat (Microchiroptera)
B. A fruit-eating bat (Megachiroptera)

Primates stands first in the animal kingdom in brain development. Most species are arboreal, apparently derived from tree-dwelling insectivores. Primates represent the end product of a line that branched off early from other mammals and have retained many primitive characteristics. It is believed that their tree-dwelling habits of agility in capturing food or avoiding enemies were largely responsible for their advances in brain structure. As a group they are generalized with five digits (usually provided with flat nails) on both forelimbs and hind limbs. All except humans have their bodies covered with hair. Forelimbs are often adapted for grasping, as are the hind limbs sometimes. The group is singularly lacking in claws, scales, horns, and hoofs.

There are two suborders;

Suborder **Strepsirhini**: (Fig. 8.9) lemurs, aye-aye, lorises, pottos, bush babies. Seven families of arboreal primates, formerly called prosimians, concentrated on Madagascar, but with species in Africa, southeast Asia, and Malay peninsula. All have a wet, naked region (rhinarium) surrounding comma-shaped nostrils, a long non prehensile tail, and a second toe provided with a claw. Their food is both plants and animals 49 species.

Figure 8.9
Primate mammals **A.** Prosimians **B.** Anthropoids

Suborder **Haplorhini**: tarsiers, marmosets, New and Old World monkeys, gibbons, gorilla, chimpanzees, orangutan, humans (Fig. 8.9). Six families, four of which were formerly called Anthropoidea. Haplorhine primates have dry, hairy noses, ringed nostrils and differences in uterine anatomy, placental development, and skull morphology that distinguish them from strepsirhine primates. Family **Tarsiidae** contains crepuscular and nocturnal tarsiers, with large, forward-facing eyes and reduced snout (five species).

New World Monkeys, (Fig. 8.10) sometimes called Platyrrhine monkeys because the nostrils are widely separated, are contained in two families:

1. **Fammily Callitrichidae** (marmosets and tamarins; 35 species). Callitrichids, which include the colorful lion tamarins, have prehensile hands and quadrupedal locomotion.

2. **Family Cebidae** (capuchin-like monkeys; 65 species). Cebid monkeys are much larger than any callitrichid. They include capuchin monkeys, Cebus, spider monkeys, Ateles and howler monkeys, Alouatta. Some cebids (including spider and howler monkeys) have prehensile tails, used like an additional hand for grasping and swinging.

Old World monkeys, termed *catarrhine monkeys* (Fig. 8.10) because their nostrils are set close together and open to the front, are placed in family cercopithecidae.

Family Cercopithecidae, with 96 species. They include mandrills, Mandrillus, baboons, Papio, macaques, Macaca, and langurs, Presbytis. The thumb and large toe are opposable. Some have internal cheek pouches; none have prehensile tails.

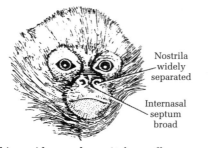

Platyrrhine spider monkey *Ateles geoffroy*   **Catarrhine rhesus monkey** *Macaca mulatta*

Figure 8.10
Difference between New World Monkeys (Platyrhina) and Old World Monkeys (Catarrhina)

Family **Hylobatidae** contains gibbons and siamang (11 species of genus Hylobates), with arms much longer than legs, prehensile hands with fully opposable thumb, and locomotion by true **brachiation**.

Family **Hominidae** contains four living genera and five species:

Gorilla (one species),

Pan (two species of chimpanzees)

Pongo (one species of orangutan), and

Homo (one species, humans).

Table 8.1: Comparison between Platyrhinii and Catarrhinii

	Characters	Platyrrhina (New world monkeys)	Catarrhina (Old World monkeys, apes, man)
1.	Distribution	Central and South America.	Africa and Asia
2.	Habitat	Mainly arboreal	Arboreal or terrestrial
3.	Size	Relatively small	Relatively large
4.	Nose	Nose flat. Internasal septum broad Nostrils widely separated and directed variously	Nose raised, Intenasal septum narrow. Nostrils close together and face downwards
5.	Check pouches	Absent	Present except in apes
6.	Dental formula	2.1.3.3. Premolars 3 2.1.3.3.	2-1.2-3. Premolars 2 2.1.2.3
7.	Tympanic bulla	Well developed	Poorly developed
8.	Bony auditory meatus	Poorly developed	Well developed
9.	Tail	Long, usually prehensile	When present, not prehensile
10.	Sigmoid flexure	Poorly developed	Well developed
11.	Ischial callosities	Ischial callosities on buttocks absent	Coloured callosities on buttocks
12.	Placenta	Non secondary discoidal	Secondary discoidal
13.	Offsprings	Usually more than one	Usually one
14.	Examples	Spider monkey (*Ateles*), howling monkey (*Aloutta*), capuchin (*Cebus*), marmoset (*Callithrix*)	Rhesus monkey (*Macaca*), langur (*Presbytis*), gibbon (*Hylobates*), baboon (*Papio*), orangutan (*Pongo*), chimpanzee (*Pan*), gorilla (*Gorilla*), man (*Homo sapiens*)

The first three of these four genera were formerly placed together in paraphyletic relationship. Family **Pongidae**; family **Hominidae** contained only humans. This separation is not recognized by cladistic taxonomy because the most recent common ancestor of the family Pongidae is also the ancestor of humans

Order 7. **Xenarthra** formerly Edentata (Fig. 8.11 A) (edentatus toothless): ant eaters, armadillos, sloths. Species of this order are either toothless (anteaters) or have simple,

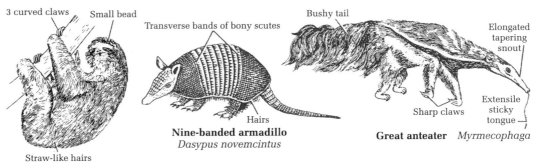

Figure 8.11
Toothless mammals and anteaters

peglike teeth (sloths and armadillos). Most live in South and Central America, although nine-banded armadillos, *Dasypus novemcinctus*, are common in the southern United States; 29 species.

Order 8. **Pholidota**: pangolins (Fig. 8.12). An odd group of mammals whose bodies are covered with overlapping horny scales that have arisen from fused bundles of hair. Their home is in tropical Asia and Africa; seven species.

Order 9. **Lagomorpha**: rabbits, hares, pikas (Figs. 8.13). Lagomorphs have long, constantly growing incisors, like rodents, but unlike rodents, they have an additional pair of peg like incisors growing behind the first pair. All lagomorphs are herbivores with cosmopolitan distribution; 81 species.

Order 10. **Rodentia** (Fig. 8.14) gnawing mammals: squirrels, rats, woodchucks.

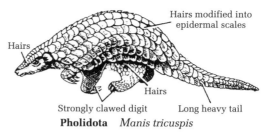

Figure 8.12
Pangolin or Scaly anteater

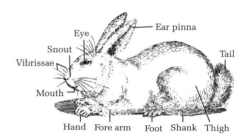

Figure 8.13
Rabbit. *Lagomorpha*

Rodents, comprising about 43% of all mammalian species, are characterized by two pairs of razor-sharp incisors used for gnawing through the toughest pods and shells for food. With their impressive reproductive powers, adaptability, and capacity to invade all terrestrial habitats, they are of great ecological significance.

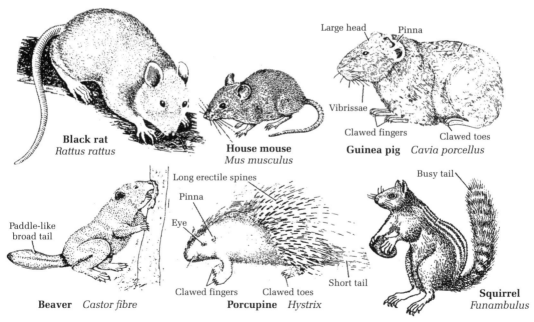

Figure 8.14
Gnawing mammals. Rodentia

Important **families** of this order are:

Family Sciuridae – squirrels and woodchucks.

Family Muridae – rats and house mice,

Family Castoridae – beavers,

Family Erethizontidae – porcupines,

Family Geomyidae – pocket gophers, and

Family Cricetidae – hamsters, deer mice, gerbils, voles, lemmings; 2052 species.

Figure 8.15 A
Carnivorous mammals. **A.** Fissipedia

Order 11. **Carnivora** (Fig. 8.15), weasels, seals, sea lions, walruses. All Carnivora, except the giant panda, have predatory habits, and their teeth are especially adapted for tearing flesh. They are distributed all over the world except in Australian and Antarctic regions where there are no native forms (besides seals).

The order Carnivora is divided into two suborders:
1. Fissipedia and
2. Pinnipedia

Suborder. Fissipedia. (Fig. 8.15 A) The suborder Fissipedia includes the terrestrial carnivores. They run very fast and may be digitigrade, plantigrade or sub-plantigrade in walking. The feet have thick pads underneath. The digits are separate and bear claws, which may be retractile. The last premolar in the upper jaw and the first molar in the lower jaw bite on other like a pair of scissors for cutting the flesh . These two teeth are called the carnassial or sectorial or shearing teeth. The Fissipedia are found all over the world.

Suborder. Pinnipedia. (Fig. 8.15 B) The pinnipedia are aquatic carnivores. They have fusiform body, reduced or no external ear, paddle-like limbs and a short tail. The digit are enclosed in the web. The Pinnipedia are greagarious animals. They feed on fish, mollusks and crustaceans. They breed on land. They inhabit the coastal seas of Temperate and Arctic regions. The common examples of the suborder are: seals, walruses and sea-lions.

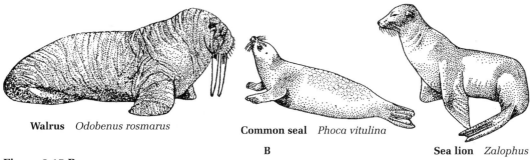

Walrus *Odobenus rosmarus*   **Common seal** *Phoca vitulina*   **Sea lion** *Zalophus*
B

Figure 8.15 B
Carnivorous mammals. **B.** Pinnipedia

Among more familiar families are

Family – Canidae (dog family), consisting of dogs, wolves, foxes, and coyotes;

Family – Felidae (cat family), whose members include domestic cats, tigers, lions, cougars, and lynxes; Ursidae (bears);

Family – Procyonidae (raccoons); and

Family – Mustelidae (fur-bearing family), containing martens, skunks, weasels, otters, badgers, minks, and wolverines;

Family – Otariidae (eared seals), containing fur seals and sea lions; 280 species.

Order 12. **Tubulidentata**: aardvark (Fig. 8.16) "Aardvark" is earth pig, a peculiar animal with a pig like body found in Africa; one species.

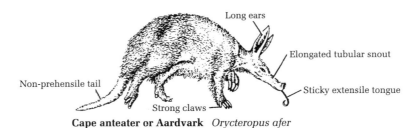

Cape anteater or Aardvark *Orycteropus afer*

Figure 8.16
Termite eating mammal

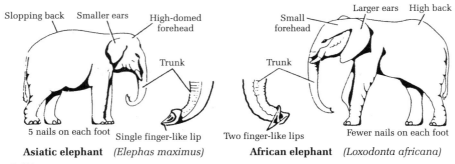

Asiatic elephant *(Elephas maximus)*   **African elephant** *(Loxodonta africana)*

Figure 8.17
Comparison between Indian and African elephant

Order 13. **Proboscidea**: (Fig. 8.17) proboscis mammals: elephants. These largest of living land animals, have two upper incisors elongated as tusks, and well-developed molar teeth. Asiatic or Indian elephants, *Elephas maximus*, have long been partly domesticated and trained to do heavy tasks. Taming of African elephants, *Loxodonta africana*, is more difficult but was done extensively by ancient Carthaginians and Romans, who employed them in their armies; two species.

Order 14. **Hyracoidea**: (Fig. 8.18) shrew, hyraxes (coneys) Coneys are herbivores restricted to Africa and Syria. They have some resemblance to short eared rabbits but have teeth like rhinoceroses, with toes and pads on their feet. They have four on the front feet and three toes on the back; seven.

Figure 8.18
Hyrax, *Procavia syriacus*

Order 15. **Sirenian**. sea nymph, sea cows and manatees. Sirenians are large, clumsy, aquatic. Mammals with large head, no hindlimbs, and forelimbs modified into flippers. The sea cow (dugong) (Fig. 8.19) of tropical coastlines of East Africa, Asia, and Australia and three species of manatees of the Caribbean area and Florida, Amazon River, and West Africa are the only living species. A fifth species, the large Steller's sea cow, was hunted to extinction by humans in the mid-eighteenth century; four species.

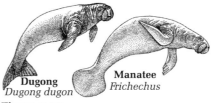

Dugong
Dugong dugon

Manatee
Frichechus

Figure 8.19
Sirenia

Order 16. **Perissodactyla**: odd-toed hoofed mammals: horses, asses, zebras, tapirs, rhinoceroses. Odd-toed hoofed mammals have an odd number of toes (one or three), each with a cornified hoof (Fig. 8.20).

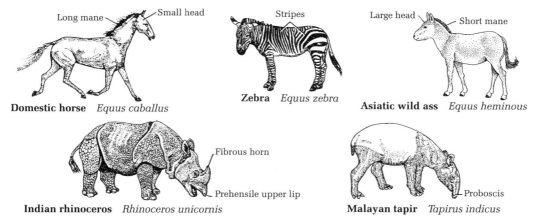

Figure 8.20
Perissodoactyla, Odd-toed hoofed mammals

Figure 8.21
Artiodactyla Even toed ungulate mammals

Both Perissodactyla and Artiodactyla are often referred to as ungulates or hoofed mammals, with teeth adapted for grinding plants. The horse family (Equidae), which also includes asses and zebras, has only one functional toe. Tapirs have a short proboscis formed from the upper lip and nose. The rhinoceros, includes several species found in Africa and South east Asia. All are herbivorous; 17 species.

Perissodactyla. It includes the horses, asses, zebras, rhinoceroses, tapirs, etc. These are terrestrial and herbivorous mammals. They have long limbs adapted for swift running. They have odd number of digits. The main axis of the limbs that divides it into two halves, passes through the third digit, which is larger than the others. The digits are enclosed in cornified hoofs. The mode of progression is unguligrade, i.e., the animal walks on the tips of the digits with the heels raised from the ground. Consequently there are three segments in the limb instead of two as in other vertebrates. There are no horns on the head.

This order is divided into Two suborders

    a) **Ceratomorpha** and

    b) **Hippomorpha**.

Suborder. Ceratomorpha. It includes the rhinoceroses and tapirs.

Rhinoceros—The Rhinoceros (Fig. 8.20). is found in South Eastern Asia and Africa. It is nocturnal and herbivorous. It is a large, heavy animal with comparatively short legs. Each leg bears three toes furnished with broad hoofs. The large elongated head bears one or two horns along the middle line of the anterior end. The horns grow from the skin and have no connection with the bones of the skull. The skin is very thick, naked or sparsely hairy, and is often deeply folded on certain parts of the body. The tail is thin and moderately long. The rhinoceros normally avoids man, but if brought to bay, proves very fierce and dangerous. Rhinoceros unicornis is the Indian rhinoceros. It has a single horn and lives in the Assam plains.

Suborder Hippomorpha. It includes horses, asses and zebras.

The Horse, Ass and Zebras (Fig. 8.20). The horse, ass and zebra are allied animals. They have a single functional digit (the third), which is enclosed in a solid hoof. The second and fourth digit are reduced to splint bones, while the first and fifth digits have altogether disappeared. The orbit is completely surrounded by bone.

In the **horse**, the tail is completely covered with long hair and bare callosities occur only on the fore-limbs. It is further characterised by larger body, smaller head, longer pendant mane, shorter ears and broader hoofs. It is no longer found in the wild state.

In the **ass**, only the lower part of the tail is covered with hair and the bare callosities occur only on the fore-limbs. It is further characterized by comparatively small body, large head, erect mane, long ears and narrow hoofs. There are no stripes on the skin. It, however, has a dark streak on the back, and sometimes another across the shoulders. The ass is found in the wild state also.

The zebra resembles the ass in all respects except that its body is fully striped.

Order 17. **Artiodactyla (Fig. 8.21)**: even-toed hoofed mammals: swine, camels, deer and their allies hippo-potamuses, antelopes, cattle, sheep, goats. Most of these ungulates have two toes, although the hippopotamus and some others have four (Figure 8.22). Each toe is

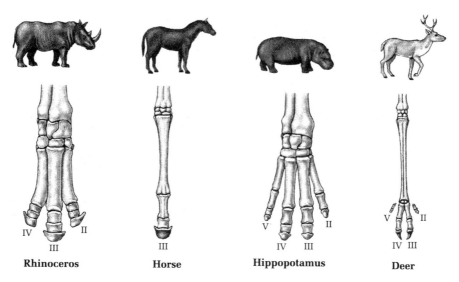

Figure 8.22
Odd-toed and even-toed ungulates. Rhinoceros and horse (order Perissodactyla) are odd-toed; hippopotamus and deer (order Artiodactyla) are even-toed. The lighter, faster mammals run on only one or two toes.

sheathed in a cornified hoof. Many, such as cattle, deer, and sheep have horns or antlers. Many are ruminants. Most are strictly herbivores, but some species, such as pigs, are omnivorous. The group is divided into nine living families and many extinct ones and includes some of the most valuable domestic animals. Artiodactyla is commonly divided into three

Suborder: The **Suina**- pigs, peccaries, and hippopotamuses,

Suborder: The **Tylopoda**- camels, and

Suborder: The **Ruminantia**- deer, giraffes, sheep, cattle; 221 species (Fig. 8.21).

Order 18. **Cetacea**: whales (Fig. 8.23), dolphins, porpoises. Anterior limbs of cetaceans are modified into broad flippers; posterior limbs are absent. Some have a fleshy dorsal fin and the tail is divided into transverse fleshy flukes. Nostrils are represented by a single or double blowhole on top of the head. They have no hair except for a few on the muzzle, no skin glands except the mammary and those of the eye, no external ear, and small eyes.

The order is divided into toothed whales

Suborder: **Odontoceti**, represented by dolphins, porpoises, and sperm whales; and baleen whales

Suborder: **Mysticeti** represented by rorquals, right whales, and gray whales.

Baleen whales are generally larger than toothed whales. The blue whale, a rorqual, is the heaviest animal that has ever lived. Rather than teeth, baleen whales have a straining device of whalebone (baleen) attached to the palate, used to filter plankton; 78 species.

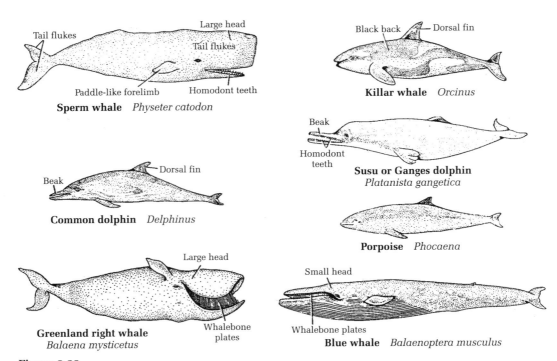

Figure 8.23
Cetacean animals

Humans and Mammals

The time when people developed agricultural, methods, they also began domestication of mammals. Dogs were certainly among the first to be domesticated. Dogs are an extremely adaptable and genetically plastic species derived from wolves. Much less genetically variable and certainly less social than dogs are domestic cats, probably derived from an African race of wildcat. Wildcats look like oversized domestic cats and are still widespread in Africa and Eurasia. Domestication of cattle, buffaloes, sheep and pigs probably came much later. Beasts of burden—horses, camels, oxen, and llamas—probably were subdued by early nomadic peoples. All truly domestic animals breed in captivity; many have been molded by selective breeding to yield characteristics desirable for human purposes.

Some mammals hold special positions as "domestic" animals.

Human Evolution

"The Descent of Man and Selection in Relation to Sex", about human evolution was written by Darwin. The idea that humans shared common descent with apes and other animals was repugnant to the Victorian world, which responded with predictable outage. Because at that time virtually no fossil evidence linking humans with apes existed, Darwin built his case mostly on anatomical comparisons between humans and apes. Darwin had this idea that the close resemblances between apes and humans could be explained only by common descent.

The search for fossils, especially for a "missing link" that would provide a connection between apes and humans, began when two skeletons of Neanderthals were collected in the 1880s. Then in 1891, Eugene Dubois discovered the famous Java man (Homo erectus). The most spectacular discoveries, however, have been made in Africa, especially between 1967 and 1977, period to which American paleoanthropologist Donald C. Johanson calls the "**golden decade**." During the same period, comparative biochemical studies demonstrated humans and chimpanzees are as similar genetically as many sibling species. Comparative cytology provided evidence that chromosomes of humans and apes are homologous. We are no longer searching for a mythical "missing link" to establish common descent of humans and apes, our closest living relatives.

Evolutionary Radiation of Primates

Humans are primates, and all primates share certain significant characteristics: grasping fingers on all four limbs, flat fingernails instead of claws, and forward-pointing eyes with binocular vision and excellent depth perception. Details of primate phylogeny are not entirely clear. The accompanying synopsis will highlight probable relationships of major primate groups.

The earliest primate was probably a small, nocturnal animal similar in appearance to tree shrews. This ancestral primate stock split into two major lineages:,

1. one of which gave origin to **prosimians**, including lemurs, tarsiers (Fig. 8.9), and lorises;
2. other to **simians**, which include the monkey (Fig. 8.24) and apes (Fig. 8.25).

Figure 8.24 A
Monkey. **A.** Olive baboons *Papio homadryas*, order Primates, an example of Old World monkeys.

Figure 8.25 B
Monkey. **B.** Red howler monkeys, *Alouatta seniculus*, order Primate family Cebidae, an example of New World monkey.

Prosimians and many simians are arboreal (tree-dwellers), which is probably the ancestral life-style for both groups. Flexible limbs are essential for active animals moving through trees. Grasping hands and feet, in contrast to clawed feet of squirrels and other rodents, enable primates to grip limbs, hang from branches, seize and manipulate food, and, most significantly, use tools. Primates have highly developed sense organs, especially good binocular vision, and proper coordination of limb and finger muscles to assist their active arboreal life. Of course, sense organs are no better than the brain processing sensory information. Precise timing, judgment of distance, and alertness require a large cerebral cortex .The earliest simian fossils appeared in Africa in late Eocene deposits, some 40 million years ago. Many of these primates became diurnal rather than nocturnal, making vision the dominant special sense, now enhanced by color vision.

We recognize three major simian clades. These are

1. **Old World monkeys** (cercopithecoids), including baboons (Fig. 8.24 A), mandrills, and colobus monkeys, and

2. **New World monkeys** of Central and South America (ceboids; Fig. 8.25 B), including howler monkeys, spider monkeys, and tamarins,

3. **Anthropoid apes** (Figs. 8.26).

Figure 8.26
Gorillas, *Gorilla gorilla*, order Primates, family Hominidae, examples of anthropoid apes.

Old World monkeys and anthropoid apes (including humans) are sister taxa, and together form the sister group of New World monkeys. In addition to their geographic separation, Old World monkeys differ from New World monkeys in lacking a grasping tail, while having close-set nostrils, better opposable, grasping thumbs, and more advanced teeth. Apes differ from Old World monkeys in having a larger cerebrum, a more dorsally placed scapula, and loss of the tail.

Apes first appear in 25-million-year-old fossils. At this time woodland savannas were arising in Africa, Europe, and North America. Perhaps motivated by greater abundance of food on the ground, these apes left the trees and became largely terrestrial. The gradual replacement of forests with grasslands in eastern Africa provided an impetus for apes to adapt to an open environment, the savannas. Because of the benefits of standing upright (better view of predators, freeing of hands for using tools, defense, caring for young and gathering food) emerging hominids gradually evolved upright posture. This important transition was an enormous leap because it required extensive redesigning of the skeleton and muscle attachments.

Evidence of the earliest hominids of this period is extremely sparse. Yet in 2001 the desert sands of Chad yielded one of the most astonishing and important discoveries of modern paleontology, a remarkably complete skull of a hominid dated at nearly 7 million years ago. Named Sahelanthropus tchadensis by its discoverer, French paleontologist Michel Brunet, this chimplike creature is by far the most ancient hominid yet discovered (Figure 8.27). Although its brain is no larger than that of a chimp (between 320 and 380 cm^3), its relatively small canine teeth, massive brow ridges on a short face, and mouth and jaw that protrude less than in most apes, confirm that the skull truly is that of a hominid. Until this skull was discovered, the earliest hominid fossil was Ardipithecus ramidus from the sands of Ethiopia, originally dated at 4.4 million years (Fig. 8.27). Ardipithecus ramidus is a mosaic of primitive apelike and derived hominid traits, with indirect (and controversial) evidence that it may have been bipedal. Between 1997 and 2001 additional fossils of A. ramidus were discovered that extend its existence back to at least 5.5 million years ago. These fossils have tentatively been assigned to a new subspecies, Ardipithecus ramidus kadabba.

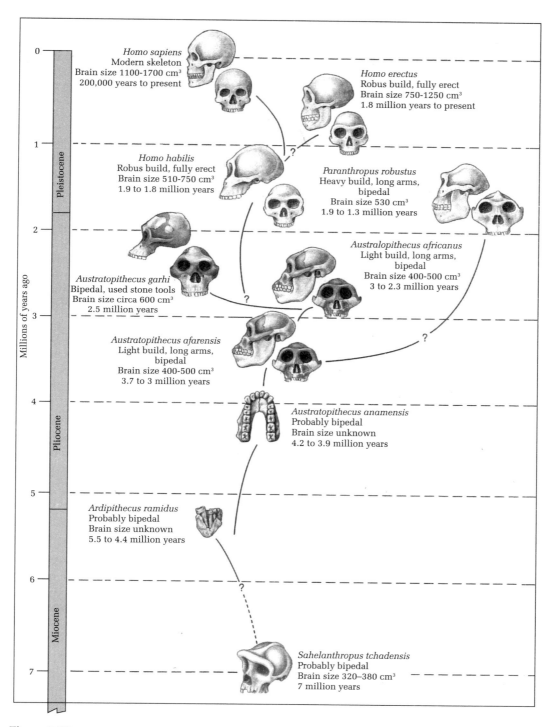

Figure 8.27

Hominid skulls, showing several of the best-known hominid lines preceding modern humans (homo sapiens).

Until the discovery of Sahelanthropus tchadensis, the most celebrated fossil was a 40% complete skeleton of a female Australopithecus afarensis (Fig. 8.27). Unearthed in 1974 and named "Lucy" by its discoverer Donald Johanson, A. afarensis was short, bipedal hominid with face and brain size resembling those of a chimpanzee. Numerous fossils of this species have now been discovered. The time range for A. afarensis is 3.7 to 3.0 million years ago.

In 1995 Australopithecus anamensis was discovered in the Rift Valley of Kenya. Many researchers believe that this species, which lived between 4.2 and 3.9 million years ago, is an intermediate between A. ramidus and A. afarensis (Lucy) (Fig. 8.28). The extremely humanlike lower leg bones of Australopithecus anamensis provide strong evidence that this species was bipedal.

In the last decade there has been an explosion of australopithecine fossil finds, with eight putative species requiring interpretation. Most of these finds are known as gracile australopithecines because of their relatively light build, especially in skull and teeth (although all were more robust than modern humans). The gracile australopithecines are generally considered direct ancestors of populations of early Homo and, by extension, of the lineage leading to modern humans. Following A. anamensis and Lucy (Australopithecus afarensis; 3.7 to 3 million years ago) there appeared the bipedal Australopithecus africanus, which lived between 3 and 2.3 million years ago. This species was similar to A. afarensis, but had a more humanlike face, slightly larger body, and a brain size about one-third as large as that of modern humans (Fig. 8.29). In 1998 a partial skull, dated at 2.5 million years ago, was discovered in Ethiopia. Named Australopithecus garhi, this species differs from other australopithecines in its combination of features, especially the large size of its molars.

Coexisting with the earliest species of Homo was a different line of large and robust australopithecines that existed between 2.5 and 1.2 million years ago. One of these was Paranthropus robustus (Fig. 8.27), which probably approached the size of a gorilla. The "robust" australopithecines were heavy jawed with skull crests and large back molars, used for chewing coarse roots and tubers. They are a side branch in hominid evolution and not part of our own lineage.

"who were the earliest species of evolutionary lineage of genus Homo , and indeed how to define the genus Homo",

The earliest known species of the genus was Homo habilis, a fully erect hominid. Homo habilis "handy man" is thought to have been about 127 cm (5 feet) tall and weigh about 45 kg (100 pounds), although females were probably smaller. This species had larger brain than the australopithecines and its brain was more humanlike in shape; one brain cast shows a bulge representing Broca's motor speech area, suggesting that Homo habilis was capable at least of rudimentary speech.

About 1.8 million years ago Homo erectus appeared, a large hominid standing 150 to 170 cm (5 to 5.5 feet) tall, with a low but distinct forehead and strong brow-ridges. Its brain capacity was around 1000 cm^3, intermediate between the brain capacity of Homo habilis and modern humans (Fig. 8.28). Homo erectus was a social species living in tribes of 20 to 50 individuals. Homo erectus had a successful and complex culture and became widespread throughout the tropical and temperate Old World.

Homo Sapiens

Recent molecular genetic studies indicate that human populations have formed a single evolutionary lineage for the past 1.7 million years. During this time, populations on different continents have shown some geographic differentiation, but they maintained at least low levels of gene exchange and all have made genetic contributions to modern humans. Several major expansions of populations out of Africa occurred during this time.

Anthropologists have named fossil species to denote spatial and temporal variation in phenotypic characters within this lineage; however, this lineage would constitute a single species according to most biological criteria of species. The earliest human remains originally classified as Homo sapiens "wise man", from 500,000 to 300,000 years ago, now are identified by anthropologists as H. heidelbergensis. One group of well-known humans, the Neanderthals, arose about 150,000 years ago. These morphologically robust humans were originally classified as a subspecies of H. sapiens, but recent evidence pointing to their distinctiveness has led anthropologists to recognize them as H. neanderthalensis. With a brain capacity well within range of modern humans, the Neanderthals were proficient hunters and tool-users. Neanderthals were not homogeneous but varied geographically in response to local conditions and the isolation of populations from one another. They dominated the Old World in the late Pleistocent epoch.

Fossil and mitochondrial DNA (mtDNA) evidence indicates that characteristics of H. sapiens, as presently defined, arose in Africa about 200,000 years ago (Fig. 8.28). About 30,000 years ago Neanderthals and remaining H. erectus disappeared approximately 10,000 years after the first appearance of H. sapiens in Europe and Eastern Asia. Modern humans were tall people with a culture very different from that of Neanderthals. Implement crafting developed rapidly, and human culture became enriched with aesthetics, artistry, and sophisticated language.

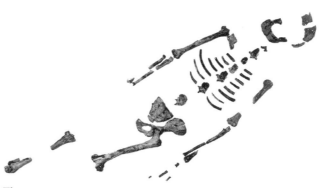

Figure 8.28

Lucy (*Australopithecus afarensis*), the most nearly complete skeleton of an early hominid ever found. Lucy is dated at 3.0 million years old.

It is important to note that recognition of species in Homo is based entirely on morphology. Recognition of three or more distinct species of Homo does not necessarily imply the occurrence of branching speciation in this lineage; it is possible that we are observing phyletic change within a single species through time, and using the species names only to denote different grades of evolution. Nevertheless, there is clearly only a single species of Homo alive today.

Homo sapiens is a product of the same processes that have directed the evolution of every organism from the time the life's originated. Mutation, isolation, genetic drift, and natural selection have operated for homo as they have for other animals. Yet we have what no other animal species has, a nongenetic cultural evolution that provides a constant feedback between past and future experience. Our symbolic languages, capacities for conceptual thought, knowledge of our history, and abilities to manipulate our environment emerge from this nongenetic cultural endowment. Finally, we owe much of our cultural and intellectual achievements to our arboreal ancestry which bequeathed us with binocular vision, superb visuo-tactile discrimination, and manipulative skills in use of our hands.

Origin and Ancestry of Mammals

Huxley (1880) advocated amphibian ancestry of mammals since there are two occipital condyles in the skulls of both Amphibia and Mammalia. However Huxley's theory is not tenable because condyles are derived from exoccipitals in Amphibia but from basioccipital in Mammalia. Apart from this, the two classes have different modes of life and display many fundamental differences .It may be hard to realize but all mammals living today have descended from reptiles. There is enough evidence from extinct reptiles and mammals for this universally accepted view that mammals had a reptilian ancestry. This view also gets strong support from the fact that the monotremes and living reptiles have close resemblances in their anatomical features, including soft as well as hard parts.

Figure 8.29
A reconstruction of the appearance of Lucy (*right*) compared with a modern human (*Homo sapiens*).

Who were these ancient reptilian ancestors of mammals?

Long before the appearance of true mammals, one group of extinct reptiles, the Synapsida, acquired several mammalian characteristics. They lived throughout the Permian and Triassic periods. These more mammal – like synapsids belonged to the order Therapsida. One of these more advanced carnivorous therapsids (suborder Theriodontia) was called Cynognathus. It lived during the early Triassic period. It was of the size of a Wolf and showed the following mammalian characters:

1. Typical upright mammalian limbs capable of generating considerable speed.
2. Skull with 2 occipital condyles, secondary palate, and enlarge lateral termporal fossa.
3. Largest bone of lower jaw was dentary.
4. Dentition consisted of incisors, canines and chewing molar.

Therapsids retained several reptile-like feature also. Their skull was intermediate between that of reptiles and mammals, having small cranium , parietal foramen, single middle ear bone, reduce quadrate and quadrato-jugal, many lower jaw bone etc. it is also not known whether therapsids were warm blooded, had hairs instead of scales and nurse their young. They were not necessarily the direct ancestors of present day mammals.

Fossil remain mainly teeth and jaws, reveal very little about the first true mammals. They were mostly creatures no bigger than rats and mice ,and were insignificant. But they could still manage to survive by exploiting different ways of life from their reptilian enemies. They were nocturnal, thus avoiding direct conflict and competition with the mostly dinning reptiles. They were either burrowing hunting for insects, or arboreal in contrast to their ground dwelling herbivorous or carnivorous contemporaries. They had a regulated high body temperature (endothermic), hairy integument, and probably carried their young in pouches for further development after birth, and safety. They were endowed with larger brains and greater intelligence.

By the end of Mesozoic, the vast majority of dominant reptiles became extinct for reasons which are still not Well known . Many ecological niches were now left open to mammals who started exploring them. By the close of Cretaceous period, placental mammals became distinct from marsupials. During Coenozoic age, all the orders of placental mammals were established, that is why this period is called the **Age of Mammals**.

Was mammalian evolution monophyletic or polyphyletic? Nothing is known about the origin of primitive Prototheria, because no fossils record older sthan Pleistocene are known. They have remained primitive probably due to isolation form rest of the world in Australia and New Zealand. During Jurassic period, there were atleast 5 well-established orders of mammals, now all extinct. According to the phylogeny, origin of mammals is polyphyletic because they were derived from at least two Triassic reptilian stocks, cynodonts and ictidosaurs. It is generally assumed that the living Prototheria possibly evolved form theriodontia, while Metatheria and Eutheria evolved independently from the pantotherians, by the end of Cretaceous period.

Part B
Comparative Anatomy

Chapter 9

Integumentary System

Integument

Integument is not only the outermost wrapping around the vertebrate body. It covers all exposed surfaces of the body, such as exposed portion of the eyeball (where it is covered by conjunctiva and is usually transparent) and external surface of ear drum. Besides this integument is also directly continuous with the lining of all passageways opening to the exterior.

Embryonic origin of integument In vertebrates integument is a composite organ having its origin from the embryonic layer such as **ectoderm** and **mesoderm**. The **epidermis** is formed by the ectoderm that remains on the surface of embryo after development of the neural tube. Next to epidermis, the **dermis**, is formed by mesoderm, i.e. the dermatome of the epimere and somatic mesoderm of the somatopleure (Fig. 9.1). The dermis contain in it the blood-vessels and nerve which grow throughout the dermis. The pigment cells migrate into dermis from the neural crest.

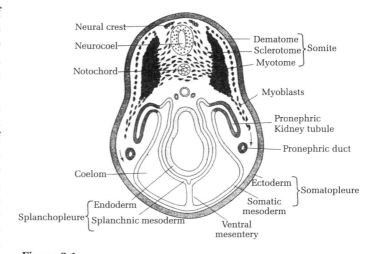

Figure 9.1
T.S. vertebrate embryo (showing fate of the mesoderm). Dark, mesodermal somite (dorsal mesoderm); Sclerotome cells move toward, notochord and form vertebra; myoblasts (arrows) form myotomal muscle; dermatome form dermis of skin

Structural details of integument

Epidermis Various layers of epithelial cells which compose the **epidermis**, arise in the deepest layer, the **stratum germinativum**. The cuboidal or columnar cell of stratum germinativum rest on a fine, fibrous basement membrane and undergo mitotic divisions throughout life. With each mitotic division, one of the two daughter cells stays in contact with the basement membrane, but the other is pushed outwards and differentiate according to species and body region (Fig. 9.2). The cells superficial to the stratum germinativum usually are connected to adjacent cells by desmosomes, which give the epidermis physical coherence. The cytoplasm of cells located peripherally may contain granules of the structural protein keratin. Some cells of stratum germinativum can differentiate into glands or other integumentary derivatives. The **stratum corneum** which is the skin's most superficial layer, consists of dead cells which contain in it the keratin. In amniotes the keratin is formed massively and exists in two molecular forms. α-Keratin, of the size of filaments 70-80 Å wide, and β-Keratin, with filament size about 30 Å wide. The β-Keratin layer is invariably hard, for example, the surface of avian and crocodilian scales, where as α-Keratin layers may be either soft as in mammalian flexible skin, or hard, as in mammalian claw. These cells of the **stratum corneum** are continually being sloughed off and replaced.

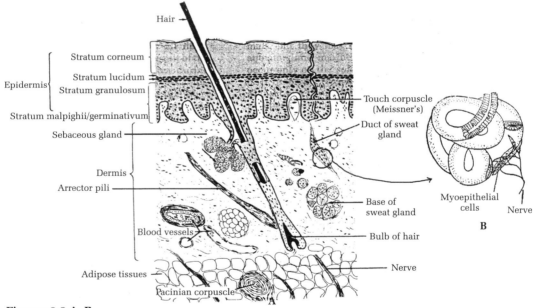

Figures 9.2 A, B
A. V.S. of mammalian skin, **B.** Magnified portion of sweat gland

The **stratum germinativum** is a mitotic layer which actively proliferates throughout life, producing cells which are pushed towards the surface and gradually differentiate, die and come to rest in the stratum corneum, and finally are lost. Whereas in the **Lepidosaurian reptiles** an entire layer of skin is shed as a unit, exuvia after a "new skin" has developed just below it.

Dermis or **corium** Consists of two portions.

  a) The **dermis proper** is a tough layer usually containing collagenous connective tissue cells and fibres, besides containing blood-vessels, nerves, pigment cells, smooth muscles, occasional skeletal muscles, and dermal bone.

  b) The **hypodermis** is made up of very loose connective tissue and may contain a great deal of fat. In some vertebrates the subcutaneous tissue has many fat layers because of which it is extremely thick e.g. the blubber of whales. The blood vessels and nerve present in the integument also supply the blood and pass through subcutaneous tissues. The thickness of different layers of skin varies considerably between species, between individuals and between body regions. In regions of mechanical stress the epidermis is much thicker, with a heavy stratum corneum, **Callouses** represents region of thickened stratum corneum that have become thickened in response to pressure or friction. In areas where skin endures lesser mechanical stress, the epidermis is much thinner. The dermis is also relatively thick in some areas and thin in others but the greatest variation is found in the fatty subcutaneous layer.

In human, it may vary from less than 1 mm on scalp to over 3cm on the buttocks and even more in woman's breast. In vertebrates, colour is usually caused by a pigment called chromatophores; they originate from neural crest and then migrate into epidermis or dermis. The most common chromatophores are melanophores which have two forms i.e. eumelanin (gives brown or black colouration) and phaeomelanin (gives yellow brown colouration).

Derivatives of the Integument

Structures derived from the integument are of two kinds; soft derivatives and hard derivatives.

Soft derivatives

The integumentary glands are the soft derivatives derived from epidermis. Their embryonic development is determined by the underlying mesoderm that induces a particular gland in a particular area of the body, even when the ectoderm above it has been transplanted from some other area. All integumentary glands are exocrine in nature. Integumentary glands can be broadly classified into (a) **Unicellular** and (b) **Multicellular**.

  a) **Unicellular glands** (or mucous glands) These are modified cells of the epidermis (Fig. 9.3 A). They are scattered throughout the body surface and produce mucin, which with water forms a slippery substance called mucus, and hence these glands are more popularly known as mucous glands. The secretion either passes out gradually over the surface of the skin or it first accumulates in the form of a droplet, and then gets squeezed out. After some time such gland cells die off and are replaced by the new ones.

  b) **Multicellular glands** Usually an invagination from the stratum germinativum extends down into the dermis to differentiate into multicellular glands. These glands may be classified into **tubular** and **alveolar**, and each of these may again be subdivided further (Fig. 9.3 B).

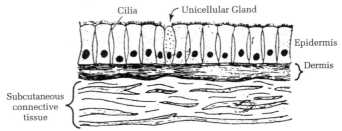

Figure 9.3 A
V.S. of integument (young Amphioxus)

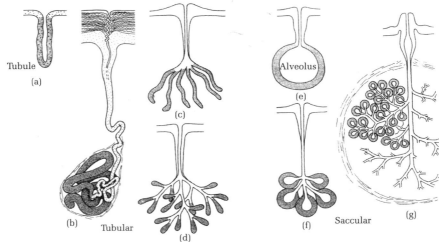

Figure 9.3 B
Multicellular glands (a) Simple tubular (b) Coiled tubular (c) Simple branched tubular (d) Compound tubular (e) Alveolus of a simple saccular gland (f) Simple branched saccular (g) Compound saccular (compound alveolar)

Tubular glands are deeper, some even extend into the hypodermis. They are lined by a single layer of epithelial cells; usually only those in the deeper part of the gland do secretion. If the tube is straight; it is called a **simple tubular gland**. If the tube coils at the end, it is called a **coiled tubular gland**; and if after entering the dermis or hypodermis the tube branches into several smaller tubes; it is called a **branched tubular gland.** Some time each branch coils in the hypodermis, forming a branched, coiled tubular gland.

Alveolar gland seldom go deep into the dermis and only rarely enter the hypodermis (Fig. 9.3 B). They have more bulk, however, being several cells thick with a small lumen. If arranged as one large clump of cells, the gland is called a **simple alveolar gland**; if in several clumps, it is called **a compound alveolar gland**.

Glands can also be classified by their **mode of secretion** (Fig. 9.4)

- **a) Merocrine** If the secretion leaves the cell without visibly damaging it and the gland cells undergo a period of rest during which they are refilling.
- **b) Holocrine** The cell must burst to release the secretion, new cells replace the discarded ones.

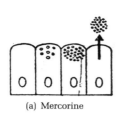 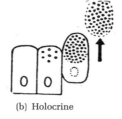 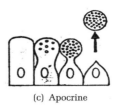

(a) Mercorine (b) Holocrine (c) Apocrine

Figure 9.4
Mode of secretion by unicellular glands

c) **Apocrine** It is an intermediate type, a small part of the cell is pinched off with the secretion but the rest of the cell survives. The injury is soon healed up and secretory activity is resumed.

d) **Eccrine** They are similar to apocrine, but differ chiefly in that the cytoplasm is broken down before secretory products are discharged.

Comparative account of soft integument derivatives

Protochordate The body wall or **mantle** or **test** or **tunic** of **tunicates** is secreted by the underlying ectodermal cells. Tunic is composed of a polysaccharide – the tunicine which is identical to plant cellulose.

Cephalochordates The integument is composed of epidermis and the dermis or cutis and is followed by muscles (Fig. 9.5). Epidermis of the Amphioxus is very thin and is composed of a single layer of cubical epithelial cells. The nucleated cells in the young or immature individual bear cilia (Fig. 9.3 A). Later these cells shed cilia and secrete a thin cuticle on outer border. Unicellular gland cells and chemoreceptor cells are scattered between the epidermal cells.

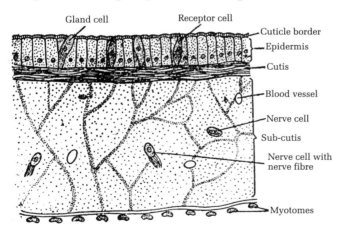

Figure 9.5
V.L.S. integument of Amphioxus

Hemichordates The integument is composed of ciliated and secretory cells (Fig. 9.6). The ciliated cells which cover the body surface are distinct and recognized by the presence of cilia and associated bundle of rootlet fibre. Secretory cells may be goblet shaped and contain a pear shaped secretory product or long granular or alveolated cell; they secrete mucus that spreads over body-wall. The **mulberry cells** have also been described, they secrete amylase.

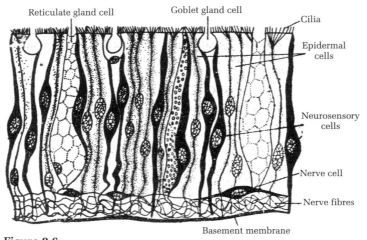

Figure 9.6
T.S. integument of *Sacchoglossus*

Cyclostomes: Lampreys and **hagfish** have prominent unicellular mucus glands, granular gland-cells and club-cells (Fig. 9.7). These glands are more numerous when scales are fewer. Mucus forms a protective layer over the body. Mucus is a complex glycoprotein. In the scaleless, hagfish, there are elongated mucous cells, "the thread cells", that are said to secrete mucus, which can coagulate muddy water.

The **mucus** (i) provides physical protection (ii) it reduces friction between the water and the body, enabling the fish to move with less energy (iii) mucus can coagulate mud. (It has been demonstrated that two drops of mucus from an eel will clear half - a- pint of muddy water in less than 30 seconds.) (iv) provides mucus layer covering because it slows the passage of water between the external and internal milieu, thus helps to maintain osmotic balance. When the water diminishes in home river the African lungfish, *Protopterus*, burrows into the mud and secretes a great deal of mucus around itself. This precipitates the mud, forming a quite impervious cocoon in which the lungfish survives the dry period.

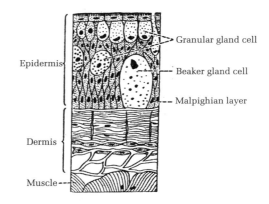

Figure 9.7
V.S. integument of a *Petromyzon*

Fish: Unicellular mucus glands are scattered all over the body. **Stickle backs** construct nests on stream bottoms, holding them together with large amounts of mucus.

In **paradise fish** the male blows-mucus-enclosed air bubbles, which float to the surface; the female deposits the eggs on the under surface of this floating nest.

In addition to these unicellular mucus glands a few fish have **multicellular** glands. In several taxonomic groups there are fish with **poison glands** viz., **string rays** among chondrichthyes, **catfish** among osteichthyes. These glands are usually associated with the spines of the fins, and they serve primarily a defensive purpose.

The **luminescent** organs are **unusual glands** of several types found mostly in deep sea fish (Porichthys) (Fig. 9.8). All fish which possess these organs produce 'cold light' and have

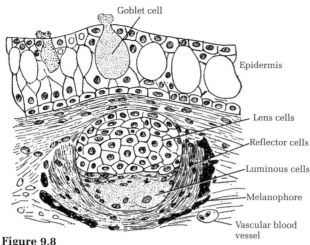

Figure 9.8
Light organ of a luminous fish

large prominent eyes, while other deep sea fish have small, often non-functional eyes. In fish where the light is produced by a true gland, the glandular portion, sink into the dermis as a sphere, the superficial hemisphere forms a transparent lens, and the deeper hemisphere forms the light producing cells, deep to this the dermis forms reflector fibres and deeper still, pigment cells. Light produced is thus reflected out to the environment. The **luminescent** organ of some of the fish are not actually a gland but contain light producing **symbiotic** microorganisms.

Amphibians: Mucus glands are better developed in the tree frogs, whose feet bear glands that secrete sticky mucus. Invariably amphibian glands are simple alveolar mucus glands (Fig. 9.9) present in the dermis. In addition to mucus glands, several anurans have serous glands also whose secretion; although proteinaceous, is more watery then mucus. Most of these **serous glands** secrete mild **toxins**, as in the **common toad** and can produce irritations in sensitive persons. *Kokoe* frog of South America, for instance, produce **batrachotoxin** which is the most powerful neural poison. Large swollen porous patches of **parotid glands** are present behind the head in **toads** and **salamanders**. **Mental glands** are tubular glands below the chin and develop as secondary sexual character in male salamander (Fig. 9.10).

Cells of **Leydig** are unicellular glands of uncertain use in tailed amphibia.

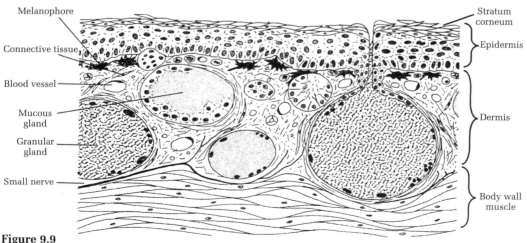

Figure 9.9
Frog integument (showing multicelluar glands) stratum corneum (thin) covers the epidermis

Reptiles: The reptilian integument contains numerous glands most of which are relatively inconspicuous and poorly understood. In some species there are small glands adjacent to the germinal epithelium. Holocrine alveolar glands are most common. Their structural details and location on the body are often species specific. In some species, larger glands are located near the cloaca. The studies suggest that their odoriferous secretion may help in intraspecific communications, such as marking territory or attracting mates.

 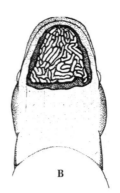

Figure 9.10
Mental gland in male salamander: **A.** ventral view of head showing position of glandular area; **B.** the same portion, with the skin below the gland removed

Hedonic glands: The second pair of glands in crocodiles is located in the region of cloacal aperture. These are related with sexual activity.

Musk glands Crocodiles of either sex have two pairs of musk or scent glands, the first pair lie in the skin of throat on the inner halves of the lower jaw, the second pair lies inside the cloaca.

Scent glands of turtles are present beneath the lower jaw and in the skin at the line of junction of plastron and carapace.

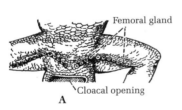 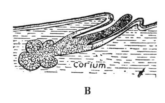

Figures 9.11 A, B
A. Femoral glands of male *Lacerta*; B. Section through a single femoral pore of a lizard, *Lacerta* (showing projecting plug of dry cells)

Femoral glands in male lizards are tubular glands on under surface of thighs (Figs 9.11 A, B). Their secretion on drying, forms hard conical projection which aids in copulation.

Stink gland occur in the cloaca of certain snakes. Function of these glands is to produce obnoxious smell.

Birds: The uropygial glands is the only integumentary glands in birds and is present at the base of the tail just posterior to the pygostyle (Figs 9.12 A, B). This is a bilobed gland with two ducts to the dorsal surface, one on each side of the pygostyle. It is a holocrine sebaceous gland, producing an oily secretion called **promatum**. It was long considered that it helps in preening the feather, hence the gland has also been called the preen gland. The function of preen gland is however doubtful because of the following reasons:

(i) Some birds such as bustards, most pigeons, parrots, and most ratite birds, do not have uropygial gland while in other it is only vestigial. However it is transiently present during development.

(ii) Some believe it to be an odoriferous gland with sexual adaptation.

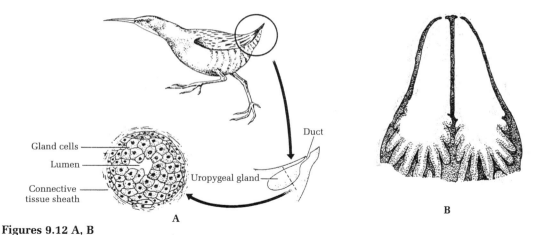

Figures 9.12 A, B
A. Section through uropygial gland in birds   B. Section bilobed uropygial gland of bird

(iii) Others believe it to be a source of vitamin D because some birds develop rickett if it is removed, but not if supplementary vitamin D is given at the same time.

(iv) In turkey oil glands of similar nature are found in ear-holes, where oil holds the dust particles.

Mammals: They have integumentary glands of great number and diversity, and all are derived from stratum germinativum.

Sudoriferous or sudorific or sweat glands   Found only in mammals, these glands secrete a watery substance containing some salts and urea. The composition of this secretion varies with the metabolic state of the animal, and it plays little role in maintaining proper osmotic balance. The evaporation of the sweat causes cooling and thus helps in maintaining a constant body temperature.

There are two major types of sweat glands **eccrine** and **apocrine**.

Eccrine sweat gland is much coiled and is long tubular eccrine or epidermal gland lined with stratum germinativum but lies deep in the dermis. Their deeper portions are vascularized and contain myoepithelial cells whose contractions force out the secretions (Fig. 9.2). Eccrine glands, which secrete sweat, extend well into the dermis and frequently into the hypodermis. Their distribution is quite variable.

Not all mammals possess sweat glands, they are absent from the skin of whales, sea-cows and the African golden mole. Very few sweat glands are found in dogs. Some species, such as humans and horses, have them profusely all over the body. Most mammals have them at least on the soles of the feet and around the lips, genitalia, mammary glands and anus.

Apocrine sweat glands extend deep into the hypodermis. Their secretion is thick and milky with the distinct odour. Apocrine glands are abundant in horses and swine. They are found only in the axillary, genital and anal regions, and around the nipples. Apocrine secretion contains a large amount of proteins, carbohydrates, iron and ammonia. Its flow is slow and less. Since it contains organic matter, bacteria frequently attacks the deposit on skin, resulting in a distinct odour and staining of clothes.

Sebaceous glands These are compound, alveolar holocrine glands and secrete an oily substance called **sebum** which is a lubricant and grooming agent for hair (Fig. 9.2). Their ducts open into hair follicles. They are common in hair-less areas also. Such sebaceous glands of man, not associated with hair follicles, are found over glans-penis, internal surface of prepuce, lips of vagina, corners of mouth and lips. Recent investigations are of the opinion that synthesis of vitamin – D by the action of sunlight occurs in sebaceous glands.

Meibomian glands These are located on the edges of the eyelids in most mammals; they produce an oily film which covers the surface of the cornea as a protective and lubricating agent. This thin oily film covers the cornea, holds below it a film of tear that keep the corneal surface moist.

Zeis glands They are sebaceous glands found in association with follicles of eyelashes. Their secretion keep the eye-lashes smooth and oily.

Genital glands Found in conjunction with the genital organs are the **preputial glands** in males and vulvar glands in females. Their oily secretion act as lubricants. Many mammals such as beavers, cats, dogs and foxes, have large hormonally controlled, compound, alveolar scent glands in anal region, in addition to and not to be confused with the apocrine glands serving the same function. In various mammals, the odoriferous secretions of these alveolar glands are used to mark territories and attract mates. In **skunks**, however, these gland get modified to produce irritating secretion.

Inguinal glands Present between the anus and the penis or vulva in rabbit.

Civet glands Certain carnivores have repellent glands opening into rectum just inside the anus. Such glands are enormous in weasles and skunks.

Musk glands These glands of certain deer are inguinal glands that produce an attractive scent. Scent glands in the antelopes are found below the eyes and in elephants over the foreheads.

Scent glands Peccary has a prominent scent gland opening on its back. Rhinoceros and certain other hoofed mammals have scent glands on the feet; on the callosities of legs in horses, and over the arm in certain lemurs. Glands on under surface of prepuce and lips of vagina in humans are also scent glands. Some sweat glands in the arm-pits of man are modified scent glands.

Wax glands (Ceruminous glands) They are found only in the mammalian external ear canal, where their secretion act with the hairs of the canal to protect the delicate tympanic membrane from microorganisms and bits of dust or dirt.

Mammary glands These glands are probably highly modified apocrine sweat glands. Their hormonally controlled secretion restricted to the female, contain water, fat, sugar, albumin and calcium and other salt and is called milk. Its precise composition varies according to species.

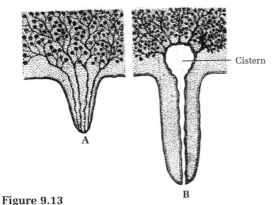

Figure 9.13
Sections: **A.** through nipple; **B.** through teat

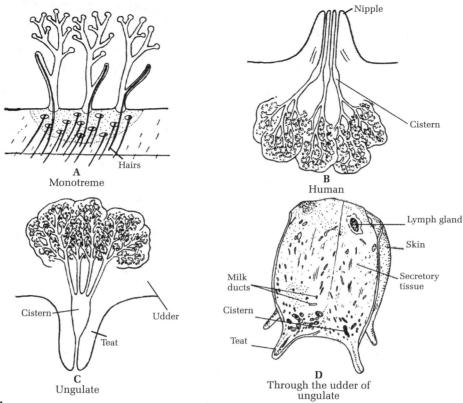

Figure 9.14

Mammary glands, ducts and nipples. The female monotreme lacks nipples, and the milk gland associated with ventral body hairs

In most **mammals** the tubules of mammary gland empty into a collecting duct which comes to the surface at the nipple, a modification of surface epidermis (Fig. 9.13).

The **prototherian** monotremes, however, have no nipples. The several branching tubules, rather than opening into a single collecting duct, open independently and secrete on to the females ventral body hairs, from which the young sucks the milk (Fig. 9.14 A).

Ungulates on the other hand, have a large storage compartment, the **udder** with the glands, opening on to its centre (Fig. 9.14 C). This is called an inverted mammary gland, as opposed to more common everted type where nipple is turned out and ducts open together at its tip.

Differentiation between nipple and teat

Nipple	Teat
1. It is actually an elevated area in the breast region.	1. A teat is a false nipple. It is in fact, an outward growth of skin in mammary area. The outgrowth is actually a large projection.
2. The ducts open directly at the tip of this raised area.	2. All major ducts first open into a cistern which is a saclike structure and lies at the base of the teat.
3. The milk is carried to the outside by ordinary ducts.	3. The milk is carried to the outside of the udder by a secondary canal or duct.
4. The milk instantly comes out as soon as the sucking stimulus is provided by the child.	4. Udder of the milking animal is not filled with milk, the milk comes out of the udder after proper pressing and massaging for sometime.

Mammary glands are located at various points along two mammary lines (mammary ridge) on the ventral side of the trunk (Fig. 9.15 B), from the pectoral to the inguinal region. In some species, such as pigs, they form all along these lines, but arboreal mammals generally have a single pair of pectoral mammae, while the udder of ungulates is in inguinal region. Mammary glands and teats may be present in both sexes but are active only in pregnant females. Gynecomastia is a condition when mammary glands in male become active. Milk is sometimes found in the teats of a newly born child, it is called *witch's milk*.

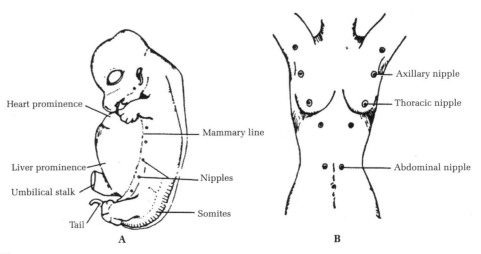

Figure 9.15
A. Milk line and nipples in pig embryo   B. Showing supernumerary nipples in human

Hard derivatives

In great majority of vertebrates, flexibility is retained because the hardenings are formed discontinuously in folds of the skin and overlap one another. This advantageous **shingle effect** provides hard layers with potential movement between them. Such dermal or epidermal hardenings formed in skin folds are called **scales**. They occur in several forms and are found in all living vertebrate classes except a few.

Keratin derivatives are found in terrestrial vertebrates whose epidermis produces a layer of cornified or keratinized cells, the stratum corneum. Scales of reptiles, birds, and certain mammals, claws, teeth, feathers, hair and epidermal fin rays are all keratin derivatives.

Scales in fish Elasmobranchs have rows of *placoid* denticles embedded in the skin over the entire body. Each denticle has a flat basal plate embedded in the dermis and a sharp, curved, protruding spine (Fig. 9.16). The basal plate and most of the spine are derived from the mesoderm of the dermis, where fibroblasts accumulate and dentine is secreted by odontoblasts. In the centre of each denticle is a hollow pulp cavity filled with blood vessels and nerves, from which fine canaliculi reach into all parts of the dentine. The spine itself is covered with a thin layer of a hard, enamel-like substance called **vitro-dentine** which is secreted by epidermal **ameloblasts** cells. In sharks both embryological and adult structure of the placoid denticles are similar to teeth and, in fact, the teeth and denticles are continuous at the edges of the mouth. In sharks the teeth are simply exaggerated placoid denticles, and both the teeth and denticles can be replaced if lost (Figs. 9.16 B, C).

Scales of ostracoderms and placoderms

Ancient fish characteristically had heavy, dermal scales and bones of three layers.

1. An inner layer of compact **lamellar bone**
2. A middle layer of spongy **vascular bone**
3. And an outer layer of **dentine** from which usually protrude tubercles or spines.

Although common in Devonian and Silurian ostracoderms and placoderms, these scales occur in no living vertebrates (Fig. 9.17).

Sarcopterygian cosmoid scales The heavy bony scales of the ancient crossopterygians and lungfish were called cosmoid scales (Fig. 9.18). They were structurally similar to those of **ostracoderms** and **placoderms**, except that instead of dentine over the two bony layers there was dentine-like cosmine substance containing radiating tubules and pulp cavities; overlying this substance, there was a very thin, extremely hard layer of **vitrodentine**. Living **sarcopterygians**, however, have much thinner scales resembling the cycloid scales of the most living **osteichthyeans**.

Ganoid scales In these heavy, bony scales the cosmine layer is overlaid not with vitrodentine but with extremely hard, multilayered **ganoin**. Its many layers and thickenings give ganoin a lustruous often **metallic sheen**. True ganoid scales, occurred in the **paleoniscoid** fish; among living vertebrates there are slightly modified ganoid scales in bichirs (Fig. 9.18 C).

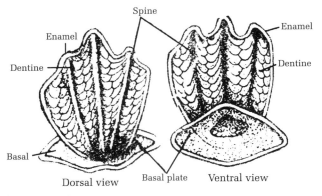

Figure 9.16 A
Placoid scale of Shark

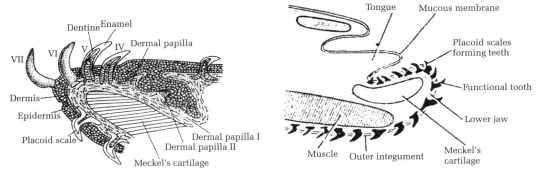

Figure 9.16 B
Placoid scale and teeth of a shark

Figure 9.16 C
Placoid scales forming teeth in shark

Scoliodon-section through lower jaw, replacement of teeth by placoid scale

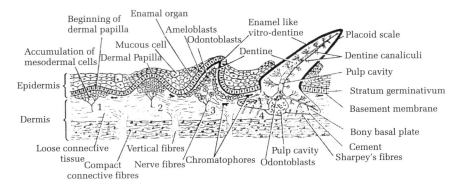

Figure 9.16 D
Development of a placoid scale. 1 Accumulation of dermal cells. 2 Formation of dermal papilla. 3 Growing spine. 4 V.S. of a fully formed scale

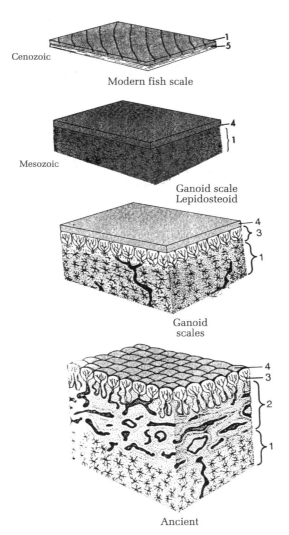

Figure 9.17
Dermal bone and dermal scales, diagrammatic. 1 Lamellar bone; 2 spongy bone; 3 dentine; 4 enamel of one kind or another, including ganoin; 5 fibrous plate characteristic of modern fish scales. The surface elevations in ancient armor are denticles consisting of layers 3 and 4. The lamellar bone in modern fish scales is acellular

Cycloid and ctenoid scales The evolutionary trend in most groups has been away from heavy armor towards other protective devices and mobility. The cycloid or ctenoid scales of most living fish have neither **cosmine** nor **ganoin**. They are composed of two thin layers: a somewhat bonelike, but acellular osteoid layer and an inner, very dense fibrous layer (Fig. 9.17). They are rather small and flexible, arranged in an overlapping single fashion and entirely embedded in the dermis. The very thin, overlapping epidermis often contains unicellular mucous glands. These scales occur in varied shapes particularly in their growth rings – which are so specific that they have sometimes been used as a tool for **taxonomy** (Fig. 9.18 F).

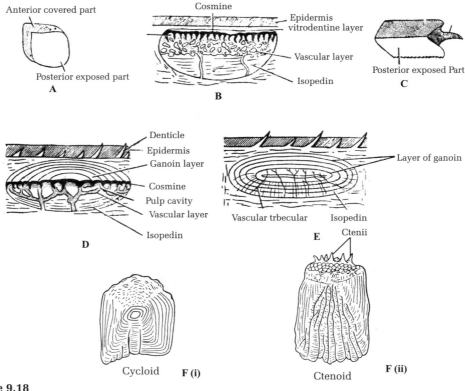

Figure 9.18
Scales of fish **A.** An isolated scale; **B.** T.S. of cosmoid scale; **C.** An isolated ganoid scale; **D.** T.S. of *Palaeoniscoid* scale; **E.** T.S. of *lepidosteoid* scale; **F.** Moden fish scales; (i) Cycloid, (ii) Ctenoid

These two scale types differ only in that the **ctenoid** scales have small stiff spines **(ctenii)** on their free posterior surfaces while cycloid scales do not.

Amphibian scales The stem amphibians, **labyrinthodontia**, possessed very heavy bony and usually cosmoid scales; during amphibian evolution these have become reduced or lost. Some modern **apodans** have tiny bony scales of dermal origin (Fig. 9.19) **anurans** and **urodeles** have none.

Reptilian scales The scales of reptiles are formed from the epidermis; their hard portion is the structural protein **keratin**. Reptilian scales can undergo various modifications. In adult turtles

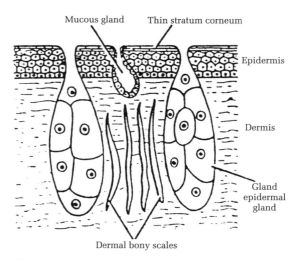

Figure 9.19
Ichthyophis. V.S. Skin showing dermal bony scales

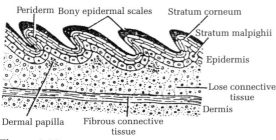

Figure 9.20
Lizard V.S. skin

and some other reptiles, the scales are underlaid with dermal bones. In the so called **horned toad**, actually a lizard, the **horns** are prominent scales on various parts of the body especially dorsally.

In **lepidosurian** reptiles (snakes, lizards, sphenodon) a complete new epidermis including its scales is formed intermittently. This growth occurs from the **stratum germinativum** and upon its completion the outer, older epidermis is shed as a unit (exuvia).

Both α and β keratin are involved in reptilian scales. **Rattle** of a rattle snake is a collection of 8-12 ring shaped epidermal scales hinged together in a long series at the tip of tail (Fig. 9.21). The scales of **crocodiles** are large **corneoscutes**.

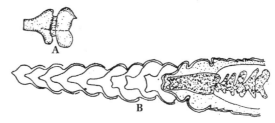

Figure 9.21 A, B
The Rattle   **A.** a single bony element of the rattle; **B.** sagittal section of end of tail of a rattle

Scales in birds Most birds have typical, epidermal keratin scales on the feet and lower leg and frequently a few small scales at the base of the beak.

Any bony spurs on the wrist or ankle of birds, like that on the **carpometacarpus** of the **Mexican-Jacana** or on the **tarsometatarsus** of **game cocks**, are covered with a sheath identical to that of the beak. The **comb** of roosters is covered with a thick warty stratum corneum.

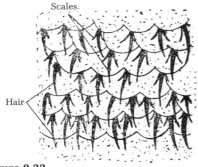

Figure 9.22
Hairs interspersed in scales of rat's tail

Scales in mammals Most mammals have no scales, but some mammals particularly **rodents**, have a few epidermal scales on the **tail** (Fig. 9.22). The *armadillos*, for instance, has an overlying layer of epidermal horny scales fused to heavy, dermal bony armor similar to the armor of turtles. The termite eating **pangolins** of South Africa and South East Asia is almost entirely covered with heavy, sharp-edged **dermal scales**. Mammals produce only α-keratin, although birds have both α-and β-keratin.

Development of placoid scale It is like the development of a tooth and in early stages resembles with that of feather or hair (Fig. 9.16 D).

First of all the mesenchymal cells collect beneath the epidermis, the nourishment is supplied by blood capillaries and soon the mesenchymal cells push the epidermis a little to form a **dermal papilla**. The mesenchymal cells in papilla get transformed and start secreting dentine and now are called as scleroblasts or odontoblast. Cells of stratum germinativum overlying

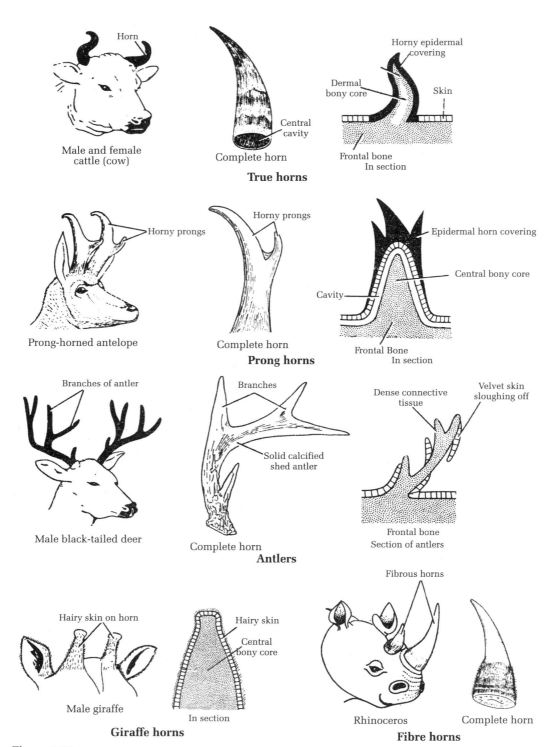

Figure 9.23
Types of mammalian horns and antlers

the dermal papilla multiply and push down around the papilla and later differentiate into an outer and inner **enamel layers**; cells between the two layers form the **enamel pulp**. The entire epidermal structure around the dermal papilla is called the **enamel organ**. Cells of the inner layer of enamel organ are called ameloblasts. Within the papilla, the **scleroblasts** lying opposite the ameloblasts, secrete a thin layer of vitrodentine followed by a thick layer of dentine. Role of **ameloblasts** is believed to be one of influencing the formation of scale and not of secreting enamel. **Dentine** though hard as bone, is pierced through by the fine canals, the **canaliculi**, into which scleroblasts send out fine processes. The middle of papilla remains soft and vascular and forms the pulp cavity. **Basal plate** is secreted by the mesodermal cells near the base of the papilla. As more and more dentine is secreted, it takes the form of a conical spine that pushes through the enamel organ and eventually cuts through the epidermis.

Horns and antlers in mammals: Among the most unique integumentary modifications are horns and antlers. Although present in some extinct reptiles and several extinct non-hoofed mammals, among living forms they are confined to the **ungulates**. They are of four basic types.

a. Hair horns or fibre horns The mid line keratin fibre horn is found only in the **rhinoceros** (Fig. 9.23). It is formed by long epidermal strands of keratin fibres held together by a sticky substance. It is an integral part of the skin, and is nerve shed. Indian **rhinoceros** has one horn, and African two, one behind the other.

b. Antlers Antlers are characteristic of the deer family, cervidae; although in **reindeer** they occur in **both sexes**, they are usually restricted to the male. An outgrowth from the dermal frontal bone forms the antlers core. This is covered by a soft layer of haired skin called the velvet. When the bony portion is fully grown, the vascularisation is restricted basally, the velvet then dies and is sloughed off. The bare, fully developed antlers remain throughout the mating season and for some time there after until they are shed regrowing the following year (except the giraffe, whose antlers remain covered with velvet throughout life and are never shed). The antlers are used chiefly as offensive weapons in intraspecific combat for females and rank. The growth of antlers and horns is hormonally controlled.

c. Pronghorn In the male pronghorn **antelope** a North American ungulate; there is an unusual horn (Figs 9.23, 9.24). Its bony core also is an outgrowth from the frontal bone, but its outer part is of heavy, hard keratin with a single prong (which is not present on the bony core). The keratin part is shed annually, forced off by a new one forming deep to it, the inner bony part is not shed.

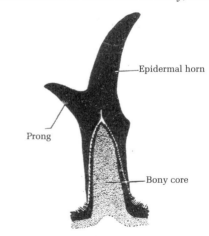

Figure 9.24
Horn of the pronghorn antelope

d. Hollow horn Members of the **Bovine** family exhibit true horns (Fig. 9.23). This family includes cattle, sheep, goats, and antelopes (except prong horn antelopes). These horns are also bony outgrowths from the frontal bone with an overlying portion of hard keratin derived from epidermis. They are not truly hollow unless the keratin portion is removed from the bony core. Hollow horns are not periodically shed.

Beaks They are characteristic of birds, but are also found in the turtles, tortoises and duck-bill moles which like birds do not have teeth. Jaws are covered with hard keratin that forms the beak.

Beaks of **duck-bill** mole resembles the beak of bird only superficially. In contrast to bird's beak, the beak of this mammal is soft, flexible, covered with moist skin richly supplied with sense organs.

Horny teeth The epidermal or horny teeth are well-known in the frog's tadpole by means of which it scrape grasses present in the water. Teeth of lamprey are similar derivatives. A small tooth like process **caruncle** or **egg tooth**, over the tip of snout is used for breaking open the egg shell and is common in *Sphenodon*, turtles, crocodilians, birds etc.

Digital tips The epidermis of the tip of tetrapod digits form hard **keratin.** The hardened keratin gets modified as claws, nails and other structures (Fig. 9.25). These structures not only protect the tips, but are often used for offense and defense. They also improve the manipulative ability of the hand.

In **amphibians**, there are only two examples of modified, keratin digital tips. In aquatic anuran **xenopus,** claws are present, the spade foot toads – **scaphiopus** have a hard keratin sheet on the hind foot, which assist in digging.

Claws: Although structural details differ between classes, all amniotes have generally similar claw made up of two curved, scale-like keratin plates. The larger, dorsal plate in the **unguis**, and the some what smaller softer, ventral plate is the **subunguis**. Together they form a sheath over the entire distal phalanx of each digits. They are derived completely from epidermis, with a thickened layer of stratum germinativum forming the claw bed in the skin of distal phalanx.

The claws in birds are usually thought of as being restricted to hind limb, one or two sharp claws may also occur at the ends of the digits of the wings, for example, ostriches, geese and some swifts. Hoatzins uses claws on the wings for climbing about on branches of trees (Fig. 9.26). The fossil **Archeopteryx** also had three claws on each wing. In some modern species there are claws in some wing digits in the nestling, these claws help the young birds move about and are lost before they leave the nest.Claw shape varies considerably and is related to habits and habitat. **Wood peckers**, for instance, have very sharp, narrow claws with which they anchor themselves firmly into tree bark during drilling. Birds of prey have long, sharp, powerful **talons** (Fig. 9.25).

The **mammalian claw** has basically the same structure as reptilian and avian claws, with two exceptions, there is usually a thickened area of epidermis underlying dermis forming a foot-pad proximal to the subunguis and the stratum corneum is thickened soft keratin.

Nails They are characteristic of the **primates** such as man (Fig. 9.25). **Unguis** or nail plate is broad and flattened and composed of stratum lucidium; **subunguis** is very much reduced and is represented by a narrow strip of less modified epidermis below the edge of the unguis. The dermis of nail bed underlying the unguis is highly vascular as is evident from the pinkish colour of the nail. At the base of nail epidermis is invaginated and supplies material for the growth of the nail. The nail groove or **sulcus unguis** and the part of nail in the groove is called the **nail root.** A thin whitish, some what broken, yet tough fold, the **eponychium** lies over the base of the nail. Nail bed is less vascular near the base producing a whitish moon or **lunula**. In **ungulates**, the hoof (Fig. 9.25) is formed by heavy unguis, the

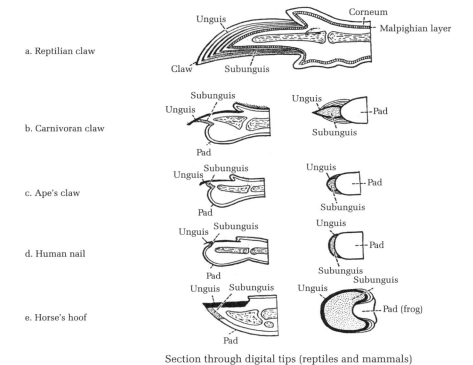

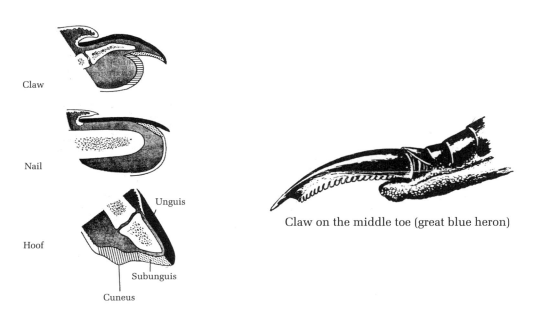

Figure 9.25
Sagittal section of claw, nail, hoof and terminal phalanx of mammals

Figure 9.26
Wing of young hoatzin shows claws on first two digits

subunguis is much softer and there is a very large, soft keratin pad (cuneus) which forms just behind the subunguis. At the other extremes are the **primates**, whose claws are so greatly modified that they are called nails, some **lemurs** have a specialized condition, with nails on all the digits except the second of the hind foot which retain a claw.

Hooves These are characteristic of horse, cow, sheep, pig, giraffe and rhinoceros etc. No part of the digit touches the ground. Unguis is large and the animal actually walks over it, and the subunguis is carried beneath it. **Fleshy pad** or **torus** of the digit lie behind the hoof and is called the **frog**. Frog or cuneus is composed of hard keratin, unguis wear away slowly.

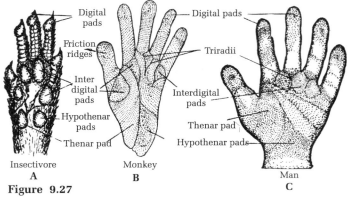

Figure 9.27
Tori and friction ridges on the hands A. insectivore; B. Monkey; C. Man

Modifications of the stratum corneum: Due to environmental stress and strain mammals exhibit various callus-like pads that develop at sites where friction occur (Fig. 9.27). The **ischial callosities** of monkeys are thick, cornified pads of epidermis, covering the ischial portion of the hip bone, on which the monkey sits. Camels have **kneepads**, which absorb some of the shock when the animal kneels preparatory to lying down.

Tori are epidermal pads on the tips of the digits (apical pads) and on the palms and soles of numerous species (Fig. 9.27). The cat is able to 'Pussy foot' by retracting the claws and walking silently upon the apical and interdigital pads. **Tori** are prominent on the human hands but they disappear before birth. **Corns** and **callouses** are thickenings of the **stratum corneum** that develop where the integument is subjected to continued friction.

Figure 9.28
Plates of Baleen hanging from upper jaw, Whale bone whale

Baleen Toothless whales have great frayed horny sheets of oral epithelium suspended from the hard plates (Fig. 9.28). As many as 370 of these sheets of **baleen** have been counted in one whale. The apparatus serves as a massive strainer, permitting only minute food-stuffs the sole nourishment of **whale bone whales**, to continue into oesophagus.

Feathers Unique of the class aves, feathers are light, strong, water proof and admirably suited to the functions of flight and insulation. They are basically an elaborate modification of β-keratin epidermal scales. There are three basic types of feathers (i) largest and most prominent are **contour feathers** or **plumae** (sing. pluma), (ii) **down feathers or plumulae** (plumula) and (iii) **hair feathers** or **pinfeathers** or filoplumes (Fig. 9.29). The **contour feathers** include the quill feathers of the body and the two largest groups of flight feathers the **remiges**–on the wing and **rectrices** on the tail.

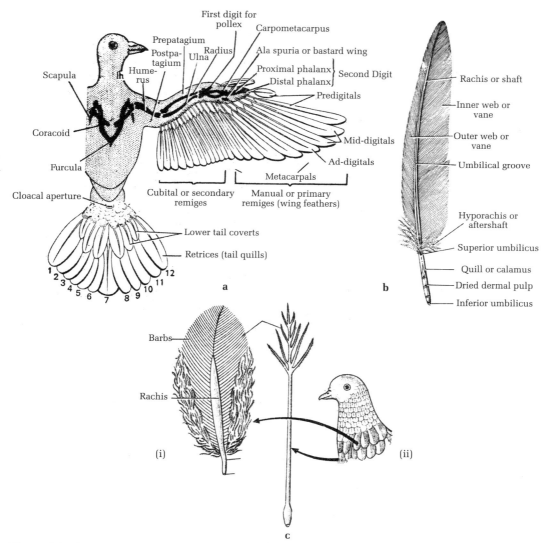

Figure 9.29 A
a. Arrangement of remiges and retrices in pigeon;   **b.** Structure of a typical flight feather pigeon;
c. Arrangement of (i) contour feather; and (ii) filoplume

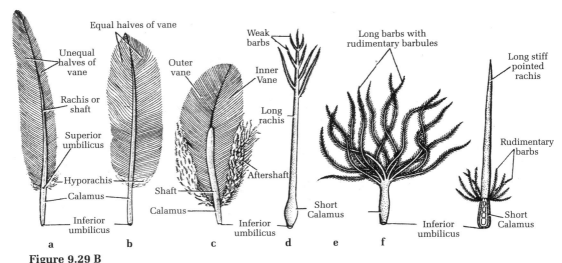

Figure 9.29 B
Kind of feathers (pigeon). **a.** Remiges; **b.** Rectrices; **c.** Contour; **d.** Filoplume; **e.** Down; **f.** Rictal bristle (not found in pigeon)

The shaft of the **contour** is composed of the **calamus**, which is embedded in the feather follicle and protrudes a short distance beyond it, and the **rachis**, which supports and is actually a part of the feathers vane. Deep in the follicle at the tip of the calamus is a hole, the **inferior umbilicus**; through which the growing feather is nourished. A second hole the **superior umbilicus**, is present at the base of the vane where calamus becomes **rachis**. From the superior umbilicus usually protrudes a 'secondary feather' the **after shaft**. The after shaft is small in most birds, but in ostrich, it may be as long as main feather itself. The vane of the feather is made up of rachis, barbs, barbules and hooklets.

Tapering rachis divides the vane into its two webs. The barbs, each attaching to the rachis are the largest elements of the web. On each barb, there are two rows of many smaller barbules; one row is along the surface of the barb that faces the feather tip and the other is along the opposite surface.

Thus the barbules from the anterior surface of one barb cross at right angles the barbules from the posterior surface of the next barb. The barbules on the side toward the tip have microscopic hooklets which firmly hold the crossing barbules. Contour feathers are important in flight. During the wings down stroke—the vanes of neighbouring wing feathers close together and present a common continuous, and almost impervious surface to the air.

During upstroke the smooth integumentary muscles attached to the feathers slightly separate the wing feathers, letting air pass through with less resistance. It is the interlocking barbules with their hooklets that make the web so impervious and yet capable of instant repair after they have separated.

Down feathers lying between the quill feathers over most of the body. These much softer feathers are particularly abundant on the breast and abdomen, and are important in the **incubation** of eggs, protection of young, and retention of body heat. Very young birds usually lack **contour feathers** and are covered at hatching by a complete coat of down. Down feathers are ancestral to the more complicated contour feathers. They have a short calamus, with a crown of barbs arising from the free end. Hooklets are lacking.

Filoplumes These superficially resemble hair, lack vanes and consist chiefly of a thread like shaft. They are usually scattered throughout the skin between the contour feathers. In peacocks they are very long, the function of these feathers, if any is not known.

Plumage The distribution of three types of feathers constitutes the birds plumage. The young birds first plumage, called the nestling down, **or neossoptiles**, contains only the fluffy tips of emerging contour features. This temporary plumage wears off as the full quill feathers develop. Gallinaceous birds (e.g. chicken) are fully covered in nestling down when they hatch, but song birds and most others develop nestling down after being born almost naked. The nestling down is replaced by juvenile plumage, **messoptiles** the first full coat of feathers that includes quill feathers. It normally lasts through the first winter and is usually less brilliant in colour than subsequent plumages. Shortly before the first mating season the juvenile plumage is replaced by the nuptial plumage. This has the colouration of the adult bird and is often quite brilliant (**teleoptiles**). A post-nuptial plumage develops after the breeding season. Some birds develop both a nuptial and a post-nupital plumage each year; others retain the post-nuptial plumage unchanged after the first year of life.

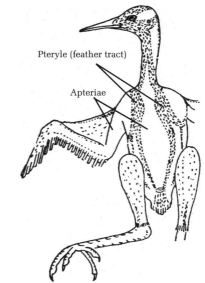

Figure 9.30
Feather tracts

Although from a casual glance it appears that the **contour feathers** occur over most of the birds body, they are actually attached in line, called **pterylae**, between which are areas called **apteriae** that have only down and hair feather (Fig. 9.30). Apteriae also occur immediately around movable joints, where heavy contour feathers would obstruct movement.

Development of Feather

Development may be followed in chick (Fig. 9.31) when the epidermis at certain places (the pterylae) show tiny projections, the rudiments of feather papilla. Within each papilla dermis extends and maintains a supply of material necessary for the development of feather. Feather papilla soon grows into a conical cylinder, at the same time, it skins into a depression, the annular groove, around the base of the papilla. It skins further to form the feather follicle. Papilla has in the centre a core of dermal pulp surrounded by the stratum germinativum and the cornified cells of the stratum corneum. Cornified layer is called the periderm or epitrichium and the developing feather at this stage is called **feather germ**. Actively dividing cells of the germinativum become arranged into longitudinal columns or ridges that will change into barb. A feather germ with rudiments of barbs in called a pin-feather and projects well above the skin. Activity in the stratum germinativum gradually comes to a stand still as the dermal pulp becomes progressively less nutritive. The periderm splits from the top and the barbs spread out. Part of the pin-feather within the feather-follicle becomes the quill or calamus maintaining a connection with the papilla below by the inferior umbilicus. Dermal pulp within quill dries into pith.

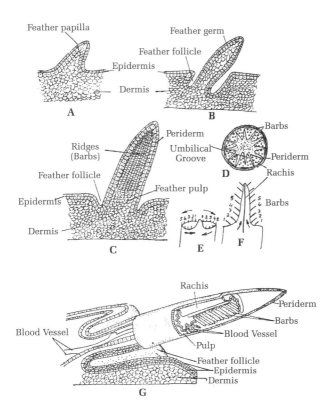

Figure 9.31
A, B, C, D, E, F, G are stages in the development of a contour feather

Hair A unique modification of the stratum corneum is found only in **mammals** and no where else in the animal kingdom. The hair may form a dense, furry covering over practically the entire body. It is represented by only two bristles on the upper lip in some whales.

The chief advantage of hair seems to be its insulating effect. In porcupines, hedgehogs, and spiny ant eaters it is stiffened into sharp quills and protects against would-be predators as well. The hairs are innervated basally, and thus serve as sense organs, this is particularly true of the specially stiffened hairs called **vibrissae** or **whiskers** in the facial region of many mammals i.e. cat.

Phylogenetic origin of hair The phylogenetic origin of hair is not known though it has been proposed that a hair may be a modification of an epidermal bristle that emerges from certain sense organs called apical pits in reptiles (Fig. 9.32). These sensory pits lie at the apices of epidermal scales singly or in groups as do hairs and the bristles emerge from the pits to extend beyond the surface of the skin. The similarity in the arrangement of sensory pits and hairs, and the sensory function of each, supports the **protothrix** theory of the origin of hair. Hair follicles are not distributed equidistant throughout the skin, but occur in isolated groups of two to a dozen or more. The number depends on the species. Certain **monkeys** exhibit group of three, **apes** five and **man** from three to five. In **armadillo** hairs actually lie in linear arrangement between rows of scales (Fig. 9.33).

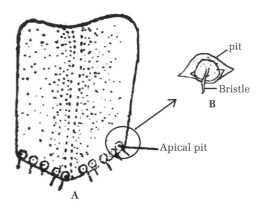

Figure 9.32
A lizard scale. Apical pits and bristle **A**. entire scale; **B**. a single pit enlarged

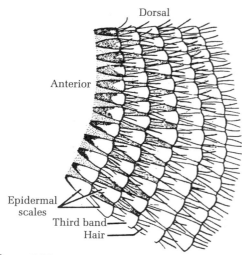

Figure 9.33
Armadillo showing epidermal scales and hair

Structure of hair The epidermal keratin shaft which projects from the skin is only part of the hair (Fig. 9.2). Its root is sunk into a deep pit on follicle, extending into the dermis. Opening into follicle is a sebaceous gland, and below the root, at the follicles base, is a small dermal papilla. Although they extend into the dermis, both follicle and hair are epidermal structure.

Inserting on the wall of each hair follicle in the dermis is a tiny smooth muscle the **arrector pili** or elevator of the hair (Fig. 9.2). When the arrectoris pilorium contract, the follicles and hair shafts are drawn toward a vertical position. The skin around the base of each hair is pulled into a tiny mound causing *goose flesh* or *chill bumps'*.

The erection of the hairs in many mammals makes the animal appear larger and thus more threatening. It also increases the insulating capacity of the fur.

Modifications of hair An extreme modification of hair is found in the toothless, whale or whale bone whale (Fig. 9.28). From the mouth edges hang huge, heavily keratinized, modified hairs (baleen) which act as strainer. The coarse hair on pig's back are modified to form stiff bristles, which have been used in the manufacture of brushes. The **quills** or spines of **porcupines** and **spiny ant-eaters** are rigid, needle like hairs. Two spectacular modification of hair are horns of rhinoceros and scales of scaly ant-eaters.

Pelage Just as birds have several plumages, mammals have more than one pelage. The first is the **lanugo** – hair which is frequently developed and usually also lost in the uterus. Hence most mammals are borne quite naked except for vibrissae.

1. Next is the juvenile pelage, often not the same colour as the adult pelage. At about puberty, the juvenile pelage is replaced by the adult pelage.

2. Adult pelage appears often in an orderly moult which start at one end of the animal, and progresses to the other, and is marked along its way by a distinct line between the new and the old hair.

Development of Hair

Hair develops from the epidermis nearly the same way as a feather. A small thickening of the epidermis appears and soon pushes down into the dermis as a column of cells called hair germ (Fig. 9.34). Dermis beneath the hair-germ becomes active, its mesenchymal cells concentrate and push an inverted cup shaped outgrowth that becomes the hair-papilla. Nourishment of the developing hair is brought about by blood capillaries penetrating the papilla. Layer of stratum germinativum lying immediately over the papilla actively produces new cells that cornifies as they are pushed up. At the same time a split occurs in the hair germ, separating a central core of cornified cells from an outer layer of stratum-germinativum.

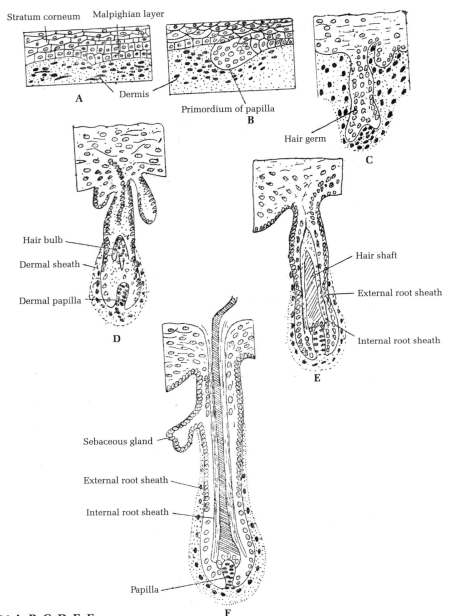

Figures 9.34 A, B, C, D, E, F
Various stages in the development of hair

Cornified cells are destined to from the shaft and the outer layer of stratum germinativum to form the hair follicle sheaths. The shaft grows as more cornified cells push it from below, eventually it pierces through the skin surface. Base of the shaft immediately over the papilla remains bulb-like and active and is responsible for the growth of the shaft. Some mesenchymal cells of the dermis outside the follicle become differentiated into smooth muscle fibres – the arrector pilli.

Functions of integument

1. **Protection** of the organism form injurious external influences. Few examples are as follows.

   a) **From mechanical injuries** Skin and its derivatives such as scales, feathers and hair, and sub-dermal fats offer protection from injuries resulting from pressure, friction and blow.

   b) **From the access of bacteria and fungi** Connective tissues of the dermis contain a variety of cells that remove foreign bodies, fight infection, take part in the healing of the wounds and the replacement of damaged tissues. Saline sweat in mammals does not favour the survival of bacteria and fungi. Of similar use is the mucus in fish and amphibia.

   c) **From the loss and intake of water** The water is of prime importance for the land vertebrates, in them almost impermeable stratum corneum makes the skin water proof and checks not only the loss of water from the body but also does not permit absorption through integument.

   d) **From the enemies** The derivatives such as osteoscutes, scales, hair and their modifications and feathers, to some extent, protect the body from the attacks of enemies.

   e) **Mimicry** Colour of the skin and the derivatives may conceal the animal in the environment or blend its outline. Changes of colour are caused primarily by position of the pigment granules within the highly branched pigment cell (Fig. 9.35). When concentrated around the centre of the cell near its nucleus, the pigment forms too small a spot to be perceived by the naked eye. However, when it is dispersed far out into the cell, it occupies a much larger area and thus can be easily detected. These colour changes are controlled by stimuli, such as light, hormones, or by a combination of these factors. In birds and mammals the change of colour cannot take place reflexly because the pigment granules within the cells cannot aggregate and disperse quickly. These animals can change their colour only when

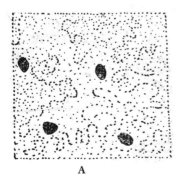
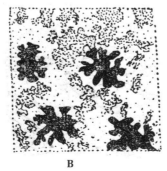
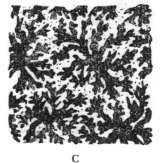

A   B   C

Figure 9.35
Patterns of pigment concentration and dispersion in chromatophores of vertebrates **A.** pigment concentrated and localized in the cell; **B.** and **C.** pigment highly dispersed throughout cytoplsm

pigment granules are synthesized or when these feathers or hairs replace the old ones, these are morphological colour changes whereas dermal or chromatophores as in chameleons are responsible for rapid colour change which is called physiological colour change. Uses of colour in fish are (i) **cryptic** – these are concealing colours which help fish in assimilation with background and conceal effectively from predators and enemies (ii) **sematic** – these are warning colours which make these animal look prominent because such colours usually are associated with unpleasant taste or special defence mechanism. (Such colours give warning to the enemies or predators.) (iii) **epigamic** – these are sex colour usually found in males and make them more brilliant during breeding season which helps them in attracting females.

2. **Exteroception** Free nerve-endings and special cutaneous receptors in the integument give warnings of heat, cold, contact or damage. Mucus membranes in the nostrils and over the tongue give a variety of informations.

3. **Respiration** Nearly all amphibia respire through skin, a thin film of water is retained over the surface of the skin with the help of mucus; exchange of gases takes place through the water film. In certain fish, such as eels and periophthalamus, skin is an additional organ of respiration.

4. **Regulation of body temperature** : Integument conserves heat, loses the excess and regulates body temperature. In cold-blooded (poikilothermous) animals such as frogs, evaporation of water from the surface helps keeping the body temperature from rising above the tolerable limits. Cold makes the frogs dormant and forces them to hibernate. The same is true of reptiles.

To keep themselves active warm-blooded animals, the birds and the mammals, produce and conserve heat and prevent its loss through various means.

Hair and **feathers** tend to trap a layer of air which insulates the body, reducing heat exchanges with the environment. Interlocking mechanism developed in feathers and in hair such as those of sheep and some rabbits increases the heat-retaining capacity.

A layer of nonconducting fat in the dermis serves the same purpose in mammals (whale, sea-cows), which have a few hair or no hair. Equally important are the means and ways for producing heat. Heat is chiefly produced by muscular activity. In all mammals, fall in external temperature induces extra muscular activity by shivering. Liver and other organs give out heat as they work. Excess heat is lost with the help of the following.

(i) The **sweat glands** When blood supply is increased and sweat output is increased, evaporation of sweat from the surface involves heat loss.

(ii) The **lungs** Heat is lost with the expired air, it is for this reason that dogs which have few sweat glands breathe faster to get rid of excess heat from the wet surface as tongue and upper respiratory tract.

(iii) **Special areas** When it is hot, flow of blood is increased in the ear pinnae of elephant which are then raised and flapped. An elephant increases its surface area this way by nearly one-sixth.

The case of **insectivorous** bat is different as they have no mechanism of control of the amount of heat produced. Every time a bat hangs up and goes to sleep its temperature falls and within an hour it reaches that of the surroundings.

Hibernation is practiced by several mammals too. Ground squirrels, chipmunks, marmots undergo hibernation. During hibernation all bodily metabolism drops to a low level the respiration and heart beats are slowed, and the body temperature falls considerably.

5. **Integument stores food** Large amount of fat is deposited in the sub-dermal parts of the skin. Sub-dermal fat of the whale, called blubber is about 45 cm thick. Besides insulating the body, fat serves as reserve food.

6. **Integument helps in locomotion** Wing membranes of bats, flying lizards, flying frogs, flying squirrels and lemurs, webs in the feet of frogs and ducks, are folds of integument that help in propulsion or locomotion.

7. **It contributes to sexual selection** Bright colour and bold patterns are related to sexual habits and apparently attracts mates and perhaps establish territories.

8. **Integument** also gives rise to various other modifications such as the several forms of **claws, hooves, horns, antlers and even rattle snakes** warring rattle, all confer advantages in the struggle for existence.

Chapter 10

Skeletal System

The term skeleton refers to the framework of the animal body. In vertebrates it is composed of cartilage, bone or a combination of the two and serves for support, attachment of muscles, and the protection of delicate vital organs which it more or less completely surrounds. The skeletal tissues are of three types the notochord, cartilage, and bone. Each type has a distinct structural characteristics, but mainly of living cells in the matrix secreted by them. New skeletal structures can always be formed and reabsorbed, and the shape of those formed can be changed in response to the external and internal stresses.

Notochord

A flexible, rodlike structure, is the primary axial skeleton in early development of all vertebrates. By the pharyngula stage (Fig. 10.1), the mid line tissue which lies just dorsal to the endoderm and ventral to the nerve cord gets differentiated into notochordal tissue. Unlike other skeletal tissue most of the notochord is cellular with a peripheral portion consisting of the collagen elastic fibres and has tough outer connective tissue sheath (Fig. 10.2). Large turgid, vacuolated cells fill its central core, and together with the connective tissue sheath give the notochord its characteristic stiff flexibility. In a few adults the notochord is the primary axial skeleton, but in most adult tetrapods it gets replaced to a greater or lesser extent by the vertebral column.

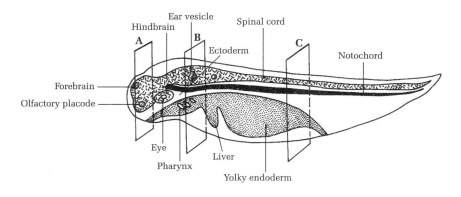
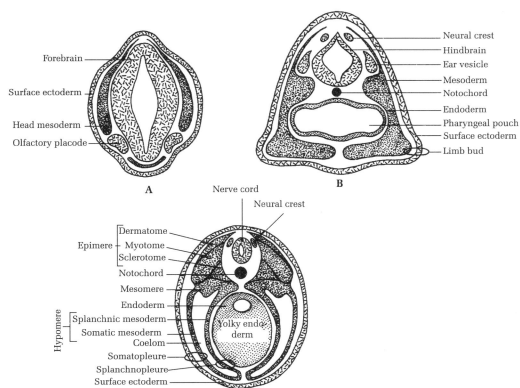

Figures 10.1 A, B, C
Pharyngula stage amphibian
A. T.S. forebrain **B.** T.S. region ear vesicle **C.** T.S. middle region of the body

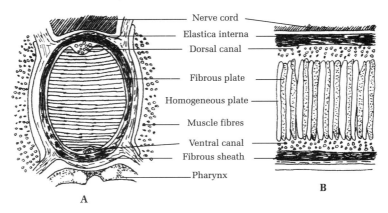

Figures 10.2 A, B
Amphioxus A. T.S. Notochord B. L.S. Notochord

Cartilage

Cartilage develops embryologically in all vertebrates, but in the adult, except agnatha and chondricthyes it is replaced to a greater or lesser extent by bone. It is a milky, clear-to-opaque substance moderately flexible and elastic. The primary or parenchymal cells of cartilage are chondrocytes. They are embedded in a gelatinous matrix of acellular material which is secreted by chondrocytes. The matrix, contain sulfated mucopolysaccharide which are acidic in nature and called chondroitin sulfate.

Types of cartilage

There are four types of cartilages:

1. **Hyaline cartilage** It has a matrix of almost pure chondroitin sulfate (Fig. 10.3 A).
2. **Fibrous cartilage** It has a dense, tough matrix containing collagen fibres. It is typically found in areas of considerable stress, such as at the attachment points of ligament (Fig. 10.3 B).
3. **Elastic cartilage** It has elastic fibres rather than collagen fibres in the matrix; it is best known in the mammalian pinna (Fig. 10.3 C).
4. **Calcified cartilage** Sometimes when cartilage is under great compression forces, as in some of the large chondricthyean fishes calcified cartilage is formed. The cartilaginous matrix becomes infiltrated with calcium salt, making the entire cartilage very hard and brittle (Fig. 10.3 D). Although it superficially resembles bone, it is technically cartilage because it contains chondrocytes and chondroitin sulfate. Unlike most other tissues, cartilage has no direct vascular supply. Blood vessels run in the thin connective tissue covering the perichondrium. The useful chemical substances and nutrient reach the chondrocytes by diffusing through the matrix.

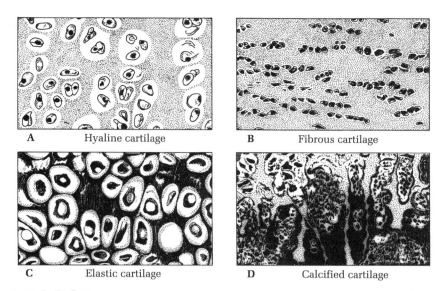

Figures 10.3 A, B, C, D
Cells of all cartilage are chondrocytes. It is the matrix which determine the type of cartilage

Origin of Cartilage   The cartilage can develop from two sources:

I. **Neural crest origin**   In the pharyngula head region, the neural crest element i.e. visual skeleton and anterior neurocranium is especially large, and much of it migrates to the anterior splanchnic mesoderm (Fig. 10.1). This tissue, is called mesectoderm and differentiates into the cartilaginous visceral skeleton which supports the gill skeleton and its derivatives.

II. **Mesodermal origin**   However, most of the cartilage has mesodermal origin, some differentiates from epimeric mesoderm in the head region; next to the anterior notochord or in the prechordal area anterior to the notochord (Fig. 10.4). Some of these cartilages form protective capsules around the sense organs, and others form the floor, walls, and some of the roof of the brain case. Post cranially, the epimeric mesoderm adjacent to the nerve cord and notochord is called the sclerotome, which gives rise to cartilage for vertebrae. The more lateral mesoderm may also give rise to ribs, and the more ventral to the sternum. Cartilages of the appendages and their girdle support form as budlike enlargements of the somatic mesoderm in the presumptive limb regions.

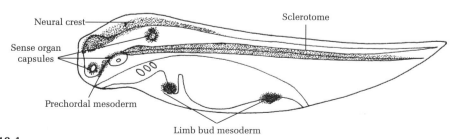

Figure 10.4
Pharyngula stage showing – differentiation of structure into cartilage

Functional limitations of cartilage

Although cartilage develops embryologically as one of the earlier skeletal tissues and is more abundant in fish than in tetrapods, it should not be regarded as a primitive tissue. Indeed in many situations it is highly adaptive, because it is flexible, it is also resistant and less apt to fracture. On the other hand, it has two potential draw backs:

1. Normal cartilage, being less resistant to compression than bone, cannot adequately protect the soft parts of animals beyond a certain critical size. Though calcified cartilage can, but calcified cartilage is heavier and more brittle than bone, and therefore less adaptive. Interestingly, the critical size differs for land and water dwelling animals. The land animals, must support and propel themselves without the buoyancy of water and hence, require larger muscles and limbs than aquatic animals of equal size. Not surprisingly, then, it is in two classes of fish which possess entirely cartilaginous skeleton, and calcified cartilage is found only in the largest Chondrichthyeans.

2. The second major draw back to cartilage is that it cannot act as a reservoir for calcium and phosphate. This is not important for marine animals which have an abundant supply of these minerals in their external milieu, but it is important for freshwater and terrestrial forms, and is no doubt another reason why in most classes bone replaces cartilage as the primary skeletal element.

Bone

Bone is hard, rigid tissue of limited elasticity; which is opaque unless thin. It is composed of parenchymal cell, the osteocyte, embedded in a secreted acellular matrix. This matrix is composed of complex inorganic salts (phosphate and carbonate), which combine with calcium to form apatite crystals, and of much organic material which consists primarily of collagen. Within this matrix the cells lie in geometrically organized lacunae or spaces (Fig. 10.5). Bone is highly vascularized, both centrally, where the bone marrow frequently manufactures red blood cells, and peripherally, where blood vessels enter the bone in canals called nutrient foramina. Thicker bone are organized into internal columns called **Haversian canal systems.** In the centre of each column is a Haversian canal in which travel blood vessels and sometimes nerve fibres that enter and leave through transverse canals of **volkman.** The osteocytes within their lacunae remain in nutrient communication with the Haversian canals and with other osteocytes by way of tiny anastomosing canaliculi. This concentration of vascular and neural tissue is responsible for the high metabolic rate and plasticity of bone. In a section of any bone one

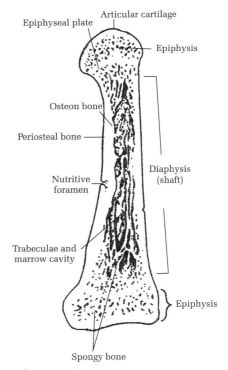

Figure 10.5
L.S Metatarsal bone

finds an outer portion called compact or **lamellar bone** and an inner portion called **cancellous bone.**

Compact or lamellar bone It is dense, hard, and solid, accounting for most of the bone's weight, mass, and rigidity. Externally it is composed of broad layers parallel to the surface; having no Haversian canal systems. There are typically two such layers, at right angles to one another, called the outer and the inner circumferential lamellae. Deep to these layers, the compact bone is organized in long cylindrical Haversian canal systems, or osteons. The osteons are tied together by less regularly shaped columns of compact bone called interstitial lamellae. Cancellous bone is more hollowed out, with trabeculae between which are spaces occupied in the living animal by bone marrow. A careful study of the trabeculae demonstrates that they are oriented much as bridge struts giving maximal strength to the bone.

Development of bone

All adult bones are structurally similar, they are classified according to their histogenesis as dermal bone, cartilage replacement bone, and heteroplastic bone.

Dermal bone Dermal bones are formed in the presumptive deep layer, or dermis of the skin by a process called intramembranous ossification, e.g., clavicle and flat bones of skull. In this process, embryonic mesodermal cells differentiate into **osteoblasts** and **fibroblasts**. The fibroblasts then lay down a network of collagen and reticular fibres, and upon this network the osteoblasts secrete the appetite crystals of calcium, phosphate and carbonate. It is this secretion of the appatite crystals by the osteoblasts that is called **ossification**. Each osteoblast continue to secrete appatite crystals until it is completely encased in a matrix; then its active secretion is suspended and it is called an osteocyte. By the continued activity of fibroblasts and osteoblasts, spicules of bones are formed in broad, flat, plate like structures within the presumptive dermis. Soon these differentiates from the dermal mesenchyme into another types of cell, the **osteoclast**, which is multinuclear in nature and is capable of reabsorbing or dissolving the matrix of the bone. The osteoblasts lie mostly along the surface of the bone, forming the inner portion of the periosteum which covers the bone. The osteoclasts lie deeper. The bone develops and continues to grow as the osteoblasts lay down new layers of circumferential lamellae, and the osteoclasts reabsorb the matrix of former outer circumferential and inner circumferential lamellae, reforming these lamellae into osteons, or Haversian canal systems; deeper into the bone, osteoclasts erode the bony matrix of former osteons, forming marrow cavities lined with fine bony trabeculae. Thus the major growth of **membrane**, or dermal bone occurs on the outer surfaces, with resorption and remodelling of bone occurring in the deeper portions. This growth along the surface is called periosteal ossification.

Cartilage replacement or endochondral bone It comprises parts of the brain case, visceral arch skeleton and, post cranially most of the vertebral column, ribs, sternum, girdles, and appendages skeleton. Its embryonic development involves a cartilage precursor: a miniature model of the bone which is to be, usually including its processes, turns and shapes. Cartilage cells and cartilage matrix are never transformed into bone; rather they are replaced by bone cells and bone matrix of appetite crystals. In a typical long appendage element such as the humerus, the replacement process begins centrally as the chondrocytes expand and the matrix begins to calcify (Fig. 10.6). When calcification is complete, the chondrocytes die. The area is then invaded by blood vessels, and with them, mesenchymal cells, some of which differentiate into osteoblasts and then start laying down bony spicules; at the same time the calcified cartilage is resorbed by other cells of the invading vascular tissue. The first product of cartilage replacement process is a spongy cancellous bone in the centre of the shaft. This bone increases in diameter, and in time the outer portion of the shaft becomes compact bone in the layers already described, and growth in girth then continues by periosteal ossification as it does in membrane bone. The two ends of a long bone, such as humerus remain cartilaginous until much of the shaft is ossified. Then calcified cartilage forms centrally in each end, this calcified cartilage is invaded by vascular tissue and the laying down of cancellous bone and replacement of the cartilage occurs. The extreme tips, or articular facets form the joints that do not ossify. The ossifying ends of the bone are called the **epiphyses**, and the shaft is the diaphysis. Between the diaphysis and each epiphysis, where the epiphyseal centre of growth and the diaphyseal centre of the growth meet, is the metaphyseal plate (Fig. 10.6). This is the only area where the bone can grow in length, which it does by the process of endochordal ossification. In this process, the cartilage along the metaphyseal plate becomes calcified; the cartilage is eroded; osteons of bone are laid down; and new cartilage is formed along the plate. The metaphyseal plate itself remains cartilaginous as long as the bone is capable of growth.

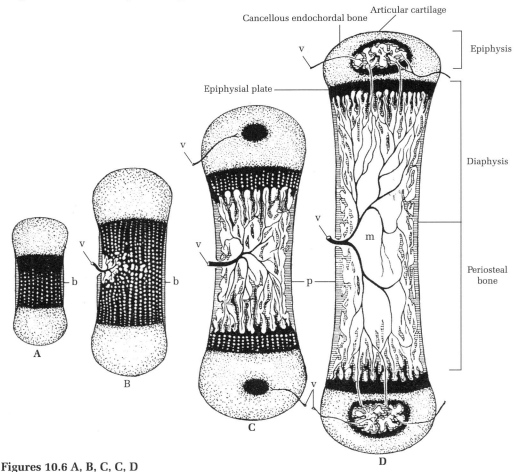

Figures 10.6 A, B, C, C, D
Formation of a mammalian long bone—Black portion—calcified cartilage, Stippled—hyaline
v – Vessel, m – Marrow cavity, b – Osteoblast, p – Periosteal ossification

Heteroplastic or Heterotopic bones

These are formed within the other already differentiated tissues. Some heteroplastic bones are abnormal apparently being produced by the physiological response of fibrous connective tissues to extreme stress. In fact, under proper stimulation bone can form in almost any tissue including heart and kidneys. Heterotopic bones, and also cartilages are especially common in mammals (Fig. 10.7). They are usually missing from routine skeletal preparations. Among heterotopic bones are the **os cordis** in the interventricular septum of the heart of the ungulates and the **baculum** (Os priapi or Os penis) embedded between the spongy bodies in the penis of bats, rodents, marsupials, carnivores, insectivores, bovines and lower primates. The baculum reaches a length of 60 cm in walruses. An **os clitoridis** is embedded in the female penis (clitoris) in others, several rodents, rabbits and numerous other female mammals. **Osseous** or cartilaginous tissue forms in the wall of the gizzard in some doves. The syrinx of birds often develops an internal skeletal element, the **pessulus**. At least one species of bat has bone in the tongue. Bone develops in the gular pouch of South American lizard, in the muscular diaphragm of camels, and in the upper eyelid of crocodilians (adlacrimal, or palpebral bone). A similar plate of connective tissue, the tarsus develops in man. A rostral bone develops in the snout of a number of mammals including pigs, and a cloacal bone in the ventral wall of the cloaca of some lizards. Epipubic (marsupial) bones are embedded in the ventral body wall of some monotremes and marsupials. Other heteroplastic bones are normal to a species; these are called sesamoid bones, e.g., the **patella** (knee cap). **Splint bones** are heteroplastic bones formed in the tendons or ligaments of older animals; in horses and some others, they are normal, sesamoid bones, while in human beings they are abnormal, and can be quite troublesome, and sometimes are surgically removed.

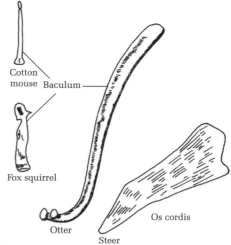

Figure 10.7 Heterotopic bones

Where two skeletal elements meet, a joint is formed. Some times, the skeletal elements are held together by connective tissue bands known as **ligaments** (as opposed to tendons, which join a muscle and a skeletal element). Tendons and ligaments are made of thick closely packed bundles of collagen between which are present connective tissue cells called fibroblasts.

Tendons connect muscles with bone; when a muscle contracts, the pull on the bone is exerted through the tendon. In accordance with this mechanical requirement the collagen is in parallel bundles and the fibroblasts lie in rows between them. **Ligaments** connect bone to bone, and the collagen bundles have a less regular arrangement. Ligaments and tendons that become flat and very wide are sometimes called **aponeuroses** (The term ligament is sometimes applied to fibrous connective tissue membranes or cord that hold visceral organs in place, for example, the **falciform ligament** and round ligament of the ovary. (However these are not skeletal structures.)

In some species tendons and ligaments normally become mineralized in one or more locations. Turkeys for example, have ossified tendons in their legs. Ornithischian dinosaurs had ossified tendons millions of years ago. **Sesamoid** bones or cartilages (so named because they resemble sesamoid seeds) are nodules of bone or cartilage that form in tendons or ligaments. Best known is the patella or knee cap which is endochondral in some species and intramembranous in others.

Joints

A joint is the site where two bones (or cartilages) meet. There are three different types of joints, classified by the degree of movement possible between the elements (Fig. 10.8).

Synarthrosis If there is no movement, as is usually the case in the skull, the joint is called a synarthrosis (Fig. 10.8 A). The two elements are usually connected by heavy, collagenous fibres, or some times by fibrous cartilage. However sometimes bones meet and become united by collagen and hydroxyapatite crystals that obliterate the suture. This condition is said to be an **ankylosis.** The premaxillary and maxillary bones of the human embryo ankylose and as a result, the premaxillary cannot be distinguished as a separate bone in adults.

Amphiarthrosis or symphysis is a joint in the mid line, in which two bones are separated by fibro cartilage and in which movement is severely limited (Fig. 10.8 B). Such joints are found in the vertebral column and at the symphysis pubis. The latter joint become a bit more movable by hormonal dissolution of the fibro cartilage shortly before labour pain begins in female mammals.

Diarthrosis If the joint is movable in more than one plane (Fig. 10.8 C). It is in this joint the two elements usually meet in cartilage covered articular facets, whose shape helps to determine the direction and extent of movement. It is also determined by the structure and the placement of the ligaments. In many diarthrotic joints an epithelial tissue called the synovial membrane lies within the joint cavity; it secretes and encloses the synovial fluids that lubricate the joint. It is not unusual to have a lemniscal fibrous cartilage within the synovial cavity as well, acting

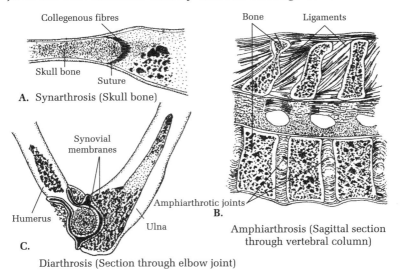

Figures 10.8 A, B, C
Structure of joints. **A.** Synarthrosis **B.** Amphiarthrosis **C.** Diarthrosis

as a pad for the joint. Synovial membranes are not limited to joints. Tendons which extend for long distances are frequently enclosed in synovial membranes and are lubricated by the synovial fluids.

Skeleton and its components : The skeleton may be divided into:

1. Axial Skeleton and
2. Appendicular skeleton

1. **Axial skeleton** As the name implies, the axial skeleton comprises bones which form the main axis of the body. It is made up of (a) the **skull,** (b) the **vertebral column,** (c) the **ribs** and (d) the **sternum**. Because of its topographical relationships sternum is considered to be a part of the axial skeleton although its embryonic origin suggests that it is a portion of the appendicular skeleton. The **visceral or branchial** skeleton, consisting of elements which in fish support the gills and jaws, contributes to the skull in higher forms. For these reasons it is included in the axial skeleton although some authors, for various reasons, make a distinction between the visceral skeleton and the somatic skeleton, which includes all the remaining endoskeletal structures.

2. **Appendicular skeleton** It is composed of the skeletal elements of the paired anterior pectoral and posterior pelvic appendages together with their respective girdles by means of which the appendages are connected directly or indirectly with the axial skeleton.

Skull

Cephalization A characteristic of vertebrates is cephalization, i.e., the concentration of sense organs, brain, food-gathering apparatus, and much of the respiratory system in the head. The concentration of so many vital organs in one area increases their potential vulnerability and their need for protection. Too much protection, however, would deprive the animal of the very values of cephalization. To be adaptive, the head skeleton must protect the delicate organs; must allow the appropriate stimuli to reach the eyes, ears, and nose; and must facilitate food gathering and oxygen intake. It is not surprising that this is the most complex and intricate portion of the skeleton exhibiting much structural diversity because natural selection has shaped specific features to specific environments. Simply stated, the head skeleton has three (Fig. 10.9) structural and functional components.

1. The **neurocranium** or primary brain case or endocranium which supports the brain and the sense organs.
2. The **dermatocranium** which protects deep-lying delicate tissues.
3. The **splanchnocranium** or visceral skeleton which supports the respiratory and food-gathering apparatus.

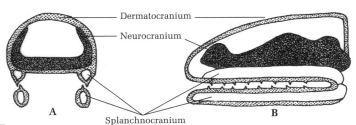

Figures 10.9 A, B
Components of skull— **A.** T.S. Skull **B.** Sagittal section

Neurocranium

Neurocranium is that part of the skull which protects the brain and certain special sense organs, it arises as cartilage and is subsequently partly or wholly replaced by bone except in cartilaginous fish. The neurocranium in all vertebrates develops in accordance with the basic pattern described below:

The cartilaginous stage The neurocranium commences as a pair of parachordal and prechordal cartilages underneath the brain. Parachordal cartilages parallel the anterior end of the notochord beneath the mid brain and hind brain (Fig. 10.10 A).

Prechordal cartilages (also called trabeculae cranii) develop anterior to the notochord underneath the forebrain. The parachordal cartilages expand across the mid line towards each other and unite. In the process, the notochord and parachordal cartilages are incorporated into a single, broad, cartilaginous basal plate. The pre-chordal cartilages likewise expand and unite across the midline at their anterior ends to form an ethmoid plate (Fig. 10.10 C).

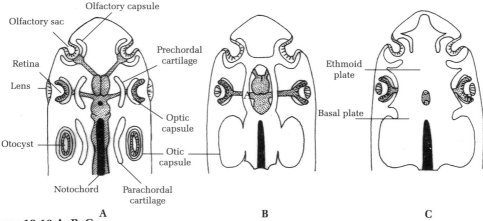

Figures 10.10 A, B, C
Development of a cartilaginous neurocranium (ventral view)
A. Notochord underlying the midbrain and hindbrain **B.** Notochord incorporated into caudal floor **C.** Cartilaginous floor completed beneath the brain

Sense capsules While parachordal and prechordal cartilages are forming, cartilage also appears in two other locations:

(i) As an **olfactory capsule** – partially surrounding the olfactory epithelium and

(ii) As an **otic capsule** – completely surrounding the otocyst, which is the developing inner ear.

The olfactory capsules are incomplete anteriorly since water (in fish) or air (in tetrapods) must have access to the olfactory epithelium. The walls of the olfactory and otic capsules are perforated by foramina that transmit nerves and vascular channels.

(iii) An **optic capsule** forms around the retina but it is not the orbit, or skeletal socket, in which lies the eyeball. Although this capsule is fibrous in mammals, cartilaginous or bony plates very often form a scleral ring within the sclerotic coat. The ring helps to maintain the shape of the eyeball. This is an ancient condition, having been present in crossopterygians and extinct amphibians and reptiles. Because the optic capsules does not fuse with the neurocranium, the eyeball is free to move independently of the skull. Therefore, the sclerotic coats is not conventionally considered a part of the neurocranium.

Completion of floor, walls and roof The expanding ethmoid plate fuses anteriorly with the olfactory capsules, and the expanding basal plate fuses with the otic capsules that lie lateral to the hind brain. The ethmoid and basal plates also expand toward one another until they meet to form a floor on which the brain rests. Further development of the cartilaginous neurocranium involves construction of cartilaginous walls alongside the brain and in lower forms, a cartilaginous roof over the brain. The cranial nerves and blood vessels are already present by this time, and the cartilage is deposited in a manner that leaves foramina for these structures. The largest is the foramen magnum in the rear wall of the neurocranium (Fig. 10.11). In cartilaginous and lower bony fishes the brain is completely covered by a

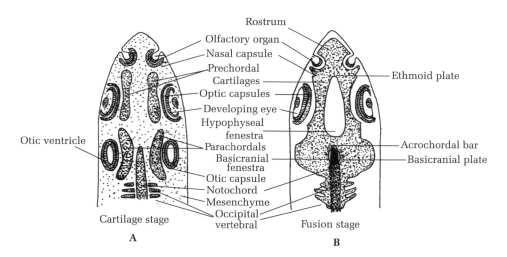

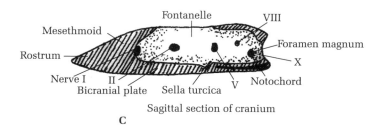

Figures 10.11 A, B, C
Development of skull **A.** Cartilage stage—chondrocranial cartilage surrounds the special sense organs **B.** Fusion Stage-all cartilages except optic capsules have fused to form chondrocranium **C.** Sagittal section of cranium

cartilaginous roof. But in teleosts and tetrapods the brain is never completely roofed over by cartilage. The preceding pattern of development recurs throughout the vertebrate series and produces a cartilaginous neurocranium that protects much of the embryonic brain, the olfactory epithelia, and the inner ear. The blastema that gives rise to the cartilaginous neurocranium is a contribution chiefly of ectomesenchyme with sclerotomal contributions in occipital region.

Adult cartilaginous neurocrania

Cartilaginous fish retain a cartilaginous neurocranium sometimes called a chondrocranium, throughout life. The neurocranium in these fishes completely encloses the brain, and the otic and olfactory capsules are fused into it along with the notochord. Dorsally, an endolymphatic fossa is perforated by endolymphatic and perilymphatic ducts and there is an opening, the pineal (epiphysial) foramen, that is occupied by the pineal body (Fig. 10.12). Among bony fish the lungfish and most ganoids retain a highly cartilaginous neurocranium throughout life. In order to see it, the membrane bones that lie over it must be stripped away.

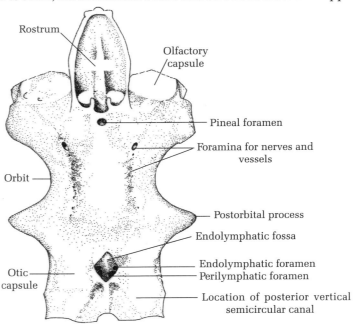

Figure 10.12
Neurocranium of *Squalus* (Dorsal view)

In cyclostomes the several cartilaginous components of the embryonic neurocranium remain in adults as more or less independent cartilages. An olfactory capsule (median, protecting the median olfactory sac), otic capsules, a basal plate, a notochord (not fused with the basal plate), and other cartilages not homologizing with those of gnathostomes can be identified. The roof above the brain remains fibrous.

Neurocranial ossification centres

In bony vertebrates the embryonic cartilaginous neurocranium is mostly replaced by replacement bone. The process of notochordal ossification occurs more or less simultaneously at numerous separate ossification centres. Although the specific number of such centres varies in different species, four regions are universally involved (Fig. 10.13). These regions are occipital, sphenoid, ethmoid, and otic regions.

Occipital centres The cartilage surrounding the foramen magnum may be replaced by as many as four bones. One or more endochondral ossification centres ventral to the foramen magnum to produce a basioccipital bone underlying the hind brain. Centres in the lateral walls of the foramen magnum produce two exoccipital bones. Above the foramen a supraoccipital bone may develop. In mammals all four occipital elements usually fuse to form a single occipital bone (Fig. 10.13 A). In modern amphibians, one or more of these may remain cartilaginous, although they were bony in stem amphibians.

The neurocranium of tetrapods articulates with the first vertebra via one or two occipital condyles. Stem amphibians had a single condyle bone, chiefly on the basioccipital bone. Living reptiles and birds still have a single condyle. Modern amphibians and mammals diverged from the early tetrapod condition with two condyles, one on each exoccipital.

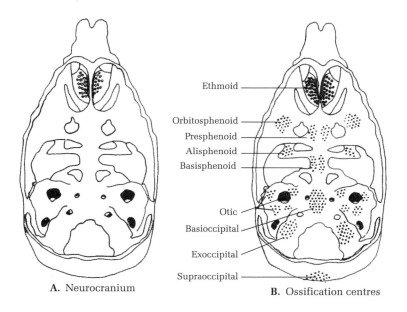

Figures 10.13 A, B
A. Cartilaginous neurocranium of (embryo) Pig   B. Chief endochondral ossification centre

Sphenoid centres The embryonic cartilaginous neurocranium underlying the mid brain and pituitary gland ossifies to form basisphenoid bone and a presphenoid. Thus a bony platform consisting of occipital and sphenoid bones underlies the brain (Fig. 10.13). The side walls above the basisphenoid and presphenoid ossification centres form other sphenoid elements (orbitosphenoid, pleurosphenoid and others) and these may remain separate or unite with basisphenoids and presphenoids to form a single adult sphenoid bone with wings. The pituitary rests in the **sella turica** of the basisphenoid region. Since the embryonic neurocranium is usually incomplete dorsally, no replacement bones lie above the brain.

Ethmoid centres: The ethmoid region lies immediately anterior to the sphenoid and includes the ethmoid plate and olfactory capsules. Of the four major ossification centres in cartilaginous neurocranium (occipital, sphenoid, ethmoid, otic), the ethmoid more than the others tends to remain cartilaginous. Ossification centres in this region become the cribriform plate of the ethmoid, perforated by olfactory foramina, and several of the conchae, or turbinal bones (ethmo turbinals) in the nasal passageways of crocodilians, birds and mammals. Mesethmoid bones ossify in cartilaginous median nasal septum. In the anurans the sphenethmoid is the sole bone arising in the sphenoid and ethmoid regions.

Otic centres: The cartilaginous otic capsule is replaced in lower vertebrates by several bones with such names as protic, opisthotic, and epiotic. One or more of these may unite with adjacent replacement or membrane bones. For example, in frogs and most reptiles the opisthotic fuse with the exoccipitals and in birds and mammals the protic, opisthotic, and epiotics all unite to form a single periotic or **petrosal bone** (Fig. 10.14 B). The petrosal, in turn, may unite with the squamosal to form a temporal bone.

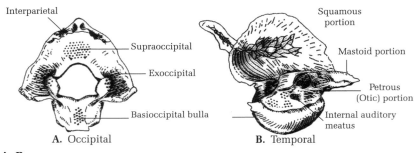

Figures 10.14 A, B
Endochondral ossification centre (dots) and intra membranous ossification centre (black reticulum) **A.** Caudal view **B.** Medial view

Dermatocranium (osteocranium)

In addition to the chondrocranium, which appears uniformly throughout the vertebrate series, several dermal plates or membrane bones contribute to the formation of the skull. Collectively, these dermal bones of skull constitute the **dermatocranium**. For convenience, the dermatocranium will be discussed under the following heads: (I) Bones that form a roof over the brain and contribute to the lateral walls of the skull.

(a) Roofing bones   (b) Bones of upper jaw
(c) Bones of the palates   (d) Opercular bones

(a) **Roofing bones :** In crossopterygians a series of paired and unpaired bones extend along the mid-dorsal line from the nares to the occiput, overlying the olfactory capsules and brain. In labyrinthodonts the unpaired bones were lost and a series of paired bones—**nasals, frontals, parietals, dermoccipitals** (post-parietals) resulted (Fig. 10.15 B). In the mid line between the two frontal or parietal bones was a parietal foramen. It is still present in many fishes, amphibians, and reptiles and houses the third eye. Forming the walls of the orbit in the basic pattern were lacrimal, prefrontal, post frontal, post orbital, and infraorbital bones (of which the jugal is one). At the posterior angle of the skull were intertemporal, supratemporal, tabular, squamosal, and quadratojugal bones. The quadratojugal disappeared as an independent bone in lizards, snakes, and mammals.

Roofing bones overlie the neurocranium when the latter is complete above the brain. When the neurocranium is incomplete dorsally, soft spots (fontanels) can be felt in the head until the membranes under the skin have ossified. In mammals a small bregmatic bone sometimes ossifies in the frontal at the junction of the coronal and sagittal structures. It is sometimes present in man, and is called the antiepileptic bone.

(b) Upper jaw bones: The first upper jaw skeleton that a vertebrate embryo develops is the pterygo-quadrate (palato-quadrate) cartilage on each side. It is a part of the visceral skeleton and the only upper jaw that cartilaginous fishes develop (Fig. 10.15 A). In bony vertebrates this cartilage becomes covered (ensheathed) by dermal bones that lock it into the skull. These dermal bones, the pre-maxillae and maxillae, comprise

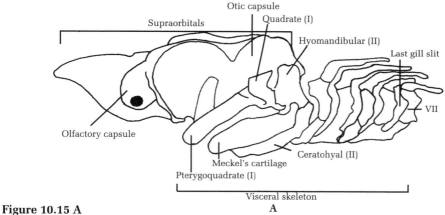

Figure 10.15 A
Skull and visceral skeleton of shark showing skeleton of I, II and VII pharyngeal arches

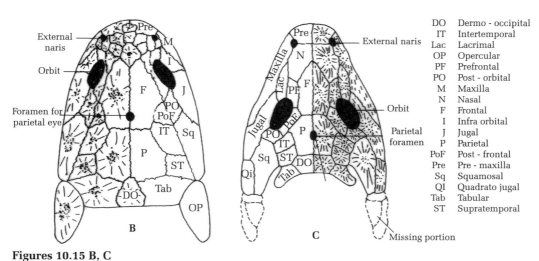

Figures 10.15 B, C
B. Skull of rhipidistian fish **C.** Skull of carboniferous labyrinthodont. Broken lines indicate missing opercular bones

the adult upper jaw of most bony vertebrates and usually bear teeth. In birds the premaxillae are elongated and form a part of the beak. Premaxillae are not seen in adult human skulls because they unite with the maxillae very early in development.

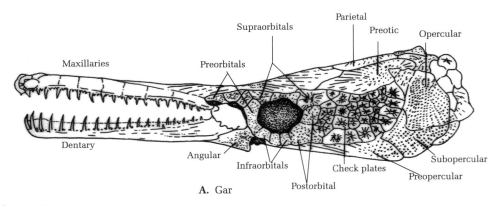

Figure 10.16
Skull of neopterygian (gar) ancient fish

Garfish have a long series of scale like bones that have been named maxillaries (Fig. 10.16). Bones called maxillae in teleosts may be toothless, reduced, or not in the upper jaw. They may not be homologous with the maxillae of older fish and tetrapods.

(c) **Palatal bones :** The floor on which the brain rests is at the same time the roof of the oral cavity in fish and amphibians. This part of the skull is the primary palate. In sharks it is cartilaginous. In bony vertebrates membrane bones form in this location. These are vomers beneath the olfactory capsules, palatines, endopterygoids and ectopterygoids beneath the pterygoquadrate cartilages, and a median parasphenoid beneath the sphenoid region of the neurocranium. Primitively, teeth were formed on all these bones, and many persists even today in lower vertebrates (Fig. 10.17).

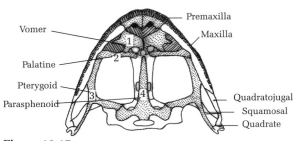

Figure 10.17
Primary palate *Rana*

In birds and mammals and some reptiles a secondary (false) palate develops. It is a horizontal partition separating the primitive oral cavity into nasal and oral passageways. Membrane bones in the secondary palate usually include palatal process of premaxillae, maxillae, and palatines. In crocodilians the pterygoid bone also contributes.

Palatal process arise as horizontal shelves of the foregoing bones, which grow toward one another in the roof of the embryonic oral cavity. Failure of these processes to meet results in cleft palate. This is a normal condition in turtles and birds, but a congenital abnormality in mammals.

(d) Opercular bones : The operculum is a fold of the hyoid arch that extends back over the gill slits in holocephalans and bony fishes (Fig. 10.18). In the latter it is stiffened by dermal opercular bones. In many bony fish a series of branchiostegal rays forms in a ventral flap of the operculum but there are no vestiges of opercular bones in tetrapods.

Splanchnocranium or visceral skeleton

The visceral skeleton, or splanchnocranium, is the skeleton of the pharyngeal arches. In fishes, therefore, it is the skeleton of the jaws and gill arches. In tetrapods this skeleton has become modified to perform new functions on land.

The blastema that give rise to the visceral skeleton come from neural crests and they first secrete cartilage. Later, the cartilage may be partly or wholly replaced by bone. In the first arch only much of it is ensheathed by dermal bone. We will look first at shark, in which no bone forms and in which we can see the visceral skeleton in its primitive capacity that of supporting the jaws and gills.

Succession of visceral skeleton and type of jaw suspension

Fish

Cartilaginous fish: The visceral skeleton of a shark consists of seven sets of paired cartilages (Fig. 10.19) in the seven visceral arches and a series of mid-ventral cartilages, basihyal and basibranchials in the pharyngeal floor (Fig. 10.19 A). All seven pairs conform to a basic pattern, but the mandibular and some of the hyoid cartilages are modified for feeding.

The skeleton of **mandibular** or first visceral arch consists of two cartilages on each side, a pterygoquadrate (also called palatoquadrate) cartilage dorsally and Meckel's cartilage ventrally (Fig. 10.19 B). The left and right pterygoquadrates meet in the mid dorsal line to form the skeleton of the upper jaw, and the left and right Meckel's cartilage meet ventrally to form the skeleton of the lower jaw. Delicate labial cartilages embedded in a position corresponding to lips articulate with the upper and lower jaw cartilages at the corners of the mouth, but they have no known phylogenetic significance. The skeleton of the hyoid arch consists of paired hyomandibular cartilage dorsally and ceratohyals laterally. The latter bears a demibranch and articulates with the basihyal cartilage. In embryo the basihyal is paired. The cartilages of the remaining visceral arches are essentially alike and all but the last pair support gills. Hypobranchials, unless absent, articulate ventrally with a basibranchial.

At the corner of the Meckel's cartilage, the pterygoquadrate cartilage, and the hyomandibular cartilage articulate in a movable joint that participates in operation of the jaws. The hyomandibular is bound by ligaments to the otic capsule, and it therefore suspends the jaws and branchial skeleton from the skull.

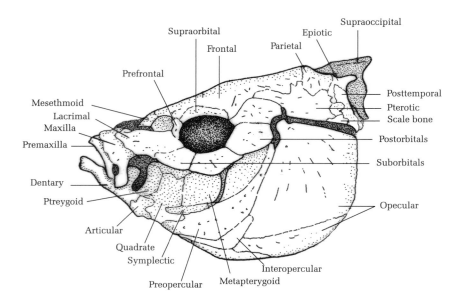

Figure 10.18
Skull of Carp black portion shows unossified cartilage

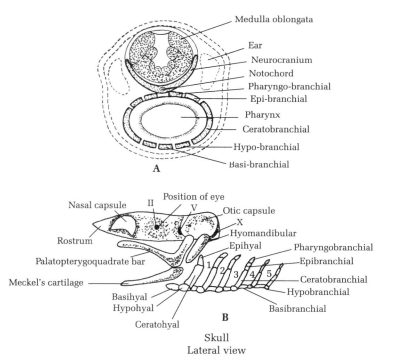

Figures 10.19 A, B
A. Cross section of primordial neurocranium and splanchnocranium.
B. Skull and visceral skeleton of the shark (*Squalus*)

Cyclostomes: The visceral skeleton of cyclostomes is quite unlike that of jawed fishes (Fig. 10.20). For example, *Myxine* has no identifiable pterygoquadrate or Meckel's cartilages. It does have a dental plate or lingual cartilage that forms a V-shaped trough in the floor of the oral cavity, and beneath this, an immovable basal plate to which the muscles of the dental plates are attached. There is evidence that these may be derived from the first visceral arch, if so, the rasping tongue like structure of which they are a part may be considered a type of lower jaw. With regard to an upper jaw a careful study of the visceral skeleton of a hagfish led to the conclusion that a rudimentary upper jaw is fused with the neurocranium. The rest of the visceral skeleton of cyclostomes consists of cartilages of unknown homology, including a basket like cartilaginous framework immediately under the skin surrounding the gill slits.

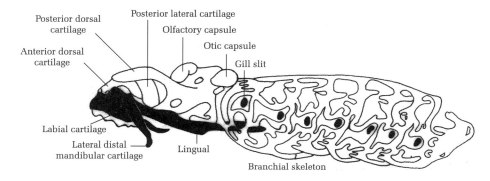

Figure 10.20
Chondrocranium and branchial basket of Lamprey (*Petromyzon*) (lateral view)

Jaw Suspension

Most important role of the splanchnocranium in jawed vertebrates is to suspend the lower jaw with the cranium (Fig. 10.21). The splanchnocranium is attached to the neurocranium in five principal ways:

(i) The **autodiastyle** type is found in the earliest gnathostomes Acanthodians. In this case the jaw is supported by direct attachment to the cranium by ligaments both at its hind as well as the front end. The hyoid arch is a typical branchial arch, not modified in any way. The spiracle is a fully functional gill slit.

(ii) The **amphistylic** type of jaw suspension occurs in early elasmobranchs such as *Hexanchus*. In this case both the pterygoquadrate and hyomandibular directly articulate with the neurocranium. Usually the pterygoquadrate articulates with the neurocranium at two points. (i) basal process at the level of the posterior end of the embryonic trabecula and (ii) otic process articulating with the side of the otic capsule.

(iii) The **hyostylic** arrangement have the upper jaw not fixed to the cranium but is slung from the latter by the hyomandibular and a prespiracular ligament. Thus, hyomandibular, a part of the hyoid arch, acts as suspensorium. This naturally allows a wide gap for swallowing large prey. This type of suspension occurs in the dogfish.

(iv) The **autostylic** peculiar feature of this type of jaw suspension is that the upper jaw is completely fused with the cranium. It is the lower jaw which is articulated with cranium and this articulation is effected in different ways and is found in dipnoans, amphibians, reptiles and birds and is of various types.

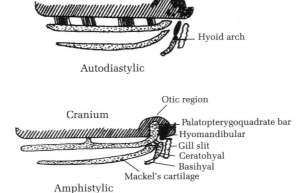

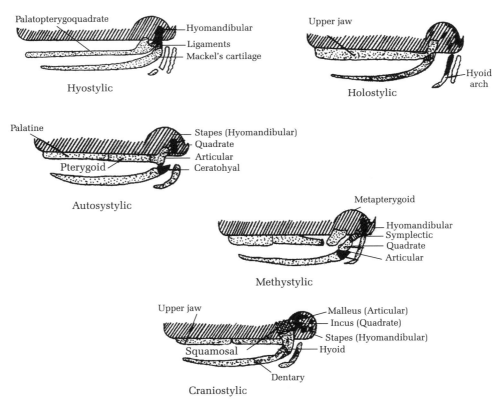

Figure 10.21
Jaw suspensorium in vertebrates

(a) **Holostylic** In this case the upper jaw articulates with the neurocranium through two or more pairs of processes without the intervention of the hyomandibular. These articulations are usually of the immovable type. The hinge occurs between the articular of the lower jaw and the quadrate of the upper jaw. Jaws of *Chimaera* are also similar but in this case the upper jaw fuses with the cranium rather completely and the hyoid arch is free.

(b) **Autosystylic** is found in the lungfish and tetrapods except mammals; here the hyoid arch breaks up and is no longer in the service of jaws; hyomandibular becomes reduced and forms the ear ossicles, the stapes. Lower jaw is articulated with the quadrate part of the upper jaw, thus, quadrate becomes the suspensorium.

   (i) **Monimostylic:** In this type the upper jaw is fused with the cranium and thus is immovable. The lower jaw articulates with the quadrate. It is mostly seen in amphibians.

   (ii) **Streptostylic:** In this types of jaw suspension, the quadrate is movable at both the ends and not firmly fixed with the skull with this arrangement, the lower jaw has two movable joints (Fig. 10.36 C). Here the articulation infact, takes place between the quadrate of the cranium and the articular of the lower jaw. Quadrate is not firmly fixed. This type of suspension is found in snake lizards and birds.

(c) **Methystylic** is characteristic of teleosts (bony fish) and a variation of autostylic suspension. Anterior part of the upper jaw is reduced and its place is taken up by membrane bones such as palatines and pterygoids, posterior part is replaced by metapterygoid and quadrate bones derived from the hyomandibular.

(v) **Craniostylic** suspension is characteristic of mammals some consider it as a modification of autostylic where quadrate of the upper jaw separates and forms the second ear-ossicle, the **incus**. A part of the lower jaw, the **articular**, separates to form the third ear ossicle called **malleus**, and the lower jaw articulates directly with a membrane bone of dermocranium, the squamosal.

Role of visceral arches in tetrapods

With pulmonary respiration and life on land, the ancestral visceral skeleton, which was necessary in gill-bearing vertebrates, underwent profound adaptive modifications. Some previously functional parts were deleted, and those that persisted perform new and sometimes, surprising functions.

Not only have changes in the visceral skeleton taken place during the evolution of tetrapods; they occur also during the ontogeny of every gill-bearing amphibian that undergoes complete metamorphosis. For example, larval frogs have six pairs of visceral cartilages, and the last four (III to VI) support gills (Fig. 10.22). These branchial cartilages unite ventrally in a hypobranchial plate. During metamorphosis visceral cartilage VI regress and disappear, the hypobranchial plate enlarge and along with the first basibranchial, becomes incorporated into a broad plate (body of the hyoid) in the buccal and pharyngeal floor. The ceratohyal cartilage (arch II) is reduced to slender anterior horn of the hyoid. And the cartilage of arch IV becomes a posterior horn. Other changes take place with the result, that a visceral skeleton initially adapted for branchial respiration (aquatic) becomes converted, in the span of a few short days, to one suitable for life on land. Perennibranchiate amphibians, on the other hand,

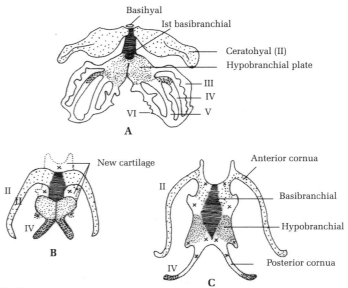

Figures 10.22 A, B, C
Changes occurring in hyoid and branchial skeleton during metamorphosis in frog **A.** Larval Skeleton II-VI visceral arches, III-VI bear gills **B.** Late metamorphosis **C.** Hyoid apparatus of a young frog

retain a branchial skeleton throughout life. In the visceral skeleton of tetrapod following bones undergo modifications:

(i) Pterygoquadrate cartilage of palatoquadrate cartilage

(ii) Meckel's cartilage

(iii) Hyomandibular cartilage or columella or stapes

(iv) Hyoid apparatus

(v) Laryngeal skeleton

(i) Pterygoquadrate cartilages are the embryonic upper jaw cartilages. In amphibians, reptiles, and birds their posterior end, which is the quadrate region, usually undergoes endochondral ossification to become the quadrate bone at the caudal angle of the skull. There it forms a joint with the articular bone of the lower jaw. Thus, in tetrapods, as in sharks, the caudal ends of Meckel's cartilages (now ossified) rock against the caudal ends of the pterygoquadrate cartilages (also ossified) (Fig. 10.23). In mammals the same joint exists; but the quadrate bone has become surrounded by the middle ear cavity, and has separated from the rest of pterygoquadrate cartilage and become the incus of the middle ear. The evolutionary transition from a jaw bone to ear ossicle was gradual (Fig. 10.24). The intermediate steps are seen in mammals like reptiles. The anterior part of the pterygoquadrate cartilage becomes ensheathed by dermal bones including the premaxilla and maxilla and by some of the membrane bones of the palate. As a result, this part of the visceral skeleton is locked into the skull. In perenibrachiate amphibians some of it remains cartilaginous and contribute to the palate. It has been claimed that bony-ring (annulus tympanicus) to which the tympanic membrane of anurans is attached is a derivative of the pterygoquadrate cartilage.

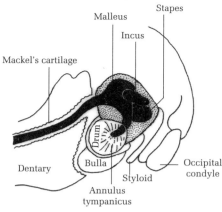

Figure 10.23
Meckel's Cartilage surrounded by developing middle ear cavity

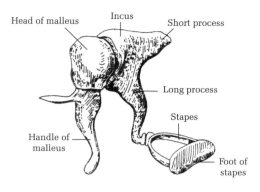

Figure 10.24
Bone of middle ear

(ii) **Meckel's cartilage** Parts of the embryonic Meckel's cartilage become replacement bone, a part remain cartilaginous, and much of it is ensheathed by dermal bones. In adult birds and mammals Meckel's cartilage fails to grow beyond the embryonic state and few or no remnants remain with in the adult mandible. The posterior tip of Meckel's cartilage is not ensheathed, rather below mammals it becomes the articular bone of the lower jaw, which articulates with the quadrate of the upper jaw. In anurans the articular portion remains unossified. In mammals the articular portion is projected into the middle ear cavity and later separates from the rest of Meckel's cartilage to become the malleus, a middle ear ossicle. The malleus still forms a joint with the quadrate (incus), but in the middle ear instead of at the end of the jaw. An ossification centre sometimes develops in Meckel's cartilage on either side of the mandibular symphysis, giving rise to a mentomeckelian bone. In some species, however the mento-mecklian is a membrane bone.

(iii) **Hyomandibular cartilage** It will be recalled that the hyomandibular of sharks is interposed between the quadrate region of the upper jaw and the otic capsule containing the inner ear. Studies have shown that the hyomandibular of tetrapod embryos ossifies to become part of the columella, or stapes, of the middle ear. It still articulates with the otic capsule but now transmits sound waves. Furthermore, in mammals it exhibits its ancient relationship, extending between the otic capsule and the quadrate bone (incus) (Fig. 10.24).

(iv) **Hyoid apparatus** The term hyoid apparatus is used here to designate the skeletal derivatives of the hyoid arch other than the columella or stapes, and the derivatives of the more caudal visceral arches other than those contributing to the larynx. The hyoid apparatus consists of a median plate, the body of the hyoid, derived from basihyal and basibranchial elements, and two or more horns, or cornua (Fig. 10.25). The anterior horns arise from arch II and are homologous with the cerathoyals of fishes. The more caudal horns arise from arches III and frequently IV.

In lizards and birds the body of the hyoid is narrow, and an elongated process extends forward into the tongue as an entoglossal bone. In some male lizards

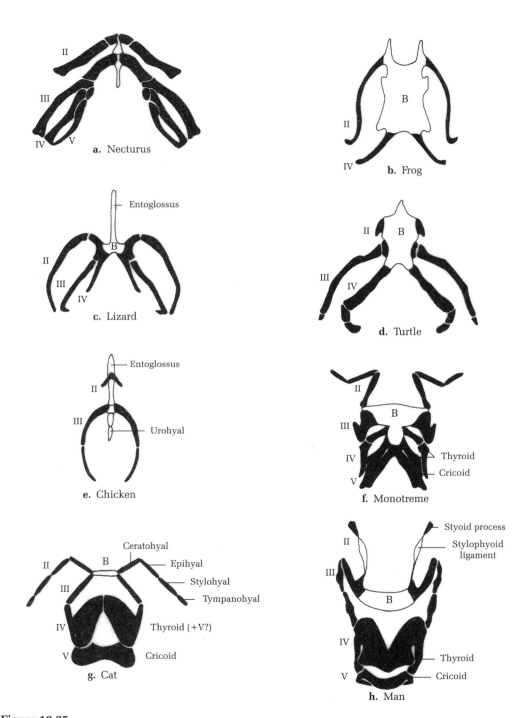

Figure 10.25
Derivatives of II, III, IV and Vth pharyngeal arches in selected tetrapods (a–h). B – body of hyoid, II-V – derivatives of arches. Projections from the body (B) are horns of the hyoid

(anoles and related genera), a similar process extends posteriorly into the gular pouch, or dewlap. The process is flexible, and bent like a bow, and during mating, the dewlap protrudes far out from the underside of the neck, where its many blood vessels and pigment cells display in sunlight a bright coloration. In snakes the entire branchial skeleton is vestigial.

The hyoid apparatus of mammals has two horns, an anterior pair from arch II and a posterior pair from arch III. In cats the anterior horns are longer (greater horns) and are composed of four segments. The dorsal most, or tympanohyal, ends in a notch in the tympanic bulla. In man the anterior horns are shorter (lesser horns), a stylohyoid ligament represents the middle segment, and the stylohyoid process of the temporal bone is equivalent to the tympanohyal. In rabbits too, the anterior horn is shorter, and a slender stylohyal bone embedded in the tendon of the stylohyoideus minor muscle close to the skull is equivalent to the tympanohyal of cats. These are few examples of the kind of variations found among mammals.

Branchiomeric and hypobranchial muscles approach the hyoid from many directions and get inserted on it. Because of this it is part of the buccopharyngeal pressure pump in anurans. In amniotes, tongue muscle and muscles used in swallowing also attach to it.

Laryngeal skeleton Nearly all tetrapods have cricoid and arytenoids elements, and crocodilians and mammals have thyroid elements as well (Fig. 10.26). Thyroid cartilage, arise from the mesenchyme of arch IV and perhaps V; cricoid and arytenoids cartilages may be products of arch V. Since the posterior visceral arch has been subject to reduction during evolution, it is not surprising that problems are encountered in relating laryngeal cartilages to specific arches. Also, there may be instances in which the laryngeal cartilages are not visceral arch derivatives.

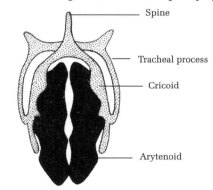

Figure 10.26 Larnyngeal skeleton of frog

Comparative account of skull in different classes of vertebrates

Cyclostomes

The skull of hagfish is very primitive consisting only of floor of cartilage with side walls and roof of connective tissue. Parachordal cartilages are united with the otic capsules. The pre-chordal cartilages terminate abruptly at the anterior end.

In lampreys (Fig. 10.20) the skull is better developed. Dorso-lateral extensions of the pre-chordal cartilages are present, forming side walls and a roof over the brain in this region. The remainder of the skin is enclosed in fibrous connective tissue. Pre-chordal and parachordal cartilages as well as optic capsules are homologous with similar structures in higher forms. Other cartilages of doubtful homology are present; supporting the round, suctional mouth, the olfactory organ, and the tongue. Posterior and anterior dorsal cartilages are located anterior to the chondrocranium. They too appear to be without homologues in higher vertebrates.

In lampreys the gill region is supported by a peculiar cartilaginous branchial basket. The cartilage of branchial basket is continuous and not divided into separate parts. Lateral openings for the gill slits are also present in addition to the openings present above and below. The branchial basket lies just beneath the skin. Its posterior end furnishes support for the pericardium. It does not appear to be homologous with the visceral skeletons of the other vertebrates. In hagfish the visceral skeleton is much reduced, but its hyoid arch are recognizable. They are located far with in the head and are interpreted as true visceral arches.

Skull of fish

Elasmobranch: Neurocranium and splanchnocranium only constitute the skull of sharks, there being no dermocranium. Neurocranium is in the typical form, consisting of the cartilaginous brain-box to which nasal and otic capsules are fused. Otic capsules are completely fused and closed, nasal capsules remain open on the outer and lower sides, The floor of the cranium called **basis cranii**, is formed of the parachordal and prechordal cartilages, which leave no gap between them expect a pair of transverse grooves, the **carotid canals** for the carotid arteries (Figs. 10.27 A, B). There is some calcification in the basis cranii, but the roof of the brain box is incomplete in the anterior part forming a big anterior fontanels. The anterior fontanelle (pre-cerebral cavity) is covered by a membrane in the natural condition. Only the occipital region (posterior) is completely roofed over by cartilages and is pierced by

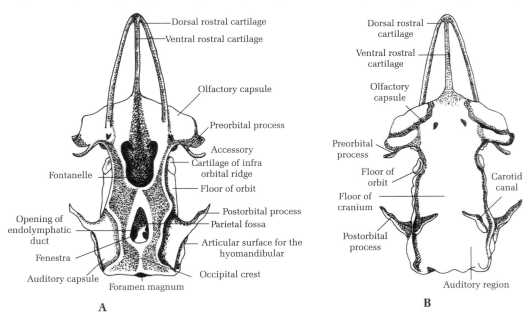

Figures 10.27 A, B
A. Skull of *Scoliodon* dorsal view **B.** Skull of *Scoliodon* ventral view

a large opening, the foramen magnum, meant for the exit of the spinal cord. Below the foramen is a concavity which opposes a similar concavity of the first vertebra. Concavity of the skull is flanked on the sides by **occipital condyles**. Head cannot be moved over the vertebral column.

Otic capsules are distinctly demarcated from occipital region, floor and roof of the cranium by three ridges, one each for three semicircular ducts of the membranous labyrinth. Over the roof of the **cranium** between the two otic capsules there is an oval depression, the **parietal or endolymphatic fossa**, in which lie two pairs of openings. During life the fossa is covered by a membrane. A slight ridge, the **crista occipitalis** extends between the foramen magnum and the endolymphatic fossa. Below the ridge of the horizontal semicircular duct lies the **post-orbital groove** and further below lies the articular surface for the hyomandibular cartilage which act as suspensorium (Fig. 10.28).

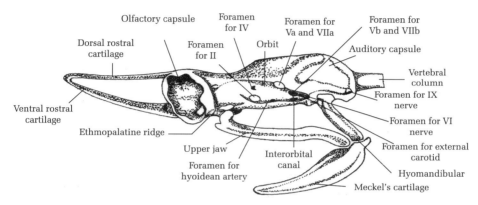

Figure 10.28
Skull of *Scoliodon* lateral view

The orbits lodging the eyeballs are in the pre-chordal region. Boundary of the orbit on each side of the cranium is marked by **supraorbital ridge** (dorsal), **infraorbital ridge** (ventral), Pre-orbital process on anterior and post orbital process on posterior, infra orbital ridge gives off an anterior orbital process to which upper jaw is attached by means of ligaments.

Ethmoid plate, formed by the union of the pre-chordals in the anterior part, has the large nasal capsules fused with it and is itself produced into a median and a **ventral rostral cartilages**. A pair of dorso lateral rostral cartilages arising from the roof of nasal capsules meet the ventro–median rostal cartilages, the three form the skeleton of the snout.

A number of openings and foramina are seen in the neurocranium. There is a large foramen outside occipital condyle of each side for the exit of X cranial nerve Of the two pairs of openings in the endolymphatic fossa, anterior small pair is for the passage of the invagination or endolymphatic canals of the internal ears, and the posterior larger pair called perilymphatic fenestrae, are openings of the perilymphatic space of each of the otic capsules. Behind and below the ridge of the horizontal semicircular duct, on each side, is a large foramen the exit of IX cranial nerve. Inside the orbit are separate foramina for the superficialis branch of VII, the III, profundus branch of V into the orbit, main branches of V and VII and for VI, also for interorbital canal, hyoidean artery II, IV, produndus of V from the orbit and for the posterior opening of orbito-nasal canal.

Immediately behind each nasal capsule is the anterior opening of the orbito nasal canal. In front of the fontanelle are the dorsal exits of the profundus branches of the V nerve.

The **visceral skeleton** of elasmobranchs is typically composed of seven cartilaginous visceral arches surrounding the anterior part of the digestive tract and furnishing a firm support to the gills (Fig. 10.29).

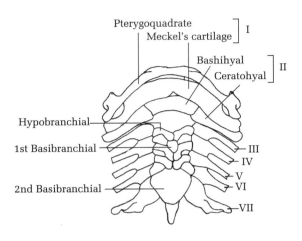

Figure 10.29
Visceral skeleton of *Scoliodon* (Ventral view)

The first or **mandibular**, arch is larger and more conspicuous than the rest. From it are derived both upper and lower jaws. On each side the mandibular arch divides into a dorsal **palato-pterygoquadrate** bar and a ventral Meckel's cartilage. These articulate with each other posteriorly resulting in hyostylic type of jaw suspension. The palato-pterygo-quadrate bars of the two sides make up the upper jaw. They have heavy cartilages which unite with each other anteriorly by means of ligamentous connection. The anterior portion of palato-pterygo quadrate bar bears a palatine, or orbital process which extends upwards along the inner wall of the orbit. The two Meckel's cartilages are united anteriorly by a ligament to form lower jaw. Both palatopterygoquadrate bar and Meckel's cartilage bear teeth. Small labial cartilages of doubtful homologies are embedded in the tissue just outside the upper and lower jaws near the anterior end.

The second or **hyoid** arch is composed of three parts. An upper **hyomandibular** cartilage articulates with the chondrocranium behind the orbit in the otic region. Below this is the **ceratohyal cartilage**. The ceratohyals of the two sides are joined together by means of a medium plate, the **basihyal**. The posterior edges of both hyomandibular and ceratohyal cartilages give off numerous slender cartilaginous gill rays which support the tissue bearing the gill lamellae composing the first or hyoid, **hemi branch**.

The number of visceral arches posterior to the hyoid varies in elasmobranchs but usually there are five. Each consists of several cartilages on each side (Fig. 10.19). These are arranged from dorsal to ventral borders in following manner – (i) pharyngobranchial (ii) epibranchial (iii) ceratobranchial (iv) hypobranchial.

The pharyngobranchials slope posteriorly and terminate just below the vertebral column. They are attached to the vertebral column by fibrous bands of tissue. The hypobranchials may meet their partners in the mid-ventral line or are connected by unpaired **basibranchial cartilages**. The epibranchials and ceratobranchials bear the gill rays which support the remaining gills. Gill rays are absent on the last visceral arch, which does not bear a gill. The pointed cartilaginous projections called gill rakers extend into the pharynx from the inner edges of the visceral arches. They are covered with mucous membrane and are used to strain food particles from the water.

Osteichthyes

The bonyfish belonging to class osteichthyes appeared first. In evolutionary terms they are much older than members of the class **chondricthyes**. Paleontological evidence indicates that very early in their history the bonyfish had already divided into two branches, the **ray-fins** or actinopterygii and **lobe-fins** or sarcopterygii. The subclass **sarcopterygii** contains members of the order **crossptergyii**, primitive relatives of which were the ancestors of tetrapods. The skulls of the ray-finned fishes with the exception of **polydon**, the **spoon bills** or **paddle fish** are invested with numerous membrane bones, several of which are homologous with similarly placed and similarly named bones of **crossopterygians** and **tetrapods**.

Among the skulls of fishes one can find all the transitional stages which gave rise to skulls which were completely cartilaginous to those which are almost completely bony. In **holocephalians** the skull is composed of cartilage with a noteworthy feature that the entire palato-pterygoquadrate bar is immovably fused to the cranium. The lower jaw is suspended from the quadrate region, and the hyoid arch plays no role in suspending the jaws. This is a variation of the autostylic method of jaw suspension. The skull articulates with the vertebral column by means of a movable joint. The hyoid arch is developed only slightly more than the remaining visceral arches. The ceratohyal component, with its gills rays supports the operculum, which first makes its appearance in this group of fishes.

The sturgeon fish *Acipenser*, has a skull almost completely formed of cartilage. No cartilage bones are present although ossification appear in the otic and orbitosphenoid region. Several well-developed membrane bones have made their appearance. On the dorsal side they include-frontals, postfrontals, parietals, supra orbitals, and others, which form a protective shield of dermal scales over the cartilaginous cranium beneath. On the ventral side are two other membrane bones, an anterior **vomer** and a more posterior parasphenoid. The rostrum covered with scales, is prominent. The method of jaw suspension in the sturgeon is **hyostylic**.

Maxillary and dentary bones are associated with the jaws, and dermal opercular bones lie over the hyomandibular cartilage. Jaws and visceral skeleton are poorly developed. In **Polydon**, a close relative of the sturgeon, membrane bones are less prominent. Nevertheless, the snout which is actually the enormously elongated rostrum is covered with dermal scales. The ancient ray fin, *Polypterus* exemplifies an advancement over the primitive condition toward the more highly developed teleostean skull. The cartilaginous skull is retained for the most part, but a veritable armor of numerous membrane bones cover it. Among these are frontals, parietals, nasals, supra temporals, post temporals, and dermal supraoccipitals. The palatopterygoquadrate bar is fused to the cranium, the method of jaw suspension is **autostylic**. Several important membrane bones are developed in connection with the palatopterygoquadrate bar. They include palatines and a number of Pterygoids. Vomer bones are also present. The outer arch of membrane bones includes premaxillaries, maxillaries, and jugals. In connection with Meckel's cartilage are developed dentary, angular and splenial bones. Other membrane bones appear on the operculum.

In the garpike, *Lepidosteus*, and in the bowfin, *Amia*, both cartilage bones and numerous membrane bones are present. Ex-occipitals, pro-optics, and lateral ethmoids are among the cartilage bones which appear. In the garpike, teeth are borne by the maxillary bones in the upper jaw (Fig. 10.16). Teeth are present in *Amia* on the palatines, pterygoids, and vomer as well as on the premaxillaries and maxillaries. The lower jaw of the gar pike consists mainly of the tooth-bearing dentary. Small angular and articular bones are also present. In *Amia* there are several splenials, an angular, suprangular, coronoid, and articular. The dentary, bearing teeth, however is the chief bone of the lower jaw. The **hyostylic** method of jaw suspension is found in these fish.

A rather wide variation is to be observed among the teleosts. Their skulls have more bones than those of any other vertebrates. Numerous membrane bones are present, but most characteristic is the presence of several cartilage bones which have formed from ossification centres in the primitive **chondrocranium**, which still persists to some degree. These bones include basioccipitals, exoccipitals, supraoccipitals, basisphenoids, pleurosphenoids, or alisphenoids of (mammals), orbitosphenoid, pro-otics, epiotics, and opisthotics. Sphenotics and pterotics present in addition, are derived from the auditory capsules. Mesethmoid and lateral ethmoids develop anteriorly. Among the more prominent membrane bones on the roof of the skull are parietals, frontals, and nasals. A series of circumorbital bones affords protection to the eyes.

The skull in **dipnoi** is not so advanced as that of teleosts. It is cartilaginous, for the most part, but both cartilage bones (exoccipitals) and a few large membrane bones are found. The palatopterygoquadrate bar, like that of holocephalians, is indistinguishably fused to the cranium. No cartilage bones are derived from it. The method of jaw suspension is **autostylic**.

In the evolution of the skull from primitive fish to mammals, there has been a gradual reduction in number of separate bony elements by elimination and fusion. As many as 180 skull bones are present in certain primitive fish. The human skull consists of 28, including auditory ossicles. In some case cartilage bones and membrane bones fuse together to form a single structure. The **autostylic** method of jaw suspension is the rule among the tetrapods.

Amphibians: Skull of early tetrapods show considerable deviation from their fish like ancestor. Among the most important changes are a reduction in the number of bones and a general flattening of the skull. In addition, the length of the skull has apparently been reduced, particularly in the occipital region. The otic capsule bears a ventral opening, the fenestra ovalis, into which a cartilaginous or bony plug fits. This is the stapedial plate of the columella which, it will be recalled, is derived from the hyomandibular cartilage. It is possible that the columella proper is derived from the symplectic. The columella (stapes) has developed in connection with the evolution of the sense of hearing and with the change from the hyostylic to the autostylic method of jaw suspension in which the hyomandibular bone loses its significance as a suspensorium.

The embryonic chondrocranium remains to a considerable extent in amphibians, but some of it has been replaced by cartilage bones. The basioccipital and supraoccipital regions are not ossified. Articulation of the skull with the atlas vertebra is accomplished by means of a pair of occipital condyles, i.e, projections of the exoccipital bones. Basisphenoid and pre-sphenoid regions also are not ossified. Pro-otics, and in some cases opisthotics, are ossified and fused to the exoccipitals. Much variation in the ossification of this region is encountered. Membrane bones form the greater part of the roof of the skull. They are no longer closely related to the integument and occupy the deeper position in the head than they do in fishes. In some of the **extinct labyrinthodonts** grooves for the accommodation of branches of the lateral line canal were present on some of the membrane bones of the skull giving evidence of their dermal origin. A rather large membrane bone, the parasphenoid, covers the ventral part of the chondrocranium. In the **labyrinthodonts** the dorsal surface of the skull was completely covered by bone, leaving only openings for the nares and eyes. In some extinct forms, however, another opening, the **interparietal foramen**, was evident. The **inter-parietal** foramen has disappeared in living species. Modern apodans, or cecilians, also show great solidity of the skull. There has, however, been a reduction in the number of skull bones. In other amphibians large spaces are present on both dorsal and ventral sides of the skull.

In tetrapods, the autostylic method of jaw attachment is the rule. The quadrate, in amphibians, is fused to the otic region of the cranium. Palatine and pterygoid membrane bones, which form about the anterior portion of the **palato-pterygo-quadrate** bar, are well developed. The outer arch of membrane bones is represented by pre-maxillaries and maxillaries. Anurans have a quadratojugal in addition. Pre-maxillaries, together with the vomer and sometimes the palatines, bear teeth in most species. The lower jaw consists of a core of Meckel's cartilage surrounded by membrane bones. In early amphibians as many as 10 bones formed Meckel's cartilage on each half of the jaw. In modern forms the number has been reduced.

Urodeles: In the tailed amphibians the skull is less well developed than in anurans. The chondrocranium proper becomes ossified in only two regions. Anteriorly there is a pair of elongated orbitosphenoids, each bearing a foramen for the passage of the optic nerve. Posteriorly are found the paired exoccipitals which are used with the otic bones. Two foramina in each exoccipital near the condyle provide space for the passage of the glossopharyngeal and vagus nerves. Placed farther anteriorly are two additional foramina for the facial and trigeminal nerves. The membrane bones on the dorsal side of the cranium consists of parietals, frontals, and prefrontolacrimals. A pair of nasal bones develops in the olfactory region. An unpaired parasphenoid bone, often bearing teeth, lies on the ventral surfaces. Anterior to this is a pair of vomero-palatines, usually bearing a row of teeth. The choanae open just posterior to the vomerine portion. The upper jaw consists of pre-maxillary and maxillary membrane bones. The premaxillaries in some cases are united and articulate with the vomeropalatines on the ventral side and with the nasals, and frequently with the frontals, on the dorsal side. Laterally the pre-maxillaries articulate with the maxillaries. The latter terminate freely at their posterior ends, failing to meet the quadrates. Jugals and quadratojugals are lacking. The quadrate bone, representing the remainder of the palato-pterygoqudrate bar, is fused to the otic capsule

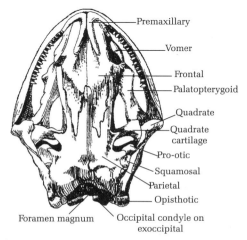

Figure 10.30
Skull of *Necturus* (dorsal view)

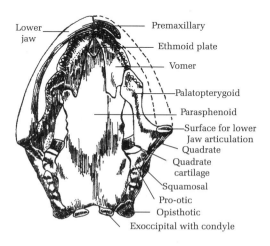

Figure 10.31
Skull of *Necturus* (ventral view)

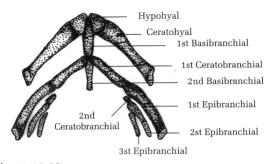

Figure 10.32
Hyoid apparatus of *Necturus*

and articulates with the stapes. It serves for articulation with the lower jaw. The squamosal, a membrane bone, invests the outer surface of the qudrate. A cartilage extending anteriorly from the qudrate toward the maxillary is invested ventrally and laterally by the pterygoid. In some cases the pterygoid may form a cartilaginous connection with the maxillary. The lower jaw of urodeles usually consists of two tooth-bearing membrane bones, and anterior dentary and a posterior splenial, which encase Meckel's cartilage, and a single cartilage bone, the articular, which articulates with the qudrate of the upper jaw. Another membrane bone, the angular, is also frequently present.

In **Necturus** (Figs 10.30, 10.31, 10.32) the skull retains many primitive features. The nasal capsules are not united with other parts of the skeleton. Maxillary bones are lacking. A palatopterygoid membrane bone with teeth on is anterior portion unites the vomer with the quadrate on each side. The remaining portion of the visceral skeleton in urodele amphibians is reduced in comparison with fish. Larval forms generally possess portions of hyoid and four other visceral arches, but these are reduced at the time of metamorphosis. In adults the hyoid arch (Fig. 10.32) usually consists of two pieces on each side, a lower hypohyal and an upper ceratohyal, which is frequently ossified. The hypohyals are in most cases united ventrally to a median basibranchial or copula. The next two arches consists of proximal ceratobranchials and distal epibranchials. The ceratobranchials unite medially with the basibranchial, which is usually composed of a single piece. After metamorphosis the last two arches are much diminished or disappear altogether. The hyoid is not always the best developed of the post mandibular visceral arches. In *Salamander* eurycea, one of the branchials is better developed than the hyoid. The entire apparatus serves to support the tongue and regions posterior to the tongue.

Anura: The broad, flat skull of anuran amphibians is noteworthy for the fact that the elements composing the jaws are so widely separated form the cranium. The cranium retains much of its original cartilaginous character (Fig. 10.33). Part of skull is replaced by cartilage bone; and a part is covered with membrane bone.

The cartilage bones consists of –

(i) an unpaired **sphenethmoid**, a ring like bone which encircles the anterior portion of the cranial cavity.

(ii) Two **exoccipitals** – forming the margins of the foramen magnum, each bearing a condyle for articulation with the atlas, and

(iii) Prootics of irregular shape, situated anterolaterally to the exoccipitals. The fenestra ovalis, a well defined opening in the cartilage lying ventral to the pro–otic, is plugged by the minute statpedial plate of the columella. The epiotic and opisthotic regions of the auditory capsule are united with the exoccipital, of which they are often considered to be a part.

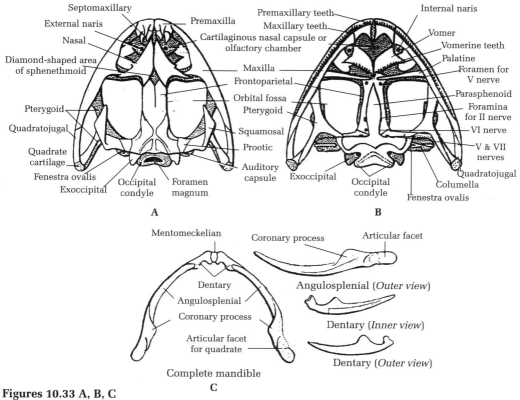

Figures 10.33 A, B, C
Frog skull   A. Dosal view   B. Ventral view   C. Bones of mandible

The membrane bones of the cranium include paired fronto-parietals on the dorsal side and an unpaired parasphenoid, located ventrally. There is no real evidence that the fronto-parietal bones of anurans represent a fusion of frontal and parietal bones. Each develops as a single entity. Three openings, or **fontanelles**, are present in the roof of the **chondrocranium**, but these are covered over by the fronto-parietals.

The nasal capsules, although united with the cranium, usually remain unossified. Two pairs of membrane bones, however, form in connection with the nasal capsules. They are the nasals above and the vomer below, the latter bearing vomerine teeth. The posterior lateral borders of the vomers bound the medial borders of the internal nares.

The upper jaw consists largely of an outer maxillary arch articulating anteriorly and posteriorly with the rest of the skull but being rather widely separated from it in the middle. The large space thus formed on each side is the orbit which accommodates the eye ball. The upper jaw is composed of series of membrane bones. From anterior to posterior ends these include the premaxillaries, maxillaries and quadratojugals. The latter are lacking in the **aglossae**. Only the first two bones of this series bear teeth. At the posterior angle of the upper jaw is a small, unossified area, sometimes called the **quadrate**, with which the quadratojugal articulates. The quadrate is connected to the cranium by a short bar of cartilage. To this, in turn, is attached a cartilaginous ring, the tympanic annulus which encircles and supports the tympanic membrane. The tympanic annulus appears to be derived from the palatopterygo quadrate bar and is therefore, not homologous with the dermal tympanic bone of higher forms which has a different origin. The columella, derived from the hyomandibular cartilage of fishes, meets the tympanic membrane distally. Its base articulates with the small stapes (stapedial plate) and also with a slight depression in the pro-otic bone. Three other bones, the pterygoid, palatine and squamosal complete the upper jaw.

The lower jaw consists of a very small pair of anteriorly located cartilage bones, the **mentomeckelians**, together with two membrane bones the dentary and angular, on each side, which form a sheath around Meckel's cartilage (Fig. 10.33 C). Teeth are lacking on the lower jaw. There is no separate articular ossification which in other vertebrates forms the point of articulation with the quadrate. Articulation is by means of an **articular cartilage**.

In anurans the remainder of the visceral skeleton or hyoid apparatus, is greatly modified. In larval forms it is much like that of urodeles. In adults; however, the hyoid apparatus, consists of a broad, cartilaginous basilingual plate with anterior and posterior projections, the cornua (Fig. 10.22 C). The basilingual plate represents the united ventral portions of the visceral arches. The anterior cornua are homologous with the ceratohyals of lower forms. They curve anteriorly, then backward and upward to articulate with the auditory capsules near the fenestra ovalis. The posterior cornua or thyrohyals, corresponding to the fourth visceral arches, are ossified. The chief difference between the skulls of urodele and anuran amphibians are as follows (i) the latter possess an unpaired sphenethmoid, homologous with the paired orbitosphenoids of urodeles (ii) the frontals and parietals which in urodeles are separate bones, have fused to form the fronto-parietals in anurans, (iii) the maxillary bone in anurans is connected posteriorly with the cranium by means of a quadratojugals, (iv) small cartilage bones, the mentomeckelians, are present at the tip of the lower jaw in anurans, (v) the skeletal elements supporting the middle ear are present in anurans, in which a middle ear appear for the first time.

Reptiles: The stem reptiles or cotylosaurs, from which all reptiles extinct and modern, are believed to have evolved, bore such close resemblance to labyrinthodont amphibian that it has not been easy to distinguish the two groups. The general pattern of the roof of the skull is essentially the same in both. Descendants of the cotylosaurs, in general, show a loss or reduction of bones of the skull roof, and numerous changes in structure are apparent.

The chief difference to be noted in the reptilian skull as compared with that of amphibians is the greater degree of ossification and the increased density of bones. Little remains of the embryonic chondrocranium which has become well ossified. Only the ethmoid region has retained its primitive cartilaginous character. All four parts of the occipital complex are ossified, but all do not necessarily bound the foramen magnum. A single occipital condyle is present formed from the basioccipital and usually by a contribution from each exoccipital. A cartilaginous or bony interorbital septum, forming a partition between the orbits, is present in many reptiles. In others in which the brain extends farther anteriorly, the anterior part of the cranial cavity lie in this region and the auditory capsule is ossified and usually remains separate from epiotics and opisthotics. The parietal bones are paired in some species but frequently are fused. The same is true of frontals. An interparietal foramen is prominent in sphenodon and many lizards (Fig. 10.34) but is lacking in other modern reptiles. Membrane bones are more numerous in reptiles than in amphibians. Prefrontal, post frontal, post orbital and lacrimal bones are often well developed.

Chelonia: The skulls of turtles and tortoises are characterized by their solidity (Fig. 10.35). The roof of membrane bones is unusually complete. Teeth are lacking with in the group,

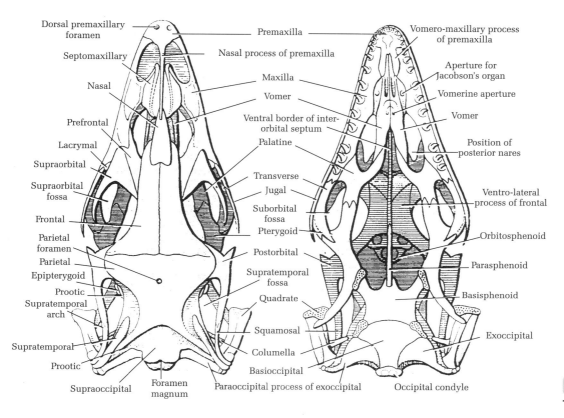

A. Dorsal view

B. Ventral view

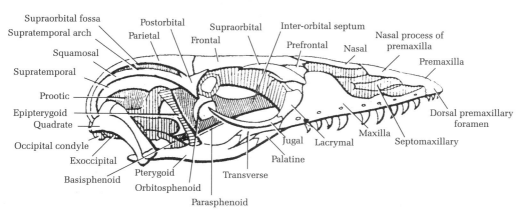

C. Lateral view

Figures 10.34 A, B, C
Varanus skull   **A.** Dorsal view   **B.** Ventral view   **C.** Lateral view

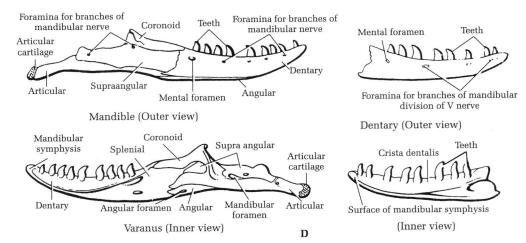

Figure 10.34 D
Bones of mandible

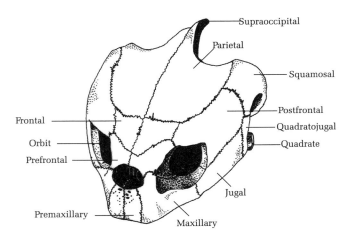

Figure 10.35
Dorso lateral view of skull of sea turtle

although vestiges have been described in embryos of soft-shelled turtle. Otherwise the jaws are covered with a sharp horny beak. In **chelonians** the single occipital condyle is formed from contributions of the basioccipital and the two exoccipitals. The latter bound the side and greater part of the floor of the foramen magnum. All four parts of the occipital segment form cartilage bones. Only the basisphenoid region of the remainder of the chondrocranium is ossified, however, the presphenoid and orbitosphenoid regions remaining cartilaginous. Membrane bones of the frontal segment consists of frontals, prefrontals, and postfrontals. In some primitive chelonians a false roof is located above the true cranial roof, the two being separated by the wide gap. The false roof is composed of portions of the post frontals, parietals, and squamosals. The three elements of the otic capsules are ossified. Only one cartilage bone is present in the upper jaw. This is the quadrate, united with exoccipital, opisthotic, and pterygoid. It joins the squamosal dorsally and the quadratojugal anteriorly. Pre-maxillaries are small. The bones of the lower jaw are generally fused into one piece, and the two sides are firmly ankylosed. As many as six pairs of bones enter into the formation of the lower jaw. The remainder of the visceral skeleton is represented by components of the hyoid and next two visceral arches. The latter may be considerably larger than the part contributed by the hyoid.

Rhynchocephalia: The skull of *Sphenodon*, the only living representative of this ancient group of reptiles, is peculiar in that it possesses two temporal fossae or fenestrae, on each side (Fig. 10.36). The other noteworthy features of the skull of *Sphenodon* include the presence

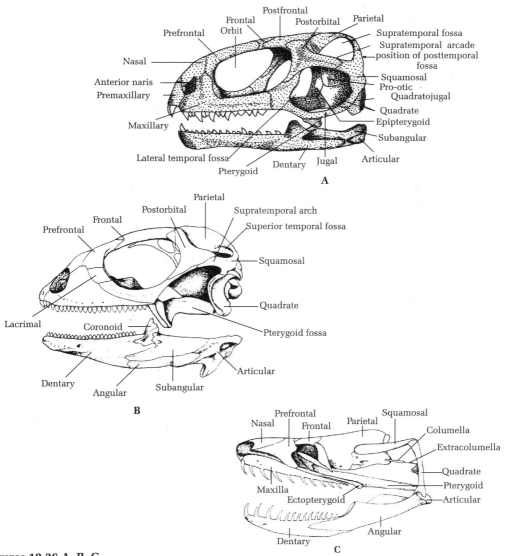

Figures 10.36 A, B, C
Lateral view **A.** Skull of *Sphenodon* **B.** Skull of *Iguana* **C.** Skull of *Boa*

of a rather large interparietal foramen, and teeth on pre-maxillary, maxillary, palatine, vomer, and dentary bones. The premaxillary teeth of the adult are large and appear much like the chisel-shaped incisors of mammals.

Temporal vacuities or fossae: Among the specialization in reptilian skulls are the presence of temporal fossae. The fossae provide working space for the stout muscles that arise on the temporal region of the skull and insert on the lower jaw. Amphibians and fish accommodate their jaw muscles besides the brain case and under the superficial roof bone of the skull which may be continuous or notched. This condition persisted in the most ancient group of reptiles (Cotylosaurs). Subsequent reptilian linage developed more powerful jaws (Fig. 10.37).

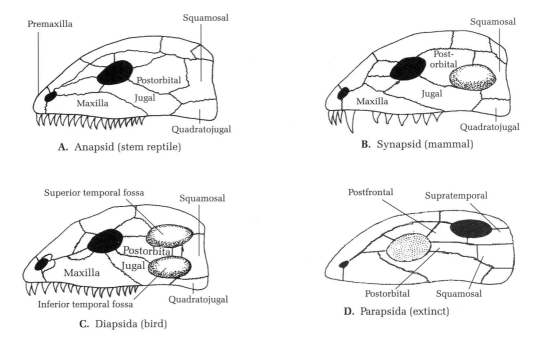

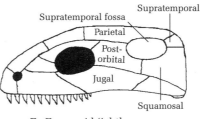

Figures 10.37 A, B, C, D, E
Temporal vacuities in reptiles, birds and mammals

Presumably in response to the resultant increase on altered stresses, the cranial vault strengthened in some places and weakened in others. Ultimately one or two pairs of openings formed in the temporal region of the skull, and jaw muscles then passed through them and got inserted on the lower jaw. It is evident that this process occurred independently several times, because the fenestration evolved at different places on the skull in different group.

Tetrapod skulls, developed one or two vacuities, or fossae, bounded by bony arches in temporal region. On the bases of presence and absence of temporal fossae, the skulls have been classified into :

(i) The **anapsid** condition, in which no temporal (Fig. 10.32 A) fossa or vacuity exist, as in the ancient stem reptiles and turtles. Today only turtles lack temporal fossae, and they are placed, with misgivings, in the subclass anapsida, along with cotylosaurs.

(ii) The **diapsid** condition, such as exists in *Sphenodon* and crocodilians in which supratemporal and lateral (inferior) temporal fossae are present with a supratemporal arcade passing laterally between the two (Fig. 10.37 C). The first birds arose from diapsid reptile and they like crocodilian have diapsid skull.

(iii) The **synapsid** condition is found in mammals and mammals like reptiles in which only the lower fossa persists, this single temporal fossa bounded by the post-orbital, squamosal, and jugal bones, the last two forming an under lying zygomatic arch (Fig. 10.37 B). This synapsid skull was transmitted to mammals.

(iv) The **parapsid** (3.37 D) condition is found in a few extinct forms, in which only the upper fossa remains. The ichthyosaurs and plesiosaurs had one dorsally located temporal fossa which may or may not have been equivalent to the superior temporal fossa of diapsids.

(v) The **euryapsid** condition no longer exists (Fig. 10.37 E). The skull had a single dorso-lateral temporal opening as was found in extinct forms either side bounded below by post-orbital and squamosal bones.

Squamata: An infratemporal arcade is lacking in squamata. The quadrate, except in chameleons, forms a movable union with the squamosal, and a quadratojugal is lacking. The nares are separate. In addition to the internal nares which are anterior in position, large spaces, the palatal vacuities are present in the roof of the mouth. All four parts of the occipital complex surround the foramen magnum.

Palates: Development of secondary palate occurred first in reptiles. In crocodilians it is exceptionally long because of the very long facial region. Palate processes of the premaxilla, maxilla, palatine and pterygoid bones all unite in crocodilians to form a completely bony secondary palate with the internal nares pushed to the rear. In most other reptiles the palatal processes of these bones do not meet, and the secondary palate is incomplete. The primary palatal bones of snakes and some lizards, the quadrate bones, and sometimes certain bones of the upper jaw are movable as independent units, so that some snakes can open their mouth wide enough to swallow objects larger than their own head. This condition is known as **cranial kinesis** (Fig. 10.38). Kinestism was present in crossopterygians and in early amphibians leading to reptiles. It has been transmitted to birds but has been lost in modern amphibians and mammals. The movable quadrate in reptilian precursors of mammals may have facilitated the transition of the quadrate from a bone of the upper jaw to a bone of the middle ear.

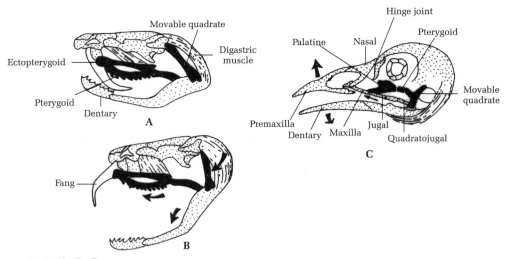

Figures 10.38 A, B, C
A, B. Cranial kinetism in the palate of snakes   C. Birds. The arrows indicate movements of components

Acquisition of importance by the dentary bone   In ancient reptiles evolving to mammals the dentary became increasingly prominent. A ramus developed on the dentary and extended upward towards the temporal region. It served for the attachment of the increasingly massive lower jaw muscles. Meanwhile, the other bones of the mandible became reduced, signalling the loss in mammals of all mandibular bones except the two dentaries.

Sauria: The presence of an imperfect interorbital septum characterizes the lizard skull. In only one small family of limbless lizards (Amphisbaenidae) does the cranial cavity extend forward between the orbits. Most lizards possess an interparietal foramen. The secondary palate in lizards is not at all complete. As in all other reptiles a single occipital condyle is present on the upper jaw, teeth are borne by pre-maxillaries and maxillaries. Occasionally they are also present on pterygoids and palatines.

The lower jaw is formed of six pairs of membrane bone surrounding Meckel's cartilage. The anterior dentary bears all the teeth of the lower jaw. The coronoid forms an upward projecting coronoid process immediately behind the last tooth (Fig. 10.36 B). The coronoid process extends into the temporal fossa. The two halves of the lower jaw are firmly united by a mandibular symphysis.

Serpentes: The skull of snakes (Fig. 10.36 C) differs somewhat from that of other reptiles in correlation with their habit of swallowing large prey. Many of the facial bones are rather loosely connected thus permitting a considerable degree of adjustment during the act of swallowing. The quadrate connects indirectly with the skull by means of the squamosal, which is interposed between quadrate and parietal. The pterygoid articulates posteriorly with the quadrate and anteriorly with the palatine. Jugals and quadratojugals are absent. The maxillary bone is connected indirectly with the pterygoid by the transpalatine (ectopterygoid). It is in the arrangement of the premaxillary and maxillary bones that the skulls of poisonous and non-poisonous snake differ mostly. In venomous snakes the premaxillary is very small and does not bear teeth. The maxillary is also small in venomous forms. Anteriorly it unites by a movable articulation with the lacrimal, which in turn is loosely connected with the frontal. In such poisonous snakes as vipers and pit vipers, these bones are so arranged that when the mouth is closed the fangs on the maxillary bone lie back against the roof of the mouth. This position is assumed because of traction exerted by the pterygoid on the transpalatine and hence on the maxillary. When the snakes strike, its mouth is opened and the quadrate is thrust forward. This in turn causes the pterygoid and transpalatine to be forced forward. As a result, the position of the maxilla changes so that its toothed surface turns downward and is in the most effective position for striking. In cobras, coral snakes, and sea serpents the fangs are permanently erect. In the snake skull the cranial cavity extends between the orbits so that interorbital septum is reduced. Snakes lack both supra-temporal and infratemporal arcades as well as interparietal foramina. Nevertheless, this is considered to be a modified diapsid condition. Lower jaw consists of several component bones surroundings Meckel's cartilage. In most species the parts of the original hyoid apparatus are lost except for an occasional remnant.

Crocodilia: The crocodilian skull is dense and massive, with a firm union between the separate bones. The outer surface of the skull bones is unusual in being rough and pitted. Of the four occipital bones only the basioccipital and exoccipitals bound the foramen magnum. Crocodilians, like sphenodon, have two temporal fossae on each side. This, is the diapsid condition. No unusual features are to be observed in connection with the sensory capsules. The upper jaw is very well developed in crocodilians. The two halves of the lower jaws are firmly united at the symphysis. Each is composed of six separate bones. The dentary, which unites with its mate to form the mandibular symphysis, is the largest and bears the teeth of the lower jaw. The hyoid apparatus of crocodilians is represented by a cartilaginous basilingual plate and pair of cornua.

Birds: It is generally agreed that birds arose in evolution from the branch of archosaurs, or "ruling reptiles" which included dinosaurs with bipedal locomotion. The skull of birds has not deviated far from the reptilian type (Fig. 10.39). It resembles that of lizards in a number of ways. Some differences in structure of the palate serve to distinguish members of the super orders paleognathae and neognathae. A single occipital condyle is present which, instead of being located at the posterior end of the skull, lies somewhat forward along the base. The skull articulates with the vertebral column almost at a right angle, thus accounting for the more anterior position of the condyle. With the greater development of the brain in birds, the cranial cavity has increased in size, mostly by lateral expansion. The orbit is large to accommodate the relatively massive eye. A complete supratemporal arcade is lacking, but the infra-temporal arcade, though slender, is completely formed. This condition in bird is essentially diapsid, although modified.

In correlation with their adaptation for flight, birds have skulls unusually light in weight. The cranial bones are fused together to a high degree. All four components of the occipital complex surround the foramen magnum. When the supraoccipital join the parietal, a prominent crest, the lambdoidal ridge, is present. Most conspicuous is the great development of the premaxillary bones, which together with the maxillaries form the upper portion of the beak, or bill. Although, **Archaeopteryx**, **Hesperornis**, and **Ichthyornis** possessed true teeth, the teeth are of course, lacking in modern birds, the beak being covered with a horny sheath derived from the epidermis. Contrary to the condition in crocodilians, there is no complete bony secondary palate. Since the quadrate is movable and the palatopterygoid can slide along the rostrum of the basisphenoid, the upper jaw can be raised or lowered within limits.

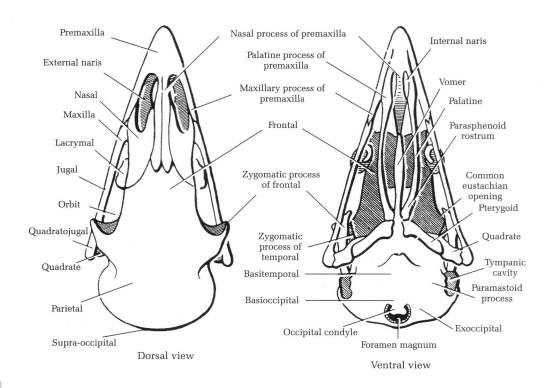
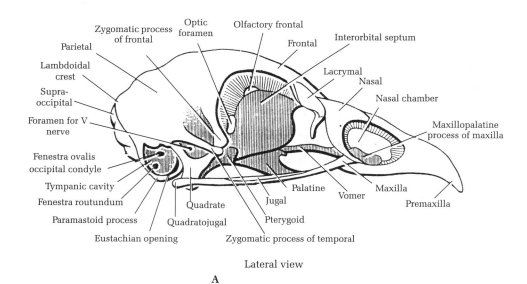

Figure 10.39 A
Fowl. Skull in dorsal, ventral and lateral views

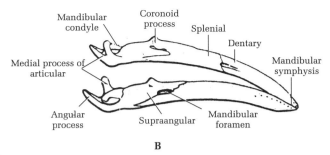

Figure 10.39 B
Fowl. Mandible (complete)

The infratemporal arcade may aid in brining about such movements. Movement of the upper jaw is best seen in parrots (Fig. 10.40). In these birds, the upper beak is not firmly united with cranium but forms a flexible attachment, so that a movable joint is present. The nasals, and the loose connections of the premaxillary bones, attach the beak to the cranium.

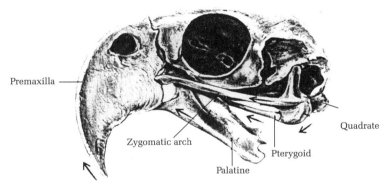

Figure 10.40
Kinetic skull in the cockatoo. The arrows indicate how the skull moves to raise the upper jaw.

The lower jaw consists of two rather flattened halves, united anteriorly and articulating with the quadrate posteriorly. Actually each half is composed of five bones fused together (Fig. 10.39). Only the articular is a cartilage bone. The anterior dentary makes up the greater part of the lower jaw. In birds the hyoid apparatus consists of a median portion composed of three bones placed end to end, and one or two pairs of cornua of horns. The three median bones are called respectively the entoglossal, bashiyal and urohyal. The anterior endo-glossal, which extends into the tongue, represents the fused lower ends of the ceratohyals. The median basihyal corresponds to the basihyal of lower forms. The posterior urohyal is probably the basibranchial of the third visceral arch (Fig. 10.25). The anterior cornua, which project from the entoglossal, are the free ends of the ceratohyals. The long and posteriorly directed posterior cornua, which articulate at the junction of basihyal and urohyal, represent the ceratobranchials and epibranchials of the third visceral arch. The posterior cornua are extremely long in wood peckers in correlation with the feeding habits of these birds in which the tongue, supported by the hyoid apparatus, can be thrust out for some distance.

Mammals: Mammals like birds, have sprung from ancestral reptilian stock but their history indicates a very early divergence from the main reptilian line. The cynodont, or mammal-like, reptiles are generally considered to be in the direct line of ascent of mammals but even they were rather far removed in time from the oldest known reptiles, the primitive pelycosaurs. The skulls of the latter differed little from those of the most ancient amphibian but already showed a trend toward mammalian evolution unlike most reptiles, which have a single occipital condyle. The cynodonts possessed two occipital condyles, as do amphibians and mammals. Although much variation exists in the skulls of different groups of mammals, there are certain features which they share in common and which serve to distinguish them from lower forms (Fig. 10.41). Most noteworthy is the increased capacity of the cranium, which has expanded in dorsal and lateral directions in correlation with the much greater size of the mammalian brain. Two occipital condyles, formed mostly by the exoccipitals, articulate with the atlas vertebra. In higher apes and man the occipital condyles are decidedly ventral in position. Articulation of mandible and upper jaw is at the glenoid, or mandibular fossa of the squamous region of the temporal rather than at the quadrate. This is the craniostylic or amphicraniostylic method of jaw suspension. Teeth of the upper jaw are borne only by pre-maxillary and maxillary bones. The pre-maxillaries frequently fuse together. The thecodont method of tooth attachment is the rule. Except in the toothed whales, which have homodont dentition, the teeth of mammals are heterodont. The number of bones in mammalian skull is less than in most lower forms. There are usually about 35 of them. This is the result of loss of certain bones and fusion of others. In most mammals a single occipital bone containing the foramen magnum is present. This also represent a fusion, although the four separate elements of which it is composed can easily be identified during development. A supratemporal arcade is lacking, but the infratemporal arcade, or zygomatic arch is invariably present. As a result, mammals possess but a single temporal fossa, an example of synapsid condition.

A hard or bony secondary palate is typically found in mammals. In whales and certain edentates the pterygoid form a portion of the hard palate, but these bones are much reduced in most mammals, so that the choanae are usually bordered by the palatines above. The nasal cavities of mammals are large in correlation with the highly developed olfactory sense.

Many interesting variations of mammalian skull are to be observed. Among these is the presence of hollow bony projections of the frontals in many members of the order artiodactyla. They are covered with horn derived from the epidermis. The antlers of deer, which develop under the organizing influence of the integument in response to stimulation by certain hormones, when fully developed are naked projections of the frontal bones.

Large air spaces, or sinuses, in communication with the respiratory passages, are present to varying degrees in the skulls of mammals. The most common are the frontal, ethomoid, sphenoidal, and maxillary sinuses, the last being often referred to as the **antra of Highmore.** The lower jaw, or mandible, of mammals differs from that of other vertebrates in its articulation with the squamosal rather than with the quadrate.

The mammalian hyoid apparatus shows much variation (Fig. 10.25). Typically it consists of a body, or basihyal, and two pairs of cornua or horns the hyoids apparatus serves to support the tongue and furnishes an area for muscle attachment. Despite the fact that mammalian skulls are, in general, built upon the same pattern, there are some very evident differences which distinguish monotremes, marsupials and the placental mammals. Monotremes more closely resemble the ancient reptilian stock, being the only mammals which have retained the post temporal fossa and paired vomer bones. Pterygoids and palatines, unlike those of higher mammals, contribute to the cranium proper. The cranial cavity of the marsupial is relatively very small in comparison with that of the placental mammals. Openings or fenestrae, are present in the palate; the lower jaw, or mandible, has a peculiar angle.

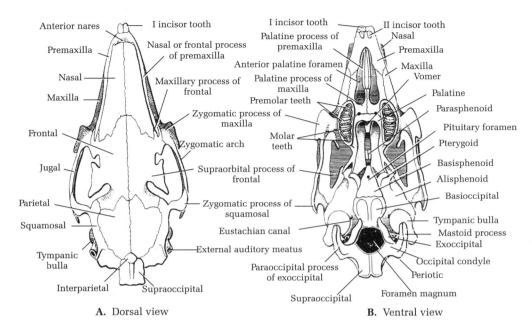

A. Dorsal view

B. Ventral view

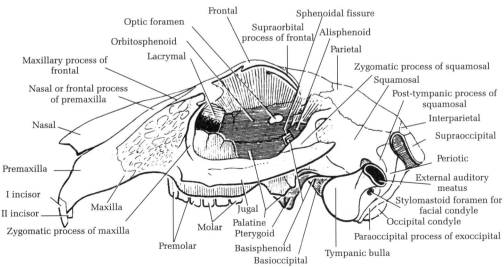

C. Lateral view

Figures 10.41 A, B, C
Rabbit Skull

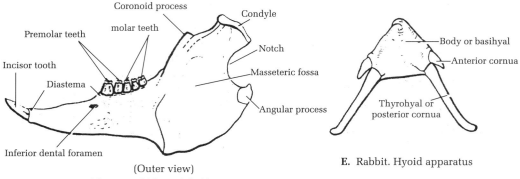

E. Rabbit. Hyoid apparatus

(Outer view)
D. Rabbit. Mandible. Left half

Figures 10.41 D, E
Rabbit   **D.** Mandible   **E.** Hyoid apparatus

Vertebral Column

The primitive axial skeleton consists of the notochord, which is present during early development in all chordates but later get replaced by the vertebral column which extends posteriorly from the base of the skull and continues to the tip of the tail. The portion of the axial skeleton which protects the spinal cord is, in most chordates composed of a series of segmentally arranged skeletal structures, the vertebrae.

Typical vertebra   A typical mammalian vertebra (Fig. 10.42) is composed of a solid, ventral cylindrical mass, the centrum or body, above which is the neural arch surrounding the neural (vertebral) canal. The ends of the centrum are smooth and joined to the centra of adjacent vertebrae by means of intervening fibrocartilaginous **intervertebral discs** (Fig. 10.44). The lateral portions of the neural arch consists of two upright pedicels or neural processes, fastened to the dorsal part of the centrum on either side. In most forms a flattened plate of bone, the **lamina**, extends medially from the upper part of each pedicle to form the roof of

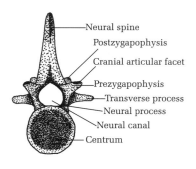

Figure 10.42
Typical mammalian vertebra (anterior view)

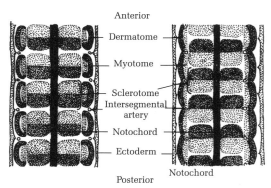

Figure 10.43
Vertebrate embryo showing somite and differentiation of sclerotomes and vertebrae

the neural arch. A dorsal median projection, the neural spine, or spinous process, is formed as the result of fusion of two laminae. Anterior and posterior articular processes, the pre-zygapophyses and postzygapophyses, are projections which serve to join the neural arches of adjacent vertebrae. The pre-zygapophyses of one vertebra articulate with the postzygapophyses of the vertebra immediately anterior to it. Smooth articular facets are present on both prezygapophyses and postzygapophyses; those on the prezygapophyses usually are on the dorso-medial surface, where as those of the postzygapophyses are generally on the ventro-lateral surface, when examined a single vertebra, it is thus possible to distinguish anterior and posterior ends merely by noting the position of the articular surfaces of the zygapophyses. Various other projections, often referred to as **transverse processes**, extend outward from the lateral sides of the vertebrae. These processes are of several types:

(i) **Diapophyses** arising from the base of the neural arch, serve for attachment of the tuberculum, or upper head of the two headed rib (Fig. 10.58).

(ii) **Parapophyses** from the centrum, serve for attachment of the lower or capitular, head of a Rib.

(iii) **Basapophyses** extending ventro-laterally from the centrum are believed to be remnants of the haemal arch.

(iv) **Pleurapophyses** projecting laterally, are extensions with which ribs have fused.

(v) **Hypapophysis** is another type of projection which is found in certain forms and extend ventrally from the median part of the centrum.

The anterior and posterior ends of the neural arches are notched so as to form anterior and posterior intervertebral notches. When two vertebrae are in apposition, anterior notch of one vertebra matches with the posterior notch of the other Vertebra and it forms an intervertebral foramen (Fig. 10.44). This serve for the passage of a spinal nerve as it emerges from the neural canal.

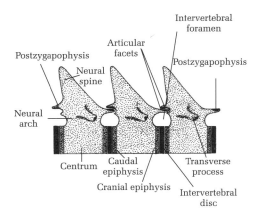

Figure 10.44
Typical vertebrae from young mammal showing epiphyses

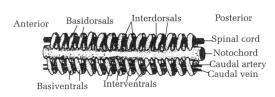

Figure 10.45
Arrangement of cartilaginous arcualia derived from the sclerotomes in relation to spinal cord, notochord, caudal artery and caudal vein

Development of a vertebra It develops from the mesenchymal cells derived from the paired segmentally arranged dorsal part of the mesoderm which eventually becomes differentiated into segmental myotomes and dermatomes (Fig. 10.43). Entire development follows in three steps.

(a) **Mesenchymal stage** Masses of mesenchymal cells budding off from the dorsal mesoderm (epimere) first collect on either side of the notochord to form the skeletogenous layer. Mesenchymal cells soon take the form of segmental masses called sclerotomes (Fig. 10.43). Myotomes from which segmental muscles are formed, lie opposite the sclerotomes and are likewise segmental. Soon each sclerotome is divided into halves called sclerotomites. These become differentiated into an anterior or cranial half and a posterior or caudal half, the latter is denser. Anterior and posterior sclerotomites separate and rearrange such that the cranial sclerotomite of one sclerotome unites with the caudal sclerotomite of the sclerotome in front of it. This arrangement brings the sclerotomes in the intersegmental positions in relation to other segmental structures, particularly, the myotomes which remain segmental and now alternate with the sclerotomes. Such a rearrangement is essential to enable the myotomes to stretch from one vertebra to the other and provide movements between the vertebrae. It may once again, be emphasized that because of this rearrangement, each vertebra develops from parts of two sclerotomes.

Mesenchymal cells of the sclerotomes on either side of the notochord now begin to surround the notochord, forming the rudiment of the centrum, some more cells take dorsolateral positions along the nerve cord above, eventually surrounding it and laying the rudiment of the neural arch. Another arch of cells passing down the ventro-lateral sides of the notochord forms the rudiment of the haemal arch or ribs.

(b) **Cartilage stage** From the mesenchyme of each half of sclerotome develops 4 cartilages, the arcualia, two dorsal and two ventral (Fig. 10.45). Dorsal arcualia are called basidorsals and interdorsals, similarly, basiventrals and interventrals are the ventral arcualia. Basidorsal and basiventrials develop from the dense cranial part (original caudal) and form the anterior parts of the vertebra. The interdorsals and interventrals develop from the less dense caudal part of the sclerotome (the original cranial part) and form the posterior parts of the vertebra. Basidorsals of a given sclerotome arch over the spinal cord, giving rise to neural arch and neural spine, the latter formed by the fusion of the basidorsals. Basiventrals in the tail region give rise to haemal arch and haemal spine. Anterior to the tail region basiventrals contribute to the centrum and give rise to ribs. Interdorsal and interventral cartilages of a sclerotome largely contribute to the centrum.

(c) **Bone stage** Only in elasmobranchii, vertebrae remain cartilaginous through out the life, in all other cartilages are replaced by bone. Endochondral ossification of the vertebrae is best seen in the mammalian vertebrae where the main part of the body represents the bony diaphysis and the two ends are cartilaginous. Separate centres of ossification appear in the cartilages to form the epiphyseal discs after quite sometime.

Formation of centrum There are two modes of centrum formation: (i) In the **elasmobranchii** and **Dipnoi** fish, mesenchyme cell leave the arcualia and invade the notochord, penetrating the notochordal sheath and becoming established in the latter. **Chondrification** (changing into cartilage) of the mesenchyme leads to the formation of a cartilaginous investment, the perichordal tube, within the limits of notochord. Not only the notochordal sheath is invaded, a considerable part of the chordal tissue is also invaded and changed into cartilage. Notochord is finally converted into **centrum** of the vertebra in which it is constricted and reduced to a thread, however, it remains thick between two adjacent centra (Fig. 10.49). Centra formed by the invasion of the notochord are called **chordal centra** and such vertebrae that have them are called **chordocentrous**. Notochord in these vertebrates presents a beaded appearance (monoliform).

(d) **In other vertebrates, the centrum develops**, not by invasion of the notochord, but the mesenchymal cells from the **arcualia** or the mesenchyme lying between the **arcualia** and around the notochord (perichordal mesenchyme) derived directly from the sclerotome, or generally both, changes into cartilage without penetrating the sheaths of the notochord. Thus, a cylinder of cartilage is laid around the notochord to be known as perichordal centrum and such vertebrae as **arcicentrous vertebrae**. In **teleosts** and **amphibia** perichordal centra are derived largely from the perichordal mesenchyme, **arcualia** contribute nominally.

Direct ossification often takes place in the mammalian vertebrae, there being no cartilage stage, such centra are called **husk centra**. Chordal centra in the true sense are rare, usually, some perichordal mesenchyme also takes part in their development. Perichordal centra of bony fishes and other higher vertebrates are so compact that they obliterate the notochord. In mammals the mesenchyme between the vertebral centra gives rise to intervertebral discs of fibro-cartilage, obliterate the notochord, however, it persists as the nucleus pulposus or pulpy nucleus of the inter-vertebral disc.

Double centrum theory Embolomerous fossils of most primitive land vertebrates, the amphibia, give the evidence of a double centrum in each vertebra, an anterior **hypocentrum**, believed to have been derived from the basiventrals, and a posterior pleuro-centrum, believed to have been formed from the basi-ventral and interventrals (Fig. 10.45). Homologies of the two centra are not yet clear. Some palaentologists believe them to have been derived from the perichondral mesenchyme, and later on hypocentra fused with the basiventrals, and pleurocentra with the interdorsal, vertebrae of such earliest known fossil amphibia are termed **embolomerous** (Fig. 10.46). Evolution of centra from this primitive forms seems to have been along two different lines, one leading to the condition found in later extinct and present amphibia, in which the hypocentrum had become the main centrum and pleurocentrum was reduced or completely lost, such vertebrae are known as **stereospondylous**. The other leading to the condition found in the reptiles, birds, and mammals, in which pleurocentrum is the main centrum, hypocentrum is reduced, such vertebrae are called **gastro-centrous** (Fig. 10.47).

Since the centrum of amphibia represents the hypocentrum (basi ventrals) the ribs are attached to the centrum, while in the amniota, ribs are not attached directly to the separate elements called intercentra representing the hypocentra. Intercentra and haemal arches together are spoken of as **chevron bones**.

Vertebral structure in different vertebrates

In living cyclostomes, dipnoi, and the coelocanth (*Latimeria*), the adult notochord is not

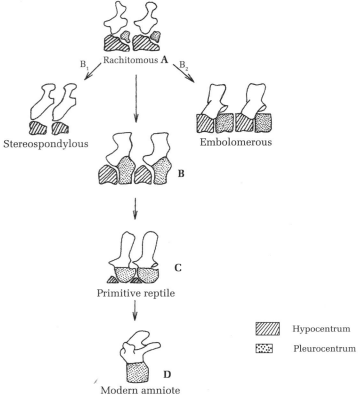

Figure 10.46
Modification from labyrinthodonts to modern amniote vertebrae (B_1, B_2 - from other labyrinthodonts)
B - labyrinthodonts from reptilian line

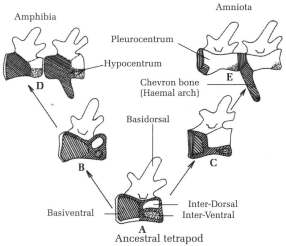

Figure 10.47
Development of amphibian and reptilian vertebrae from ancestral tetrapod
A → B → D typical amphibian   A → C → E typical amniote

constricted by the ventral column and is the primary axial skeleton, preventing telescoping and allowing lateral undulations. There are some inter segmental cartilages and bones in *Latimeria*.

In other fish each centrum encircles the notochord. This prevents further expansion of the notochord which continues to grow. The centrum also continues to grow, enlarging anteriorly and posteriorly and surrounding these expanded intervertebral portions of the notochord. The result is a centrum with concave anterior and posterior surfaces and a tiny hole in the centre filled with notochord. This is **amphicoelous centrum** (Fig. 10.48). The characteristic of **chondrichthyes** and **actinopterygii**, allows limited movement between centra. In urodeles and in the neck region of hoofed mammals the posterior surface of each centrum is concave and the anterior surface convex; this centrum shape is called **opisthocoelous**. The reverse is true in anurans and in the majority of reptiles, where the anterior face of the vertebra is concave and the posterior convex, a condition known as **procoelous** (Fig. 10.48).

In mammals, with the exception of the neck region of hoofed mammals, both anterior and posterior surfaces of the centrum are flat, a condition known as acoelous or **amphiplatyan**. In contrast, the centra faces are complex in birds, being roughly saddle-shaped; considerable movement is possible between these **heterocoelous** vertebrae, particularly in the neck region.

Differentiation and regionalisation of the vertebral column

In *Branchiostoma* notochord persists in the most typical form and no vertebrae are formed. Sclerotomes, however, give rise to form connective tissue sheaths that surround the notochord and nerve cord.

In **fishes** vertebral column of chondrichthyes shows remarkable advancement over that of the cyclostomes. **Fully** developed vertebrae are found throughout the length of the vertebral column and there is a clear differentiation into trunk and tail vertebrae. Development of vertebrae stops at the cartilage stage and vertebrae are **chordocentrous** and **amphicoelous** i.e. having both the anterior and posterior faces of the centrum concave and the notochord is **monoliform** (Fig. 10.49). A typical trunk vertebra of *Scoliodon* has an amphicoelous centrum surrounding the notochord which, here, is reduced to a fine thread, a dorsal neural arch enclosing the spinal cord, a short blunt neural spine and a pair of transverse processes arising ventro-laterally from the centrum, hence probably representing the remnants of haemal arch. Notochord remains un-invaded between two successive vertebrae. Centra are not purely cartilaginous, calcification takes place and fibro cartilage develops in the whole of the centrum except at four wedge-shaped places. A centrum cut across, show the calcified areas reaching a deep into the centre and presenting the appearance of the cross, popularly known as **maltese cross**. Centra in which the calcified area present a cross or starlike appearance, are called **asterospondylous.** In other forms of centra such as **cyclospondylous** the calcified area is confined in the middle of the centrum and around the centre. In **tectospondylous**, where the central calcified ring is surrounded by alternate rings of calcified and fibrous material.

Vertebral column of *Scoliodon* is primitive in the sense that still there are separate inter dorsals forming inter neural plates. Neural plates is perforated by a foramen for the passage of the ventral root of a spinal nerve and the interneural plate bears a foramen for the passage of the dorsal root. Tail vertebrae have longer neural spines and the transverse processes are absent. Strongly developed haemal arches and haemal spines of tail vertebrae are derived from the basiventrals. Haemal canal encloses the artery and vein of the tail region. In the region of the tail fin haemal spines (also called hypurals) are flattened at the extremities and

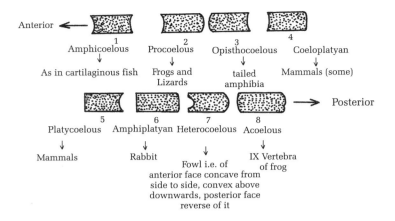

Figure 10.48
Types of centra in vertebrates

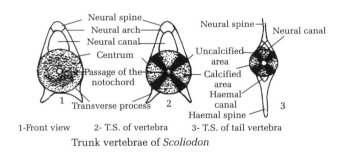

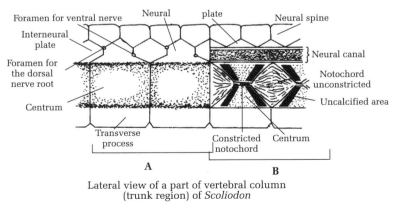

Figure 10.49
Vertebrae of *Scoliodon*   1, 2 Trunk vertebra   2. T.S. vertebra   3. T.S. caudal vertebra
bottom   **A.** Left - two vertebrae entire   **B.** Right - two vertebrae in L.S.

support the ventral lobe of the tail fin (Fig. 10.49). Dorsal lobe of the tail fin is similarly supported by the neural spines of the tail vertebrae.

Primitive actinopterygii represented by sturgeons, garpikes and bowfins, show some outstanding features. In sturgeons, notochord is unconstricted, basi and interdorsals, and basi and interventrals are present as separate elements, thus, centra are completely lacking, In the garpike, centra are opisthocoelous. In the tail region of the bowfin (*Amia*) there are two kinds of vertebrae: (a) vertebrae with a single centrum, a neural arch, and a haemal arch and (b) vertebrae with a double centrum. Posterior centra with neural and haemal arches are the basidorsals and basiventrals, and those without, are inter dorsals and interventrals. Presence of a double centrum is a condition called **diplospondyly** as against the normal monospondyly in which there is one centrum in each myotomal segment.

Vertebrae of teleosts are compact, bony, chordocentrous and usually amphicoelous. Opisthocoelous centra are found in some cases and in some teleosts vertebrae have articular processes, the zygapophyses.

Tetrapods: Vertebral column is divisible into regions corresponding to neck, thorax, sacrum (abdomen) and the tail. Vertebrae are provided with articular processes for allowing movement between them. First neck vertebra is always modified to carry the head and is known as the atlas. At least one vertebra articulates directly with the pelvic girdle, and is called sacral.

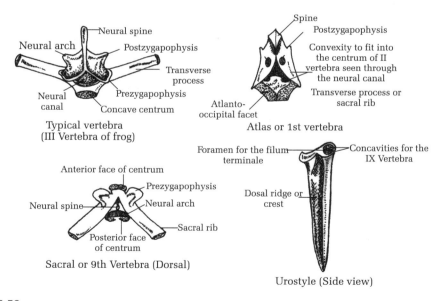

Figure 10.50
Typical vertebrae of frog

Vertebral column in amphibia: Frog and tail-less amphibia in general have shortest vertebral column (Fig. 10.50). Indian frog *Rana tigrina* has 9 vertebrae and a long bone the urostyle. First vertebra is the only **neck or cervical** vertebra. Vertebrae from II-VII are typical trunk vertebra. VIII is the presacral vertebra and IX is the sacral vertebra. Vertebrae of frog are stereospondylous, since hypocentrum alone supports the neural arch. All typical vertebrae are **procoelous**, having a concavity in front and a convexity behind, VIII is **amphicoelous**, IX **acoelous** or biconvex and the first acentrous. A typical vertebra has a centrum supporting a neural arch produced into a neural spine and the neural arch bears pre and postzygapophyses, and a pair of transverse processes. Cartilaginous tips of the transverse processes represents the ribs. Atlas vertebra has no transverse processes or prezygapophyses. It bears on its front face a pair of articular facets for articulation with the occipital condyles of the skull. The VIII centrum receive the convexity of the VII in front and convexity of the IX behind. Sacral vertebra has two convex projections over the posterior face of the centrum, which fit into corresponding concavities of the urostyle. Stout backwardly directed transverse processes or sacral ribs articulate with the ilium bones of the pelvic girdle forming the sacroiliac joint. Urostyle or tail piece has two concavities for receiving the IX centrum, a crest or ridge on the dorsal side, a pair of small openings near the anterior end for the exit of nerves, and a narrow neural canal in which lies the filum terminale.

Tailed amphibians They have a large number of vertebrae, in their column as many as upto 250 in some. The vertebral column shows distinct cervical, thoracico lumbar (between the cervical and sacral vertebrae) sacral and a large number of caudals in the tail. All caudal vertebrae, except the first, bear haemal arches and haemal spines. Vertebrae of the abdominal and sacral regions have ribs articulating with the flattened and divided transverse processes. Vertebrae of *Triturus* are long and cylindrical, they have well developed centra and neural arches. Tail vertebrae have haemal arches also. All trunk vertebrae have lateral transverse processes with which the ribs articulate. In the Indian newt, the sharp free ends of ribs penetrate the integument, apparently as a protective device. There are cartilaginous pads on the front and rear surfaces of the centrum of vertebra and the notochord is much constricted. First vertebra or atlas bears-paired articular surfaces for the two occipital condyles of the skull. The anterior face of the centrum gives off a small projection called odontoid process; it lies between the articular surfaces of occipital condyles. Sacral ribs are attached to the ilium bones of the pelvic girdle. Limbless amphibia have amphicoelous centra, very short; caudal region and no sacrum.

Vertebral column of reptiles: *Varanus* and other lizards have long vertebral columns, 83-85 vertebrae in *Varanus* and 56-60 vertebrae in *Uromastix*, with distinct regions, such as cervical (neck), thoracico-lumbar, sacral and caudal (Fig. 10.51). Pleurocentrum is the main centrum, the vertebrae are therefore, **gastrocentrous**. However, reduced hypocentrum lingers on in some vertebrae. The centra are procoelous.

The cervical vertebrae (9 in *Varanus* and 8 in *Uromastix*) except the first two, have a procoelous centrum, a neural arch with well developed anterior and posterior zygapophyses, a neural spine and a pair of transverse processes. Cervical ribs are very short and do not reach the sternum. Hypocentrum lingers on the ventral side of the centrum as a bony nodule called **hypapophysis**.

First cervical vertebra or atlas is ringlike and formed by three separate ossification, a ventral and two dorso-lateral, the latter do not meet above but are covered with a membrane. Ventral piece bear an anterior facet for receiving the single occipital process of the second vertebra. A ligamentum transversum divides the canal of the atlas into an upper spinal canal for the spinal cord and the lower odontoid canal for the odontoid process. Prezygapophysis, transverse processes and centrum are absent.

Second cervical vertebra, the axis or epistropheus, is like a typical cervical vertebra bearing both pairs of zygapophyses, a laterally compressed neural spine and well developed

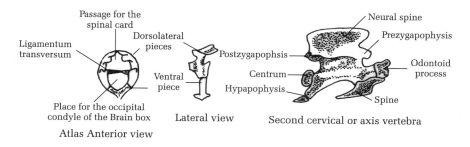
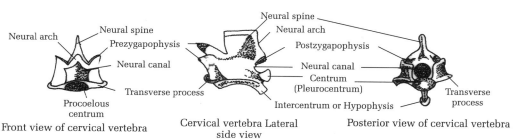
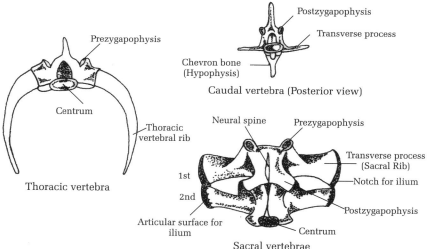

Figure 10.51
Vertebrae of *Varanus*

procoelous centrum to which hypapophysis is attached. Anterior face of the vertebra bears a tooth-like, bony, roughly rectangular odontoid process. It may be recalled that atlas has no centrum; odontoid process is really the pleurocentrum of the atlas, and a bony backwardly directed process below the odontoid process is the hypocentrum (intercentrum) of the atlas. Odontoid process serves as a pivot, which permits considerable freedom of movement of the head.

Thoracico-lumbar vertebrae are 21 in *Varanus* and 16 in *Uromastix*. They resemble the typical cervical but do not have hypapophyses and the neural spines are short. All have a pair of single headed (holocephalous) vertebral ribs attached to the vertebrae at the junction of the centrum and neural arch in capitular facets. Ventral ends of the ribs of first five vertebrae reach the sternal ribs. Sternal ribs are made up of calcified cartilage. Last seven vertebrae have small ribs. **Sacral** vertebrae, only two, are stoutly and firmly held together. They have short procoelous centra, without hypapophyses. Transverse processes are strongly expanded and notched at the outer ends for the ilium bones of pelvic girdle with which they articulate. Fusion of the two sacral vertebrae forming the sacrum provided strength and rigidity in this part of the body.

Caudal vertebrae (49-51 in *Varanus* and 30-34 in *Uromastix*) are recognized by the presence of a pair of small pegs (hypocentra) below the centrum and known as intercentra, attached to these are Y-shaped haemal arches and spines collectively called **chevron bones.** Caudal vertebrae have procoelous centra, well developed neural spines, and transverse processes. Posterior caudal vertebrae gradually become smaller and smaller till they are left only with elongated rod-like centra. Caudal vertebrae in lizards including the wall lizard, possessing the quality of autotomy in self-defence, have a narrow transverse and an unossified zone traversing the centrum. A vertebra readily breaks at this zone.

Normal division of the vertebral column seen in lizards are found in crocodilians too, but in the **turtles**, **snakes** and limbless lizards such clear divisions are wanting. In turtles, there is no true lumbar region, the cervicals and all caudal vertebrae except the first, are the only movable vertebrae. Ten thoracic vertebrae (2nd to 9th inclusive) do not have transverse processes, and the thoracic ribs and neural spines are firmly united with the carapace. Thoracic vertebrae themselves are immovably united with one another by means of fibro cartilaginous inter-vertebral discs. Two sacrals and first caudal are also fused with the carapace. In **snakes** and **limbless lizards** there are only two divisions of the vertebral column, the pre-caudal and caudal regions. The vertebrae of snakes certain lizards and of *Sphenodon*, have additional articular surfaces which coexist with the normal zygapophyses (Fig. 10.52). Additional articular surfaces are a pair of **zygosphenes** at the anterior face of the neural arch, and a pair of **zygantra** at the posterior face. Zygosphenes of one vertebra articulate with the zygantra of the next anterior vertebra. Vertebrae of snakes have well developed vertebral ribs. The hood of cobra is formed when the long cervical ribs are raised. A sacrum in snakes and limbless lizards is weak or absent.

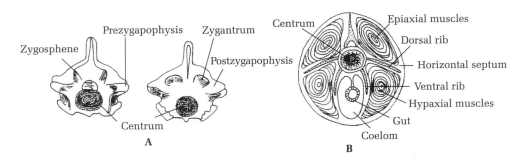

Figures 10.52 A, B
A. Anterior and posterior view of vertebrae of snake   B. T.S. Body of a vertebrate showing the ribs

Ribs are developed in cervical and thoracic vertebrae, in the caudal region they are replaced by haemal arches. But in *Sphenodon*, crocodiles and turtles there are caudal ribs which become fused with the vertebrae, and in addition to thoracic ribs there are abdominal ribs. Each rib of *Sphenodon* and crocodilia has a backwardly directed uncinate cartilage connected with it.

Vertebral column of birds: It is remarkable for rigidity and compactness in every part except the neck which is long and highly mobile (Fig. 10.53 A). Vertebrae have full complements of basidorsals, basiventrals, interdorsals and inter ventrals and are gastrocentrous i.e. having pleuro centra as the main centra. Centra are saddle shaped or heterocoelous—anterior face is convex from side to side and concave from above downwards; the posterior face is just the opposite of it, being concave from side to side and convex from above downwards.

There are 13 cervical vertebrae in the fowl (14 in pigeon), the first and second being modified into atlas and axis respectively. A typical cervical vertebra any one from 3rd to 9th is elongated, with a heterocoelous centrum, well developed zygapophyses and transverse processes pierced by **foramina transversaria** or vertebraterial canals for the passage of vertebral arteries and

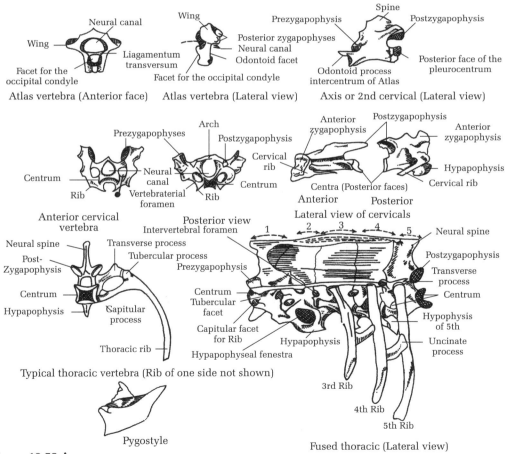

Figure 10.53 A
A. Vertebrae of Gallus

veins. Last three cervicals differ from other cervicals in having double-headed vertebral ribs. One of the heads (capitular) articulates with the centrum, the other (tubercular) articulates with the transverse process. These ribs do not reach the sternum. In all cervicals except first two, ribs are present but not well developed; they are small backwardly-directed processes and their heads are fused with the parts of vertebrae with which they articulate. Posterior cervicals have compressed **hypapophyses**.

Atlas vertebra is ring-shaped, and separate pieces which make it, are firmly united. There is a large rounded facet on its anterior face for the single occipital condyle of the skull. A thin ligament divides the cavity as in the atlas of varanus. Axis vertebra has the same morphological features as in *Varanus*, having at its anterior end a peg-like odontoid process or dens. There are seven thoracic vertebrae, first being the one whose ribs reach the sternum. Ribs are distinctly double-headed and the vertebrae show a great degree of fusion. Each thoracic rib and the rib of last cervical has a backwardly-directed bony uncinate process overlapping the rib next behind.

Hypapophyses for the attachment of flexor muscles of neck are present in the anterior thoracic vertebrae. First thoracic vertebra is free, second to fifth are fused, their neural spines are united to form a compressed plate for the attachment of muscles of the back. The sixth thoracic vertebra is free while the seventh is united with the first lumbar vertebra to assist the formation of synsacrum. Thoracic ribs of the first five vertebrae meet corresponding sternal ribs which are bony.

Thoracic region is followed by the lumbar, sacral and caudal regions. Last thoracic, five or six lumbars, two sacral and five anterior caudal vertebrae are firmly united in a composite structure which is fused with the ilia of the pelvic girdle, the whole thing forms a synsacrum. It is easily analysed when examined from the ventral side, the first vertebra with free ribs is definitely thoracic, next six do not have free ribs, hence looked upon as lumbars, their transverse processes arise high up on the neural arches and the ligament uniting them is ossified so that the lumbar region is a continuous bony plate. Next two vertebrae with their transverse processes arising from the neural arch have a pair of processes each, the sacral ribs, which abut against the ilium bones of pelvic girdle, hence regarded as sacrals. The remaining five with flat transverse processes must be the caudal vertebrae.

Synsacrum is followed by four or five free caudal vertebrae with somewhat amphicoelous centra and ill-defined parts. Five or six posterior caudal vertebrae are fused to form the plow share bone or pygostyle.

The vertebral formula of Pigeon = $C_{14}T_1+(4)+1, L_5S_2, Cd_{5+4}+(6)$ pyg = 42

Birds in general have heterocoelous centra, penguins, parrots and some others have opisthocoelous centra, and the vertebral centra of archaeopteryx were amphicoelous. Cervical vertebrae are freely movable and their number is variable from 8-25. Swans have the largest number, 25 cervical vertebrae. Thoracic vertebrae always show a great degree of fusion but all are not fused. Posterior thoracic, lumbar, sacral and some other anterior caudal vertebrae are fused with the pelvic girdle to form the synsacrum and some posterior caudals forms the pygostyle. Posterior cervicals and anterior thoracic vertebrae have hypapophyses. Number of sacral vertebrae varies from one to five. Vertebral ribs are always double-headed (dichocephalous) and thoracic vertebrae have bony uncinate processes.

Vertebral column of mammals: It is divided distinctly into five regions i.e. the cervical, thoracic, lumbar, sacral and coccygeal (Fig. 10.53). Centra are pleurocentra and have both surfaces more or less flattened (amphiplatyan). Vertebra grows by means of epiphyses which later fuse with the main body of vertebra. Between the two centra, there are intervertebral discs of fibrocartilage, the central portion of disc called, nucleus pulposus, represents the remnant notochord and the disc itself probably represents the hypocentrum.

Various regions of the vertebral column are developed to varying extent according to the way the weight is distributed. In rabbit, weight is carried in two different ways. When standing much of the weight is carried on the fore legs and when springing forward, it is mostly carried on the hind legs.

Vertebral formula of Rabbit is: $C_7 T_{12-13} L_{6-7} S_{(1+3)} Cy_{16}$

Remarkably there are seven cervical vertebrae in **rabbit**, rat, and in all mammals, of course, there are some exceptions. length, and not the number of cervical vertebrae, determines the length of neck. Neck is fairly movable and allows movement of the head in several directions. First and second cervical vertebrae are modified into **atlas** and **axis** respectively. A typical cervical is broad and has a small centrum, large neural canal and a small neural spine. Zygapophyses are well developed. The transverse processes arise from two roots, one from the base of the neural arch and the other from the centrum. Between the two roots on each side lies the **foramen-transversaria** or the vertebraterial canal. Cervical ribs are more or less incorporated in the vertebra, the sixth cervical vertebra of rat shows some peculiarity.

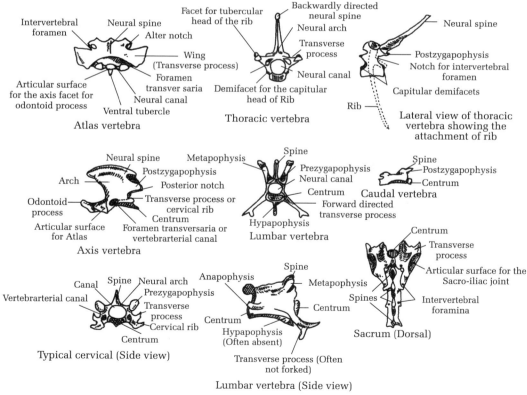

Figure 10.53 B
Vertebrae of Rabbit

Extending ventrally and slanting backwards from the transverse process is a thin plate of bone, called carotid or **Chassaignac's tubercle**. It lies directly ventral to the vertebrarterial canal. The tubercle is actually an enlarged part of the cervical rib. Seventh cervical of rabbit and rat differ from others in having more elongated spines, in the absence of foramina transversaria and in the presence of small concave semilunar facets at the posterior edges of the centra.

Atlas and **axis** are responsible for the extensive movements of the head. Atlas vertebra is very wide, larger than other cervicals and has a thin neural arch but no centrum, transverse processes are broad and long for the attachment of muscles that hold and rotate the head and neck. Occipital condyles of the skull fit into large concave facets on the front face of atlas making the atlanto - occipital joint, which allows movement in the sagittal plane, as in nodding the head. Foramen transversaria pierces the transverse process of each side. Neural canal of atlas is divided by a transverse ligament, the lower parts is occupied by the odontoid process. Axis is narrow in the transverse plane and bears the odontoid process and a pair of smooth articular surfaces, which form the atlanto-axial joint. It allows rotation of the head and atlas on the neck in a variety of planes. Neural spine of axis presents a flat surface for the attachment of muscles running forward to the skull and backwards to other neck vertebrae.

Thoracic vertebrae, 12 or 13 have double headed ribs and long, backwardly-directed neural spines. Tubercular process of a rib is attached to the transverse process which is short, stout, placed high on the side of the neural arch and bears the tubercular facet for receiving the tubercle of the rib. Capitular head of the rib is situated at the junction of the centrum and neural arch and the attachment of the rib is such that the capitular facet for the rib becomes divided into two demifacets, one at the posterior face of the corresponding centrum and the other over the next adjacent centrum. A demi facet for the first thoracic rib is present over the last cervical vertebra. All, except the last 5 thoracic ribs, meet the sternum below, hence are called true ribs; last five ribs are the floating ribs.

Lumbar vertebrae, 6 or 7, are large, strongly-built and immediately recognized by their short and broad neural spines, forwardly-directed large stout transverse processes and additional processes, called **mammillary processes.** one such pair is dorso-lateral to the prezygapophyses and the transverse processes and is called matapophyses, the other between the postzygapophyses and the transverse processes, is called **anapophyses.** Neural spines, transverse processes, and mammillary processes provide surfaces for the attachment of muscles. Movement allowed between the lumbar vertebrae are mainly in the sagittal plane, causing arching or straightening of the back and some rotation. Two or three anterior lumbar vertebrae have well developed hypapophyses, in others, it is represented merely by a ridge. **Sacral** vertebrae are firmly united to form a firm synsacrum. There are 4 vertebrae in the synsacrum but only the first in rabbit and first two in rat are really sacral because only these have great expanded transverse processes, to which sacral ribs are indistinguishably fused for articulation with the articular surfaces of ilium bones of the pelvic girdle. Remaining three vertebrae must, therefore, be anterior caudals. Anterior vertebrae resemble the lumbars but do not have hypapophyses and anapophyses; metapophyses are present but are comparatively smaller; the spines are low and vertebrae have expanded upper surfaces for muscle attachment. Foramina for the sacral nerves are present on both the dorsal and ventral sides. **Synsacrum** is not fused with the pelvic girdle and some movement is provided at the **sacroiliac** joint. Articular surfaces of the sacrum and ilium are partly covered with cartilage and roughened for the attachment of storing interosseus sacroiliac ligament, which, together with bands of fibres above and below the joint, holds the parts together. Sacroiliac joint provides additional strength to the pelvic girdle and the vertebral column for throwing the body forward when the hind-limbs are straightened.

Caudal vertebrae, 16 in rabbit, and from 27-30 in rat, are small, only some anterior caudals have neural arches, spines and zygapophyses. These parts gradually diminish in size towards the end of the tail till near the end of the tail only elongated centra are left. In the tail of **rat, chevron bones** are present in all the caudal vertebrae except the last five or six. Muscles attached to the anterior caudal vertebrae provide movements of the tail in many planes.

In other **mammals**, vertebral column has the same divisions except the whales, which do not have a sacrum. Cervical vertebrae are always seven except in some edentates, for instance, in **two-toed sloth** there are 6, in the **tamandua** 8 and in the **three-toed sloth** 9. Cervical ribs are short and fused with their attachment and all except a few posterior cervical vertebrae have vertebrarterial canals. Normally, the centra are amphiplatyan but in odd-toed hoofed mammals, cervical centra are opisthocoelous. Posterior cervicals and all thoracic vertebrae may be any where between 9-25. Centra bear demifacets, in rabbit, and fullfacet in cat. Nearly in all mammals the thoracic neural spines are directed backwards and the lumbar spines forward. There is usually one vertebra between the thoracic and Lumbar regions with a straight spine, this is called the anticlinal vertebra. Lumbar vertebrae are very strong and are usually 4-7 vertebrae, which abut against the pelvic girdle, and considered as sacrals, remaining vertebrae of the synsacrum are decidedly caudals. Largest number of caudal vertebrae is 50, found in the scaly anteater. Caudal vertebrae of man 3 or 4, are fused into a single mass, the coccyx. Caudal vertebra may vary from 4 in man to 50 in scaly ant eater (*Manis*) and insectivore (*Microgale*).

Functions of vertebral column

a) It forms a strong beam, with which the viscera is suspended by means of mesenteries.
b) It carries the weight of the body both in motion and when the animal is standing.
c) It allows flexion and bending of the back and the body without injuring internal organs.
d) It protects the nerve cord.
e) Its sacroiliac joint prevents fracture under sudden forces.
f) It allows free-movement of the head with the help of anterior cervical vertebrae.

Ribs

Ribs may develop in the hypomeric mesoderm anywhere from the posterior edge of the skull through the tail region. Ribs are long and short, cartilaginous or bony myosepta articulating medially with vertebrae and extending into the body wall. In *Polypterus* and some **teleosts** (Fig. 10.54) have two pairs of ribs for each centrum of the trunk.

A **dorsal rib** passes laterad into a horizontal septum between epaxial and hypaxial muscles, and a **ventral rib** arches ventrad into the body wall just external to the parietal peritoneum. Dorsal and ventral ribs may be primitive. Most teleosts have ventral ribs only, Sharks and some other fish develop dorsal ribs only. **Aganathans** have no ribs. This may be correlated with absence of centra. In the tail the paired ventral ribs frequently meet underneath the centrum to form haemal arches.

The exact nature of the amniote rib is not clear. It would seem a simple matter to determine whether it develops embryologically between muscle masses or next to the body cavity (Fig. 10.55). However, the rib develops with one surface against the dorsal muscle mass and the other against the body cavity, while the ventral muscles develop between the ribs. Therefore whether amniote ribs are homologous with the dorsal ribs of teleosts, or possibly both (which could account for their biciptal nature) is not known.

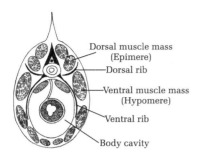

Figure 10.54
T.S. Through trunk teleost (showing position of dorsal and ventral ribs)

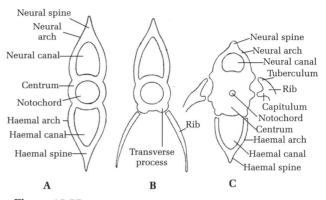

Figure 10.55
Structure of a vertebra **A. B.** Vertebra of a fish **C.** Vertebra of higher vertebrate

In some vertebrates the ribs have evolved in spectacular ways. In turtles they are inseparably fused with the dermal bones of the carapace (Fig. 10.56). In snakes their ventral edges are attached to the ventral scales so that the intercoastal muscles, which move the ribs, are also the primary muscles of locomotion. In one lizard, *Draco volans*, the ribs pierce through the body wall and support a large flap of skin that aids the animal in gliding from tree to tree. Although in most tetrapods the ventral portion of the ribs remain cartilaginous, in birds both dorsal and ventral portions ossify, and a joint is formed between them. Adjacent ribs are overlapped and held together by the small uncinate processes (Fig. 10.57). This firm rib cage, movable at the dorso-ventral joint, is adaptive in respiration.

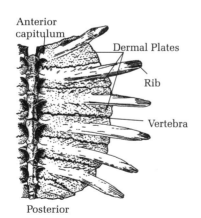

Figure 10.56
Under side portion of carapace (young turtle showing fusion of ribs and vertebrae)

Tetrapod ribs: There is some dispute among comparative anatomists as to whether the ribs of tetrapods are homologous with the dorsal or with the ventral ribs of fish. Most believe that they are the homologues to dorsal ribs. Some investigators are of the opinion that in tetrapods there has been a fusion of dorsal and ventral rib, thus accounting for the two headed or bicipital, condition, typical of tetrapod ribs.

Tetrapod ribs usually form articulation both with the centra and neural arches of vertebrae (Fig. 10.55). In these bicipital ribs the upper head of the rib or tuberculum articulates with a dorsal process, the diapophysis, coming from the neural arch. The lower head or capitulum, joins a projection of the centrum, the parapophysis. The bicipital condition is considered to be primitive in tetrapods, since it exists in some of the oldest fossil amphibians and reptiles. It undoubtedly aids in terrestrial locomotion by strengthening the trunk.

In higher forms the diapophysis may be represented by a smaller tubercular facet on the ventral side of the transverse process; the parapophysis in such cases may consist of nothing more than a small facet on the centrum for the reception of the capitulum. The capitulum may articulate at the junction of two centra, in which case each centrum may bear a half facet or demifacet, at either end. Sometimes tetrapod ribs have but a single head. This may represent either the tuberculin or capitulum or a fusion of the two.

In the cervical region where ribs are fused to the vertebrae, the space formed between the two articulations of the rib with the vertebrae is known as the foramen transversarium or vertebrarterial canal. It serves as a passageway for the vertebral artery and vein and a sympathetic nerve plexus. In the lumbar, sacral, and caudal regions the ribs are fused with the vertebrae (pleurapophyses), and there is little or no evidence of the primitive bicipital condition.

Rib in different classes of vertebrates

Fish: In **elasmobranchs**, small ribs, attached to the ventro-lateral angles of the centrum (basapophyses), extend laterally into the skeletogenous septum. It would at first sight appear as though these are dorsal ribs, but it is generally agreed that they more probably represent ventral ribs which have moved secondarily into this position. The caudal vertebrae bear haemal arches with which the ventral ribs are homologous.

In most of the higher fish only ventral ribs are present. However the chondrostean **polypterus**, as well as the **salmon** and a number of other teleosts, possess both dorsal and ventral ribs.

Amphibians: In some of the fossil labyrinthodont amphibians the ribs were strong structures which extended ventrally around the body for some distance. In modern amphibians they have been reduced and are small and poorly developed.

In most urodele amphibians all the vertebrae from the second to the caudal bear ribs which are typically bicipital. The dorsal tubercular head forms a union with the diapophysis, or base of the neural arch; the lower or capitular, head articulates with the parapophyses, a projection of the centrum. In a few primitive **salamanders** the pointed ribs may protrude through the skin as small, naked projections of a protective nature. The strong sacral ribs are attached to the pelvic girdle. The caudal vertebrae lack ribs. The ribs in anurans are lacking entirely except on the sacral vertebrae, or else they take the form of minute cartilages attached to the transverse processes.

Ribs in **caecilians** are somewhat better developed than in other amphibians. They occur on all except the first and a few of the more posterior vertebrae and are attached by means of a double or single head, as the case may be. Ribs of amphibians never form connections with the sternum.

Some of the extinct fossil amphibians possessed ventral, dermal **ribs** or **gastralia.** These consisted of V-shaped, rib-like, dermal bones located on the ventral abdominal region and extending in postero-dorsal direction on either side. **Gastralia** should not be confused with true ribs, which are endoskeletal structures.

Reptiles: Rib development in reptiles has deviated far from the condition observed in amphibians. Ribs may be borne by almost all the vertebrae in trunk and tail regions. Attachment of ribs to a sternum is first observed in the class **reptilia**.

Turtle lack ribs in the cervical region. The 10 trunk vertebrae bear ribs which are flat and broad and like the vertebrae in this region, are immovably fused to the underside of the carapace (Fig. 10.56). The carapace is thus formed partly of dermal skeletal bony plates and partly of endoskeletal structures derived from vertebrae and ribs. It will be recalled that epidermal scales overlie these skeletal elements. Each rib has but a single head, the capitulum, for articulation with the vertebral column. The point of articulation is usually at the boundary of adjacent vertebral centra. The two sacral and first caudal vertebrae are also united to the carapace. The ribs of the sacral vertebrae form a union with the pelvic girdle. Those of the anterior caudal vertebrae consists of small projections, their union with the vertebrae being indicated by the presence of distinct sutures.

The **ribs** of *Sphenodon* are extensively developed, being present even in the caudal region. A few of the anterior thoracic ribs join a typical sternum. Most of the remaining ribs, however, join a median ventral **parasternum** which extends from sternum to pubis. The parasternum is an endoskeletal structure derived from cartilages which develop in the ventral portions's of the myocommata. They should not be confused with **gastralia**, or **dermal ribs**, which are dermal derivations and which in sphenodon are present in the same region. Each rib is typically composed of three sections, an upper ossified, **vertebral section**, a ventral cartilaginous **sternal** or costal section, and an intermediate section also composed of cartilage. In *Sphenodon* each rib bears a flattened, curved uncinate process which projects posteriorly from the vertebral section, overlapping the next rib and thus providing additional strength to the thoracic body wall.

Small cervical ribs occur in **lizards** except on the atlas and axis. In the **geckos**, however, even the first two vertebrae bear ribs. In most lizards a few of the anterior thoracic ribs curve around the body to meet the sternum, Each of these ribs is divided into two or three sections, but only the vertebral section is bony in most cases. Articulation with the vertebra is by capitulum alone. The tuberculum being much reduced. Uncinate processes are lacking. The ribs of the flying dragon, *Draco volans* are of particular interest, since the posterior ribs are greatly extended, supporting an extensive fold of skin on each side of the body. This provides a wide surface which, when extended, supporting an extensive fold of skin on each side of the body. This provides a wide surface which, when extended, enables the animal to soar. When at rest, the "wings" are folded against the sides of the body.

In **snakes** all trunk vertebrae except the **atlas** and **axis** bear ribs. They articulate loosely by means of a single head, the capitulum, since no sternum is present in these reptiles, the ribs terminate freely. The lower ends however, have muscular connections with ventral scales which are used extensively in locomotion.

In **crocodilians** the ribs are typically bicipital. All five regions of the vertebral column are rib-bearing, although the ribs are much reduced in the cervical, lumbar and caudal region. Even the atlas and axis bear ribs in these reptiles. The cervical ribs increase in length as they progress in an anteroposterior direction. The two heads of the cervical ribs unite with the vertebrae so as to surround an opening, the foramen transversarium. Eight or nine thoracic ribs connect with the sternum. These are composed of vertebral, intermediate, and sternal sections, of which only the vertebral section is completely ossified. Uncinate processes are present on the vertebral section. The last two or three thoracic ribs are floating ribs with no sternal connection. Only the vertebral section is present in these. The two pairs of sacral ribs are strong projections which articulate with the ilium of the pelvic girdle. Some of the anterior caudal vertebrae bear ribs which are fused to the transverse processes, of which they appear to be parts. Dermal gastralia similar to those of sphenodon, are well developed in the ventral abdominal regions of crocodilians.

Birds: In *Archaeopteryx* the ribs resembled those of lizards more closely than those of modern birds. They were slender, articulated by means of a single head, and lacked uncinate processes. *Archaeopteryx* also possessed **gastralia**, not encountered in birds of today. In modern birds some of the posterior cervical vertebrae bear movable ribs. Strong flattened, bicipital ribs connect most of the thoracic vertebrae with the sternum. Each is typically composed of vertebral and sternal sections, both ossified. A prominent uncinate process from each vertebral section (Fig. 10.57, Fig. 10.58) overlaps the rib behind, furnishing a place for muscle attachment and increasing the rigidity of the skeleton, an important factor in flight. In the posterior thoracic, lumbar, sacral, and anterior caudal regions the ribs are fused to the large synsacrum.

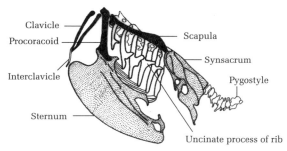

Figure 10.57
Skeleton of Trunk, tail and pectoral girdle of Pigeon

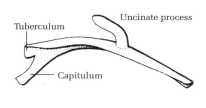

Figure 10.58 A
Fowl. A vertebral thoracic rib

Mammals: Mammalian ribs are usually bicipital. In the cervical region the transverse processes are composed of the remained portion of ribs which have fused with the vertebrae, the **foramina transversia** present only in cervical vertebrae, represent the original space between the points of attachment. Thoracic ribs are well developed in mammals. The tubercular head articulates with the tubercular facet (diapophysis) on the ventral side of the transverse process; the capitulum typically articulates at the point of junction of two adjacent vertebrae, the centrum of each bearing a coastal demifacet. In some posterior ribs, only the capitular head remains. Each rib consists of two sections: an upper, bony vertebral section and a lower, usually cartilaginous sternal section (Fig. 10.59). The latter joins the sternum directly or indirectly and is commonly referred to as the costal cartilage. The ribs of monotremes are composed of three sections as in primitive reptiles. Mammalian ribs do not bear uncinate processes.

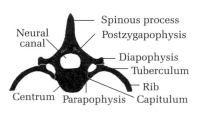

Figure 10.58 B
Showing articulation of tetrapod rib with vertebra

The part of the rib between the two heads is called the neck, the remainder is the shaft (Fig. 10.59). The area where the shaft curves most markedly is the angle of the rib. Those ribs which makes a direct connection with the sternum are called true ribs. **False ribs** are located posteriorly, and their costal cartilages either unite with the costal cartilages of the last true rib, or they terminate freely. The latter are termed floating ribs. Mammals lack ribs in the lumbar and caudal regions although the transverse processes of the lumbar vertebrae are, in all probability, pleurapophyses. The number of ribs in mammals shows considerable range, from 9 pairs in certain whales to 24 in the sloth. True ribs, however, range in number only from 3 to 10 pairs.

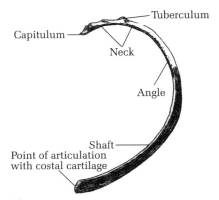

Figure 10.59
Vertical section of human rib

Sternum

Functionally the sternum gives skeletal continuity between the ventral portions of the ribs on the right and left sides of the body. Its phylogenetic origin is unclear; it probably evolved independently in several groups whose environmental situations gave significant adaptive value to this skeletal connection. It is present in only a few fishes and in a limited number of amphibians. It is largest in birds (Fig. 10.57) but is also prominent in modern flying mammals (bats) as it was in the flying reptiles (pterosaurs of the Mesozoic). The **sternum** does two things for a flying animal (i) it makes the thoracic cavity more rigid and (ii) it provides a large surface area for the attachment of powerful flight muscles. So important are these functions in fact, that the surface area of the birds sternum-also called the keel- is to a great extent proportional to its flying ability, being relatively largest in the humming bird and absent in flightless ratitae birds such as the Ostrich.

Origin of sternum

Several theories have been advanced to explain the origin, of sternum, none of view are wholly acceptable.

(a) **According** to one view the earliest indication of a sternum is to be found in the perennibranchiate (salamander, *Necturus*) in which it is referred to as the **archisternum** (Fig. 10.61). In this amphibian, cartilages occur in a few of the myocommata in the ventral thoracic region. The largest and most prominent of these irregular skeletal elements is found at precisely the same region in which, in higher salamanders, a sternal plate overlaps the coracoids of the pectoral girdle. Such a theory of origin of the sternum, is plausible, relating it to the parasternum of such forms as sphenodon, lizards, and crocodilians. The parasternum is a posterior continuation of the sternum.

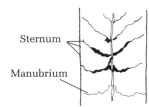

Figure 10.60
Sternum of *Necturus*

Fusion of some of the anterior archisternal or parasternal elements may thus possibly account for the origin of the true sternum. If this theory is correct, the connection of ribs to the sternum must be secondary.

(b) According to another theory the sternum arises as a fusion of the ventral ends of some of the anterior ribs. This type of sternum found in amniotes, has been spoken of as **neosternum.** This theory is based upon the fact that although sternum is present in most amphibians, in no case are ribs associated with it. It is possible, however, that the ribs have become shortened secondarily in amphibians in the course of evolution and have lost their sternal connections. There is no palentological evidence for this.

(c) Another theory seeks to derive the sternum from the median portion of the **pectoral girdles of fishes**. This concept has least support of all from a palentological or an embryological point of view.

Some authorities are of the opinion that the archisternum of amphibians and the neosternum of amniotes, with its rib connections, are in no way related and are of independent origin. Since, however, there is considerable evidence from embryology that the ribs of amniotes connect with the sternum secondarily, there is no particular reason for assuming a lack of homology between the two types of sterna.

Recent evidence from experiments in which marking technique were used shows that the sternum of birds and mammals is a derivative of the lateral plate mesoderm, where as ribs, along with the vertebrae, are clearly somite derivatives. This indicate that the sternum should properly be classified as a portion of the appendicular skeleton rather than of the axial skeleton.

Amphibians: In urodele amphibians the sternum appear for the first time and in its most primitive form (*Necturus*). In most **salamanders** it consists of little more than a small, median, triangular plate, lying behind the posterior, medial portions of the coracoids of the pectoral girdle.

In **anurans** it is better developed. In common frog, for example the anterior clavicles and posterior **coracoids** of the pectoral girdle are separated from their partners of the opposite side only by narrow cartilaginous strips, the **epicoracoid**. Anterior to the junction of the clavicles with the epicoracoid cartilages lies the bony **omosternum**, with an expanded cartilaginous **episternum** joined to it anteriorly (Fig. 10.69). Posterior to the junction of coracoids and epicoracoids lies the **sternum proper**. An expanded cartilage, the **xiphisternum** is attached to it posteriorly. The sternum is thus considered to be composed of four median elements. It is not clear which part corresponds to the sternum in urodeles, but most anatomists consider the section referred to as the sternum proper to be the actual homologue.

Reptiles: The snakes, most limbless lizards and turtles lack a sternum, but in many other reptiles this portion of the skeleton is more fully developed than it is in amphibians. In turtles, the plastron, composed of dermal plates covered with epidermal scales, is closely associated with the pectoral girdle. The membrane bones of the plastron are usually considered to be homologous with the gastralia of such forms as *Sphenodon*, crocodilians, certain other reptiles, and *Archaeopteryx*.

In *Sphenodon* a few of the anterior thoracic ribs join a typical mid ventral sternum. Most of the remaining ribs, however, join the median ventral parasternum which extends from the true sternum to the pubis. It will be recalled that this endoskeletal structure is derived from cartilages which appear in the mid-ventral portions of the myocommata in this region.

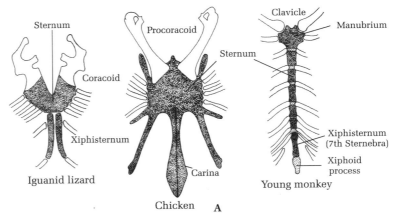

Figure 10.61 A
Tetrapod sternum of vertebrates

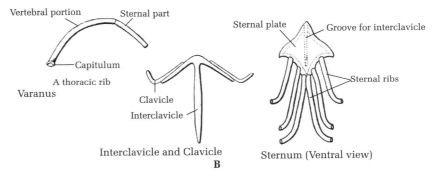

Figure 10.61 B
Varanus. Thoracic rib, sternum and bones of pectoral girdle

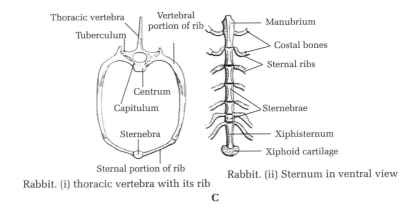

Figure 10.61 C
Ribs of Rabbit

In **lizards** (Fig. 10.61) the sternum usually consists of a large, cartilaginous, flattened plate to which the sternal sections of several anterior thoracic ribs are united. A membrane bone, referred to as the episternum, of interclavicle, in some cases lies ventral to the sternum. It is not considered to be part of the sternum.

The sternum of crocodilians is a simple cartilaginous plate which splits posteriorly into two **xiphisternal cornua**. The sternal sections of the thoracic ribs are attached to the sternum and to the xiphisternal cornua. Numerous gastralia of dermal origin are present posterior to the cornua but form no connections with them.

Birds: The sternum of birds is well developed bony structure to the sides of which the ribs are firmly attached (Fig. 10.61). In flying birds and **penguins** it projects rather far posteriorly under a considerable portion of the abdominal region. The ventral portion is drawn out into a prominent keel or carina which provides a large surface for attachment of the strong muscles used in flight it is of interest to note in this connection, that the extinct flying reptiles, the pterosaurs, had carinate sterna as do bats among mammals. A sternum in *Archaeopteryx* is unknown, from the posterolateral borders of the sternum extends a pair of elongated xiphoid processes. Running birds of the super order paleognathae have rounded sterna which are not carinate. In swans and cranes a peculiar cavity of the anterior end of the sternum houses a loop of the trachea.

Mammals: In mammals the sternum is typically composed of a series of separate bones arranged one behind the other (Fig. 10.61). Three regions may be recognized: an anterior Presternum, or manubrium, a middle mesosternum consisting of several **sternebrae**; and a posterior **metasternum** or **xiphisternum** to the end of which a xiphoid cartilage is attached. The costal cartilage of the ribs articulate with the sternum at the points of union of the separate bones. In bats the presternum is conspicuously keeled, as is the mesosternum in some cases. The number of sternebrae is variable in mammals, as is the number of true ribs which articulate directly with the sternum. In some cases, as in certain **cetaceans**, **sirenians**, and **primates**, the separate elements may fuse. Thus is human being, the entire structure, which is considerably flattened, consists of only three parts: an anterior **manubrium**, a body (gladiolus), and a small posterior xiphoid process. The last is long and thin and essentially cartilaginous. In older individuals the proximal portion may become ossified. The clavicles, and the costal cartilage of the first pair of ribs, join the manubrium, the second pair of costal cartilages articulating at the junction of the manubrium and body. Five more pairs of ribs join the sides of the body of the sternum at rather evenly paced intervals. No ribs articulate with the xiphoid process. Study of development of the sternum in man shows it to be composed of separate structures, each derived from separate centres of ossification. These fuse together at various times from puberty to old age. The mammalian sternum arises quite independently of the ribs which unite with it secondarily. In such aquatic mammals as whales and sirenians the sternum has lost much of its importance. It is composed of a single bone which forms articulations with only a few of the most anterior ribs.

Median – unpaired fins

In discussing the axial skeleton, certain unpaired, median appendages must be taken into consideration. Only aquatic vertebrates possess median appendages, which are always fin like in character. They are found in **cyclostomes**, **fish**, larval amphibians, some adult urodele amphibians and cetaceans. Typically they are present in dorsal, anal and caudal regions. In some forms the median fins are continuous (Fig. 10.62), but most frequently the continuity is interrupted and the various fins are separated by gaps. In some of the **newts** the dorsal fins of male becomes developed only during the breeding season. At other times it is rather inconspicuous. Fin development and regression in these forms are controlled by the endocrine activity of the testes. In cyclostomes and fish the median fins are supported by skeletal structures (fin rays) and special muscles, but in higher forms they are merely elaborations of the integument. Dorsal and anal fins are used chiefly in directing the body during locomotion. The caudal fins also helps in this respect in addition to being the main organ of propulsion.

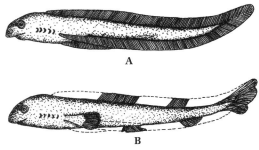

Figures 10.62 A, B
Fin fold theory of origin of paired appendages   A. Undifferentiated condition   B. Manner in which permanent fins might be formed from continuous fin fold

The skeleton supporting the unpaired dorsal and anal fins in fishes consists of radial cartilaginous or bony pterygiophores supporting slender fin rays at their distal ends. Union of several pterygiophores may take place at the base of the fin to form one or more basipterygia or basalia. These are ossified in bony fishes. Pterygiophores and basalia may form secondary connections with the neural spines of the vertebral column. There may or may not be a segmental correspondence between the skeletal elements of the fin and the vertebrae. Distal to the basalia the pterygiophores continue as cartilaginous or bony radialia. From the radialia, extend numerous fin rays of dermal origin, supporting the greater part of the fin. Frequently the fin rays connect directly with the basalia, the radialia being absent. Various types of joints permit the fin to be moved in a complex undulating manner or merely to be raised or lowered, as the case may be. In the males of many fish the anal fin and its skeletal elements are modified to form a gonopodium (Fig. 10.63), which serves to aid in the transport of sperm from male to female in internal fertilization. Development of gonopodium has been shown to be under control of the male hormone, testosterone.

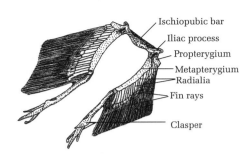

Figure 10.63
Pelvic fin and girdle of male *Squalus*

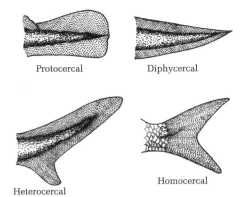

Figure 10.64
Four types of fish tails

Caudal fins or tail fins:

1. **Protocercal** Caudal fins are of several shapes (Fig. 10.64) presumably the most primitive type is the protocercal-tail of adult cyclostomes. The notochord is straight and extends to the tip of the tail, the dorsal and ventral portions of which are of practically equal dimensions.

2. **Diphycercal** It is found in dipnoi, (*Latimeria* and *Polypterus*). It appears at first sight to be similar to the protocercal type. However, palentological and embryological evidence indicates that the diphycercal tail may really be a secondary modification.

3. **Heterocercal** is the modification of third type. Here skeletal axis bends upward to enter the dorsal flange of the tail, which thus becomes more prominent than the ventral flange. A heterocercal tail is typical of elasmobranch and lower fishes in general.

4. **Homocercal tail** it is the most common type of tail. Superficially it appears to be composed of symmetrical dorsal and ventral flanges. The posterior end of the vertebral column, the urostyle, is however deflected into the dorsal flange. This type of tail, therefore, is not far removed from the heterocercal type.

In caudal fins the fin rays on the dorsal side may connect with basal pterygiophores or with neural spines directly. On the ventral side, however, they form connections with modified hemal spines which are spoken of as **hypurals**.

Functions of tail

1. as a powerful organ for swimming in fishes, tailed amphibia, crocodilians and whales;
2. as a balancing organ in animals that walk or leap mainly on the hind legs such as kangaroos and squirrels;
3. as a fly-whisk in cattle and horses;
4. as a prehensile organ in the American monkeys, oriental langurs, chameleons and sea horses;

5. for self defence in lizards;
6. as a seat, in kangaroos;
7. for protecting a sensitive under surface in monkeys;
8. as a signal of warning of dangers in rabbit, who show the under side of the tail at such times;
9. for storing food (fats) in Heloderma (gilamonster);
10. for sexual display in birds of paradise and peacocks;
11. for carrying the young, in some shrews and opossums;
12. for steering the flight in birds;
13. as a covering for the anus and genitalia;
14. as a blanket in squirrels that hibernate in winter;

Appendicular-skeleton

The appendicular skeleton includes the pectoral and pelvic girdles and the skeleton of the fins and limbs. Agnathans, caecilians, snakes, and some lizards have no appendicular skeleton, and it has been much reduced in certain other vertebrates.

Origin of paired appendages and girdles Several-theories have been advanced, some of which have been discarded in light of new palentological discoveries. It is universally agreed that the limbs of tetrapods have arisen in evolution from the fins of fishes. The **fin-fold** theory of the origin of both unpaired and paired vertebrate appendages seems to be one of the most plausible and in the past has been quite generally accepted.

According to some authorities, the fin fold theory goes back to amphioxus as a starting point (Fig. 10.62), In this animals the single dorsal fin continues around the tail to the ventral side as far forward as the atriopore. At this point it divides in such a manner that a metapeural fold extends anteriorly on either side almost to the mouth region. Gaps appearing in the dorsal fins and in the metapeural folds, for one reason or another, may have resulted in the appearance of median and paired fins. Certain rather valid objections have been advanced which would make it seem improbable that the metapeural folds of amphioxus were in any way concerned with the origin of the paired appendages of vertebrates. Among these are the fact that the folds terminate at the atriopore, a structure without no skeletal structures other than a notochord. The so called "fin rays" of Amphioxus; which support the fins are small rods of gelatinous connective tissue not homologous with the cartilaginous or bony fin supports of fishes. Usually the cyclostomes are dismissed in this connection as a specialized, limbless, divergent side off shoot of the ancestral vertebrate stock.

According to another idea, fin folds similar to the metapeural folds of amphioxus were present in some hypothetical ancestral fish but terminated at the anus rather than at the atriopore. In some of the very primitive acanthodian sharks, a row of six or seven spiny fins extended on each side between pectoral and pelvic fins. They have been interpreted as a remnants of fin folds. Traces of lateral fin folds can be observed in embryos of certain elasmobranchs as mesodermal proliferations which develop extensively only in pectoral and pelvic regions as metameric myotomic buds. The pectoral and pelvic fins may thus be regarded as persistent remnants of primitive fin folds. Furthermore, the basic skeletal structure of the paired fins is essentially like that of the unpaired fins, indicating that the two have a common origin. The structure of the fins of the extinct shark, **cladoselache** has been cited as providing additional support to the fin fold theory (Fig. 10.65). The fins are broad at their bases and contain numerous parallel pterygiophores which by their very arrangement suggest a primitive **fin-fold-like origin**.

The pterygiophores are primitively metameric, appearing in the fin along with metameric myotomic buds. Each pterygiophore is subdivided into several small pieces. Those at the base, the **basalia**, show a tendency to fuse, the **radialia** may form two or three rows of short cartilages distal to the **basalia**. Dermal fin rays in turn are distal to the radialia. According to one theory, fusion of the anterior basalia with their partners in the midline resulted in the formation of a transverse base, This is the rudiment of the pectoral or pelvic girdle, as the case may be.

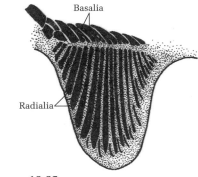

Figure 10.65
Pectoral fin of cladoselache

Another more recent theory explaining the origin of the paired appendages and girdles does not assume a primitive fin-fold origin. According to this idea, the appearance of paired appendages goes back to the ostracoderms. Ostracoderms may have been the remote ancestors of existing cyclostomes. Some of the early ostracoderm had a cephalothorax covered with a shield of plates with posteriorly directed cornua. The plates in some cases are known to have contained bone cells. Certain forms possessed lateral fleshy lobes, projecting from each side, medial to the cornua of the thoracic shield. A bony, skeletogenous septum passed behind the gill chamber and behind the pericardium. The fleshy lobes may have become pectoral fins, and the endoskeletal part of the pectoral girdle may have developed as an ingrowth of basal portions of the pectoral fins. The dermal bones of the pectoral girdle may have been derived from the bony plates of the thoracic shield.

Another group of ostracoderms is known to have possessed a paired row of dermal spines on the ventral side of the body not unlike the extra fins of spines of the **acanthodian sharks** referred to above. The spines may have been used merely as an aid in clinging to the bottom of rapidly flowing freshwater streams. Loss of armor and reduction of spines, but their persistence in pectoral and pelvic regions, might well account for the origin of the paired appendages from **ostracoderm** ancestors. With the growth of muscles from the metameric myotomes into the bases of the spiny fins, their movement may first have been accomplished. Appearance of radial skeletal elements, the pterygiophores, could have accompanied the segmental muscles as they spread outward in a fan-like manner, in opposing groups, to the peripheral parts of the fin. According to this concept the basal pieces of the endo skeleton of the fin may have pushed farther into the body to give rise to the girdles. This idea holds that the paired fins were not originally parts of fin folds but were, from the beginning, separately spaced ridges supported by spines. Much paleontological evidence has been cited in support of the ostracoderm theory.

Paired fins and girdles of fish

Fish: Pectoral fins and **girdles in** certain extinct sharks; such as **cladoselache**, possessed the simplest fins known to occur in vertebrates. The skeletal structures of which they are composed consists of several **basalia** from which **radialia** extend in a fan-like manner. The basalia of the two sides did not unite as in a true pectoral girdle.

In existing elasmobranchs the pectoral girdle is a U-shaped cartilage in the form of an arch, located just posterior to the branchial region and open dorsally (Fig. 10.63). It is not connected with the axial skeleton except in skates in which the upper portion or **suprascapula**, joins the vertebral column on each side. The girdle consists of one piece, the scapulocoracoid cartilage. The ventral portion is the coracoid bar from which a long scapular process extends dorsally on each side beyond the glenoid region where the pectoral fin articulates. Each scapular process has a suprascapular cartilage attached at its free end. The fin itself consists of three basal cartilages (basalia) which articulate with the girdle at the glenoid region. They are an anterior, lateral propterygium, an intermediate mesopterygium, and a posterior, medial metapterygium. Numerous segmented radial cartilages extend distally from the three basal cartilages. They are arranged in rows. From the radialia many dermal fin rays extend peripherally.

In bonyfish the pectoral fins themselves exhibit much variation in their detailed structure. In the **chondrostean, polypterus,** the fin is much like that of elasmobranchs except that ossification of the separate elements has begun to take place. In most bony fishes the number of separate skeletal pieces is reduced, the dermal finrays taking over much of the supporting function. Many separate basalia, however, may articulate with the pectoral girdle. In the dipnoan *Epiceratodus* (Fig. 10.66) the pectoral fins consists of a main segmented metapterygial, radial axis from the sides of which numerous segmented secondary rays project. Dermal fin rays extend from the secondary rays to the periphery. In **protopterus** and **lepidosiren** only the main radial axis persists. The type of fin seen in *Epiceratodus* (Fig. 10.66 C) is some times referred to as the **archipterygium**. Since it was formerly thought to be the archetype from which the tetrapod limb evolved. This view is no longer held. Instead, the type of fin found in living dipnoi is interpreted as being a specialized derivative of the primitive, tribasic, multiradial fin of elasmobranchs. The link between the fins of fishes and the limbs of tetrapod is sought among the fossil **crossopterygians**, only a few specimens of which are known. Study of the fin structure of *Eusthenopteron* (Figs. 10.66 A, B) strongly suggests that the group will provide the clue to the true explanation of the origin of tetrapod girdles and limbs.

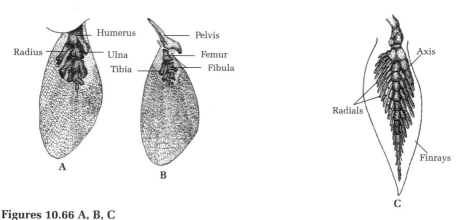

Figures 10.66 A, B, C
Skeletal structure **A.** Left pectoral **B.** Left pelvic crossopterygian fish, *Eusthenopteron*
C. Pectoral fin of *Epiceratodus*

Pelvic fins and girdles: The pelvic fins and girdles of fish are usually more primitive and simple in structure than those of the pectoral region. In **cladoselache** there is little difference between the two, although the pelvic structures are somewhat reduced. The sturgeon shows little change over the condition in **cladoselache**. The pelvic girdle of fishes is always free of the axial skeleton. Fusion of the anterior basalia of the two sides occurs in elasmobranch to form the **ischiopubic bar** (Fig. 10.63) comparable to the coracoid bar of the pectoral girdle. Attachment of the fin is at the acetabular rather than at the glenoid region. A small iliac process, projecting dorsally to slight degree at either end of the ischio-public bar, corresponds to the scapular region of the pectoral girdle. In the Dipnoi the pelvic girdle gives off anterior and posterior medial and lateral process. The anterior medial projection, spoken of as the epipubis, has homologies with similar structures of higher forms. The posterior lateral processes serves for fin articulation. The other processes offer a surface for muscle attachment. It is rare to find any membrane bones-in the pelvic region of fishes comparable to those associated with the pectoral girdle. The pelvic girdle of teleost fishes is reduced to a small element on each side with no connections between the two. It may even be lacking, as in the eels which have no pelvic fins. Often the pelvic fins have shifted so far forward that they come to lie just behind the pectoral fins and are attached at their bases to the **cleithra**. In elasmobranchs each pelvic fin typically consists of two basalia: a medial, posterior metapterygium and a small anterior propterygium. The latter may be absent.

Numerous radialia extend outward from these two basal pieces. Dermal fin rays complete the skeletal structure of the fins. In male elasmobranch the skeleton supporting the clasper is a continuation of the metapterygium which is broken up into several segments. In the dipnoi the pelvic fins are of the same archipterygial type as the pectoral fins. The central axis in some cases is split into two. Teleosts possess rather degenerate pelvic fins. Usually a single basal bone gives off a few poorly developed radials from which the fin rays project. Often the fin rays arise directly from the basalia, cushion situation exists in certain fishes, such as the cod, in which the pelvic fin have pushed forward so as to lie anterior to the pectoral fins and have become attached to the throat region. The fossil crossopterygian *Eusthenopteron* shows a better developed pelvic fin structure than is to be found in forms living to day. Although somewhat smaller in size than the pectoral fins. The basic skeletal structure is similar. A strong pelvic girdle is also present.

Girdles and limbs of tetrapods

Several theoretical possibilities have been advanced in the past to explain the origin of tetrapod girdles and limbs from piscine ancestors. There is general agreement among authorities today that connecting links are to be sought among fossil crossopterygians of the extinct suborder rhipidistia. The theory of **Gegenbaur**, driving the tetrapod limb from the archipetrygium of the dipnoi, has been generally discarded. A more recent theory, based upon an exhaustive study of the fossil remains of the rhipidistian crossopterygian *Eusthenopteron* (Figs. 10.66 A, B) rather convincingly points out that the various skeletal parts of the fins of this primitive fish can be closely homologized with those of the pentadactyl limbs of tetrapods.

Each fin of *Eusthenopteron* consists of a chain of bones along the postaxial side, from which a series of radials comes off as shown in figure (Figs. 10.66 A, B). It is believed that the first skeletal element at the base of the fin is equivalent to the humerus (femur) and the second element to the ulna (fibula). The first radial is comparable to the radius (tibia). The rest of the radials are believed finally to have become carpals (meta tarsals) have risen as new distal outgrowths from the margin of the fleshy, muscular portion of the paddle like fin. *Eusthenopteron* itself is not considered to be the direct ancestor of tetrapods. Some close **rhipidistian** relative, rather, was probably the tetrapod progenitor. Joints are believed to have developed during the evolution of such limbs at points in the appendages where bending was required to produce effective locomotion.

It has already been mentioned that in the evolution of the pectoral girdle of tetrapods there has been a loss or reduction of those portions derived from the dermal skeleton. The clavicle, however, persists in most forms although in mammals it is frequently much reduced or absent. A new element, the interclavicle, a membrane bone, has been added in many tetrapods. This is an unpaired, median portion of the skeleton, lying between the coracoids just posterior to the clavicles and joining the anterior end of the sternum. It is best developed in lizards but appears also in birds as the median part of the **furcula** or wish bone, an inter clavicle is also present in monotremes among mammals. The elements derived from the original, cartilaginous pectoral arch are of most importance in the evolution of the pectoral girdle. Scapula and usually the coracoid persists. In most mammals, however, the coracoids are reduced. Precoracoids, except in mammals, are usually present but frequently remain unossified, the clavicles taking their place. The glenoid fossa is usually at the point of scapula, precoracoid and coracoid persists, but in higher mammals the precoracoid is lost and the coracoid is represented only by a small coracoid process attached to the scapula.

The **pelvic girdle**, in fishes is poorly developed and not attached to the vertebral column, is a much more important structure in tetrapods in which, in most groups, it forms a firm union with the sacrum. It has been suggested that the pelvis of tetrapods could have logically been derived from the primitive pelvic girdle of such a fish as Eusthenopteron. Changes may have involved a shifting of the acetabulum to the lateral surface, an elongation of the ischial region, and a dorsal extension of the iliac process on each side. A gradual widening of the ischial processes with the resultant formation of the ischial symphysis, posterior to the pubic symphysis, was a probable development. The union of the iliac processes with sacral ribs was the next important change to occur but probably did not take place until a later geological era. Ossifications in pubic, ischial and iliac region apparently resulted in the appearance of the three bony elements making up each half of the pelvic girdle.

Amphibian girdles and limbs: In some of the primitive amphibians the pectoral girdle was very similar to that of piscine ancestors except for the addition of interclavicle. In modern amphibians the dermal elements have been lost except for the clavicle which persists in anurans. Limbs and limb girdles are much better developed in anurans than in urodeles, but in both groups the basic structural plan is similar. Caecilians lack limbs and limb girdles.

The limb and girdles of urodeles are small and weak. In siren both pelvic limb and girdle are absent altogether. The urodele pectoral girdle is a simple structure (Fig. 10.67) and is apt to remain cartilaginous except in the region of the glenoid fossa. The coracoids, which are united on each side with the scapula to from a single piece are broad and overlap each other medially in a loose manner just anterior to the sternal cartilage. A precoracoid process may project anteriorly from each coracoid. Clavicles are absent. Each scapula is connected dorsally with a broad supra scapular cartilage.

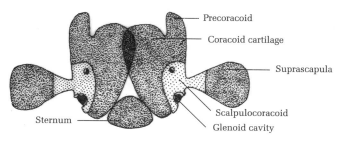

Figure 10.67
Pectoral girdle of *Ambystoma* (Salamander)

The urodele pelvis is firm, the irregular plate-like puboischia, being united in a median symphysis. An iliac process on each side unites with the transverse process of the sacral vertebra. Ossification centres appear in ilium and ischium but not in the pubis. A Y-shaped, (Ypsiloid) cartilage lies anterior to the pubo-ischium but is not generally regarded as part of the pelvic girdle. This cartilage and the muscles attached to it have been shown to be concerned with the hydrostatic function of the lungs.

In the urodele forelimb (Fig. 10.68) humerus, radius, and ulna are distinct: Carpal bones are reduced in number by fusion. No more than four digits are present, the first, or pollex (thumb) probably being the one that has been lost along with its carpals and metacarpals. In some species, only two or three digits remain separate, and in most cases the primitive pentadactyl condition is retained. In a few species, however, only two or three digits persists.

The anuran pectoral girdle show several modifications with in the group. In frogs (firmisternia) the two halves are firmly united in the middle and are closely related to the sternum (Fig. 10.69). In toads (*Arcifera*) the two halves overlap in the middle. Coracoids, precoracoids, clavicles, scapulae, and suprascapulae are present. The clavicles are fused to or cover the cartilaginous precoracoid in the firmisternal type, whereas in toads the cartilaginous precoracoids and coracoids join their partner medially. All stages between the two types of girdles-have been identified with in the group. The pelvic girdle of anurans is V-shaped and consists of on each side of a long ilium and a small ischium and pubis. The ilium is attached anteriorly to the transverse process of the sacral vertebra (Fig. 10.69). All three bones join at the acetabulum. The limbs of anurans are more specialized than those of urodeles. Humerus

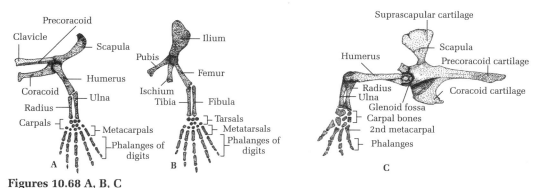

Figures 10.68 A, B, C
Tetrapod limbs and girdle **A.** Pectoral girdle and limbs **B.** Pelvic girdle and limbs **C.** Pectoral girdle and limbs of *Necturus* (Lateral view)

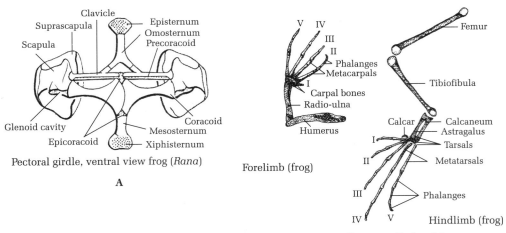

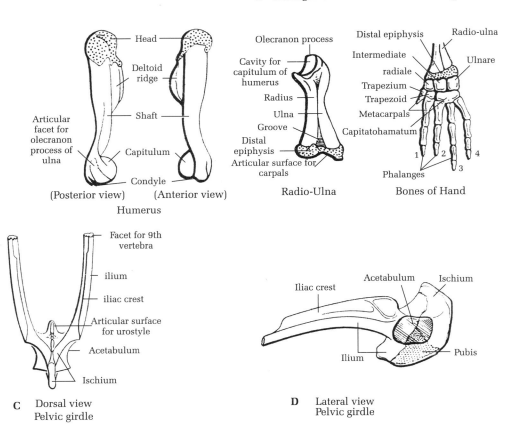

Figures 10.69 A, B, C, D
Girdles and limbs of Frog **A.** Pectoral girdle **B.** Forelimb and hindlimb bones of frog
C. Pelvic girdle (Dorsal view) **D.** Pelvic girdle (Lateral view)

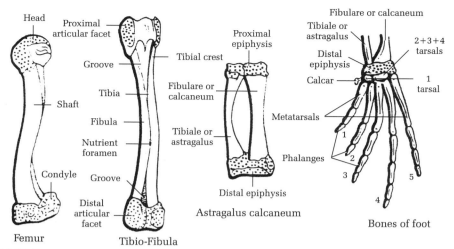

Figure 10.69 E
Hind limbs bones (frog)

and femur are typical, but radius and ulna in the forelimb, and tibia and fibula in the hind limb, usually has only four digits. The tarsal bones are modified and consist of two rows. The proximal row contains two long bones, an inner **astragalus** (talus) and an outer calcaneum. The distal row of tarsals is reduced two or three small cartilaginous or bony pieces. Metatarsals and phalanges of the hind limbs are long and webbed. A small additional bone, the prehallux, or calcar, occurs on the tibial side of the tarsus in most anurans. It is sometimes interpreted as being the rudiment of an extra digit. It is more probable, however, that the prehallux represents an additional tarsal bone.

Reptilian girdles and Limbs: The girdles and limbs of reptiles are well developed except in **snakes** and limbless lizards. In snakes there is no trace of pectoral girdle or forelimb, but vestiges of a pelvis and hind limb are found in members of the families **glauconiidae** and **boidae**. In the former group the three components of the pelvis and even the femur, are represented, In the boidae the pelvic vestiges appear externally in the form of small spurs on either side of the cloacal opening. Remnants of the girdles persists to some extent in most of the limbless lizards, but in a few no traces remain.

Reptilian pectoral girdles usually possess the typical three cartilage bones: coracoid, precoracoid, and scapula. Two membranes bones, clavicles and interclavicle, are frequently present (Fig. 10.70). In turtles the location of the pectoral girdle is peculiar in that it lies within the arch formed by the ribs or is ventral to them. A sternum is lacking in turtles. Coracoids and precoracoids are not united medially except by the fibrous bands. Clavicles and interclavicles are lacking unless certain ossification in the plastron represent these structure. Lizards have a well developed pectoral girdle united medially by the sternum. Clavicle and interclavicles are present except in chameleons. Crocodilians, on the other hand, have an incomplete pectoral girdle. Clavicles and precoracoids are lacking. A small interclavicle lies between the coracoids ventral to the sternum but projects beyond it anteriorly. The supra scapula is represented by the small cartilaginous dorsal border of the scapula.

Pelvic girdles of reptiles are typically composed of the usual three bony elements which retain their integrity throughout life. The ilium is firmly united with the two sacral ribs.

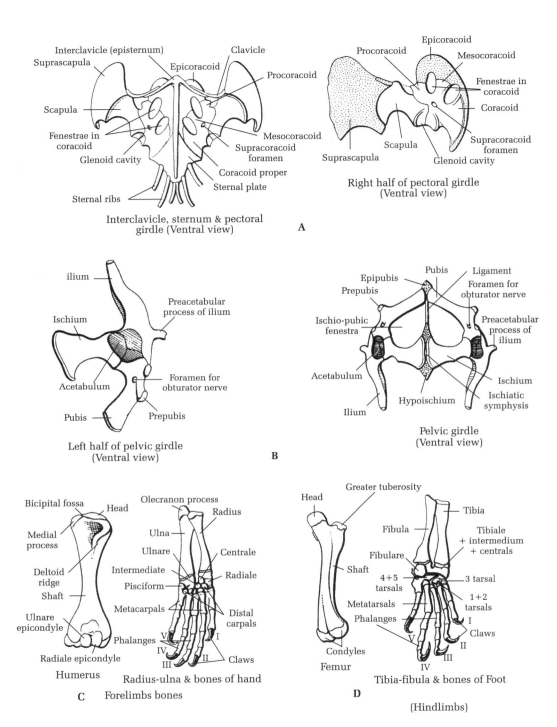

Figures 10.70 A, B, C, D
Girdle and limbs bones of Varanus **A.** Pectoral girdle **B.** Pelvic girdle **C.** Forelimb bones **D.** Hindlimb bones

Both pubic and ischial bones form median symphysis. A small foramen in the pubic bone, serving for the passage of the obturator nerve, was the only opening or foramen in the pelvic girdle in primitive amphibians and reptiles. Next, as in *Sphenodon*, a puboischial foramen appeared between pubis and ischium. In most reptiles these two foramina have become confluent and a new name, obsturator formen, is then applied (Fig. 10.70). in crocodilians only an ischiatic symphysis is present. The two pubic bones are reduced and a pair of epipubic bones projects ventrally. These are separated by a membranous area and do not form a symphysis. The epipubic and pubic bones are frequently confused. According to some authorities the epipubic bones are actually pubic bones which do not enter into formation of the acetabulum and which form a movable articulation with the rest of the pelvis. In most **lizards**, and **turtles**, a median fibrocartilaginous bar connects the rather widely separated pubic and ischial symphyses and separates the obturator foramina of the two sides. In these groups of reptiles a median, cartilaginous epipubis lies anterior to the pubic symphysis. A posterior prolongation from the ischial symphysis in *Sphenodon* and in many lizards and turtles is called the hypoischiac process, or cloacal bone.

The limbs in **reptiles** show no unusual features. They are typically pentadactyl, even in sea turtles, the limbs of which are in the form of paddle-like flippers. The chief differences in the limbs of the various groups are to be observed in the carpal and tarsal bones, in which the number may be reduced because of fusion. In *crocodilians* three tarsal bones unite just distal to tibia, to form an astragalus (talus). A single bone, the calcaneum, lies distal to the **fibula** The calcaneum bears a heel like projection which appears, for the first time in phylogenetic history, in this group of reptiles. A feature of the reptilian hind limb is the presence of an intratarsal joint between the two rows of tarsal bones (Fig. 10.70 D). Movement takes place here rather than at the junction of the tarsals with tibia and fibula. A patella, or knee cap, which is a sesamoid bone, appears in certain lizards for the first time.

In some of the prehistoric reptiles interesting modifications are present. The paddle-like limbs of **plesiosaurs** and **ichthyosaurs** have a very large number of phalanges (hyperphalangy). As many as a hundred phalanges to a digit have been observed. In some instances additional rows of phalanges are present in excess of the usual pentadactyl number. In the flying pterosaurs, humerus, radius and ulna are of normal proportions, one heavy metacarpal and three slender bones lie distal to radius and ulna. The slender metacarpals support the first three digits which are small and free and bear claws at their tips. The phalanges of what is presumably the fourth digit are enormously elongated. They supported the large integumentary fold which extended outward from the body from shoulder to ankle and was used as a wing. A spur like **sesamoid** bone, the pteroid, which is not a modified digit, projects toward the shoulder from the base of the metacarpals. It is believed that the pteroid provided additional support to the wing. The Flexure of the pterosaur wing occurred at the junction of the fourth metacarpal and the first phalanx of that digit.

Girdles and limbs of birds: The appendicular skeleton of birds shows a remarkable uniformity within the group. The pectoral girdle consists on each side of a large coracoid, a thin, narrow scapula, and a slender clavicle (Fig. 10.71). The two clavicles which are fused medially to a small interclavicle form the furcula or 'wish bone'. The precoracoid has practically disappeared. The coracoids form a firm union with coracoid grooves on the sternum. The glenoid fossa is formed by an imperfect union of scapula and coracoid. In the **paleognathae** the pectoral girdle is relatively small the clavicles being much reduced or absent. In the extinct **moas** a pectoral girdle seems to have been missing altogether, and in the kiwi it is extremely small.

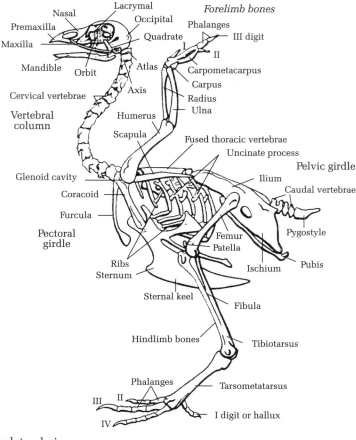

Figure 10.71 A
Fowl. Complete skeleton in lateral view

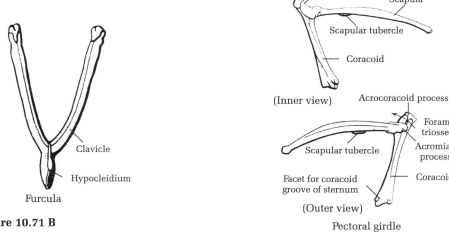

Figure 10.71 B
Fowl. Furcula   B. Fowl. Pectoral girdle. Right half

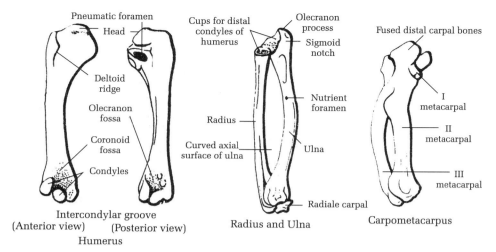

Figure 10.71 C
Fowl. Forelimb bones

Except in *Archaeopteryx* the three bones of the very large pelvis are fused together to form the innominate bone. The ilium is large and projects forward and backward from the acetabulum for some distance. It is completely fused to the synsacrum along the entire length of the latter (Fig. 10.57). A suture, however, indicate the line of fusion. The synsacrum, it will be, recalled, is the single bony mass composed of the posterior thoracic, lumbar, sacral and first few caudal vertebrae, all fused together (Fig. 10.57). Embryological studies make it clear that only two of the vertebrae are truly sacral. The ischium extends posteriorly, paralleling the ilium and fusing with it throughout the greater part of its length. A gap, the ilio-ischial foramen, varying in size in different birds, separates the two bones for some distance. Ischia and pubis do not form symphyses. The two pubic bones, projecting in a postero-ventral direction, usually terminate freely, thus permitting passage of the relatively large eggs of birds. In a number of birds the distal end of the pubis unites with the distal end of the ischium. Only in the ostrich is there a true pubic symphysis. An ischial symphysis occurs in the *Rhea*. *Archaeopteryx* may have had an ischial symphysis. The large space between pubis and ischium, whether or not these bones unite distally, represents the obturator foramen. The pelvic girdle of birds is much like that of some extinct dinosaurs.

The fore limb (wing) of birds is much modified from the primitive condition. In flying birds, humerus, radius and ulna are exceptionally well developed, since in strong fliers the wings are much longer than the hind limbs. Modification of carpals, meta carpals, and phalanges is primarily responsible for the specialized condition of the endo skeleton of the wing (Fig. 10.71). Carpal bones, due to fusion, are at first reduced in number to four. These are arranged in two rows of two each. The two distal carpals become fused to the corresponding meta carpals, forming the carpometa carpus, thus leaving the proximal carpal bones free. They are referred to as the radials and ulnare. Three metacarpals persists in birds, but there is a difference of opinion as to which three are actually represented. In order to avoid confusion, we shall refer to them merely as metacarpals, I, II and III. Metacarpal I is much reduced. All three are united proximally and number II and III unite distally in addition.

The first metacarpal bears one phalanx, the second bears two and the third only one, except in the ostrich in which two phalanges are present on the third digit. In *Archaeopteryx*, claws were present at the ends of all three digits, in which the meta carpals remained separate. In other birds, claws are sometimes encountered on digit I. The young **hoatzin**, **opisthocomus** has claws on the first two digits, but they disappear in the adult animals.

The various bones of the forelimb form a base supporting the large wing feathers which are called **remiges.** There are three sets of remiges: (1) **Primaries** or **metacarpodigitals**, attached to the various bones of the wrist and hand. (2) **Secondaries** or **cubitals**, attached to the ulna, and (3) **Humerals;** supported by the humerus (Fig. 10.71). The primaries are further subdivided into groups, depending upon the particular bone to which they are attached.

The hindlimbs of birds are modified for bipedal locomotion in contrast to the wings, which have become adapted for flight (Fig. 10.72). In addition to their use in locomotion, the feet of various birds are adapted for swimming perching, wading, running, scratching, nest building, clinging, fighting, and sundry purposes. Although there is so much variation, nevertheless a striking uniformity exists in the basic structure of the hindlimbs of birds. A strong femur articulates with the pelvis at the acetabulum. The fibula is usually much reduced and often represented only by a small bony splint (Fig. 10.72). The tibia is strong and fused to the

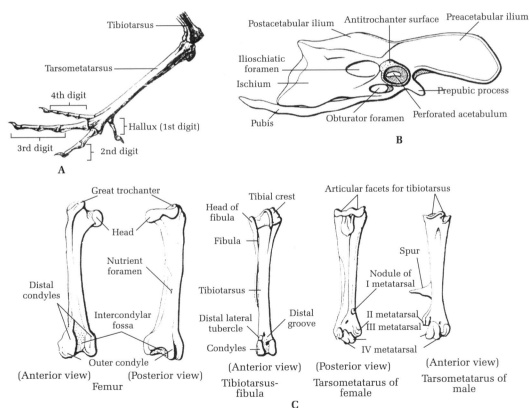

Figures 10.72 A, B, C
Fowl. **A.** Hindlimbs of chicken **B.** Pelvic girdle Right Os innominatum in outer view **C.** Hindlimb bones

proximal tarsal bone to form a tibio-tarsus. A sesamoid bone, the patella, is found in most birds anterior to the junction of femur and tibia. The distal tarsal bones unite with the second, third, and fourth metatarsals, forming a single tarso meta tarsus, which is thus a compound bone. Grooves at its distal end indicate its composite origin.

Because of the fusion of the tarsals with other bones, the ankle joint is said to be intratarsal in position. An oblique bar of bone crosses the anterior surface of the distal end of the tarsometatarsum. It is lacking in the ostrich. The first meta tarsal is represented by a free projection from the distal end of the tarsometatarsus. The spur of the domestic fowl is a bony projection of the tarsometatarsus and is directed posteriorly. It is covered with an epidermal cap of horny material. No more than four digits are present in birds, the fifth meta tarsal and the accompanying phalanges being lacking. The first digit (hallux) is usually directed backward, and the other three are directed forward. Some variation is met within the direction in which the toes point. The typical number of phalanges, going from the first to the fourth digit is 2, 3, 4, 5. The terminal phalanges bear claws (Fig. 10.72).

A few variations from the above description are encountered in birds. **Tibia** and **fibula** in *Archaeopteryx* were distinct and of approximately equal length. In **penguins**, too, the fibula is complete. Members of the super order Paleognathae, with the exception of the **kiwi**, have only three toes, the first, or hallux, being absent. In **ostriches** the second toes is also lacking, of the two remaining digits in the ostrich, the third is by far the larger. It bears a claw, but the fourth digit is claw less.

Mammalian girdles and limbs The appendicular skeleton of mammals shows considerable range from a primitive, reptile like condition to one of a very high degree of specialization. As would be expected, the monotremes are the most primitive in this respect.

In Monotremes the cartilage bones of the pectoral girdle consists (on each side) of scapula, coracoid and precoracoid (epicoracoid). The coracoids form a ventral connection with the manubrium (pre-sternum). They furnish the greater portion of the glenoid fossa. Precoracoids join a median episternum which lies anterior to the sternum. Clavicles and inter clavicle are present, representing the dermal portion of the reptilian pectoral girdle. In most mammals, however, the interclavicles and precoracoids are lost and the coracoid is reduced to a small coracoid process on the scapula adjacent to the glenoid fossa. (Fig. 10.73) The latter lies entirely on the scapula. In some mammals the clavicle persists as a strong bony arch from scapula to manubrium. In other it is lost or else remains as a small, unimportant bony vestige embedded in muscle. Persistence of the clavicle is correlated with freedom of movement of the pectoral limb. In the absence of the clavicle all direct connection between axial skeleton and pectoral girdle is lost. The scapula is thus the most important part of the mammalian pectoral girdle.

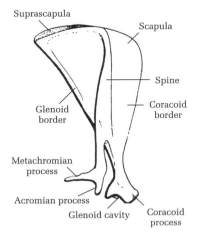

Figure 10.73 A
Rabbit. Pectoral girdle A. Right half

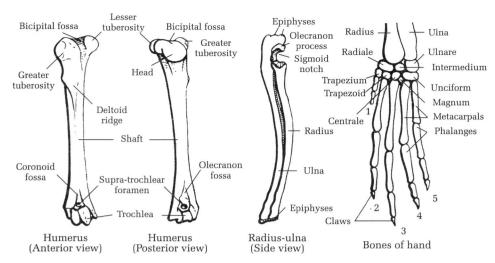

Figure 10.73 B
Rabbit   B. Forelimb bones

The pelvic girdle of mammals is made up of the usual three bony elements which, in most cases are united at the acetabulum (Fig. 10.74). Although the three parts are distinguishable in the young, they are fused in adults to form a single innominate bone on each side. A small acetabular (cotyloid) bone is frequently present in the acetabulum, substituting for the pubis and sometimes the ilium in forming the acetabular depression. The ilium and sacrum are firmly ankylosed. Both pubis and ischia, in many mammals, form symphysis, but in others the ischia do not meet ventrally, so that only a pubic symphysis is present. Strong posterior projections of the ischia may develop. These support the body when in a sitting position. Even a pubic symphysis is lacking in certain mammals. In the female **pocket gopher** it undergoes resorption under the influence of the estrogenic hormone of the ovaries, thus facilitating passage of the young from the reproductive tract at the time of parturition. The presence of a large obturator foramen on each side, bounded by pubis and ischium, is characteristic of mammals. The ilium in primates, particularly in man, has become broad and flat in connection with the assumption of an upright posture. The weight of the body is then supported by ilia, sacrum and the two femurs where they join the innominate bones. Pelvic girdle and hind limbs are lacking in **whales** and **sirenians**. However, paired remnants of pubis and ischium remain in much reduced form.

In **monotremes and marsupials** an additional pair of bones, preformed in cartilage, extends forward from the pubis in the ventral wall of the abdomen. They are the marsupial, or epipubic bones. A movable articulation occurs between them and the pubis. The homologies of these bones are uncertain. Some authorities state that they are sesamoid bones which develop in the tendons of the external oblique muscles of the abdomen. Apparently they have no homologues in other vertebrates although it is possible that they represent ossifications of the epipubic cartilages of such reptiles as lizards and turtles and even are comparable to the epipubic bones of **crocodilians**, if needed these reptiles actually have epipubic bones.

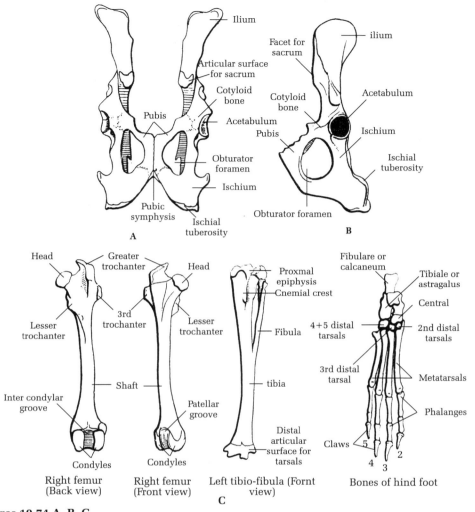

Figures 10.74 A, B, C
Rabbit. Pelvic girdle **A.** Complete girdle in ventral view **B.** Left half **C.** Hindlimb bones

The forelimbs of **mammals**, which in many cases are highly specialized, usually, deviate less than the hind limbs from the primitive pentadactyl form. A mammalian humerus may bear a supra **condyloid foramen** near its distal end. This serves for the passage of the branchial artery and median nerve of the arm. The branchial vein does not pass through the supra condyloid foramen. In man, the foramen is usually lacking but may be encountered in an occasional individual. Radius and **ulna** connect with the distal end of the humerus by a hinge joint which permits movements in only one plane. In primates and a few other mammals with prehensile forelimbs, radius and ulna are not arranged in a fixed position but, rather, articulate in such a manner that the distal end of the radius can rotate about the ulna so as to turn the hand in either a prone (Pronator) or supine (supinator) position. Articulation with the humerus is chiefly through the ulna, which bears a notch and a process for junction with the **olecranon fossa** of the humerus (Fig. 10.73 B). A projection from the proximal end of the ulna is called the olecranon process, commonly spoken of as the elbow. In many cases **radius** and ulna are fused to some extent. The ulna is most important in forming the elbow joint. The radius is chiefly concerned in forming a support for the hand.

In **Primates** the pollex is usually quite independent of other digits and more freely movable. The fact that it can be brought into opposition to the remaining digits and to the palm of the hand makes it possible for animals with opposable thumbs to pick up minute objects and to use their hands for a variety of purposes. This is one factor which has led to the superior position held by man among his contemporaries.

Carpal bones in mammals consists of several separate elements, although some fusion may occur. Meta carpals are elongated usually only two phalanges are present on the first digit with three in each of the remainder. Reduction in number of digits is common in mammals, the tendency being toward reduction in the following order 1, 5, 2, 4. In the horse and its close relatives **perissodactyl** only the third digit remains it is called canon-bone (m3) and the total weight of the body is borne by this digit m3. Such a foot is called **mesaxonic** foot. Remnants of the metacarpals of digits 2 and 4 remain as the small meta carpals, or splint

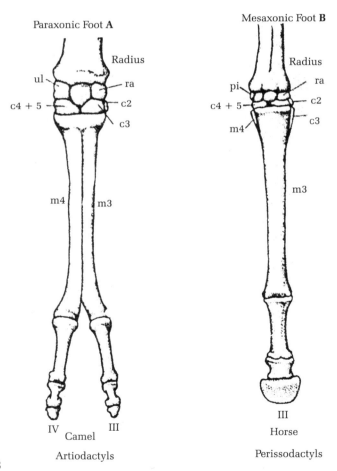

Figures 10.75 A, B

A. Right manus of an artiodactyl (camel) paraxonic foot m3 and m4 Canon-bone **B.** Right manus of a perrisodactyl (horse) mesaxonic foot. m3 in horse is "canon bone"

bones or middle (digit). The evolution of the limb of the horse from a primitive pentadactyl ancestor has been traced with a degree of completeness. Almost without parallel in paleontological investigations. In **artiodactyl** camel there is one large and one small metacarpal. The larger of the two is actually a fusion of the third and forth meta carpals, the line of fusion being clearly indicated by a groove. There are united to form Canon bone. The small meta carpal is a vestige of the fifth digit, only digits 3 and 4 are developed. Canon bone (m3, m4) bear the body weight on two parallel axis. Such foot is called Paraxonic foot. Each consists of three phalanges, the terminal phalanx bearing the hoof in these split-hoofed animals. Several sesamoid bones may be present in addition to the bones just explained.

Among the many interesting modification of the mammalian forelimbs is the condition found in **Bats**. Most of the elements are elongated. The ulna is very much reduced, only the proximal portion being present in most cases. Its distal end is fused to the radius, which is thus the main skeletal element of the lower arm. The first digit is short and free and bears a long claw. The meta carpals and phalanges of the remaining digits are greatly elongated and support the web, or wing membrane. The third digit is the longest. Comparison of the wings of bats and of pterosaurs is of interest. The same principle is involved in the development of wings in these two groups, but the manner in which it is accomplished differs greatly.

The pentadactyl limbs of **sirenians and cetaceans** are webbed to form flippers. The pentadactyl nature of the flipper is very evident in the developing embryo although not superficially apparent in the adult. The skeletal structure differs essentially from other mammals only in the excessive number of phalanges in the central digits.

Mammalian pelvic appendages in general show more variation, than is found in the pectoral region. A patella or knee cap, is present in most mammals (Fig. 10.74). The tibia is the chief bone of the lower leg, the fibula usually being much reduced in size or else fused to the tibia as a small splint. Tarsal bones are distinct entities. A projection of the calcaneum forms the heel. In some of the primates the large toe, or hallux, is opposable in the same manner as the pollex. Unlike the condition in most mammals, the head of the femur of the **orangutan** is not attached by **ligamentum teresfemoris** to the acetabulum. This permits greater freedom of movement of the hind limbs but makes them less strong. These apes walk with difficulty.

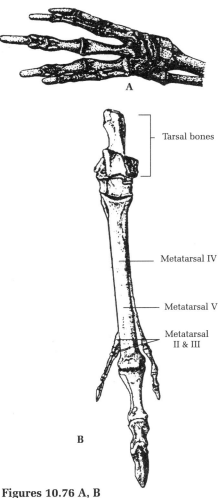

Figures 10.76 A, B
Kangaroo   **A.** Hand of kangaroo
　　　　　 B. Hind foot of kangaroo

In mammals, which show a reduction in number of digits, the hind limbs, in most cases, show a greater reduction than do the fore limbs. This is well shown (Fig. 10.76) in the **Kangaroo** in which the hand is pentadactyl but the foot has lost the hallux. In addition, meta carpal 2 and 3 are very slender. The fourth toes greatly developed and bears formidable claw at its tip, which is of great value in defense. The fifth toe is less pronounced. The meta carpals of digits 2 and 3 are closely applied to each other, and their basal portions are enclosed in a single integumentary sheath. These two modified toes are used by the animal as a sort of comb in grooming its fur.

Foot Posture in Mammals: Among mammals, three types of foot posture are recognized (Fig. 10.77).

1. **Plantigrade:** The most primitive is the plantigrade posture, observed in the hind limbs of man, in bears, and certain insectivores. In plantigrade animal the entire foot is in contact with the ground during locomotion.
2. **Digitigrade:** Animals such as cats and dogs place only their digits on the ground, the wrist and ankle being elevated.
3. **Unguligrade:** (Horse, cow, deer). Only the hoof is in contact with the ground. This type of foot posture is the most specialized of the three, since it deviates most from the primitive condition and is found only in the most swiftly running mammals.

The diverse modifications of the appendicular skeleton include both general and specific adaptations. Among the general adaptations is the firm attachment of the pelvic girdle to the axial skeleton so that the power stroke of locomotion, coming from the pelvic limbs, is passed directly to the axial skeleton. Another adaptation is the loose or non-existent skeletal connection of the pectoral girdle to the axial skeleton, which allows the fore limbs and pectoral girdle to act as a shock absorber for locomotion. This freedom of the fore limbs also facilitates greater movement and manipulation. Specific adaptations occur primarily in the distal portions of the limbs, examples are the reduced toes and lengthened meta carpels in horse and other ungulates; the grasping ability of the digit in **man, opossum** and **raccon and the short, stout, powerful digging appendages of moles.**

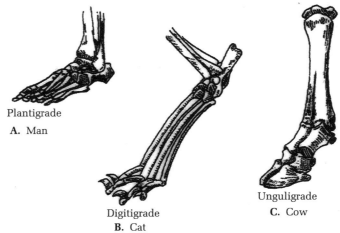

Plantigrade
A. Man

Digitigrade
B. Cat

Unguligrade
C. Cow

Figure 10.77
Types of foot posture in mammals

In the appendicular skeleton, therefore, there are general adaptations proximally and specific adaptations distally. In fact, the structural and functional adaptations in the distal portion of the limbs is that portion which is in the most intimate contact with the environment and are an effective mirror of the environment.

Chapter 11

Muscular System

The movement of the body as a whole from one place to another or movement of different part of the body with respect to body itself is brought about by muscles. In addition to above, vital processes such as contraction of heart, constriction of blood vessels, breathing movements and peristaltic movement in digestive tract are also performed with the help of muscles. The muscles are essentially machines which convert chemical energy into mechanical work. Muscles are the active part of the motor apparatus of higher animals.

The vertebrate body has striated, smooth and cardiac muscles. The microscopic study reveals that all types of muscle cells **smooth, striated** and **cardiac** contain the contractile proteins—actin and myosin.

Smooth Muscle

The smooth muscle is incorporated into every organ system except the skeletal, the nervous, and the muscular system itself; it controls the movement of food through the alimentary canal, the secretion of glands, the contractions of the urinary bladder. In general, the smooth musculature regulates a vertebrates's internal environment. It acts more slowly than other muscles, but can sustain contraction for long periods. These are also called **unstriated** or **involuntary** muscles and are devoid of any cross-striation. They are widely distributed in the visceral organ systems; so usually referred to as **visceral muscles**. They are composed of uninucleated cells constituting the fibres. A single fibre is about 5 to 10 μm in diameter and 30 to 200 μm long. It is spindle shaped and contains a centrally placed nucleus (Fig. 11.1).

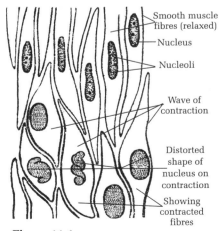

Figure 11.1
Smooth muscles

Skeletal Muscle

Skeletal or striped muscles are, also called striated muscles. Most of these muscles are responsible for movement of the skeleton, others insert on the skin and cause it to move. In general, striated muscles move large parts of the body in response to and in relation to the external environment. They act much more quickly than do smooth muscles, with rapid, strong contractions hence they fatigue more readily. The skeletal muscle fibres are elongated cylindrical cells of variable size. The length of these fibres may vary from a few millimeters to a few centimeters while their thickness may vary from 10 μm to 100 μm in diameter and depending on location, the length of a muscle fibre may extend from few microns to centimeters. Each muscle fibre is bounded externally by a tough, exceedingly thin elastic membrane called the **sarcolemma**. Inside the sarcolemma is present semifluid cytoplasm called the **sarcoplasm**.

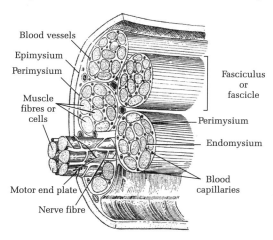

Figure 11.2
Sectional view of skeletal muscle

The ultrastructure of striated muscle shows that striated muscles have complex structure having a large number of muscle fibres of different length up to 12 cm (Fig. 11.2). The muscle fibres are usually found parallel to each other and are grouped in bundles or **fasciculi**. Each muscle is composed of many such bundles. The fasciculi are arranged in a particular pattern to allow the movements of a limb or any other region of the body. The muscle fibres, fasciculi and the muscle as a whole are invested by connective tissue that forms a continuous framework. The outer part of this framework, the **epimysium**, covers the whole muscle. Thin collagenous partitions extending inward to surround the individual fasciculi comprise the **perimysium** and the delicate network of connective tissue that invests the individual muscle fibres is the **endomysium**. The connective tissue serves to bind these individual contractile fibres together and to integrate their action. The skeletal muscles are attached to bones by means of tendons which are found at the ends of muscles. A tendon is composed of dense, white, fibrous (inelastic) connective tissue. An expanded tendon consisting of a fibrous or membranous sheet is called an **aponeurosis**. Each striated or skeletal muscle has blood vessels and afferent (sensory) and efferent (motor) nerves. Blood passes along the blood vessels, delivers nutrients to the muscles and carries away their waste products. The nerves link the muscles and the central nervous system.

Cardiac Muscle

The cardiac muscle are found only in the vertebrate heart and are characterized by its rhythmic activity, which can be modified by neural or hormonal stimulation. In general its structural and functional characteristics are intermediate between those of smooth and skeletal muscle. The fibres are involuntary in nature as their activities are not under the control of the animal. The fibres of these muscles are both longitudinally and transversely striated, like striated or skeletal muscles, but its fibres exhibit branching and contain myofibrils and filaments of actin and myosin which are arrayed similar to those in skeletal muscles (Fig. 11.3). Likewise, the mechanisms of contraction are essentially the same as those in skeletal muscle but these muscles differ histologically from striated muscle of vertebrates in many ways:

(i) The cardiac muscles are uninucleate whereas striated muscles are multinucleated.

(ii) The cardiac muscles are not directly innervated as a result there are no motor end plate.

A part of intercalated disc serves to transmit the electrical excitation and they interdigitate to form what are known as intercalary discs (Fig. 11.3). These are actually cell membranes that separate individual cardiac muscle cells from each other. Thus, cardiac muscle is not a syncytium. Despite, lack of syncytium, cardiac muscle fibres transmit impulses from one fibre to another throughout the muscle without the need of booster stimuli from nerves. This suggests that cardiac muscle fibres have pseudosyncytial arrangement through which the cardiac muscle is allowed to contract sequentially from fibre to fibre and, thus, the cardiac muscle is said to be in functional syncytium.

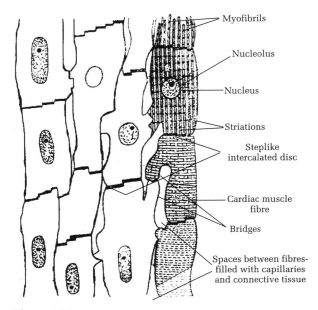

Figure 11.3 Cardiac muscles

From the above account, it becomes clear that the cardiac muscles have one character each of both voluntary and involuntary muscles i.e. being striated (voluntary) and not under the control of animal (involuntary).

Skeletal muscles maintain the vertebrates's posture and cause its movements. In addition, to this their energy is dissipated as heat which is circulated by the blood to all parts of the body. In homeothermic vertebrates this helps to maintain a constant temperature; in poikilotherms it provides a change in temperature, warmer than ambient temperature which facilitates the metabolic reactions of the body. The skeletal muscles comprise about one-third to one-half of the animal's body bulk.

Some time it is appropriate to classify muscles as somatic or visceral. The **somatic** muscle orient the animal (body or soma) with respect to environment. They enable the animal to go deeper into an environment to search food or mate and to withdraw from the environment that is hazardous. The chief role of visceral muscle is to maintain an appropriate internal milieu. It is the muscle of hollow organs, vessels, tube and duct, the intrinsic muscle of the eye ball, erector of hair and feathers and striated muscle of the jaws and remaining visceral arches, lumen of bladder or heart, causes peristalsis and serve as sphincter and dilator muscles in other locations.

Kind of voluntary muscles

On the basis of embryonic origin Wilder has classified the voluntary muscles into groups such as **metameric, branchiomeric** and **integumental**. The metameric (or somatic) group include the axial and appendicular muscle (Table 11.1). The branchiomeric muscles are associated with splanchnocranium and its derivative, while the integumental group is associated with integument rather than skeleton.

(i) **Axial muscles** are skeletal muscles of the trunk and tail. The axial muscles are segmental because of their embryonic origin. They arise from segmental mesodermal somites.

(ii) **Appendicular** muscles help in movement of fins or limbs. Appendicular muscles of fish serve as stabilizers and contribute little to forward locomotion. In tetrapod these muscle provide leverage for locomotion on land.

(iii) **Branchiomeric** muscles are associated with the pharyngeal arches of vertebrates. In fish these muscles help in operation of jaws and gill arches. In tetrapods they continue to operate the jaws.

(iv) **Integumental** muscles are well differentiated in mammals only. In humans more than 30 different muscles are known to cause gestures of grief, smiling, wrinkle the forehead, close the lid, raise the eye brow, draw milk on sucking stimulus, or dilate nostrils. These are extrinsic integumental muscles. The intrinsic integumental muscle develop with in the skin, in the dermis. These are arrectores plumarum that insert on feather and ruffle the feather and elevate the fur and arrectores pilorum that insert on hair follicle and elevate the hair.

Table 11.1: Vertebrate skeletal muscles

	Muscle	Derivation	Innervation
Somatic		Somatic mesoderm	Somatic motor nerves
Axial	1. Extrinsic ocular	Preotic head Myotomes	Somatic motor cranial Nerves III, IV, VI
	2. Hypobranchial	Postotic head Myotomes	Occipital nerves (Anamniotes), Hypoglossal nerve (Amniotes)
	3. Trunk		
	Epaxial	Dorsal portion of myotomes	Dorsal-rami of spinal nerves
	Hypaxial	Ventral portion of myotomes	Ventral rami of spinal nerves
Appendicular	1. Intrinsic	Limb bud mesenchyme	Ventral rami of spinal nerves
	2. Extrinsic	Limb bud mesenchyme and/or myotomes	Ventral rami of spinal nerves
Branchiomeric		Anterior splanchnic mesoderm	Special visceral motor cranial nerves V, VII, IX, X, XI.

Metameric muscles: Axial

Extrinsic ocular muscles: The anterior most somatic muscles are the extrinsic ocular muscles, which move the eyeball (Table 11.2). These are phylogenetically the most conservative muscles of the vertebrate body. They are derived from the three preotic myotomes, and except for the relatively simple condition of cyclostomes, six of them are present and are the same in all vertebrates. There are four rectus and two oblique muscles, they originate from various parts of the bony or cartilaginous orbit, and their insertions are onto the wall of the eyeball, working in coordination, they can move the eyeball freely and smoothly (Fig. 11.4).

Table 11.2: Extrinsic ocular muscles

Muscle	Myotome	Cranial nerve and number
Superior rectus	First	Oculomotor (III)
Inferior rectus		
Medial rectus		
Inferior oblique		
(Levator plapebrae superioris)	First	Oculomotor (III)
Superior oblique	Second, dorsal part	Trochlear (IV)
External rectus	Second and third ventral parts	Abducens (VI)
(Retractor bulbi)	Second and third Ventral parts	Abducens (VI)

Two others, the retractor bulbi and the levator palpebrae superioris, are variably present. The retractor bulbi pulls the eyeball directly into the orbit, a protective mechanism for animals with slightly protruding eyes. It has no antagonist, instead, the elasticity of a large fat pad at the back of the orbit pushes the eyeball towards the surface when the retractor bulbi relaxes. The levator palpebrae superioris inserts on the upper eyelid and raises it. Its antagonist is not a somatic muscle but a branchiomeric muscle, the orbicularis oculi, which also attaches to the eyelid and lowers it.

Hypobranchial Muscles

The dorsal part of the third myotome plus the entire fourth, fifth and sixth myotomes never differentiate in vertebrates, but are reabsorbed during development of the otic capsule. Postotic myotomes form just behind this region and migrate ventrally, medially, and finally anteriorly, coming to lie ventrally between the gill arches and the floor of the mouth and pharynx. These myotomes then differentiate into the hypobranchial muscles. In anamniotes they are innervated by the occipital nerves, and in amniotes, by the hypoglossal nerve

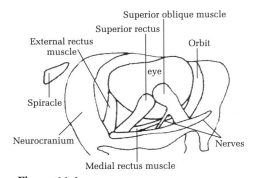

Figure 11.4
Extrinsic ocular muscles in dogfish (Dorsal view)

(cranial nerve XII) which is a coalescence of the occipital nerves. In fish the hypobranchial muscles insert on the ventral parts of the gill arches and the jaw apparatus and play a major role in depressing them. In chondrichthyes they have their origin on the coracoid (ventral) portion of the scapulocoracoid. The most superficial pair extends all the way to Meckel's cartilage and forms the coracomandibular muscles. Deep to this is a pair of coracohyoid muscles inserting on the basihyoid. Deeper still are coracoarcuals and coracobranchial going to the more posterior gill arches (Fig. 11.5). In teleosts, although the hypobranchial muscles are more diverse, they still go primarily to the jaw apparatus and the gills, and have generally depressor functions. In living amphibians the hypobranchial muscles can be divided into a prehyoid and a posthyoid group (Table 11.3).

Table 11.3: Hypobranchial musculature of Lissamphibians

Muscle	Position
Prehyoid	
Geniohyoid	Symphysis of lower jaw to hyoid
Genioglossus	Symphysis into tongue
Posthyoid	
Rectus cervicus	Inserting caudally onto hyoid
Omoarcuals	Coracoid portion of pectoral girdle
Pectoriscapularis	Scapula

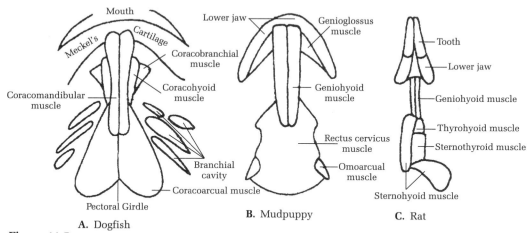

Figure 11.5
Hypobranchial muscles lie ventrally between the level of the pectoral girdle and lower jaw

In amniotes the hypobranchial musculature (Table 11.4) forms the major intrinsic and some extrinsic musculature of the tongue, and also some of the superficial ventral neck muscles controlling gross movements of the hyoid and thyro-hyoid apparatus. In contrast to anamniotes, amniotes have no pectoral girdle attachments of hypobranchial muscles (Fig. 11.5).

Table 11.4: Hypobranchial musculature of amniotes

Muscle	Position
Hyoglossus	Hyoid apparatus to tongue
Geniohyoid	Symphysis to hyoid
Genio glossus	Symphysis to tongue
Sternohyoid	Anterior part of sternum, to basihyal
Sternothyroid	Sternum to thyroid cartilage
Thyrohyoid	Thyroid cartilage to basihyal

Trunk-musculature

In **fish** the post cranial, axial musculature remains segmented in the adult, with connective tissue – myoseptae between adjacent myotomes and a horizontal septum between epaxial and hypaxial musculature (Fig. 11.6).

The myotomes run dorsally to ventrally in a zigzag line whose angles are greatest in the most active fish. Further, the myotome are slanted sharply anteriorly from the body surface to the vertebral column, with the fibres converging towards the centra of the vertebrae. Thus a transverse section through the trunk musculature of an active fish cuts across several myoseptae, which appear as concentric circles as the myotomes taper in toward their insertions; the myotomes are stacked like cones with their apices pointing forward (Fig. 11.6). With this type of arrangement a larger percentage of the muscle fibres exert their force on or close to the vertebrae, efficiently producing the characteristic lateral undulations of fish locomotion.

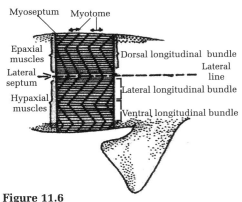

Figure 11.6
The epaxial and hypaxial musculature of dogfish (showing zig-zag arrangement of myotome)

Tetrapods: Tetrapods that locomote on four limbs sustain quite different stresses in the axial region and have undergone much more regional differentiation of the axial musculature both anterior to posterior and dorsal to ventral. In general, the epaxial (dorsal) musculature, rather than being segmented, forms long, cable like muscles. These prevent the vertebral column from sagging between the appendage and girdles, allow free movement of the tail, and strongly support the head. The hypaxial (ventral) musculature differentiate as laminae of broad, flat muscles which protect the body cavities and, fore and aft of the girdles, support the neck and tail. The stresses differ in the epaxial and hypaxial regions, and the two regions have evolved quite separately.

Tetrapod epaxial musculature

Amphibians: The epaxial musculature of urodeles is quite fishlike in some respects. Even in the adult, myoseptae separate individual myomeres, and in the tail there are typical myotomes producing lateral undulations. In the trunk region, however, the epaxial part of the myomere is divided into a very small medial portion, the interspinalis, and a large lateral portion, the dorsalis trunci (Fig. 11.7).

In **anurans** the transverse myoseptae are lost and long cable like muscles extend the length of the trunk. The medial muscle here named the **longissimus** dorsi, extend all the way from the ilium to the posterior part of the skull. The lateral muscle, here called the iliolumbaris, runs from the ilium on to the transverse processes of more anterior vertebrae. The caudal vertebrae are fused to form a single bony rod, the urostyle.

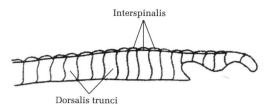

Figure 11.7
Epaxial musculature (dorsolateral view) mud puppy

A single dorsal muscle, the coccygial iliacus, goes from the ilium caudally and obliquely to the urostyle, extending the urostyle when it contracts.

Reptiles: Since the vertebral column in **turtles** is fused to the carapace, bending of the trunk is impossible. Not surprisingly, there has been an almost complete loss of axial trunk muscles.

Among other living reptiles (squamata, rhynchocephalia, crocodilia) several strong, cable like epaxial muscles are formed in the trunk region. Most medial is the transversospinalis, connecting the neural spines and arches of adjacent vertebrae. Lateral to this there is the broader longissimus dorsi muscle, running from the ilium and sacrum to the transverse processes of a variable number of more anterior vertebra. Lateral to the longissimus dorsi is the iliocostalis, arising from the lateral part of the ilium and adjacent fascia and inserting on the ribs. In the cervical region the epaxial musculature is more complex, with many muscles controlling the movements of the neck and head. The most intricate of these movements are caused by the occipital muscles, derived from the transversospinalis. More superficially, the longissimus dorsi partially inserts onto the spines of cervical vertebrae, and a separate muscle, the **cervicus capitus**, runs from the neural spines to the posterior part of the skull and controls the gross movements of head and neck (Fig. 11.8).

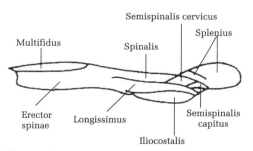

Figure 11.8
Epaxial musculature of a rat (dorsolateral view)

Birds: There is little epaxial musculature of the trunk in birds, for the fusion of thoracic, lumbar, sacral and anterior caudal vertebrae into one rigid avian structure has eliminated much of the selective pressure for it. In the cervical region, however, where complex and elaborate movement is necessary, the epaxial musculature includes many fine, small muscles that control movement between individual vertebrae and that between vertebrae and skull.

Mammals: In generalized mammals, the epaxial musculature is similar to that of the more generalized reptiles, although split into more parts anteriorly (Fig. 11.8).

Tetrapod Hypaxial Musculature

Amphibians: In urodeles the myoseptae between myotomes persist, and in the caudal region the myotomes are typical. In the trunk hypaxial region, however, each myotome differentiates into four muscles. Superficial fibres, forming the **external** oblique muscle, run at right angles to intermediate fibres, forming the internal oblique muscle; the former runs antero-dorsally to postero-ventrally, and the latter caudo-dorsally to ventro-anteriorly Deeply embedded to these two is a third, very thin muscle, the **transversus**, whose fibres run transversely (Figs. 11.9 A, B)

Mid-ventrally there is a line of connective tissue called the linea alba (Fig. 11.10). On either side, the rectus abdominous muscle runs from the pubis anteriorly to the cartilage of the pectoral girdle. This segmented muscle is the fourth trunk **hypaxial muscle.**

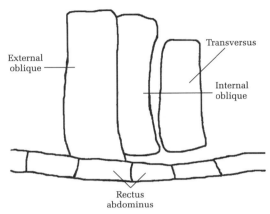

Figure 11.9 A
Hypaxial musculature of mud puppy (ventrolateral view)

Anurans retain more of the myoseptae in the hypaxial than in the epaxial musculature. For instance, distinct myoseptae divide the rectus abdominous into segments. The other two hypaxial muscles, however, are broad and sheetlike with no segmentation; the external oblique, superficially, and the transversus deep to it. There is no internal oblique.

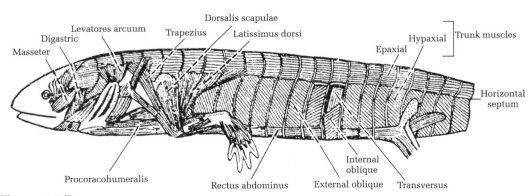

Figure 11.9 B
The musculature of a salamander

Reptiles: In reptiles (except chelonia) the hypaxial musculature differs regionally. In the lumbar region there are typical sheetlike, unsegmented muscles—the external oblique, internal oblique, and transversus; ventrally and medially is the segmented rectus abdominous. In the thoracic region, between the prominent ribs that reach ventrally to the sternum, are two sets of muscles whose fibres run at right angles—the external intercoastal and underneath

Table 11.5: Major tetrapod shoulder muscles

Embryonic origin	mammal	lizard	salamander	labyrinthodont
trunk to girdle = "sling"				
Hypaxial	Levator scapulae ventralis	Levator scapulae	Levator scapulae	Levator scapulae
	serratus ventralis	serratus ventralis	Thoraciscapularis	Thoraciscapularis
	Rhomboideus	—	—	—
Branchiomeric	Trapezius complex Sternocleidomastoid complex	Trapezius and Cleidomastoid	Cuccularis	Cuccularis
Hypobranchial	— —	— —	Pectoriscapularis Omoarcual	— —
Trunk to humerus				
Limb Bud	Latissimus dorsi Teres major Cutaneous maximus (Part)	Latissimus dorsi	Latissimus dorsi	Latissimus dorsi
	Pectoralis major Pectoralis minor Pectoantibrachialis Xiphihumeralis Cutaneous maximus (Part)	Pectoralis	Pectoralis	Pectoralis
Girdle to humerus				
	Deltoid { Clavodeltoid, Spinodeltoid, Acromiodeltoid }	Clavodeltoid Scapulodeltoid	Procoracohumeralis Longus Scapulodeltoid	Deltoid
	Teres Minor	Scapulohumeralis Anterior	Procoracohumeralis Brevis	Scapulocoracohumeralis
	Subscapularis	Subcoracoscapularis Scapulohumeralis Posterior	Subcoracoscapularis	Subcoracoscapularis
	Supraspinatus Infraspinatus	Supracoracoideus	Supracoracoideus	Supracoracoideus

them are the internal intercostals. A third group lies even deeper, running between the ribs and the transverse processes of vertebrae—the levatores costarum muscles. In the cervical region fine muscles run from the cervical vertebrae to the skull and depress the head. Other muscles lower the neck; chief among these is the longissimus coli which is attached along the ventral border of the centra from the thoracic through the cervical region.

Birds: The avian axial musculature is that of a modified reptile. Because the trunk is fused and the sternum and appendicular musculature expanded, the hypaxial musculature is much reduced in the trunk region. It is, of course, extremely elaborate in the cervical region.

Mammals: In mammals the abdominal hypaxial musculature is almost reptilian, consisting of the external and internal obliques, the transversus, and the rectus abdominus. In the thoracic region the external and internal intercostals are developed further than they are in reptiles, with the external intercostals pulling the ribs forward and the internal intercostals pulling them backward (Fig. 11.10) (Table 11.7). Variable numbers of supercostals and subcostals aid in respiration as does a migrated cervical, hypaxial muscle and the diaphragm, Finally, a pair of longitudinal thoracic hypaxial muscles, the scalene muscle, run from the ventro-lateral portion of the ribs up to the cervical vertebrae; depending upon which skeletal part is fixed, these can either bend the neck or raise the rib cage.

Matameric muscles (Appendicular)

Fish: In chondrichthyes and also apparently in teleosts the muscles of the paired fins arise as direct buds from the hypaxial myotomes. Connective tissue holds the girdles in their position on the trunk. Small muscles running from the girdle to the proximal part of the fin form its extensors and flexors. In fish there are a few muscles which are completely contained within a fin, and movement is usually of the fin as a whole rather than of its parts. Since the primary function of paired fins is stabilization rather than locomotion or manipulation, fine movements of parts are unnecessary. The male chondrichthyean pelvic fin provides significant exceptions, as they get modified as the intromittent clasper organ, they have complex muscles and elaborate movements.

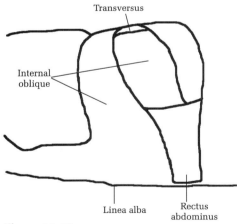

Figure 11.10
Hypaxial musculature of a rat (Ventrolatral view)

Skates and rays, however, are flattened dorsoventrally and hence are incapable of lateral undulations. Their primary locomotor organs are huge pectoral fins which have very large muscles continuous with the myotomes of the hypaxial region. Nervous coordination permits rhythmic, anterior to posterior contractions which provide the impulses for locomotion.

Tetrapods: In tetrapods there are extrinsic and intrinsic appendicular muscles. The extrinsic muscles relate the girdles to both the axial structure and the limbs. A functional category, it includes muscles derived from the axial musculature, the branchiomeric musculature, and the limb musculature, which itself is derived from limb bud mesenchyme. The intrinsic muscles, however, are totally derived from limb bud mesenchyme; they are defined as muscles whose origins and insertions are both within the limb itself. The extrinsic limb musculature, involved with proximal parts of the skeleton, is conservative compared to the more distal intrinsic musculature which, like its associated skeleton, has undergone gross modifications in different tetrapod groups. Generally, **intrinsic appendicular** muscles are simple in living amphibians than in reptiles and most complex in birds and mammals and, of course, they are correlated with the degree of variation in the distal part of the skeleton and the degree of fine movements of parts.

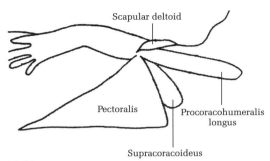

Figure 11.11
Ventral superficial musculature of pectoral girdle of mud puppy (ventral view)

Pectoral musculature

In those tetrapods whose clavicle and sternum are not articulated, the pectoral appendage has no skeletal connection to the rest of the animal skeleton but is held to it by an elaborate muscle sling (Fig. 11.11). This is advantageous for two aspects of the animal behaviour. First, the girdle can be moved relative to the trunk, and therefore, the limb is more mobile. Second, the muscles that form the interface between the pectoral and axial skeletons act as shock absorbers when the animal lands on its front legs, as during locomotion or a predatory leap.

Mammals: The three major hypaxial muscles contribute to this pectoral girdle sling — the levator scapulae ventralis, the rhomboideus, and the serratus ventralis (Table 11.5, Fig. 11.12). The levator scapulae ventralis is a band like muscle running from the posterior part of the skull to the spine of the scapula. The rhomboideus is heavy muscle coursing from the base of the neural spines to the dorsal border of the scapula. A slip of this muscle, the rhomboideus capitus, arises from the back of the skull. The serratus ventralis is the largest of the three. Many separate fascicles originate on the ribs and then converge, inserting on the dorso-medial surface of the scapula. Lizards lack a rhomboideus muscle but have a large levator scapulae and a serratus ventralis that is somewhat smaller than that of mammals (Table 11.5). Salamanders have the same two muscles, but the latter is called the thoracic scapularis muscle (Fig. 11.11).

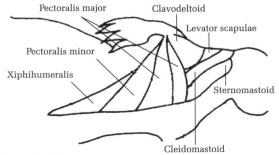

Figure 11.12
Ventral superficial musculature of petoral girdle of a rat (ventral view)

The branchiomeric muscles also contribute to the **pectoral girdle sling** (Table 11.6). In mammals there are the trapezius complex and the sternocleidomastoid complex (Fig. 11.12). The trapezius muscles originate from the back of the skull and from the neural spines as far caudally as the thoracic region. The anterior portion of the complex inserts on the clavicle and the more posterior portions on the spine and fascial surface of the scapula. The cleidomastoid runs from the clavicle to the mastoid process of the temporal bone; the sternomastoid, as its name implies, does not attach to the pectoral girdle (Fig. 11.12). In lizards the cleidomastoid and trapezius are a continuous muscle originating on the back of the skull and neural spines and inserting onto the clavicle and scapula. In urodeles, these muscles are represented as a small muscle, the cuccularis, inserting onto the front of the scapula (Fig. 11.13, Table 11.5). As mentioned, two hypobranchial muscles are part of the shoulder musculature of amphibians. With this exception; the remaining shoulder muscles are all derived from limb bud mesenehyme and go from either the trunk fascia or the pectoral girdle to the humerus. Those from the trunk to the humerus are, in all tetrapods, two large superficial muscles-the latissimus dorsi and the pectoralis complex (Fig. 11.14). The latissimus dorsi, which

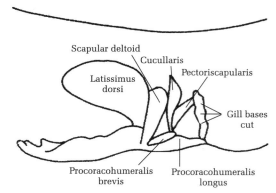

Figure 11.13
Dorsal superficial musculature of pectoral girdle of a mud puppy (dorsolateral view)

has its origin from the dorso-lateral portion of the trunk behind the pectoral girdle, is a major muscle pulling the humerus caudally. It is small in urodeles, larger in reptiles, and still large in mammals. In mammals two other muscles are derived from the same embryonic mass as the latissimus dorsi; these are the teres major (Fig. 11.15 B) which runs from the ventral border of the scapula to the humerus and a portion of the cutaneous maximus which inserts onto the skin of the trunk and acts to twitch it (Table 11.5).

The pectoralis complex, which has its origin from either the sternum or mid-ventral fascia (the linea alba) courses medially to insert onto the humerus; thus it is a major adductor of the forelimb (Fig. 11.8). Like the latissimus dorsi, the pectoralis complex is smallest in urodeles, larger in reptiles and largest in mammals. In most mammals this complex is divided into separate muscles, in the cat, these are the pectoralis major, pectoralis minor, pectoanti-branchialis, and xiphihumeralis (Fig. 11.15 A). The complex also gives rise to the portion of the cutaneous maximum not contributed by the latissimus dorsi.

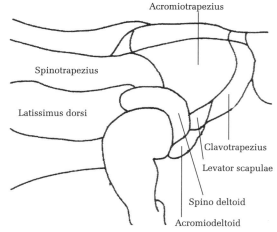

Figure 11.14
Dorsal superficial musculature of pectoral girdle of a rat (dorsolateral view)

Those muscles derived from limb bud mesenchyme and going from the pectoral girdle to the humerus include the deltoids, teresminor, subcaspularis, and supra- and infraspinatus. The deltoid musculature undergoes considerable variation both within and between classes (Table 11.7). In mammals there may be a single deltoid originating from the clavicle, scapular spine, and acromian process and inserting on the humerus or there may be three separate muscles, the clavodeltoid, spinodeltoid, and acromiodeltoid (cat) (Fig. 11.15 A).

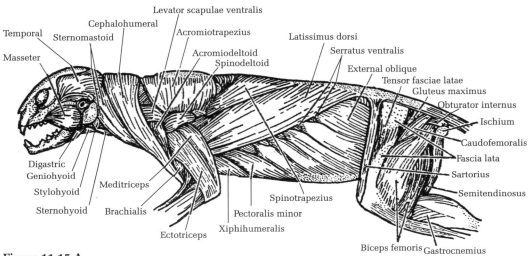

Figure 11.15 A
The musculature of a cat

Lizards usually have a clavodeltoid (clavicle to humerus) and scapulodeltoid (scapula to humerus). Salamanders also have a scapulodeltoid, but since they lack a clavicle, the other deltoid muscle runs from the procoracoid to the humerus, this muscle is named the procoracohumeralis longus (Fig. 11.13). In mammals there is a small muscle deep to the teres major called the teres minor, which runs from the scapula to the humerus. The same muscle is relatively larger in lizards and is called the scapulohumeralis anterior. In urodeles this muscle runs from the procoracoid to the humerus and is called the procorachohumeralis brevis (Table 11.5).

In **mammals** the large subscapularis is a strong adductor. It originates from most of the medial surface of the scapula and inserts on the medial surface of the proximal portion of the humerus (Table 11.7). In lizards two muscles form from the same embryonic anlage-the subcoraco scapularis and the scapulo—humeralis posterior (Table 11.7). As with the mammalian subscapularis, each runs from scapula to

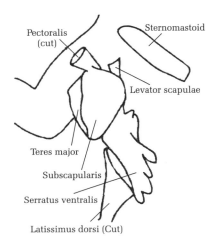

Figure 11.15 B
Pectoral girdle musculature of a rat (ventral view)

humerus (Fig. 11.15 B). Urodeles have a small subcoracoscapularis but no scapulohumeralis portion. Finally, in mammals, there are two adductor muscles which lie on the lateral surface of the scapula, just above and below the scapular spine, and are named the supraspinatus and infraspinatus, respectively (Table 11.7). Both insert on the proximal end of the humerus. Embryologically these muscles form in the ventral portion of the limb bud and then migrate dorsally. In the **lizard** and **salamander** a single muscle, the supracoracoideus, is homologous to these two, not surprisingly it is located on the ventral surface of the shoulder girdle, the procoracoid, and runs to the humerus (Table 11.5).

Pectoral musculature The common ancestors of today's living mammals, lizards, and salamanders were the stem amphibians, labyrinthodontia. From the fossil record we know that the labyrinthodont pectoral girdle was composed of an endochondral scapula and procoracoid and dermal cleithrum, clavicle and interclavicle, Faint muscle scars on those fossil bones indicate some of the points of muscle attachment. Putting this together with own knowledge of its living descendent; inferences can be made about its shoulder musculature.

With the loss of gills in adult amphibians the muscles which had acted as levators of the gill apparatus lost their adaptive value. At the same time, as amphibians began to walk on land, there was a strong adaptive value for anything that would strengthen the shoulder girdle. Thus, it is not surprising that the branchiomeric gill levators became attached to the pectoral girdle. This apparently happened quite early, muscle scars on the labyrinthodont clavicle and cleithra are probably from the cuccularis (Figs. 11.16 A, B).

The fact that among living vertebrates only amphibians have hypobranchial muscles involved in shoulder movement the pectoris capularis and omoarcuals (Table 11.5) suggest that these evolved after the lissamphibians had split off from the evolutionary group giving rise to the amniotes. The diversity of deltoid musculature in living tetrapods strongly suggest that they are the derivatives of a single, broad deltoid muscle, not unlike that in humans, and they evolved differently in independent tetrapod lines.

Pelvic Musculature

In all tetrapods the pelvic girdle is firmly attached to the axial skeleton by way of the sacrum; therefore it cannot move independently of the axial skeleton, and so needs no "sling" muscles. The extrinsic pelvic muscles, therefore, run only from the girdle and associated vertebrae onto the femur. There are also intrinsic muscles.

Branchiomeric muscles

Pharyngeal pouches form embryologically in all vertebrates and the splanchnic mesoderm between the pouches differentiate into a mass of skeletal muscles which form the branchiomeric muscles. Homologies are easy to determine, since in all classes each branchial pouch is innervated by a specific, special visceral nerve from the brain. On the basis of their ontogeny and innervation these muscles can be considered in groups; with each group composed basically of levators and dorsal and ventral constrictors (Table 11.6).

Table 11.6: Chief branchiomeric muscles and their innervation in *Squalus* and in tetrapods

Pharyngeal nerve arch innervation	pharyngeal skeleton in squalus	Chief branchiomeric muscles		Cranial
		Squalus	Tetrapods	
I Mandibular arch	Meckel's cartilage	Intermandibularis	Intermandibularis Mylohyoideus (anterior part) Digastricus (anterior part)	V
		Adductor mandibulae	Adductor mandibulae Masseter Temporalis Pterygoidei Tensor tympani	
	Pterygoquadrate cartilage	Levator maxillae superioris		
II Hyoid arch	Hyomandibular Ceratohyal Basihyal	Epihyoideus (dorsal constrictor) Interhyoideus (ventral con-stictor)	Stapedius Stylohyoideus (anterior part) Mylohoideus (posterior part) Depressor mandibulae Digastricus (posterior part) Sphincter colli (gularis) Platysma Mimetics	VII
III Glosso-pharyngeal	Gill cartilages	1. Constrictors 2. levators 3. Adductors 4. Interarcuals	Stylopharyngeus Stylohyoideus (Posteriorpast)	IX
IV to VI Vagal	Gill cartilages	1. Constrictors 2. And levators 3. Adductors 4. Interarcuals	Striated pharyngeal muscles Thyroarytenoideus Cricoarytenoideus Cricothyroideus	X
VII	Seventh visceral cartilages	Cucullaris (probably derived also from dorsal constrictors 3 to 6)	Trapezius Sternomastoideus Cleidomastoideus	Occipitospinal nerves in Shark; spinal roots of XI in amniotes

Fish: In the dogfish, the jaws are formed by the mandibular arch and are moved by its muscles (Fig. 11.16 A). The levator is the levator palato-quadrate, running from the neurocranium to the upper jaw; as its name implies it raises the upper jaw. The dorsal constrictor muscles are a large adductor mandibulae mass, which is broken up into a small pre-orbitalis and quadratomandibularis; these close the mouth. The ventral constrictors are the paired intermandibularis muscles, running from the linea alba to meckels cartilage, they function in unison to raise the floor of the mouth.

The hyoid arch has a very small levator, the spiracularis, when contracted it does not raise the hyoid arch but constricts the spiracle, which is a small gill slit in the same arch. The dorsal constrictor is the epihyoidean, it is the main muscle raising the hyoid, and it also helps push the jaws forward. The ventral constrictor is the interhyoideus, lying just underneath to the intermandibularis and inserting on the ventral hyoid structures (Fig. 11.16 A).

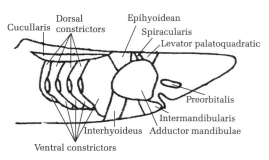

Figure 11.16 A
Superficial branchiomeric musculature of a dog fish (lateral view)

The third through the sixth arches have superficial dorsal and ventral constrictors which are numbered as the arches (3,4 etc). Their levators are fused into a single muscle, the cuccularis, raising the gill arches.

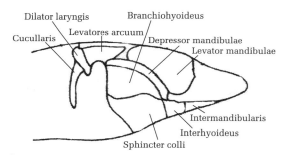

Figure 11.16 B
Superficial branchiomeric musculature of a mud puppy (lateral view)

In the quite independent actinopterygian evolution the brachiomeric muscles, like the head and gill arch skeleton have become much more complex and are intimately involved in the mechanism of kinetic jaws. In the adult many originate and insert not on the splanchnocranial elements with which they are associated in early embryology, but on dermatocranial elements, for example, the dermal bones of the teleost operculum are moved by branchiomeric muscles.

Amphibians: Since adult amphibians do not respire by gills, their branchiomeric muscles differ from those of fishes, but they can still be recognized by their innervations and embryology. As in fishes, the dorsal constrictor of the mandibular arch closes the jaws; in amphibians the muscles derived from the dorsal constrictor are the levator mandibuli muscles. The levators of the mandibular arch are fused with these muscles. Ventrally the intermandibularis muscles are quite similar to those in chondrichthyes in both position and functions (Fig. 11.16 B).

In the second gill arch the dorsal constrictor and levator are also fused and together are called the depressor mandibuli. This muscle runs from the back part of the skull to the posterior part of the lower jaw, behind its articulation. When it contracts, it pulls the back part of the jaw upwards; depressing the rest of the lower jaw and opening the mouth. The interhyoideus (ventral constrictor) runs from the midline to the hyoid apparatus. Its posterior part, the sphincter coli, becomes attached to the integument rather than to the hyoid arch and acts to tighten the skin in the ventral throat region (and therefore, is an integumentary muscle). The constrictor muscles of arches three through six are primarily concerned with the larynx and voice production. The levators are fused as the cuccularis which functions as an extrinsic shoulder muscle (Table 11.5).

Amniotes: In amniotes the mandibular arch is greatly reduced; only one branchiomeric muscle retains its splanchnocranial connections, and that only in mammals. This is the tensor tympani, a very small muscle from the dorsal part of the mandibular arch musculature (Table 11.6). When the tensor tympani contracts it pulls on a modified splanchnocranial element, the malleus of the middle ear and thus tightens the tympanic membrane.

With this one exception the amniote mandibular muscles are involved with the dermatocranium. The fused levator and dorsal constrictor, in all amniotes, are the major muscles of mastication, closing the jaws and often allowing for lateral chewing; these are the masseter, the temporalis, and the pterygoideus internus and externus (Fig. 11.16 C) (Table 11.6).

The mandibular arch ventral constrictor is the intermandibularis in birds and reptiles; in mammals it is divided into two elements. Anteriorly there is the mylohyoid, which runs to the lower jaw much as the intermandibularis does in other vertebrates. The posterior part has become attached to part of the ventral constrictor of the second gill arch, thus forming a two part muscle called the diagastricus (Fig. 11.16 C) (Table 11.6). Because of its embryonic origin, the diagastricus has a double innervation the anterior part by the trigeminal nerve and the posterior by the facial nerve. The digastricus muscle is the primary muscle opening the mammalian jaw.

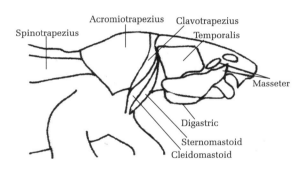

Figure 11.16 C
Superficial branchiomeric musculature of a rat (lateral view)

In reptiles and birds as well as amphibians the levator and dorsal constrictor of the second gill arch give rise to the depressor mandibuli which opens the mouth. In mammals, whose jaws are opened by the digastricus, the levator and dorsal constrictor of the second gill arch are present as the tiny stapedius muscle. This muscle is attached to a hyoid arch derivative, the stapes of the middle ear, and acts to tighten the chain of tiny bones in the middle ear; thus protecting the ear from over stimulation.

In adult reptiles and birds, as in amphibians, the ventral constrictor is present as the interhyoideus and sphincter coli. In mammals, it is present as the posterior portion of the digastricus and the muscles of facial expression, so finely developed in humans. In the neck region this musculature is called the platysma, and its more specialized regions around the eyes, nose and mouth are named "mimetic muscles" according to the part of the facial skin they move. e.g., the orbicularis oculi closes the eye (Table 11.6).

The musculature of arches three to six becomes involved primarily with the intrinsic muscles of the pharynx and larynx and with some large muscles of the shoulder (trapezius muscle), as well as muscles running from the sternum and clavicle to the mastoid at the rear of the skull (sternomastoid and cleidomastoid muscles, respectively) (Table 11.6). Thus the loss of functional gills in tetrapods frees the muscles which in early embryology were related to gill pouches, and they then become involved in facial expression, jaw movements, swallowing, vocalizing, hearing (in middle ear), and movement of the shoulder (Fig. 11.16 C). Despite these diverse developments, their homologies can be accurately traced by their embryology and innervation.

Table 11.7: Representative myotomal muscles of head, trunk and forelimbs in mammals

Head and neck		
Eyeball	**Tongue**	**Hypobranchial**
Superior oblique	Genioglossus	Geniohyoideus
Inferior oblique	Hyoglossus	Sternohyoideus
Medial rectus	Styloglossus	Sternothyroideus
Lateral rectus	Lingualis	Thyrohyoideus
Superior rectus		Omohyoideus
Inferior rectus		
Trunk and tail		
Epaxial muscles		**Hypaxial muscles**
Longissimus		Subvertebrals
L. dorsi		Longus colli
L. cervicis		Psoas
L. Capitis		Iliacus
Extensor caudae lateralis		Quadratus lumborum
Iliocostales		Oblique group
Spinales		Internal and external intercostals
S. dorsi		Internal and external obliques of abdomen
S. cervicis		
S. capitis		Diaphragrm
Transversospinales		Cremaster
Extensor caudae medialis		Transverse group
Intervertebrals		Transversus thoracis
Intertransversarii		Transversus abdominis
Interspinales		Rectus abdominis
Interarcuales		pyramidalis
Interarticulares		Abductors and flexors of tail

Forelimb	
Extrinsic muscles	**Intrinsic muscles**
Levator scapulae	(**Muscles of girdle**)
Latissimus dorsi	Girdle to humerus, proximally
Rhomboideus	*Deltoideus*
Serratus ventralis	*Subcoracoscapularis*
Pectoral group	*Subscapularis*
	Scapulohumeralis
	Teres minor
	Supracoracoideus
	infra spinatus
	Supraspinatus
	Coracobrachialis
	Teres major (derived from latissimus dorsi)
Muscles of upper arm	
	(Girdle or humerus to proximal end of radius or ulna)
	Triceps brachii (anconeus)
	Humeroantebrachialis
	Biceps brachii
	Brachialis
Muscles of forearm	
	(Humerus and proximal end of radius and ulna to manus)
	Extensors and *flexors* of wrist and digits
	Supinators and pronators of hand
Muscles of Hand	
	Flexors, extensors, abductors and *adductors of fingers.*

Integumental Muscles

The integumental, or dermal, muscles split off embryonically from the underlying skeletal muscles. While in many cases retaining their skeletal origins at one end and inserting under the skin at the other, they sometimes as in sphincter muscles, lose all skeletal connection.

Integumental muscles can be divided into:

A. **Extrinsic integumental muscles** They have their development and anatomical origin away from dermis. Their insertions are along the under surface of the dermis.

B. **Intrinsic integumental muscles** The muscles develop entirely with in the skin, in the dermis. They are smooth muscles in nearly all species and are innervated by the sympathetic nervous system.

Extrinsic integumental muscles are striated muscles and they move the skin. In **fish** and **amphibians**, slips of branchiomeric or body wall muscles insert here and there on the dermis, firmly attaching the skin to the underlying muscle; however, the slips cause little movement of the skin. The cutaneous pectoris of anurans is such a muscle.

The tight-skinned **fish** are without any dermal musculature, and amphibians have only a trace of anything of the sort in the tiny muscles that open and close the lids of the nostrils.

Snakes among reptiles use integumental muscles in locomotion, for these muscles enable the scales to get a grip on the ground. Costocuaneous muscles are hypaxial muscles that erect the ventral scutes of snakes, thereby providing friction for locomotion.

Birds fluff the feathers by means of integumental muscles, in this way changing the thickness of the layer of warm air held next to the body to regulate the body temperature. The so-called patagial muscles in the web of a bird's wing, that assist in flight, belong to the integumental group and are derived from the pectoralis muscles of the breast, together with various muscles of the shoulder and arm.

It is in mammals, however, that integumental muscles reach their greatest differentiation, serving a wide range of uses from defence to the expression of the emotions.

Under defensive skin muscles may be mentioned (1) those which cause hairs and bristles to stand on end in terrifying fashion, as on the tail of a frightened cat or on the scruff of an angry dog's neck; (2) those which erect defensive spines or quills, as in the skin of the "fretful porcupine,".(3) those which enable animals like the armadillos and the European hedgehog (*Erinaceus*) to roll up into an impregnable ball; and (4) muscles which tend to dislodge annoying insects by causing the skin to shudder or twitch, as on the neck, shoulders, and the anterior side of the horse.

Voluntary integumental muscles fall into two general groups according to their derivation; first, the panniculus carnosus group from the lattissimus dorsi and pectoralis muscles, and second, the sphincter coli group, from the branchiomeric musculature of the hyoid region, under the dominance of the seventh pair of cranial nerves.

The panniculus carnosus (Fig. 11.17) is particularly evident in mammals, although somewhat degenerate in man, and with only primitive traces showing in the lower vertebrates. It is a thin sheet like muscle that tends to wrap about the body under the skin. In monotremes it extends over the entire body as far as the cloaca, and includes a sphincter marsupii, and a sphincter clocacae.

Fragments of the enveloping panniculus carnosus remain longest in the axillary,

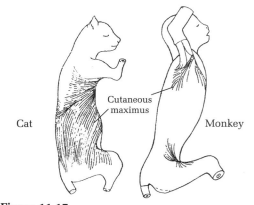

Figure 11.17
Panniculus carnosus (cutaneous maximus) of a cat and primate

inguinal, and sternal regions. These fragments make up the shuddering muscles of the horse and the muscles by which a wet dog shakes itself. There is also occasionally a sternalis muscle under the skin and superficial to the pectoralis, which is sometimes visible in man when well developed.

The progressive sphincter colli group of integumentary muscles, originally associated with the hyoid arch, is supplied by the facial (VII) nerve. In such animals as turtles and birds it is a well-developed group of muscles enwrapping the neck. During its evolution it migrated forward and expanded so as to spread over the head and down onto the shoulders, becoming differentiated into a superficial sheet of muscle designated as the platysma and the deeper lying sphincter colli proper.

With the upgrowth of the cranium that part of the platysma extending over it becomes divided into an occipital and a facial part, separated by a broad sheet of connective tissue, the galea aponeurotica, that stretches over the top of the cranium under the skin. The facial parts of the platysma and sphincter colli may be classified into four groups of muscles in close association with the underlying muscles of mastication. These are the muscles of the external ears, eyebrows, nostrils, and lips and cheeks (Fig. 11.18).

The muscles of the external ear, namely, auricularis anterior, posterior, and superior, enable an animal to turn the pinna on the external ear toward the source of sound without changing the position of the head. They are better developed in animals like dogs and horses than in man.

The eyebrow group takes in four muscles (Mimetic muscle): the frontalis; orbicularis oculi; levator palpebrae superioris; and corrugator supercilii or brow-wrinkler.

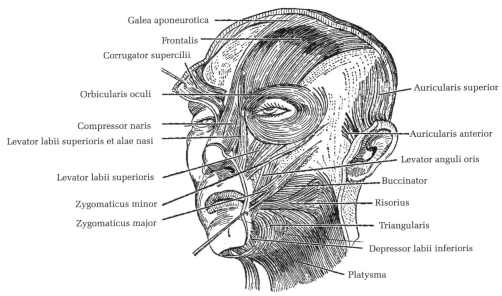

Figure 11.18
Mimetic muscles (man)

Three more or less well-developed muscles of the nostril group or levator labii superioris et alae nasi, by means of which man as well as beast sneers and snarls; the dilatores naris and the compressor naris, by means of which rabbits and men dilate the nostril and wiggle their noses.

Finally, the lips and cheeks group consists of a strong sphincter muscle, the orbicularis oris around the mouth opening, from which radiate several other muscles. Of these the risorius muscle, attached at the corner of the mouth opening and pulling laterally in opposite directions and triangularis, pulling down the corners of the mouth, serve in humankind to express the diverse emotions of laughter and tears. The buccinator makes up much of the cheek wall.

Intrinsic integumental muscles develop entirely within the skin in dermis and are found in **birds** and **mammals**. They are mostly smooth muscles that attach to feather follicles (arrectores plumarum) and to hair follicles (arrectores pilorum). These muscles permit ruffling the feathers or elevating the hairs for insulation. They are innervated by visceral motor fibres (sympathetic nervous system).

Some muscle specializations due to adaptations

Birds: One of the most extreme adaptations of the muscular system is seen in flight muscles of the class aves. Flight requires powerful adduction (when the wings are brought down in the power stroke) and then rapid abduction (when the wings are raised in preparation for another power stroke). Although small epaxial muscle give fine control and stability to the wing dorsally, the mass of both its adductors and abductors are located ventrally, along the enlarged sternum (unlike other vertebrates, whose abductors are located dorsal to the girdle). The largest of these flight muscles is also the largest muscle in the birds body often providing as much as one-fifth of the total body weight (Fig. 11.19). This is the pectoralis muscle, whose origin is from the very large sternum and its keel, from the ribs, and from the furcula. The pectoralis muscle fibres all converge to insert onto the proximal part of the humerus, so that its contraction adducts the wing powerfully and rapidly.

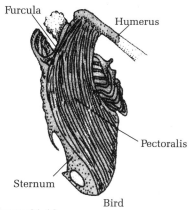

Figure 11.19
The power stroke of flight is caused primarily by a single large muscle in bird (pectoralis)

Deep to the pectoralis lies the supracoracoideus, sometimes called the deep pectoral muscle. It originates on the sternum beneath the pectoralis muscle. Its tendon passes up parallel to the procoracoid to its dorsal aspect and then passes through the foramen triosseum, which is formed by the articulation of the scapula, procoracoid, and furcula of the pectoral girdle. After passing through this foramen, the tendon inserts on the dorsal part of the proximal portion of the humerus. When the supracoracoideus contracts, the foramen triosseum act as a pulley and the wing is abducted. In this way both the abductor and adductor add weight only ventral to where the wing articulates onto the pectoral girdle, thus the bird's centre of gravity is below the axis of the wings and stability is added to its flight.

Bats: The only flying mammals are bats, which have quite different adaptations. The sternum has only a small keel (Fig. 11.20). The pectoralis musculature is divided into three parts, the pectoralis anterior, posterior, and abdominus, all of which insert on the proximal end of the humerus and provide much of the power stroke for adduction of the wing. The clavodeltoid, posterior division of the serratus ventralis, and subscapularis provide additional power and also give fine angle adjustments. Abduction involves several muscles, including the supra and infraspinatus, the acromio- and spinodeltoids, and the teres major, although all of these are located dorsally, none is large enough to make the bat "top-heavy". There are elaborate intrinsic muscles that control the patagium (diving membrane) and the digits which support it. One such muscle originates from the back of the skull and inserts distally on the wing, supporting its anterior portion.

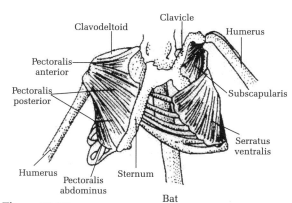

Figure 11.20
The power stroke of flight in bat coordinated by contraction of six muscles

Moles: Among fossorial (burrowing) animals there are several different adaptations, in the case of moles, it is once again an adaptation of the forelimb, particularly in the shoulder region. The teres major is the largest and most powerful muscle in the complex, and the one that does the work (Table 11.7). However, the humerus, and the pectoral girdle are rotated approximately 90 degrees, so that when the teres major contracts it does not retract the limb but pulls it directly posteriorly. The other muscles, such as the pectoralis and triceps, fix the joints.

Whales and porpoises: Quite a different locomotor adaptation is found in the wholly aquatic whales and porpoises, (order cetacea.) Their hindlimbs have been secondarily lost; the forelimbs acts as stabilizers for steering much as do the paired appendages of fishes. Also as in fishes, the caudal musculature produces the power stroke for the cetaceans swimming; but instead of lateral undulations they are dorsal ventral undulations. The tail is moved dorsally by very large dorsal iliocostalis muscles running back and inserting onto the caudal vertebrae; it is depressed by a very large ventral muscle, the ischio-caudalis, which runs from the ischia of the pelvic girdle back to the caudal vertebrae.

Red and White Muscle Fibres

Birds and mammals have in their skeletal muscles two kinds of striated muscle fibres: red or slow muscle fibres and white or fast muscle fibres.

Table 11.8: Differences between red muscle fibres and white muscle fibres

Red muscle fibres	White muscle fibres
1. They are thinner.	They are much thicker.
2. These muscle fibres are dark red in colour which is due to the presence of red haem-protein called myoglobin. Myoglobin binds and stores oxygen as oxymyoglobin in the red fibres. Oxymyoglobin releases oxygen for utilization during muscle contraction.	These muscle fibres are lighter in colour as they do not have myoglobin.
3. Red muscle fibres are rich in mitochondria.	White muscle fibres are poorer in mitochondria.
4. They carry out considerable aerobic oxidation and so aerobic contractions of the muscles without accumulating much lactic acid. Thus red muscle fibres can contract for a longer period without fatigue.	They depend mainly on anaerobic oxidation (glycloysis) for energy production and so anaerobic contractions accumulate lactic acid in considerable amounts during strenuous work and soon get fatigued.
5. These muscle fibres have slow rate of contraction.	These muscle fibres have a fast rate of contraction.
6. Red muscle fibres perform sustained work at a slow rate but for a long time.	White muscle fibres are specialized for very fast and strenuous work for a short time only.
7. Extensor muscles on the back of the human body are very rich in red muscle fibres. Some flight muscles of birds are red muscles.	The muscles for eyeball movements are very rich in white muscle fibres. Flight muscles which are used in short and fast flying such as that in sparrow, are white muscles.

Electric Organ

The **electric organs** are highly modified muscle masses which are able to produce, store and discharge electricity. Each electric organ consists of columns of electric discs called **electroplax** (up to 20,000 in the tail of one ray) arranged vertically (ray) or longitudinally (eel) (Fig. 11.21). Each disc is a modified multinucleate muscle cell separated from the next by a connective tissue partition. Motor nerve endings terminating on each disc control the discharge of electric current. These muscle masses may be derived from branchiomeric muscles or from hypaxial musculature. In the electric ray *Torpedo* the electric organ lies on each pectoral fin near the gill and is innervated by 7th and 9th cranial nerves. It is probably **branchiomeric** in origin. In another ray *Raia*, and in the electric eel *Electrophorus*, the electric organ lies in the tail and is derived from the hypaxial muscles. The charge produced by the organ in the eel amounts to several hundred volts and delivers a stunning shock. The electric organs are not distributed in a systematic manner and are probably a result of convergent evolution. In some they are not muscular, for instance, in a catfish native to Africa it is dermal in origin.

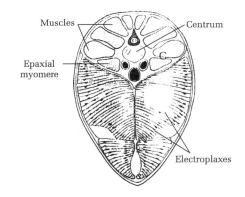

Figure 11.21
Electric organs in tail of electric eel. Each nucleated horizontal disc (electroplax) is a modified hypaxial muscle fibre

Chapter 12

Digestive System

Nutrition is a process by which the organisms procure those substances which are utilized for their growth, maintenance and for replenishing their energy requirements. Digestion is a process in which non-diffusible foods are converted into diffusible form with the help of digestive enzymes. Hence precisely the ingestion, digestion, absorption of food and the formation and egestion of faecal matter are the functions of digestive system. A few important factors discussed below have played a significant role in inducing modifications in increasing efficiency of the alimentary canal of vertebrates.

a) **One such very important factor has been the ratio between the absorptive surface area and the size of the animal.** As the ratio between absorptive surface and animal mass improves, the metabolic rate in animals can become higher. Therefore, down the years, gradually various complexities evolved in the absorptive surface to keep this ratio constant or to improve it, as species evolved to bigger size. These developments also included lengthening and folding of the alimentary canal; an increase in its diameter, and folding of the canal lining or its cells. Some or all of these modifications are to be found in most evolved species.

b) **Metabolic rate itself has been important and second crucial factor in this direction.** The higher the metabolic rate, the more rapidly the absorbed food is utilized and sooner more food is required. The vertebrates with high metabolic rate require efficient mechanical and chemical digestive processes, as well as large surface area for absorption.

c) **Adaptability:** Environment influences all types of organisms to such a great extent that a change in the environment may bring about significant changes in organisms. The change in environment gives rise to new needs, the new needs or requirements may cause significant change in habits of organisms. For instance, the gills of plankton eating fish have an elaborate strainer mechanism which traps plankton during respiration and concentrate them in pharynx, before they are swallowed.

The tendency to invade aquatic niche and its exploitation for food has affected the modifications in alimentary canal of plankton eating toothless whales, which unlike fish have evolved a quite massive strainer mechanism consisting of frayed horny sheets of oral epithelium called baleen of highly modified keratin hairs hanging from roof of the mouth. The baleen acts as a sieve during feeding. The food is filtered at baleen and swallowed where as excess water goes out though nares.

In short the vertebrate's alimentary canal usually gets adapted to available specific diet both morphologically and anatomically by several types of modifications in order to adjust with the new environment.

Alimentary Canal

Origin and Development

The lining of the entire alimentary canal is formed from epithelial tissue. An ectodermal invagination called **stomodaeum** forms the oral cavity, another invagination forms the anal canal epithelium and is called **proctodaeum**, between the two is endoderm and endodermal out-pocketing which forms the rudiments of liver, pancreas gall bladder and paired pharyngeal pouches (Fig. 12.1).

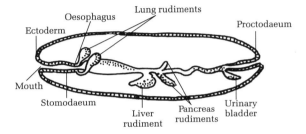

Figure 12.1
Alimentary canal and it derivatives (vertebrate embryo)

In the stomodeal ectoderm a dorsal out-pocketing forms **Rathke's pouch**, which contributes to the pituitary. From the embryonic digestive system are also derived various respiratory organs, such as lungs, gas bladder, and gills.

The epithelial lining of the alimentary canal is surrounded by tissues of mesodermal origin. During very early development the hypaxial mesoderm splits into a lateral, somatic mesoderm and a medial, splanchnic or visceral mesoderm (Fig. 12.2)

The splanchnic mesoderm gives rise to the inner connective tissue and outer muscular tissue which at any

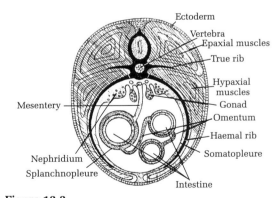

Figure 12.2
Cross section through the trunk of a vertebrate showing position of mesentries, omentum and mesoderm etc.

level make up the bulk of the alimentary canal. Between the somatic and splanchnic mesoderm develop the body cavities. Except in mammals and birds there are two body cavities a large **pleuroperitoneal** cavity, which contains most of the viscera and the lungs if present, and a smaller **pericardial cavity**, which contains only the heart (Fig. 12.3).

The body cavities together are called coelom, they are filled with coelomic fluid and lined with a shiny-serous membrane, called **mesothelium**, which is formed from epithelial cells of mesodermal origin. Between the visceral and parietal serous membranes is another continuous subdivisions, the double-walled membranes called mesentries (Fig. 12.2). These mesentries hold the viscera in place while permitting some movement. They allow movement for communication and nourishment by virtue of the blood vessels and nerves which travel through them.

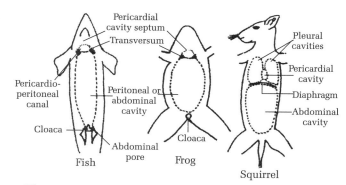

Figure 12.3
Divisions of the coelom of craniates

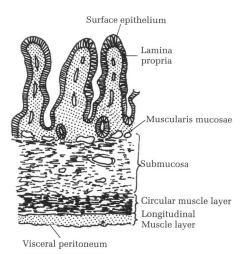

Figure 12.4
T.S. ileum, showing layered organization in alimentary canal.

Although each region of the alimentary canal (stomach, oesophagus) has specific morphological characteristics, there is a general layered organization which is common to all regions lined by endoderm (Fig. 12.4). Immediately outside the epithelial lining is a layer of vascular connective tissue called **lamina propria**, containing the capillaries through which the digested food enters the circulatory system. Lamina propria and epithelial lining are collectively called **mucosa** which is absorptive layer of the gut. Separating the lamina propria from deeper, denser and less vascular connective tissue is a thin, smooth muscle layer the muscularis mucosa. The denser connective tissue layer beyond the mucosa is the **submucosa**. External to the submucosa are two layers of smooth muscles an inner, circularly oriented layer and on outer longitudinally oriented layer. The coordinated contractions of these muscles, controlled by the autonomic nervous system is called peristalsis. Finally the gut is covered by outermost layer the **visceral peritoneum**.

Parts and their modifications

Ingestion brings food into the mouth and pharynx. Regardless of structural variation, however, the mouth and pharynx are basically homologous in all vertebrates. The other structures associated with mouth opening are the lips, tongue, glands tentacles and teeth.

Mouth

In a restricted sense, the term mouth refers only to the anterior opening into the digestive tract. In a broader sense the term is used as a synonym for oral cavity. Mouth is usually terminal in position, although in some fish, especially in elasmobranch and sturgeons it is located ventrally and often away from the cephalic tip of the head (Fig. 12.5). In cyclostomes the opening is at the vertex of the buccal funnel (Fig. 12.6) and is always open, since there are no hinged jaws or other mechanisms for closing it. In amphioxus the opening of mouth is in the velum (Fig. 12.7). In lower chordates it is sometimes encircled by papillae. As in amphioxus, larval cyclostomes (Fig. 12.6) and catfish, papillae usually have chemoreceptors akin to taste buds. Mouth of the larval tadpole (frog) is a small oral suctorial organ surrounded by horny lips and rasping papillae consisting of rows of horny teeth. These are adaptations for feeding on plants. Fleshy muscular lips surround the mouth in most mammals. Muscular lips and cheeks are adapted for sucking. In mammals movable lips help in osculation.

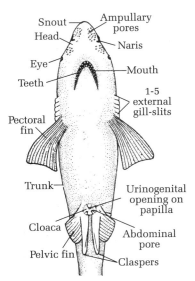

Figure 12.5
Male *Scoliodon* (dogfish) in ventral view

Oral cavity The stomodaeum develops into the oral cavity, which includes the area from the lips to the pharynx. Since, this part of the alimentary canal has intimate contact with the environment and receives food in its raw state, it has evolved with extreme modifications to diet. In adult fish the oral cavity is very short and almost non-existent. It is longer in tetrapods and culminates in mammals in a sucking and masticating organ bounded laterally by muscular cheeks (Fig. 12.8). During mastication (chewing) food squeezed between the teeth

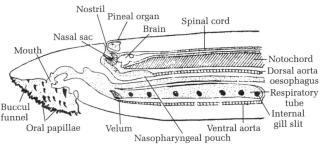

Figure 12.6
Petromyzon Sagittal section of anterior portion

falls either into the oral cavity proper or into the vestibule, from where it can be brought back between the teeth by contractions of the platysma musculature lining the cheeks. Except mammals in other vertebrates there is no distinct vestibule. The teeth are frequently on the margins of the mouth and food outside the oral cavity proper falls to the ground or into the water. Some mammals such as squirrels and many monkeys, have internal cheek pouches which are expandable dilations inside the cheeks in which they carry or temporarily store food. Not surprisingly, the buccinator muscles of these mammals are particularly well developed (Fig. 12.8).

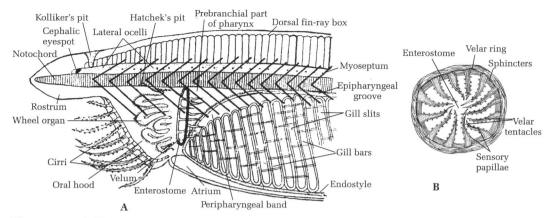

Figures 12.7 A, B
A. *Branchiostoma* L.S. of anterior region   B. Structure of velum showing tentacles (magnified)

Oral cavity proper In mammals the oral cavity is separated from the nasal chambers by the secondary palate, whose anterior portion, the hard palate, is formed by extensions of the premaxilla, maxilla and palatine bones, and whose posterior portion, soft palate, is formed by a tough but pliable connective tissue membrane. The entire secondary palate is covered by the stratified squamous epithelium of the oral cavity (Fig. 12.9). An adaptive value of the secondary palate is to extend the internal nares caudally so that they open into the pharynx rather than the oral cavity. This permits simultaneous mastication and respiration. Among **reptiles** only the crocodilians have a complete secondary palate, it is bony in its entire length. Other reptiles and birds have an incomplete soft secondary palate formed of tissue folds that do not meet medially. Anamniotes have no secondary palate.

Tongue

Most fish and larval amphibians have only a primary tongue. This structure does not have the fleshy element of most tetrapod tongues, it is a little more than a fold of mucous membrane covering the basihyal and ceratohyal elements. It is moved by hypobranchial muscles, particularly the coracohyoid. There are taste buds on the primary tongue, as well as in the pharynx and other parts of the oral cavity.

The **cyclostome** tongue bears hard, sharp keratin teeth anteriorly. It is supported by a complex group of lingual cartilages derived from the splanchnocranium and lingual muscles derived from the branchiomeric musculature. Within this arrangement the cyclostome tongue is quite mobile and can be drawn in and can be protruded out of the mouth for considerable distances. It is used to rasp away the flesh of the fishes on which the cyclostome prey, and it is a remarkable adaptation for the specific diet.

Figure 12.8
Cheek pouches of monkeys, showing complex buccinator muscles

In **amphibians** the tongue undergoes two major changes during metamorphosis. (i) As functional gills are lost there is a corresponding reduction in the branchial skeleton, some hypobranchial muscles are therefore needed to move the primary tongue (ii) A secondary fleshy tongue also develops just behind the mandibular arch in the anterior oral cavity, a glandular field develops in relationship to the basihyal.

Particularly in **frogs** and **toads** the tongue can be extended to a considerable length, the base remains in approximately the same position in the anterior oral cavity (Fig. 12.10), while the tip which normally lies back in the pharynx, is flipped over and out of the mouth. Insects are caught on its sticky-surface, and it is then flipped back into the mouth. A few frogs, for example, *Pipa americana* lack tongue.

The hyoid apparatus in amniotes has migrated caudally, and a neck region has formed. As a result, the base of the tongue lies far back in the pharyngeal region. The diversity in the shape of the amniote tongue fits it to perform different functions.

Most **reptiles, birds,** and **mammals** use the tongue for lapping up liquids, for taste, for touch, and (especially mammals) for food manipulation. Many mammals use it as a curry comb, as it bears numerous small keratin structures called filiform papillae, which improves its ability to groom the fur (Fig. 12.15). The chameleon tongue is so extensible that it can be protruded from the mouth for a distance greater than body

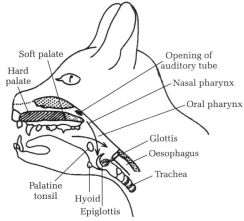

Figure 12.9
Position of soft and hard palate (in mammal)

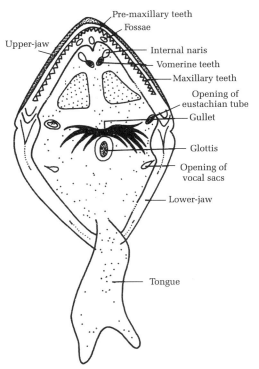

Figure 12.10
Frog: Buccuo-pharyngeal cavity of male

length. The forked tongue of snakes can be flashed in and out very rapidly and is highly sensitive to chemical and physical stimuli. Anteaters have long very narrow tongues which can reach deep into the holes of termites. Among humans the tongue is extremely important for production of the complex sounds of speech (Fig. 12.14). In most mammals the tongue is attached to the floor of the oral cavity by a ligament the **frenulum**. In man if the frenulum extends unusually forward, the individual is 'tongue-tied'. For various functions tongue needs the taste buds innervations and fine muscular control. In reptiles and mammals tongue has much intrinsic musculature. In birds, however, the tongue is a relatively hard structure built around a forward extension of the hyoid apparatus (Fig. 12.11), it bears taste buds and a few glands. Bird songs are produced by the syrinx, and the tongue has little intrinsic musculature. Although it lacks the flexibility of the mammalian tongue, the bird tongue is nevertheless highly mobile because of its extrinsic musculature, which is primarily hypobranchial in origin.

Oral glands

Labial glands open into the vestibule at the base of the teeth and secrete mucus which lubricates the food. The duct of the parotid gland, called stenson's duct, also opens into the vestibule (Fig. 12.13). The parotid is a major salivary gland which helps start the digestion of starch.

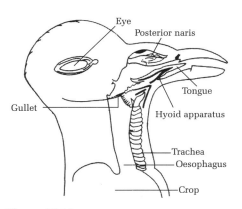

Figure 12.11
Birds tongue showing hyoid apparatus

Salivary glands open into the oral cavity proper, they are unicellular in fish and multicellular in tetrapods (Fig. 12.16). Their secretions are stimulated by autonomic innervation and pushed along the duct of the gland into the oral cavity by muscles. Most of these glands secrete mucus which acts as both a lubricant and a solvent. Only a few multi cellular glands are found in the oral cavity of fish, aquatic amphibians, aquatic turtles and birds that obtain their food from water. Their absence is no handicap, since any secretion would be constantly washed away and is unnecessary for lubrication or digestion. In fact, most oral glands in the groups primitive to mammals, contain no digestive enzymes. The goblet cells, however, are common. The males of the species of catfish carry the fertilized eggs in their mouth, and during the breeding season the oral epithelium becomes extensively folded and exhibits numerous crypts supplied with large goblet cells. These brood pouches shelter the eggs and atrophy after the breeding season.

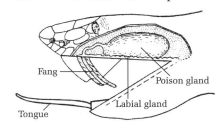

Figure 12.12
Rattle snake, head dissected to show poison gland and fangs

In land dwelling vertebrates unicellular and multicellular oral glands, chiefly mucous, are scattered about in the roof, walls and floor of the oral cavity, but the secretion does not becomes copious in the groups primitive to mammals. In tetrapods the secretion facilitates taste by dissolving food, in some cases the secretion also contains the enzyme ptyalin, which initiates carbohydrate digestion. In amphibians and many lizards special lingual glands produce the sticky mucus used for catching insects. In amniotes there are variable numbers of large salivary glands, called mixed salivary glands, which contain clumps of both mucus and serous cells, these are frequently located some distance from the opening of their ducts. Largest oral glands are the poison glands of snakes (Fig. 12.12) and the compound salivary glands of mammals. In viper the poison gland is a modified salivary gland that is enfolded in the masticatory muscles; contraction of these muscles simultaneously forces the glands secretion to the ducts termination at the base of the large hollow fangs and closes the jaws.

Oral glands are often named according to their location.

Labial glands open at the base of the lips.

Palatal glands open onto the palate.

Intermaxillary (internasal) glands lie between the premaxillary bones. In frogs they consists of up to 25 small glands, each emptying a sticky secretion by its own duct into the oral cavity. **Lingual glands** lie under the tongue. In *Heloderma* the only poisonous lizard; they secrete the toxin.

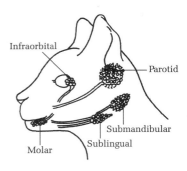

Figure 12.13
Cat: location of oral glands

Parotid gland are the largest salivary glands. They open into vestibule opposite one of the upper molars (Fig. 12.16). The submandibular in rabbit opens in the caudal floor of the orbit (Fig. 12.13).

Molar gland in cat (Fig. 12.13), under the skin of lower lip open via several ducts into mucosa of cheek.

Monotremes apparently do not have salivary gland. Saliva chiefly is a mammalian product, but Ptyalin secreting glands occur in some frogs and birds.

Tentacles

In *Branchiostoma* the mouth is a large widely open space below the oral-hood and the tentacles bordering it have sensory-papillae, which serve to filter minute particles of food. Epithelium of the oral hood is produced into a characteristic structure, the wheel organ, for drawing a current of water into the mouth (Figs. 12.7 A, B). Mouth leads into the pharynx through the enterostome a perforation in the velum bearing filtering tentacles.

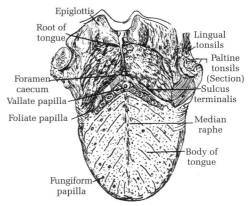

Figure 12.14
Dorsal view of human tongue

In **cyclostomes** intake apparatus of hagfish is altogether different. A small some what triangular mouth is surrounded on the sides by four pairs of sensory tentacles. A single median epidermal tooth lies above the mouth and two rows of teeth (dental plates) are located on the tongue. The hagfish burrow actively into the bodies of dead or dying fish and feed upon their tissues and organs. In the Indian carp *Labeo rohita* sub-terminal mouth is bounded by soft-fleshy lips and sensory barbels.

Teeth

True vertebrate teeth are specialized remnants of an ancient dermal armor. The fossil record does not indicate that teeth evolved from placoid denticles, rather that both placoid scales and teeth evolved from hard denticles on the outer surface of superficial dermal bone. The development of placoid scales and also of teeth, shows that a dermal papilla produces a

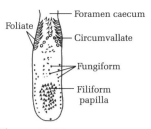

Figure 12.15
Tongue showing arrangement of papillae (in cat)

root composed of dentine; and part of this root erupts through the overlying epidermis (Fig. 12.17). An enamel organ, an ingrowth of the germinal layer of the epidermis, usually secrete enamel on that part of the tooth which erupts. The enamel organ may be functionless in sharks, and is clearly functionless in *Armadillos*, and a few other fish and mammals. In such cases any hard material on the surface of the tooth is a product of the dermis.

Development of teeth The embryological development of vertebrate teeth start with either a thickening or an inpocketing of the stomodeal ectoderm (Fig. 12.17) or, in some fish, of the pharyngeal endoderm. Deep to the thickening or inpocketing a condensation of mesoderm forms a vascularized dermal papilla. The outer part of this papilla differentiate into epithelial cells called **odontoblasts** which secrete dentine, a cellular substance harder than bone but of similar structure which makes up the bulk of the tooth. The ectodermal thickening or inpocketing forms the enamel organ; its cells differentiate into secretory cells called **ameloblasts**, which secrete hard enamel over the tooth. Depending on taxon, the enamel-capped portion of tooth erupts either during or after development. It may be held into a socket by dentine roots below the surface, or it may be attached to the dermal bone by collagenous fibres or by the secretion of hard cement like substance—the cementum.

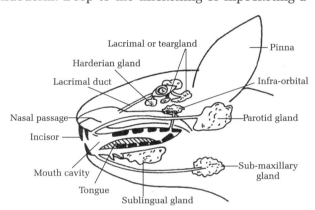

Figure 12.16
Salivary glands in Rabbit

A few toothless species are found in every vertebrate class. Agnathans, sturgeons, some toads, sirens (urodeles) turtles and modern birds are toothless. However, one species of *Terns* develops an embryonic set of teeth that fails to

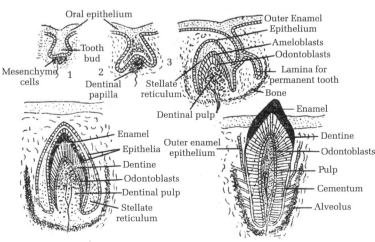

Figure 12.17
Different stages of development of a milk-tooth

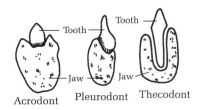

Figure 12.18
Methods of tooth attachment

erupt, and at least one genus of turtles develops an enamel organ. Toothless mammals develop a set that either do not erupt at all or, after eruption are soon lost. Teeth among vertebrates very in number, in their distribution within oral cavity, in degree of permanence, mode of attachment and in shape.

Distribution of teeth Teeth are vary numerous and widely distributed in the oral cavity and pharynx of fish. They develop in relation to the bones of the jaws, palate and even the pharyngeal skeleton. When teeth are lacking in some of these areas in fish, smaller tooth like stomodeal denticles may appear.

In early tetrapods, as in fish teeth were widely distributed on the palate, even today most amphibians (Fig. 12.10) and many reptiles have teeth on the vomer, palatine, pterygoid bones and occasionally on the parasphenoid. In crocodilians, toothed birds and mammals teeth are confined to the jaws and are least numerous among mammals. Teeth of vertebrates have therefore tended toward reduced numbers and restricted distribution. Only in mammals is there a precise number of teeth in given species.

Degree of permanence of teeth Most vertebrates through reptiles have a succession of teeth that develop from new dermal papillae, and the number of replacements is indefinite but numerous—**polyphyodont dentition**. Mammals for the most part develop two sets of teeth—**diphyodont dentition**—deciduous (milk) teeth and permanent teeth. Milk teeth usually erupt after birth, but in guinea pigs and a number of other mammals milk teeth develop and are shed before birth. A few mammals platypus, sirentians, and toothless whales develop only a first set of teeth—**monophydont dentition** and these may not erupt, if they do, they are usually shed shortly afterwards. In platypus they are replaced by horny teeth.

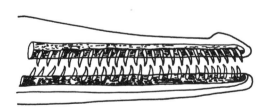

Figure 12.19
Homodont dentition of dolphin

Mode of attachment of teeth Teeth are intimately attached to the underlying skeleton by fibrous connective tissue. Jaw teeth may be attached to the outer surface or to the summit of the jaw bone, as in frogs—**acrodont dentition** or inner side of jaw bone, as in *Necturus* and lizards—**pleurodont dentition** (Fig. 12.18). Teeth may be rooted in individual bony socket or alveoli—**thecodont dentition**. Socketed teeth occur in crocodilians, in toothed birds and mammals, and in many fish.

Morphological variants of teeth Some of the teeth of a shark bear spines for tearing flesh; others exhibit flattened or rounded surfaces for sawing or crushing.

Fangs of some poisonous snakes are specialized teeth borne on the maxillary bones. For most part, in any single individual below mammals all teeth are similar in shape except for size—**homodont dentition** (Fig. 12.19). In mammals alone the teeth of each individual exhibit different shapes and different function—**heterodont dentition**. Extinct reptiles in the mammalian line such as synapsids were the first to exhibit heterodont dentition. Incisors - located anteriorly are usually specialized for cutting or cropping. In rodents and lagomorphs incisors continue to grow throughout most of life (Fig. 12.20). Enamel is deposited only on the anterior face of the incisors of teeth of these orders, and since enamel wears down more slowly than dentine, sharp chisel like edges result. Incisors may be totally absent in sloth. Lacking on the upper jaw in ox. Elephant tusks are modified incisors teeth—mastodon (Fig. 12.21).

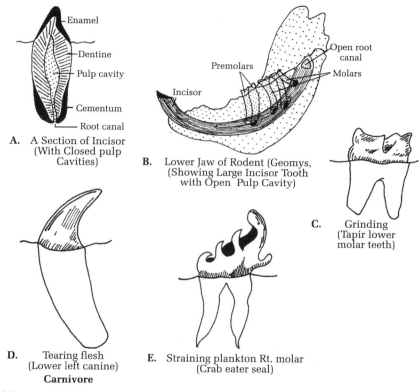

Figure 12.20
Varieties of mammalian teeth **A.** Lower molar from Tapir **B.** Lower jaw of rodent **C.** and **D.** Flesh eating canine jaguar **E.** Right molar Crab eater Seal

Canines teeth lies immediately behind the incisors. In shrew incisors and canines scarcely differ in appearance. In carnivores canines (Fig. 12.20) are spear like and are used for piercing the prey and tearing flesh. They are tusks of *Walrus* (Fig. 12.21). Canines are absent in rodents and lagomorphs and thus leave a toothless space, the diastema (Fig. 12.21) between last incisors and first cheek tooth.

Cheek teeth (premolars and molars) are used for macerating food. Molars are cheek teeth that are not replaced by a second set. In herbivorous mammals like ox, cheek teeth tend to be numerous, and there is little differentiation between premolars and molars (Fig. 12.21). In carnivores the number of cheek teeth is often reduced, and cusp are prominent. At least one pair on each jaw may have very sharp cusps for cracking bones and shearing tendons and is called **carnassial** teeth in dog (Fig. 12.21).

The cheek teeth of other mammals are more or less intermediate in shape between the extremes exhibited by herbivores and carnivores. Eruption of last cheek tooth (wisdom tooth) is delayed in higher primates, and this tooth is sometimes imperfectly formed or absent in man. The classification of both living and extinct mammal is based upon dentition. Primitive-placental mammals exhibit 3 incisors, 1 canine, 4 premolars and 3 molars. Dental formula (number of each type of teeth in a half of the upper jaw (is identical) and in a half of lower jaw of the same side is denoted in a numerical formula called dental formula. Number of teeth in dental formula multiplied by 2 gives total number of teeth. (Table 12.1)

Table 12.1: The dental formula in different animals is

	i	c	pm	m	
Rabbit	$\frac{2}{1}$	$\frac{0}{0}$	$\frac{3}{2}$	$\frac{3}{3}$	= 28
Rat	$\frac{1}{1}$	$\frac{0}{0}$	$\frac{2}{1}$	$\frac{3}{3}$	= 22
Horse	$\frac{3}{3}$	$\frac{1}{1}$	$\frac{4}{4}$	$\frac{3}{3}$	= 44
Squirrel	$\frac{1}{1}$	$\frac{0}{0}$	$\frac{2}{1}$	$\frac{3}{3}$	= 22
Mole	$\frac{3}{3}$	$\frac{1}{1}$	$\frac{4}{4}$	$\frac{3}{3}$	= 44
Dog	$\frac{3}{3}$	$\frac{1}{1}$	$\frac{4}{4}$	$\frac{2}{3}$	= 42
Cat	$\frac{3}{3}$	$\frac{1}{1}$	$\frac{3}{2}$	$\frac{1}{1}$	= 30
Cow	$\frac{0}{3}$	$\frac{0}{1}$	$\frac{3}{3}$	$\frac{3}{3}$	= 32
Man	$\frac{2}{2}$	$\frac{1}{1}$	$\frac{2}{2}$	$\frac{3}{3}$	= 32

i = incisors, c = canines, pm = premolars and m = molars

Cusp pattern (tubercle) distinguishes four kinds of molar teeth. (i) **Bunodont** When the cusps remain distinct and separated from one another as in man and monkeys—tooth is bundont (ii) **Lophodont** When the cusps are fused by means of intermediate masses of dentine to form ridges or lophs is called lophodont (Fig. 12.22 B)—horse, elephant, (iii) **Selenodont** When the cusps are cresentric—tooth is selenodont, sheep, cow and goat. (iv) **Secodont** When cusps have sharp cutting surfaces—carnivores dog. Combination of **bunolophodont** and **buno-selenodont** are not uncommon.

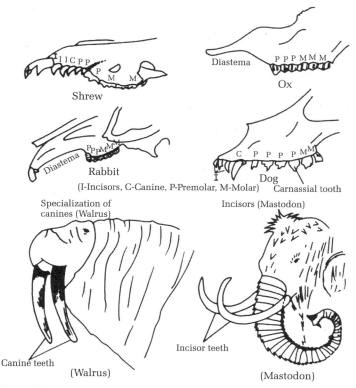

Figure 12.21
Mammalian upper teeth showing generalized pattern and specialization in ox herbivove, rabbit-gnawing animal, Canine-dog, canine of walrus, incisors of mastodon
(I-incisor, C-Canine, P-Premolar, M-Motor)

Horny teeth Aganathans lack true teeth but the buccal funnel and tongue are provided with horny epidermal teeth for rasping flesh. Horny teeth arise from the stratum corneum. The lips of the suctorial mouth of some species of frog tadpoles develop horny teeth. In turtles, sphenodon, crocodile, birds and monotremes a horny egg tooth for cracking the shell of cleidoic egg develops on the upper jaw at the tip of the beak prior to hatching. In lizards and snakes embryos that develop within an egg shell, the **egg tooth** is transitory genuine tooth of dentine. The adult duck-billed platypus has only horny teeth. Horny beaks in some turtles and birds have serrations that simulate teeth.

Pharynx

The pharynx extends from the back of the oral cavity caudally to the posterior limit of the last gill slit in fish, or to the point where the trachea begins in tetrapod (Fig. 12.23). In other words, it is the portion of the endodermally lined digestive tract which is common to both the digestive and respiratory system. In crocodilians, birds and mammals it can be divided into three regions:

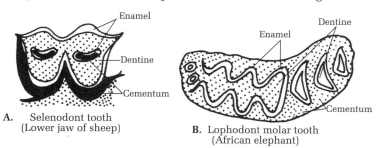

Figures 12.22 A, B
Grinding surface of selenodont and lophodont tooth

Nasal pharynx above the secondary palate

Oral pharynx below nasal pharynx

Laryngeal pharynx caudally opening into both oesophagus and larynx

In **fish** food and water passes through the pharynx. Most of the water is then passed out through the gills slits while the food is pushed into the oesophagus.

In **tetrapods** air may reach the pharynx from either the mouth or the nostril. In either case the air enters the pharynx

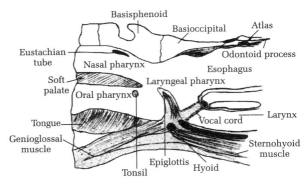

Figure 12.23
Section through larynx of a mammal

dorsally and must pass ventrally through the pharynx to reach the larynx and trachea. At the same point the food must pass dorsally to enter the oesophagus.

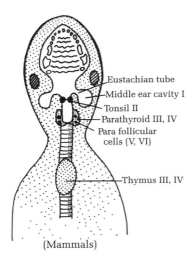

Figure 12.24
The derivatives of pharyngeal pouches in mammals. Roman numbers indicate pouch numbers

This area of the laryngeal pharynx where the air current crosses the food current, is called the **pharyngeal chiasma.** Only in whales there is a different and presumably better system; the most anterior part of the larynx, the epiglottis wraps around the pharynx and inserts into the internal nostrils eliminating pharyngeal chiasma. In all vertebrates, of course, six pairs of pharyngeal-pouches develop in the pharyngeal region (except for cyclostomes, which have many more). In fish the pharyngeal pouches (gill pouches) unite with ectodermal invagination to form gill slits which break through the surface. In tetrapods the endodermal lining of pharyngeal pouches forms several different structures (Fig. 12.24).

Ist pouch form **middle ear** cavity and with eustachian tube it retains continuity with pharynx.

2nd pouch form **palatine tonsils**—lymphatic tissue on each side of the pharynx that retain continuity with pharynx. 3,4,5 and 6th pouch lose their continuity with pharynx and the endoderm migrate caudally and ventrally. 3 and 4th contribute to thymus and parathyroid gland. 5th and 6th to ultimo-branchial and parathyroid. The thyroid gland is formed from mid ventral of the anterior pharynx. The only prominent glands in the pharynx itself are the lymphoid tissue of palatine tonsils and adenoids.

Oesophagus This is a distensible muscular tube connecting pharynx with stomach (Fig. 12.25). In a few vertebrates the oesophagus is lined with fingerlike papillae but more often the lining has longitudinal folds or is smooth. Peristalsis begins at the oesophagus. The endodermal lining of the oesophagus is frequently stratified epithelium which contains mucous glands whose secretion help lubricate the ingested food. The upper area of the

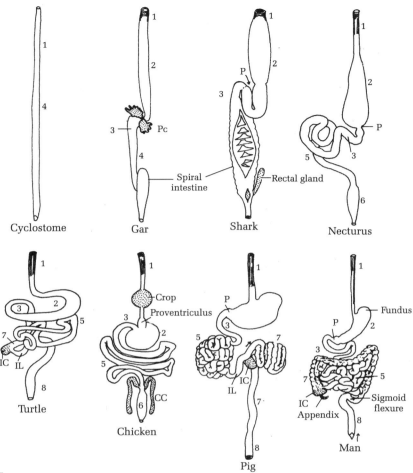

Figure 12.25
Alimentary canal of a few craniates after pharynx showing different modifications
1. Oesophagus
2. Stomach
3. Duodenum
4. Intestine
5. Small intestine
6. Large intestine
7. Colon
8. Rectum

CC Paired caeca of bird
IC Ileocolic caecum
IL Ileum
P   Pyloric sphincter
PC Pyloric caeca

oesophagus is wrapped by stratified musculature and its lower by smooth musculature except in ruminants whose entire oesophagus is walled by striated musculature. **Ruminants**, therefore, have more control over their oesophagus than do most vertebrates and can regurgitate their food—the cud which is rechewed.

The oesophagus has a good deal of elastic tissue, enabling it to stretch enough to accommodate large pieces of food. Some tetrapods such as snakes and fish can swallow pieces of food larger than their own diameter. In most birds the posterior part of the oesophagus forms a

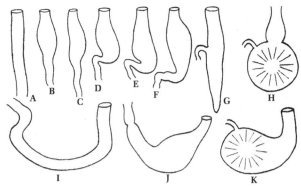

Figure 12.26
Shapes of vertebrate stomach
A. Belone **B.** proteus **C.** Tropidonotus **D.** Gobius
E. Shark **F.** phoca **G.** polypterus **H.** Fulica **I.** Testudo
J. Land tortoise **K.** owl

sac, called the **crop** which is particularly large in seed eaters (Fig. 12.26). The crop acts primarily as a storage depot; food is eaten rapidly and then passed on gradually for digestion. In pigeons (Fig. 12.29) and few other birds the crop sloughs of its squamous epithelium when there are nestlings. This cast off material forms a highly nutritious **pigeon milk** or crop milk which is regurgitated and fed to the young bird. Both males and females produce crop milk, apparently at the mere sight of nestlings.

Stomach

The portion of the stomach which is solely meant for storage is close to the oesophagus, and is called the cardiac stomach (Fig. 12.27). Here the diameter of the alimentary canal is increased-a great deal in vertebrates that eat discontinuously and much less or not all in those fish whose easily digested diet is continuously consumed. Storage also occurs in the remainder at part of the stomach. In most stomach: bands of muscular tissue in the external layer cause folds, called **rugae**, these folds provide the potential space for additional distension. When the stomach is filled, internal pressure stretches this folds and obliterates the rugae.

A large stomach contributes significantly to a more mobile lifestyle. For many terrestrial animals it is advantageous to eat rapidly and then hide from potential predators while digestion occurs. For predators it allow the animal to stock up while the kill is still fresh.

Digestion

The breaking up of food by mechanical digestion provides a greater surface area on which digestive enzymes can work. The wall of stomach is more muscular than those of other parts of the alimentary canal. There are often encircling oblique bands of muscle, which by powerful rhythmical contractions churn and break down the

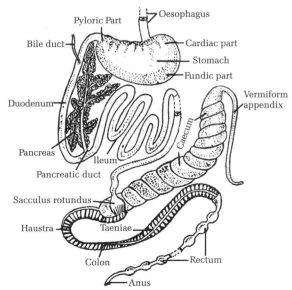

Figure 12.27
Rabbit, alimentary canal with associated glands

food. This mechanical action plus the effect of the gastric juice reduces the food to a semiliquid state.

Now chemical digestion follows immediately. Protein digestion begins in the stomach with secretions from gastric glands in the central (fundic) region (Figs. 12.27, 12.28). In mammals gastric glands have two secretory cell types (i) Chief cells secrete proteolytic enzymes collectively called pepsin. (ii) Parietal cells secrete hydrochloric acid.

In nonmammals gastric glands have only one secretory cell type which evidently secrete both hydrochloric acid and pepsin.

Mucus glands are present throughout the alimentary canal, and are particularly prominent and usually multicellular in pyloric region of the stomach. Their secretion lubricate the food and protect the delicate epithelium of stomach from mechanical damage or chemical digestion by its own enzymes.

Stomach has acquired specializations to meet diversity of diet. In the mucosa are many depressions, called gastric pits, lined with columnar epithelium (Fig. 12.28). In both cardiac and fundic region of stomach, tubular fundic glands lie within the lamina propria and contain secretory cells, called **body** cells, which evidently secrete both pepsin and hydrochloric acid. In the pyloric region instead of fundic glands there are mucus secreting pyloric glands also opening into the base of the gastric pits. The stomach is present in **protochordates** and is simple, and the stomach is poorly delineated in cyclostomes. The boundary between the oesophagus and stomach is indefinite or lacking in vertebrates below birds. The stomach is straight when it first develop in-embryo and may remain so throughout life in lower vertebrates. More often, flexure* develop producing a J shaped or U-shaped stomach (Fig. 12.26). As a result, the stomach may exhibit a concave border (lesser curvature) and a convex border (greater curvature). The stomach also undergoes torsion in higher forms so that it may finally lie across the long axis of the trunk. As flexion and torsion gets pronounced during development of mammalian stomachs; the dorsal mesentery of the stomach (mesogaster) becomes twisted and finally get suspended from the greater curvature. The part of the dorsal mesentery attached to the greater curvature is then called the **greater omentum**. The stomach of some vertebrates, especially of fish, exhibit one or more blind pouches, or caeca. The stomach of some marine teleosts is lined with a horny membrane that may exhibit some spiny projections.

Amphibians and reptiles: The stomach of amphibians and most reptiles, when examine grossly are relatively simple dilations of the alimentary canal. On detailed microscopic examination they can be divided into regions by types of gland present. In crocodilians stomach the smooth musculature is thickened and extremely powerful, it functions analogously to the avian **gizzard**, violently and efficiently macerating food.

Birds: The avian stomach is differentiated into two sections:

i) The glandular portion—**proventriculus** and

ii) Mastigatory portion—**gizzard**

Food is stored in the crop located at the distal end of the oesophagus and is periodically released in the stomach at the junction between gizzard and proventriculus

---

* In mammals, e.g. mouse a constriction in the middle part of the stomach mark off a cardic chamber from a pyloric chamber. In Man the stomach bear resemblance to this condition found in mice and is called "hour glass" stomach.

(Fig. 12.25/12.29 A, B). The gizzard, a unique modification for flight, is found only in birds, although the thickened stomach walls of crocodiles which function similarly are interesting in this regard since crocodiles descended with birds from a common ancestor. The avian gizzard has extremely thick powerful muscular walls and a glandular mucosa that produces a very hard, keratin like protein called koilin. The koilin is analogous to teeth, when the gizzard muscles contract the food is crushed. The gizzard is another means of accomplishing mechanical digestion (essential for an animal with a high metabolic rate). If instead of a gizzard birds had a mammalian type masticatory apparatus with teeth muscles and skeletal supports, the weight of the head would create disastrous aerodynamic problems. The extent of muscular development in the gizzard depends on both heredity and diet. For instance, if herring gulls are fed their usual soft fish diet their gizzard remain thin walled and relatively inconspicuous. If they are forced to eat grain, however, the gizzard walls soon proliferate and strengthen.

Mammals: Like birds, mammals have a high metabolic rate and require an efficient digestive system. There is no need for a masticatory apparatus in the stomach since mammalian heterodont teeth provide grinding surface in the oral cavity. The stomach however, must act as a large storage reservoir, and in most mammals, the cardiac portion is distensible. In small insectivorous mammals, such as the *grasshopper mouse* the cardiac portion is lined with keratinized, stratified squamous epithelium, which protects the stomach from damage by the chitinous exoskeleton of insects. The gastric glands, instead of opening through numerous gastric pits are all funnelled into one short duct. Since this duct is the only place where they contact the lumen, the opportunity for damage to the glands is greatly minimized.

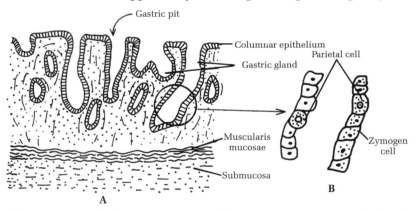

Figure 12.28
A. Histology of alimentary canal showing secretory cell types   B. Portion magnified to show cells

Ruminant stomach   The most bizarre vertebrate stomach, however, is surely that of the ruminant *Artiodactyl*, a four chambered structure with a capacity of up to 60 gallons (Figs. 12.30 A, B, C). The first and largest part of this stomach into which the food enters is the **rumen**, its simple epithelial wall is filled with symbiotic bacteria which can digest plant cellulose. The next part is **reticulum**, the mucosal lining of this chamber usually also contains symbiotic bacteria. This lining is folded in such a way which produces geometric configuration similar to honey comb.

Food is frequently regurgitated from both rumen and reticulum (Figs. 12.30 B, C) to the oral cavity, there it is resalivated, rechewed and reswallowed to pass once again through the rumen and reticulum, for further cellulose digestion by the symbiotic bacteria. During this

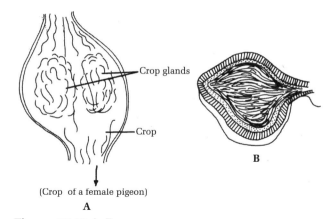

Figures 12.29 A, B
A. Crop gland of pigeon **B.** Gizzard (dissected) of a grain eating bird

process much of the water and sugar present in food are reabsorbed through the walls of the rumen and reticulum into the circulatory system. From the reticulum the food is passed to the **omasum**, a heavy muscular structure with longitudinal muscle folds. The walls of **omasum** subject the food to mechanical churning, the purpose of which is unclear. Surgical experimentation, has shown that cows suffer no ill effect if the omasum is removed. The fourth chamber is the glandular **abomasum**, and it is the site of initial digestion of protein.

At the junction of stomach and intestine the circular smooth muscles are greatly thickened, forming the pyloric sphincter. When the sphincter contracts it completely closes the alimentary canal; thus it controls the rate at which food passes from stomach to intestine. The pyloric sphincter itself is controlled by the autonomic nervous system and by endocrines.

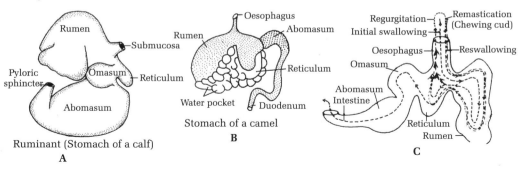

Figures 12.30 A, B, C
Diagram of Ruminant stomach **A.** Stomach of calf **B.** Stomach of camel **C.** Functional relationship. Dotted lines show initial swallowing and swallowing after chewing cud.

Intestine

The rest of the digestion as well as absorption of food into the blood stream and formation of faecal matter occur in the intestine. Food enters the intestine in a fluid or semi fluid state called **chyme**, after having undergone mechanical digestion and some chemical digestion. In the intestine are present the necessary enzymes for lipid digestion and for the final digestion of carbohydrates, and protein. Some of these enzymes are produced and secreted by the intestinal walls, others are produced in pancreas and carried to the intestines by pancreatic ducts. There are also mucus secreting cells in intestine, particularly *goblet* cells which concentrate the mucus into large globule before releasing it. As in the stomach, mucus helps protect the intestinal epithelium from **auto-digestion**.

The intestine's large surface area is richly supplied with capillaries so that absorption of nutrients and water can occur. The columnar absorptive cells of the intestine are similar in all vertebrates (Fig. 12.31) i.e., with a brush border on the free surface, The brush border of a single cells is seen to be comprising as many as 2,000 microvilli, or long, fingerlike projections from the apical end of each cell: **Microvilli** are a major and universal way of extending surface area. Absorption takes place through the cells, not between them. The digested material (primarily amino acids and simple fats and sugars) passes directly into the capillaries in the underlying lamina propria.

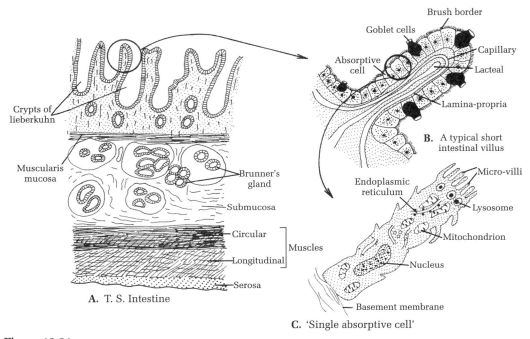

Figure 12.31
A. T.S. of intestine of a mammal   B. An intestinal villus showing goblet cells and absorptive cells and their relation to capillaries and lacteal in lamina propria.   C. A single absorptive cell with its microvilli

Fish: In contrast to mammals there is no real distinction between large and small intestine in fish. Frequently however, the proximal portion, where the pancreatic duct open, is called duodenum; the most posterior portions; where faecal matter is formed, is called the large intestine, and the intermediate portion is called the small intestine. The fish have evolved four **major mechanism** for further increasing the absorptive surface area. In cyclostomes a simple fold, called the **typhlosole**, runs the length of the intestinal wall. In elasmobranchs, holosteans, chondrostean and dipnoi the mucosa and submucosa forms a prominent spiral fold, called **spiral valve** throughout most of the intestine (Fig. 12.32 A). This fold forces the food to spiral on its course through the intestine, rather than passing directly down its length, thus prolonging its contact with the intestinal walls.

In most teleosts blind diverticuli called **pyloric caeca** (Fig. 12.32 B) (caecum), come off the intestinal lumen just past the pyloric sphincter. In some such as *mackerel* there may

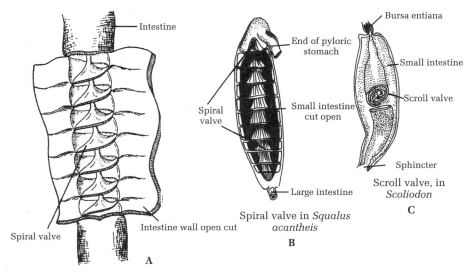

Figures 12.32 A, B, C
Small intestine of elasmobranch fish **A.** cut open to show spiral valve **B.** *Squalus* **C.** *Scoliodon*

be as many as 200 of these pyloric caeca. In other **teleosts** however, the intestine is greatly lengthened, and coiled or curved back upon itself.

Amphibian and reptiles: In amphibian and reptiles, intestinal length is also greatly increased, particularly in herbivorous forms. In herbivorous anuran tadpole for instance, the intestine is more than five times the length of the intestine in the insectivorous adult. In reptiles the epithelial lining is deeply folded as well, the folds involve not only the mucosa but the entire depth of the lamina propria and some of the submucosa. Large and small intestine are differentiated in amphibians and reptiles. Both have mucus secreting goblet cells and absorptive columnar cells with microvilli very much like those of fish, but in the large intestine there are many more goblet cells secreting mucus which helps to form faecal matter.

Birds: The intestinal epithelium of birds contains columnar and goblet cells similar to those of reptiles but also has two sharply distinguishing features. In the mucosa are extremely deep groove, the *crypts of Lieberkühn* which carry the epithelium throughout the very thick lamina propria and give the inner surface of the intestine a folded appearance. Large, branched tubular glands at the base of these crypts secrete enzymes which hydrolyse carbohydrates, proteins and fats.

All over the intestine there are also **villi**, barely macroscopic fingerlike processes, which extend out into the lumen and give a velvety appearance. A villus (Fig. 12.31 B) has a core of lamina propria completely covered by columnar epithelial and goblet cells. The lamina propria core contains a capillary network where absorbed amino acids and carbohydrates enter the blood. It also has a single blind lymph vessel called a **lacteal** which absorbs digested fat and passes them into larger lymphatic vessels from which they eventually drain into main circulatory system. At the junction of the large and small intestine there are usually paired caeca, which in herbivorous forms may contain symbiotic bacteria and masses of lymphoid tissue. In the large intestine, which is quite short in birds the number of goblet cells increases and the number of villi deceases. In **mammals** the large and small intestines are distinct. The small intestine itself can be divided into three portions: duodenum, jejunum and ileum (Fig. 12.33 A). Throughout all three portions the surface area in increased much as it is in birds by both villi and crypts of Liberkühn and also by large folds of the submucosa, called the *valves of kerkring*, which extend into the lumen. The three sections are distinguished morphologically. The duodenum has *glands of brunner* (Fig. 12.33 A), alveolar glands secreting both mucus and a proteolytic enzyme (not pepsin), which pierce through the muscularis mucosa and lie in the submucosa.

The **jejunum** and **ileum**, which lack glands of brunner, are quite similar to one another; the ileum has more goblet cells and smaller villi than the jejunum (Figs. 12.33 A, B, C). At the junction between the small and large intestine there is usually a prominent colic caecum a blind diverticulum structurally similar to the small intestine but with more lymphatic tissue and larger concentrations of goblet cells. Occasionally, there is also a small blind pouch, the **vermiform appendix** (Fig. 12.27) extending from the caecum, the vermiform appendix,

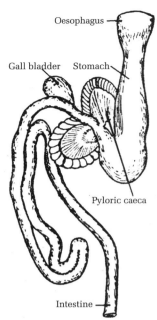

Figure 12.32 B
Pyloric caeca of a teleostean fish

lacks villi but contains many tonsil like lymphatic nodules. The large intestine is much shorter, but has a large diameter. It has more lymphoidal tissue than the small intestine, and usually no villi. The long tubular glands of Lieberkühn secrete copious amounts of mucus. The large intestine resorbs a great deal of water still in the undigested material and forms the remainder into faeces.

Anus or cloaca

The alimentary canal ends as it began, as a derivative of an ectodermal invagination. The posterior chamber, developed from the embryonic proctodaeum is the cloaca (in most vertebrates) or anus (in mammals and most teleosts). The cloaca receives not only undigested food (faeces), but also urine and reproductive discharges (Fig. 12.34) In birds it is a large tripartite chamber divided into three compartments an anterior **coprodaecum** which receives the rectum; a short middle **urodaeum** into which open the ureters and genital ducts and a large **proctodaeum** which open to outside by cloacal aperture. A small glandular blind pouch of lymphatic tissue the **bursa fabricii** lies in the dorsal wall of proctodaeum in young birds but it disappear on sexual maturity. It is also described as cloacal thymus. The anus, on the other hand, carries only faeces and open separately from the urinary and reproductive tracts.

Glands associated with digestive system

Pancreas

The alimentary canal by itself is not enough to complete the digestion. The digestion will not occur without digestive enzymes. Although several enzymes are produced

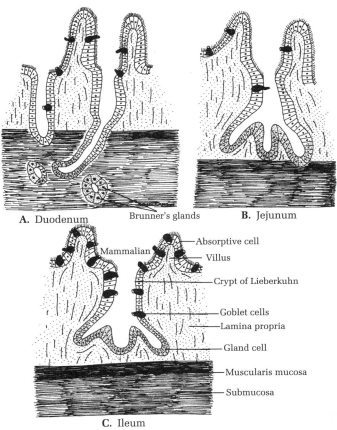

Figure 12.33
Three portions of the mammalian small intestine **A, B, C** showing similarities and difference in the villi, glands and goblet cells

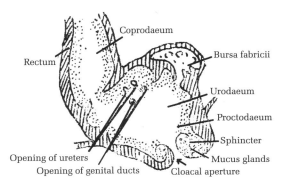

Figure 12.34
Cloaca in longitudinal section (pigeon)

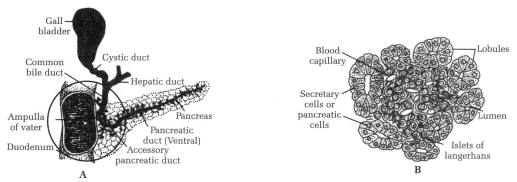

Figures 12.35 A, B
A. Gall bladder, bile ducts and pancreatic ducts in relation to duodenum (man)
B. T.S. pancreas of rabbit

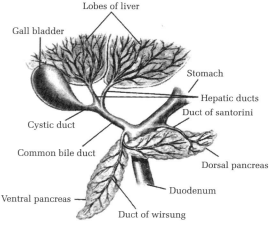

Figure 12.35 C
Development of the liver, gall bladder, and pancreas

in the alimentary canal, most enter the intestine from pancreas (Figs. 12.35 A, B) The pancreas is formed by two or frequently three outgrowths of endodermal epithelium just posterior to the presumptive stomach, which retain their embryological connections as the pancreatic ducts. In most vertebrates the pancreas is both exocrine and endocrine gland (Figs. 12.35 A, B).

The exocrine portion is a complex, tubular alveolar gland whose secretions contains enzymes for the digestion of proteins, carbohydrates and fats. In fact the pancreatic enzymes alone are capable of breaking down all these major food components into units that are small enough to be absorbed by the intestinal epithelium. In the cyclostomes and several bony fishes there is no compact pancreas, instead, microscopic bits of pancreatic tissue present line the walls of the gut and adjoining organs and are connected by ducts which empty into the intestine. In other vertebrates the pancreas is a relatively compact, two part organ sometime there is secondary fusion of the dorsal and ventral pancreas to form a single adult structure. From the dorsal pancreas there is the *duct of santorini*, from the ventral pancreas there is the *duct of wirsung*, usually a paired branched duct draining into the intestine next to or in conjunction with the bile duct (Figs. 12.35 A, B).

Liver

During development paired endodermal cords grow out of the alimentary canal, close to point of origin of the pancreas. The terminal end of these endodermal cords proliferate and form the bulk of liver and gall bladder. Although most of the embryonic ventral mesentery disappears during subsequent development but the mesentery ventral to the duodenum and stomach that was invaded by the liver bud remains as the hepato-duodenal ligament connecting the duodenum with the liver, and as the gastro hepatic ligament connecting the pyloric end of the stomach with the liver. These two ligaments are continuous with one another and constitute the **lesser omentum**. The omentum serves as a bridge transmitting the common bile duct, the hepatic artery and the hepatic portal vein (Fig. 12.36). The shape of the adult liver conforms to the space available in the coelom. In forms with an elongated trunk the liver is elongated. In animals with short trunks it is broad and flattened. It is variously subdivided in different species but usually exhibits median and lateral lobes.

The liver is digestive organ and manufacture bile, which aid in fat digestion and absorption. Other roles, associated are **homeostasis**—maintenance of a suitable internal milieu, are more vital. It helps regulate blood sugar level by *glycogensis*, *glycogenolysis* and *glyconeogenesis*, it deaminates amino acids taken in as food and hence is a source of **urea** and it manufacture several blood proteins including certain clotting substances. The fetal liver is an important source of **red blood cells**, and the adult liver excretes the breakdown

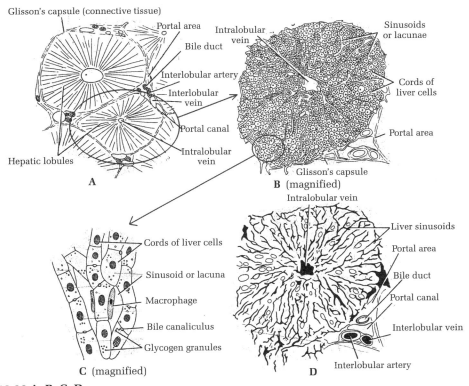

Figures 12.36 A, B, C, D
Liver of rabbit. **A.** T.S. of a group of lobules **B.** Structure of a lobule and a Glissson's capsule **C.** Liver cords magnified **D.** T.S. of a lobule showing sinusoids with veins injected

products of hemoglobin as bile pigment. **Bile** is partially an excretion; eliminating wastes such as detoxified bacteria and excess iron from red blood cells. Bile is also partially a secretion which emulsifies fats. Thus this large and important gland embryologically derived from the alimentary canal, assists in digestion and also controls a great deal of the body metabolism by its selective filtration and secretion and its unique blood supply.

Gall bladder

One spout of the liver bud expands to become the gall bladder and cystic duct (Fig. 12.35, 12.36). A gall bladder develops in most vertebrates including *hagfish*. However, no gall bladder develops in lamprey, many birds, rats, perissodactyl or whales. Since the gall bladder serves primarily to store bile for emulsifying fat. Because of its embryonic origin from liver bud, the gall bladder is embedded, on, or suspended from one of the lobes of the liver and empties into common bile duct. The terminal segment of the common bile duct is embedded in the wall of the duodenum for a short distance and is called the **Ampulla of vater**.

Chapter 13

Respiratory System

In all animals that exhibit the aerobic respiration oxygen must reach each cell of their body. But in vertebrates, the individual cells of the living tissues have no immediate access to the supplies of oxygen from outside. Hence in all coelomates, the transportation of oxygen from the source of supply (water/air) to the deeper tissues is done by the blood. In order to fulfil the respiratory needs efficiently the animals developed remarkable modifications to draw maximum benefit and obtain optimum amount of oxygen from environment (air/water) they were living.

Respiratory requirement for efficient exchange of gases is centered on or around following points.

a. Uninterrupted supply of oxygen— In oxidative metabolism oxygen should be supplied to body cells and tissues and carbon dioxide, (CO_2) should be removed from them. Unlike many small animals in vertebrates with increased metabolic rates, the gas exchange by diffusion across the general body surface area is inadequate. Instead, it is done by means of the respiratory and circulatory system.

b. Oxygen availability in respiratory media—Air contains 200 cm^3 of oxygen per litre, whereas fresh water contains not more than 9 cm^3, and extremely stagnant water contains as little as 0.02 cm^3 of dissolved oxygen per litre. Thus there is relatively little oxygen available for aquatic animals, and air breathing vertebrates would apparently have a great advantage.

c. Efficient exchange—When the gases are exchanged between the external milieu (air or water) and circulatory system, oxygen gets loaded to blood corpuscles and dissolve in blood plasma it is termed as external respiration. When the gas exchange occurs between blood and tissues. It is termed as internal respiration. The oxygen reaches tissue and CO_2 moves out of tissue.

d. Adequate respiratory surface—In order to maintain a smooth exchange of gases between the environment and the blood, the essential requirement is of a thin membrane. The membrane must be moist for the gases O_2 and CO_2 to dissolve and absorb before they can diffuse. The respiratory membrane surface area must be adequate enough so that the required large amount of gas, can be exchanged from the environment by the animal. The considerations of surface area and volume relationships are similar to the digestive system; i.e. the respiratory surface area also increase in the same ways—by causing internal folding. The membrane must also be richly vascularized so that there is enough blood to accomplish the exchange.

e. Increase capacity of blood to carry oxygen—The capacity of blood to carry oxygen, is increased because of the carrier molecule—**haemoglobin,** which loosely binds much of the oxygen in the blood. An exception to this is the *Antarctic ice fish*, which however, have no haemoglobin. They are thin fish with a large surface area compared to their mass and they live in cold water, which has a high oxygen content, enough gases can diffuse through their surface to meet their needs.

f. Effective ventilation mechanism—For the process of diffusion to occur, a higher concentration of oxygen and a lower concentration of carbon dioxide must be maintained outside the membrane than in the blood of the respiratory membrane.

Therefore the second requirement for a respiratory system is a **ventilation mechanism** to move the respiratory medium (air/water) past the respiratory membrane. In many of the vertebrates this is accomplished by a muscular pump, operating by either suction or pressure.

g. Quick elimination of carbon dioxide—Carbon dioxide elimination is equally important as oxygen intake in respiration. Carbon dioxide is transported dissolved in the plasma, usually as a bicarbonate ion, to the respiratory membrane and then eliminated. Both water and air normally have a low carbon dioxide tension but because carbon dioxide is much more soluble in water than in air its elimination occurs more rapidly in aquatic animals than in aerial tetrapods. Respiratory diffusion is limited in most aquatic animals by the rate at which the respiratory membrane can absorb the limited available oxygen. In tetrapods it is limited by the rate at which the membrane can expel carbon dioxide. In addition, the respiratory membrane in tetrapods must be kept moist. In early vertebrate embryos respiration occurs directly between the individual blastomeres and their surroundings. Later, highly vascularized membranes may serve, in some species, for respiration during embryonic life.

Respiratory devices of both the larval fish and larval amphibians (being aquatic) are gills where as terrestrial vertebrates have lungs as respiratory organs.

a) **external gills** which usually project from the surface of the pharynx.

b) **internal gills** are located in chambers in pharyngeal wall. The adult animals rely chiefly on **internal gills.**

c) **amphibians** have the most complicated respiratory organs primarily because the amphibian includes all those animals which live partly in water and partly on land. Their skin is usually soft, glandular and moist lacking in external scales. Hence amphibians respire through **skin, external gills, internal gills, lining of buccopharyngeal cavity** and the **lungs**.

d) The tetrapods rely chiefly on **lungs** for respiration. Though accessory respiratory organs may be present in some species.

Respiration by Gills

The primary and usually the only respiratory organ of most fish and larval amphibians are gills. The internal gills, characteristic of adult fish, differentiate from the endodermal lining of the pharyngeal pouches. External gills characteristic of embryonic and larval fish and larval amphibians differentiate from the ectodermal portion of the gill slit.

Chondrichthyeans In *Squalus* a series of five-external gill slits are visible on the surface of the pharynx. The slits in *Squalus* are not covered by an operculum and so they are naked gill slits (Fig. 13.1). In subclass **elasmobranchii**, because of the absence of operculum, each gill slit is covered externally by individually flapped valves. The space between the gill lamellae and the flap valve of each gill slit is called the parabranchial chamber (Fig. 13.2), it functions analogously to

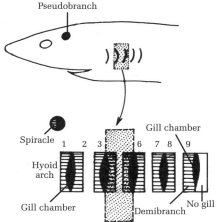

Figure 13.1
Gill chambers (black) and nine demibranches in *Squalus*

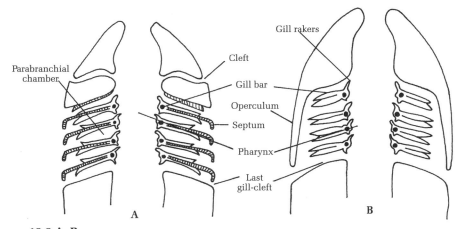

Figures 13.2 A, B
Gills and branchial chamber **A.** Elasmobranch fish (*Scoliodon* ventral view) **B.** In a teleost fish

the teleosts opercular chamber, elasmobranchs also have a **spiracle**. The modified anterior gill slit of the hyoid arch is completely closed in teleosts. It is not a respiratory structure but is used to bring water into the oropharyngeal cavity. Only oxygenated blood is brought to the spiracle, and its lamellae, which do not function in gas exchange, and hence form a **pseudobranch**. The anterior and posterior walls of the first four gill chambers exhibit a gill surface or **demibranch.** The last or fifth gill chamber lacks a demibranch in the posterior wall. The demibranch in the posterior wall of a gill chamber is post-trematic. Separating the two demibranch of a gill is an interbranchial septum. The septum is strengthened by delicate cartilaginous gill rays radiating from the gill cartilage. Gill rakers protrude from the cartilage,

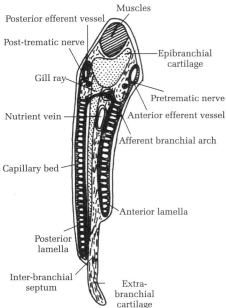

Figure 13.3
A holobranch (cross section)

and not only keep food from entering the gill slit but also help in directing food towards oesophagus. Associated blood vessels and branchiomeric muscles, nerves and connective tissues, constitute **holobranch** (Fig. 13.3). *Squalus* has four holobranchs.

The anterior demibranch of the one gill arch interdigitates with the posterior demibranch of the arch in front of it (Fig. 13.4). Water must pass through narrow space between filaments; this space is made even narrower by secondary lamellae or folds, on the dorsal and ventral surfaces of each filament. It is these secondary lamellae that constitute the vascularized respiratory membrane. Each secondary lamella is covered by their epithelial cells on a fine basement membrane (Fig. 13.4). Deep to the basement membrane are present unusual cells called **pilaster cells,** which form both the surface epithelium of the gills and the walls of the respiratory membranes forming blood channels. The blood channels so formed are large enough for the passage of one blood cell at a time and thus provide maximum surface area for gas diffusion. Further, diffusion occurs between the blood cells and the pilaster cells on both sides of the secondary lamellae. The gill filaments also contain

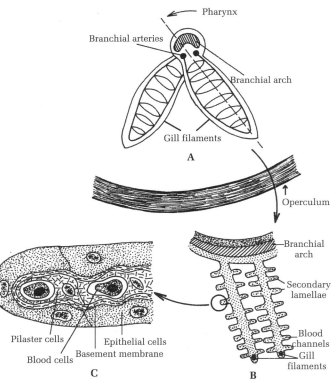

Figure 13.4
The teleost gills **A.** A horizontal section through pharynx showing two holobranches **B.** Section of two gill filaments **C.** Magnified secondary lamella

occasional acidophilic cells, called *chloride* cells. These cells have no role in respiration rather have important role in the excretion of salts. The efficiency of the gill

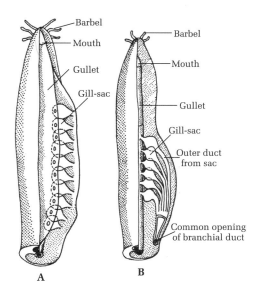

Figures 13.5 A, B
A. Bdellostoma B. Myxine
(Respiratory region dissected to show the arrangement of gill slits)

apparatus is increased by a *counter current* distribution system. The water entering the gills contains maximum amount of oxygen and least amount of carbon dioxide, blood leaving the gills also contains more oxygen and less carbon dioxide while the water leaving and blood entering the gill contain most CO_2 and least oxygen. This maintains a constant diffusion gradient between blood and water in the animal body so that as much as 80 per cent of the dissolved oxygen is diffused into the blood. If the flow of water is experimentally reversed, the efficiency is reduced from 80 per cent to just 10 per cent.

Mechanism of ventilation

A double pump involving both suction and pressure mechanisms is used to move water, past the gills, while the oesophagus is kept closed by sphincter muscles.

The **suction pump** functions when the oropharyngeal cavity is expanded by opening the mouth and by contracting the sternohyoideus muscle which lowers the floor of the cavity. This drains water into the oral cavity through mouth and spiracle with the gill flaps closed. The coracohyoid and coracobranchial muscles are contracted, further expanding the oropharyngeal cavity. This forces water past the gills and also expands the parabranchial chambers, which are highly elastic due to the cartilaginous nature of the chondrichthyeans branchial basket.

The **pressure pump** functions with the mouth and spiracle closed and gill flaps opened. Closing the mouth reduces the volume of the oropharyngeal cavity, which forces water past the gill lamellae. The volume of the oropharyngeal cavity is further reduced by the geniohyoideus muscle, which raises its floor; this forces still more water past the lamellae.

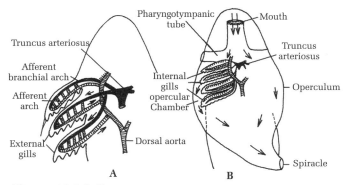

Figures 13.6 A, B
The tadpole of frog A. external gill stage B. internal gill stage

Throughout the cycle water is passed almost continuously from pharynx to the parabranchial chamber, always passing over the respiratory membrane. Of course, the double pump requires some oxygen to run it, but it is so efficient that much more oxygen is gained than is utilised. The structure is not as tightly knit in **chondrichthyeans** as it is in **teleostei**, and there is only about 50 per cent oxygen utilization. Most elasmobranchs are of the **pentanchid** type, i.e, they have five-gill slits. *Hexanchus,* another shark, has six gill slits and a spiracle. The five-gill slits in adult rays are on the underside of the flattened body but the spiracle is located dorsally behind the eyes, and is used for intake of water. The branchial region of *Chimera* is further modified. The spiracle is closed, there are only four adult gill pouches, the interbranchial septa are short and do not reach the skin, and a fleshy operculum extends backward from the hyoid arch and hides the gills. In several of these respects *Chimera* resembles bony fish.

Teleosts: The gill apparatus of bony fish is basically similar to that in *Squalus*. Five gill slits is the rule, but there are few exceptions.

    a) **Presence of bony operculum** An operculum, a bony flap arising from the hyoid arch, projects backward over the gill chamber (Fig. 13.2 B). The result is an opercular cavity that opens by a **crescentic cleft** (a small aperture in eels) just anterior to the pectoral girdle. In few fish the left and right opercular chambers open by a common aperture in the mid-ventral line. Opercular movements assists in the expulsion of water from the gill chambers. The operculum of many bony fish (*Sturgeons, Spoon bill, Polypterus, Lepidosteus* and many teleosts) bear on its inner surface either a functional gill or a pseudobranch.

    b) **Reduction of interbranchial septa** The interbranchial septa in bony fish do not reach the skin and in teleosts the interbranchial septa are shorter than the demibranchs.

    c) **Elimination of spiracle** A spiracle occurs in the relatively ancient chondrosteans, but it closes during embryonic life in modern fish and in lungfish.

    d) **Reduction in number of demibranchs** Most **teleosts** lose the demibranch on the hyoid arch. It is retained in the more ancient **chondrostei** and **holostei**. Additional demibranchs are lost in some lungfish. Since lungfish breathe partly by lungs, the loss was no disadvantage.

Cyclostomes: Living cyclostomes (Agnathans) have six to fifteen pairs of gill or pouches (Fig. 13.5). *Myxine* usually has six pairs of gill pouches but occasionally there are five or seven. The various species of *Bdellostoma* have from five to fifteen pairs and within the species. *B. stouti* the number varies between ten and fifteen, although twelve pairs are common. Among Lampreys, *Petroymzon* has eight embryonic and seven adult pairs.

The gill pouches in agnathans are connected to the pharynx by afferent branchial ducts and to the exterior by either separate or common efferent branchial ducts. *Lampreys* and *Bdellostoma* have separate ducts that open by separate apertures. In *Myxine* and its relatives the efferent duct unite to open via a common external apertures on each side. The pathway of flow of the respiratory stream differ in lampreys and hagfish. In the former the water can be taken in through the external gill slits and ejected by the same route. This is essential when the lamprey is attached by its buccal funnel to a host fish. In the hagfish the water enters via the nares and passes along the nasopharyngeal duct into the pharynx. The flow of respiratory water in hagfish is maintained by the pumping action of a **velar chamber** at the anterior end of the pharynx, into which the nasopharyngeal duct empties.

Respiratory role of larval-gills In Lampreys, and a few larval fish i.e., *Polypterus,* lungfish, some teleosts and in all larval amphibians external gills occur (Fig. 13.6). The gills arise from the exposed surfaces of the branchial arches and are typically larval and hence usually temporary organ.

In tadpoles, before the gill slits perforate, external gills develop as fingerlike elevations on the outer surfaces of visceral arches III to V. A rudimentary gill may develop on VI arch. Later, mouth perforates and pharyngeal pouches II to V rupture to exterior to form four gill slits. The first pouch typically become the middle ear cavity when the gill slits have developed, their walls become folded to form a second set of gills, and from the hyoid arch a fleshy operculum grows backward over the entire gill apparatus and encloses it in an opercular cavity. A second set of gills is referred to as internal gills because they lie within the opercular chamber. Internal gills serve throughout the most of the tadpole stage. Later on anterior limb will sprout, through the operculum. In larval **urodeles** and **caecilians,** a vestigial operculum makes an appearance, but it becomes only a small fold on hyoid arch.

In **elasmobranchs** with very long periods of development within the uterus or egg shell, the larval gills appear and project into uterine fluid or under the egg shell. They serve not only for respiration but also for the absorption of nutrients material.

Evolutionary significance of spiracles, gill pouches and gills There has been a trend towards reduction in the number of gill slits in modern vertebrates, as is evident by the following observations.

a) The spiracle, when, present, arises during embryonic life in series with the gill slits and some times bear a pseudobranch. This suggests that spiracle is a vestigial gill slit.

b) The spiracle remains open in phylogenetically ancient fish but in modern forms the first pharyngeal pouch either fails to perforate or closes during later development.

c) The last pharyngeal pouch fails to achieve an opening to exterior in some relatively recent fish.

d) Demibranch fail to develop in some gill chambers of certain fish.

e) Larval amphibian develop a larger number of pharyngeal pouches then rupture to form gill slits.

f) The number of open slits in amphibian is reduced as compared with fish.

g) Amniote embryos develop a series of four to six pharyngeal pouches that may rupture to form temporary slits in lower amniotes but tend not to rupture in higher amniotes.

From the foregoing facts we may deduce that modern fish and amphibians are descendants of vertebrates that had a large number of functional gills and that amniotes may be descendants of gill breathing vertebrates.

How much reduction in the number of gill slits may have occurred! If the number of gill chambers in generalized fish such as elasmobranchs is a reasonable criterion. It may be concluded that the number of gill pouches in jawed vertebrates may have been not much more than seven. In **ostracoderm**, the most ancient known fish, the largest number yet discovered is ten.

Did pharyngeal pouches and slits originally serve for respiration or did they have a still more primitive function in the earliest chordates! The suggestion has been advanced that their original use may have been the ensnaring of minute food stuffs from an incurrent water stream in the anterior end of the digestive tract (filter feeding). In the lowest living chordates (ammocoetes, amphioxus and urochordates) food particles collected by the ciliated branchial apparatus represent the entire diet. Such a hypothesis is attractive because it furnishes an explanation for the presence of the respiratory system in the anterior wall of the digestive tract. The hypothesis, is of a necessity, based on speculation.

Swim Bladder

In most **teleosts** a gas filled sac called the swim bladder is present in retroperitoneal position just dorsal to the serous membrane lining the body cavity, The gas bladder develops as a diverticulum of the embryonic gut, and in some teleosts it remains connected to the oesophagus by the pneumatic duct **physostomous** while in others it has no adult connection **physoclistous** (Figure 13.7).

The internal structure of swim bladder is similar to that of the post-pharyngeal alimentary canal with layers of epithelium such as lamina propria, muscularis mucosa, submucosa and muscles (Fig. 13.8). Its outer muscle layer is made up of tough, collagenous connective tissue, embedded with many *guanine crystals* which give it a silvery appearance in gross

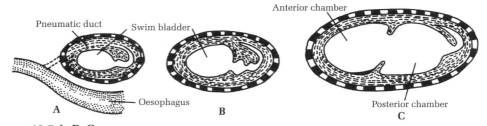

Figures 13.7 A, B, C
A. and **B.** Swim bladder showing formation of anterior chamber   **C.** Formation of posterior chamber from pneumatic duct (*Siphostoma*)

dissections (Fig. 13.8) This layer makes the gas bladder impermeable to diffusion, although it is not clear whether it does so by its guanine or because its collagenous fibres are very tightly packed. Anteriorly, the epithelium is modified into the **red gland**, which is made up of folded columnar epithelium and heavily vascularized by long loops of densely packed capillaries. This type of structure, wherever it occurs, is called a **rete mirabile**. The rete mirabile here produces gas, which the red gland secretes into the gas bladder.

In many **teleost** species this gas closely resembles air, but in some it is almost pure oxygen and in others almost pure nitrogen. In *Physoclistous* fish a posterior and usually dorsal part of the gas bladder is modified as the **oval gland**, which is surrounded by a sphincter muscle. When this sphincter is contracted the oval gland is shut off from the rest of the gas bladder; when the sphincter is open, the oval gland absorbs gas from the bladder. In physostomous fish the amount of gas in the bladder can be more rapidly decreased by contractions of the bladder musculature which force gas out the pneumatic duct, oesophagus, oropharyngeal cavity and finally mouth. Conversely, by gulping in air physostomous fish can get additional gas into the gas bladder more rapidly than the red gland can secrete it. In general, therefore, physostomous fish have no oval gland and a smaller red gland than do physoclistous fish.

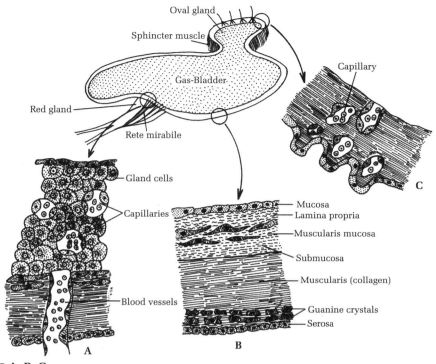

Figures 13.8 A, B, C
The sagittal section through the gas bladder of a physoclistous teleosts, showing **A.** Red gland **B.** Unspecialized wall and **C.** Oval gland

Functions The important functions of gas bladders are such as follows.

a) **Role in buoyancy:** Gas bladder functions primarily as a **hydrostatic** organ. By changing the amount of gas in its gas bladder a fish changes it specific gravity and thus its buoyancy. When it has the same specific gravity as the water around it, the fish can maintain its level without spending metabolic energy.

b) **Role in hearing:** Swim bladders in some species perform other functions such as hearing in addition to their hydrostatic role. In one suborder of **teleosts (cypriniformes),** a series of small bones, the *weberrian ossicles* connect the anterior end of the swim bladder and the sinus-impar (a projection of the perilymph cavity). Low frequency vibration of the gas within the swim bladder, evoked by waves of similar amplitude in the water, are transmitted by the ossicles to the membranous labyrinth. Therefore these fish can hear.

c) **Depth perception:** In certain **herring** like teleosts a diverticulum of the swim bladder comes into a direct contact with the membranous labyrinth, since there are no interposed bones. There is a possibility that the specialization may reflex or regulate gas pressure in the swim bladder or perhaps be related to depth perception.

d) **Sound production:** In a few fish, contractions of extrinsic striated muscles cause the swim bladder to emit **thumping sounds** or may force air back and forth between one chamber and another through muscular sphincter, as in *croakers* and *grunters*.

e) **Bioluminescence:** The outermost muscle layer of swim bladder is embedded with quinine crystals. It is believed that this silvery outer layer reflects bioluminescence through the translucent body tissue.

f) The urinary bladder of one species of marine **teleosts** from Indian waters passes through the air bladder. The advantage, if any, is not known.

Are gas bladders the precursor of lungs or there exist homology between lungs and gas bladders?

Only few fish (inhabiting) in oxygen deficient waters use either lungs or gas bladder as respiratory organs. Air breathing is apparently not a recent specialization to a stressful environment but an ancient feature. Moreover, careful studies of fossil fish impressions suggest that lungs evolved with the earliest **osteichthyes**, or possibly even in some **placodermi**. These lungs were very likely an adaptation to the oxygen depleted environments in which the earliest **osteichthyes** probably lived. It is quite possible that for fishes moving into more adequately oxygenated waters there would be a strong adaptive value not for an aerial respiratory organ, but for a hydrostatic organ, and that the teleost gas bladder represents not a precursor of the lung but a more adaptive modification of it.

Some biologists are of the opinion that the gas bladder and lung cannot be homologous because between the lungs of lungfish and gas bladders of teleosts there are apparent inconsistencies in *development*, *adult connections*, and *inter relationships* with other parts.

The fact that there is a single gas bladder while lungs are usually paired is not a major problem. Living lungfish have but a single lung that is bilobed in two genera. The lung of *Polypterus* is paired, although one is considerably larger than the other. The lungs of the *Bowfin* and *Garpike* are both single organs.

If lungs and gas bladder are homologous one would expect them to have a similar arterial supply and venous drainage.

The elementary respiratory organs of three genera of lungfish, *Polypterus, Bowfin,* and of one teleost *(Gymnarchus)* are all supplied by the artery to the last gill arch; this vessel called 'pulmonary artery' is homologous to the pulmonary artery of tetrapods. However, the arterial supply to the gas bladder in other teleost and in the *Garpike* and *Lepisosteus*, is by branches of the dorsal aorta (Fig. 13.9).

The venous drainage of the lung is the same in Lungfish as in tetrapods — a pulmonary vein going directly from the lungs to the heart. There is a chaotic distribution of veins in other air breathers none of which has a true pulmonary vein. Lungs and gas bladder also differ in microscopic structure. The adult teleost gas bladder has the typical epithelium, connective tissue, and muscle layers the same as has been described for the alimentary canal (Fig. 13.8).

In lungs of tetrapods and lungfish, however, connective tissue and smooth muscles are very scanty and are never arranged in the characteristic layers as those found in most of the alimentary canal.

As usual the best clue to possible homologies is **embryology**. Both gas bladders and lungs develop as foregut diverticuli, but they involve different areas of the foregut. In

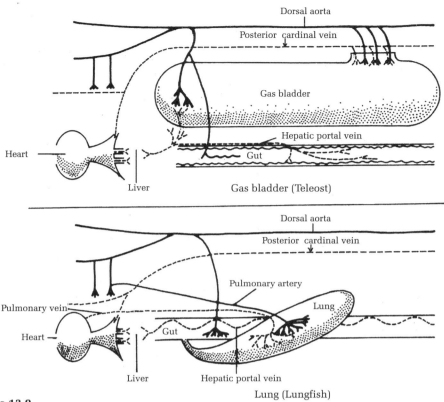

Figure 13.9
The arterial supply and venous drainage of the teleost **gas bladder,** and the lung of **lungfish**

sarcopterygians and **tetrapods**, the lungs develop from ventral diverticuli of the foregut of the pharyngeo-oesophageal border, or possibly even from a posterior pair of pharyngeal pouches.

In (teleosts) their early development, lungs are histologically more similar to the pharynx than to any other part of the alimentary canal (Fig. 13.10). In teleosts, however, the gas bladder very clearly arises as a more posterior, dorsal diverticulum from oesophagus or even from the presumptive stomach region; in the larval codfish the gas bladder even has peptic glands.

The gas bladders adult structure also resembles posterior rather than pharyngeal part of the alimentary canal. Considering all these factors, it seems likely that the earliest **osteichthyes** had lungs adaptive to an oxygen-deficient aquatic environment, and that these have been retained in the very generalized living **actinopterygians** (e.g. *Polypterus*) as well as in the **sarcopterygians** and **tetrapod**. During the evolution of those **actionopterygians** that successfully inhabited the increasing number of adequately oxygenated fresh water and salt water regions, it is also likely that an **air-bubble-lung** especially one in the ventral body cavity would be maladaptive. Natural selection would then favour individuals with a reduced lung, or even no lung. A hydrostatic organ, on the other hand, would no doubt have been a great advantage during the remarkable adaptive radiation of **teleosts**.

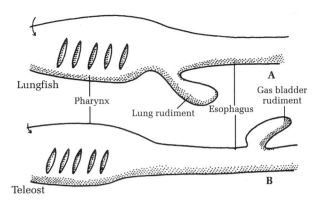

Figures 13.10 A, B
A. Location of lung rudiment. The lung develops ventrally at the border of pharynx B. Gas bladder rudiment develops dorsally and far more caudally from oesophagus

The gas bladder, then, may be neither a precursor nor a modification of the lung, but may have evolved quite independently for its own adaptive value. During subsequent radiations, as some **actionopterygians** came to inhabit areas with oxygen depleted waters, the hydrostatic organ may have secondarily evolved a respiratory function, providing another uninhabited niche to colonize.

Respiration by lungs

The lungs of the **chondrostean** fish — *Polypterus* are asymmetrically bilobed and the ducts open into the pharynx ventrally. The lining is not smooth as it is of swim bladders, and there are at least a few furrows that increase the surface in contact with the air.

The lungs of **dipnoi** may be bilobed, as in African and South American lungfish or there may be a single sac, as in Australian lungfish. The duct in dipnoans open into the oesophagus slightly to the right of mid-ventral line, but the sacs grow into the dorsal mesentry of the gut, where they lie in the adult. There is some subdivision of the lining into pockets. In both *Polypterus* and lungfish the lungs are supplied by pulmonary arteries of the sixth embryonic aortic arches but in *Polypterus* the venous return is to the hepatic veins and in **dipnoans** it is to the left atrium, as in **tetrapods.**

There are authorities who refer to any pneumatic sac in fishes as a swim bladder even though it performs a respiratory role, but this is a matter of semantic.

Tetrapod lungs have an unpaired trachea that enters the pharynx ventrally. They are supplied by pulmonary arteries arising from the embryonic sixth aortic arch and are drained by veins emptying into the left atrium. The lining is septate or pocketed. The lungs and air ducts of successively higher tetrapods exhibit increasing specializations. The lungs of **amphibians** are simplest and resemble those of lungfish.

Embryogenesis of lung Lungs in tetrapods arise as mid-ventral evagination from the caudal floor of the pharynx (Fig. 13.11). The lung-bud in **amphibian** commences as a solid outgrowth but soon develops a lumen. In higher vertebrates the bud is hollow from the beginning. The median opening in the pharyngeal floor becomes the *glottis*. The unpaired lung-bud elongates only slightly before bifurcating to form the primordia of the two lungs. The primordia, push posteriorly underneath the foregut to finally bulge into the embryonic coelom lateral to the heart. As they push into the coelom, the primordia carry along an investment of peritoneum, which becomes the visceral pleura. The lungs of **amphibians** and of many **reptiles** cease development after two relatively simple sacs have been formed within the coelomic cavity, but those of higher reptiles, birds and mammals become more complex.

Amphibians Although the respiratory organ of most adult amphibians is the lung, at different times in their life-cycles amphibians also use gills, skin and walls of the oropharyngeal cavity. External gills derived from the ectoderm are the primary respiratory organs of the larval tadpoles of all amphibians (Fig. 13.6 A). In anuran tadpoles a larval operculum covers the gills (Fig. 13.6 B), while in salamanders and caecilians there are external gills which project directly into the water and are rhythmically moved by the branchial muscles for ventilation. External gills persists throughout life in Neotenic **salamanders** (e.g. *Necturus*), which gain sexual maturity while otherwise retaining larval characteristic.

In many larvae and adults the skin is an important respiratory organ. For instance, where an air-bubble-lung would be maladaptive to locomotion, have lost the lungs in the adult; all respiration is cutaneous, facilitated by the fact that the water is cold and rapidly moving. For **salamanders** living in quiet water the lungs have some respiratory function, but are more important as hydrostatic organs similar to the **teleost** swim bladder.

Lungs As in all tetrapods, the amphibian lung develop from the floor of the alimentary canal just behind

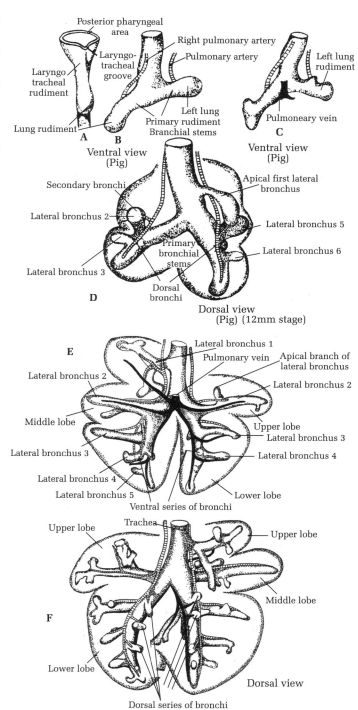

Figures 13.11 A, B, C, D, E, F
Development of mammalian lung through different stages

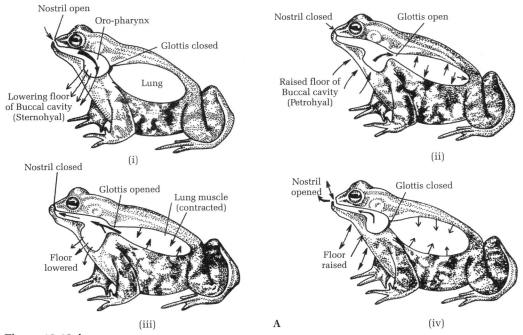

Figure 13.12 A

A. Ventilation of the lungs in a frog (thick arrow shows the inward movement of air) (Fig i, ii, iii and iv) show outward movement
  (i) The glottis is closed, lowering the floor of the orpharynx draws air into cavity.
  (ii) The nostril are closed and the floor of the orpharynx raised forcing the air into the lungs
  (iii) The glottis is opened and lungs muscle contracted forcing air out through the glottis, orophaynx into buccal cavity
  (iv) The glottis is closed and the nostrils opened, osscilatory movement of phaynx force air out through nostrils

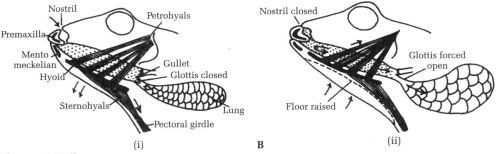

Figure 13.12 B
Muscle involved in inspiration during pulmonary breathing in frog (i) Floor of buccal cavity is lowered (Sternohyals) to draw air in through open nostrils, glottis is kept closed (ii) Nostril closed, floor is raised by petrohyals, glottis forced open, lungs inflated

the gill slit region. They are supplied with blood by derivatives of the arteries to the last gill arch and drained by the pulmonary vein which go directly to the heart. The lungs lie within the pleuroperitoneal cavity and are covered by visceral peritoneum except anteriorly where they are attached to the anterior walls of the body cavity. The lumen of each lung is continuous, either directly or through very short bronchi, with the laryngeal chamber, a mid line structure deep to the floor of the pharynx lined with muscles and small cricoid and arytenoid cartilages. The cartilages surround a slit-like opening into the pharynx, called the glottis, which is controlled by laryngeal muscles attached to the laryngeal cartilages. In most tetrapods the glottis opens in the floor of the pharynx, just anterior to the opening of the oesophagus.

In **amphibians**, however, there is a deep concavity in the floor of the oropharynx (particularly prominent in **anurans** and the position of the glottis is shifted to a posterior vertical surface of this concavity, just ventral to the oesophagus (Fig. 13.12 B).

During respiration the oesophagus is closed by sphincter muscles. The internal complexity of the lung and hence the extent of its respiratory membrane vary according to the order. Septae support the vascularized respiratory membrane and divide the lung into inter-communicating air spaces called *infundibuli*. Each infundibulum is lined by a thin epithelium with capillary bed lying deep to it in connective tissue. There are also smooth muscles just deep to the vascularized layer. In frogs and toads the septae are more complex than in **salmanders** and **caecilians**; the septae divide the lung into more and smaller air spaces (called **alveoli**), giving a larger total surface area for the respiratory membrane. **Toads**, whose keratinized skin permits little cutaneous respiration, have the most complex alveoli and the largest respiratory membrane.

Ventilation Amphibians force air into their lungs by a *pumping action* of the floor of the oral cavity in association with the action of valves that surround the external nares (Fig. 13.12 B). In **inhaling**, the external nares are opened and the floor of the oral cavity is depressed by contraction of the mylohyoid (sternohyal) and other muscles. Atmospheric pressure forces the air into the oral cavity. The nares are then closed by smooth muscle action and the floor of mouth is elevated. This action forces the air in the only direction available, through the glottis and into the lungs. In male **anurans**, opening the entrance to the vocal sacs would also force air to enter those chambers.

In **exhaling**, the body wall muscles compress the contents of the pleuro-peritoneal cavity and thus force the air up into the oral cavity here petrohyal muscle contract and raise the floor of buccal cavity and pump air out. Sometimes the air may then be pumped back into the lungs again or released out through open nostrils. **Urodeles** gulp air as well as inhale via the external nares, but the floor of mouth must still force the air into the lungs. In exhaling, some urodeles (e.g. *Siren*) with open gill slits eject the air forcefully through the slits.

Reptiles: Like other reptilian systems, the respiratory system is characterized by diverse morphology and hence diverse functions. The simplest reptilian lungs are in the ones living—rhynchocephalian, *Sphenodon* with a few, small septae and a simple general form.

In **ophidians** (snakes) the left lung is rudimentary and the right lung is long, narrow and functional, with moderately developed septae and alveoli.

The **lacertilians** (lizards) have a more complex lung, with definitive bronchioles coming off the trachea and leading to alveoli. In a few forms **(chameleons)** air sacs or thin membranes lead from the lungs to the body cavity. These structures like other structures in birds, supply additional air spaces but are completely unvascularized. In **chelonia** (turtles) although the lungs occupy much of the body cavity, the internal foldings consists of large, separated infundibuli (or alveoli), so that the total respiratory surface area is not great.

The large lungs of **crocodilians** have a more complex internal structure, their bronchioles and very small alveoli give a large respiratory surface area. Only in **crocodilians** are the pleural cavities, separated from the posterior peritoneal cavity by a transverse septum.

In all the reptiles, air passes through either the nostrils or the mouth, then through the glottis into a distinct larynx and to the trachea, which is supported by cartilaginous rings. The treachea then bifurcates, sending a bronchus to each lung.

Ventilation In all reptiles studied, ventilation of the lungs is caused by *suction* rather than by pressure, i.e., air is sucked in rather than being pushed into the lungs (Fig. 13.13). However, the mechanism of the suction pump vary considerably among reptilian groups.

In **lepidosaurians** both rhynchocephalia and lacertilia, the costal musculature moves the ribs and expands the thoracic region of the body cavity. Since the cavity is filled with uncompressible fluid, this expansion puts *negative pressure* on the contained organs, this in turn expands, the size of the lungs and thus draws air into them.

Contraction of the abdominal muscles compresses the body cavity and helps in exhalation, which is also assisted by contraction of the lungs intrinsic smooth musculature. **Ophidians**, with their long narrow bodies, have trunk musculature specializations that facilitate ventilation. A dorso-lateral muscle sheet runs from the ventral surface to the vertebrae to the medial surface

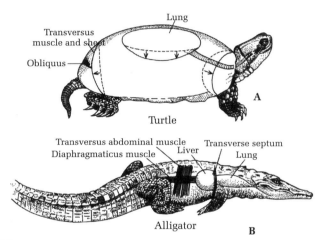

Figure 13.13
Ventilation mechanism in a turtle and an alligator **A.** In *turtle* the body cavity is expanded by moving the girdle outward and by contracting the obliqus muscle. This draw air into the lungs. Moving the pectoral girdle inward and contracting the transversus muscle compress the body cavity forcing air out of the lungs **B.** In alligators, contraction of diaphragmaticus muscle expand the pleural cavity and lungs and draw air in. Contraction of transversus muscle pushes the transverse septum forward, forcing air out of the lung.

of the ribs. A second ventro-lateral muscle sheet runs from the medial surface of the ribs to the mid-ventral skin. Contraction of these two muscles compresses the pleuro-peritoneal cavity, causing exhalation.

Inhalation is caused by *levator costalis muscle* (one per rib). Their contractions pull the ribs anteriorly and thus into a more vertical position.

In **chelonians** most of the trunk is covered by the bone of the **carapace** dorsally and **plastron** ventrally, which cannot expand or contract to change the pressure in the body cavity. However there is a **transversus muscle** attached to an externally convex connective tissue sheet, this sheet lies just deep to the skin over the openings for the hind limb and tail (Fig. 13.13). When the transversus muscle is contracted it flattens the sheet and thus reduces the volume of the body cavity. This forces air out through the glottis. Its antagonists, the **obliquus**, pulls in the other direction and increases the curvature of the skin, it therefore acts as a *suction pump* drawing air back into the lungs.

An analogous situation exists at the fore limb. The pectoral girdle is completely contained within the shell and is loosely articulated with it. By withdrawing the limb and pulling in on the pectoral girdle the volume of the body cavity is reduced and air is forced out. The opposite movements expand the body cavity and fill the lungs.

In **crocodilians** there is an even more unusual condition. Running from the pelvic girdle and inserting on the liver is a muscle called the **diaphragmaticus** (Fig. 13.13). The liver is attached to the transverse septum, which separates the pleural from the peritoneal cavity. Contraction of the diaphragmaticus pulls both the liver and the transverse septum posteriorly, compressing the peritoneal cavity and expanding the pleural cavities and thus drawing air into the lungs. The transverse abdominal muscles are its antagonists. These muscles compress the peritoneal cavity, which pushes the transverse septum forward and promotes exhalation. These movements are augmented by coordinated rib movements caused by contractions of the intercostal muscles.

Aves Because of their high metabolic rate (highest of any vertebrate group) birds require large amounts of oxygen and a rapid and efficient exchange of oxygen and carbon dioxide. They have evolved the most extraordinary respiratory apparatus. The lungs lie dorsally against the ribs in small pleural cavities. They contain the respiratory membrane, and it is here that all gas exchange takes place (Fig. 13.14 A). They are nevertheless small and compact but a fraction of the total respiratory system.

From the oropharynx air passes through the glottis, into a relatively simple larynx with laryngeal cartilages, and down through a cartilage supported trachea. The voice box is not the larynx, as it is in other **tetrapods**, but the **syrinx**, which is located at the base of the trachea where the treachea bifurcates into two primary bronchi which lead to the lungs (Fig. 13.14 B)

When a primary bronchus enters the lung it is called a mesobronchus (Figs. 13.14 C, D) and as such it continues directly through the lungs, giving off secondary bronchi on its way. The secondary bronchi are of two types; posteriorly they are called **dorsal bronchi**, although some of them come off ventrally, and anteriorly they are called **ventral bronchi**, which may come off dorsally. Some secondary bronchi terminate in thin walled air sacs, but all send branches into the lungs. The dorsal bronchi and the mesobronchus communicate with the more **posterior** air sacs, these are paired abdominal air sacs and the paired posterior thoracic air sacs (Fig. 13.14 B). The ventral bronchi communicate with **anterior air sacs**. These are paired cervical and anterior thoracic air sacs and midline interclavicular air sacs.

These air sacs, whose general positions are indicated by their names, have little vascularization and are not the site of gas exchange. They are extra air spaces required for ventilation. Diverticuli of these air sacs enter all portions of the body except the head and legs, sometimes including hollow bones. In the lungs are tertiary bronchi, called **parabronchi** (Figs. 13.14 C, D, E) which interconnect dorsal and ventral bronchi and facilitate gas exchange. Along their entire length are tiny diverticuli, called **air capillaries**. This arrangement puts

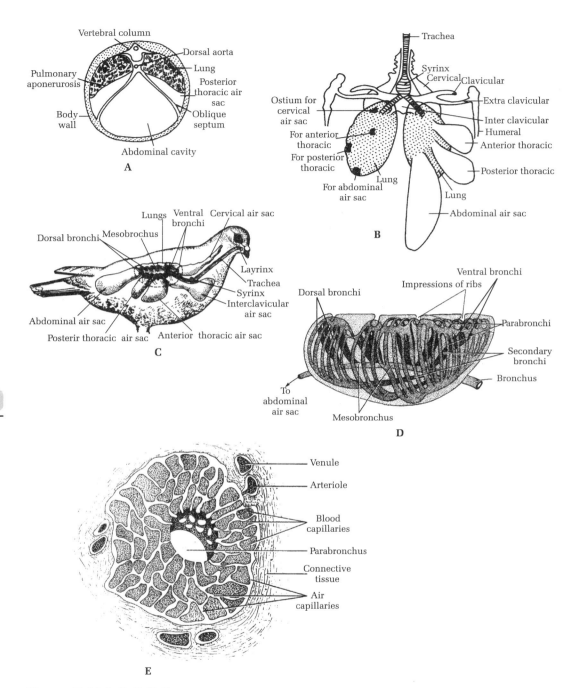

Figures 13.14 A, B, C, D, E

A. T.S. showing relation of air sac with the lungs in pigeon **B.** Pigeon diagram of air sacs, on the left air sacs have been removed to show their connection with lungs **C.** Relationship of the parts of a birds respiratory system **D.** Medial view of a chicken, showing relations of bronchus, mesobronchus, secondary bronchi, and parabronchi **E.** Section through a small segment of a parabronchus in the avian lungs.

the air capillaries into extremely close apposition to the blood stream and it is here that gas exchange occurs.

Ventilation Inhalation involves an **aspiration** pump analogous to that of reptiles (Fig. 13.15 A). Contraction of the external intercostal, subcostal, scalene and costal levator muscles straighten the ribs, which in birds are jointed mid-way between their vertebral and sternal connections. The straightening moves the sternum and precoracoid anteriorly and ventrally and expands the thoracic cavity; this creates suction and draws in air, most of which passes into the posterior air sacs. At the same time when *inhalation* draw air into the body, other air is, surprisingly, forced out of the lungs and into the anterior air sacs. Lung volume is changed

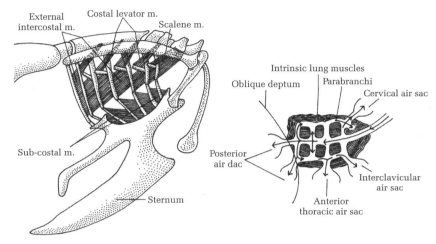

Figure 13.15 A
Movement of air flow during inspiration. Oxygen rich air passes the respiratory membrane during both inspiration and expiration

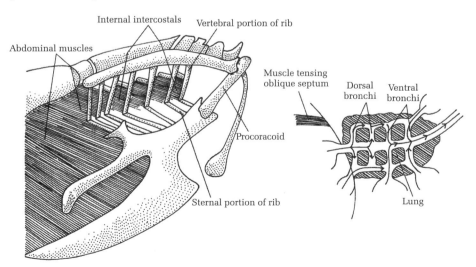

Figure 13.15 B
Ventilation in birds   B. muscles involved in expiration and movement of air flow during expiration through lungs and air.

by muscles attached to the oblique septum. Contraction of these muscles pulls on the oblique septum, providing a negative pressure around the lungs and therefore, causing the lungs to expand. Thus inhalation causes some air to move into the posterior air sacs and dorsal bronchi while other air moves out of the lungs into the anterior air sac.

In *exhalations*, processes are reversed (13.15B). The internal intercostals, contract, as do the abdominal muscles and the muscles flattening the oblique septum. This puts positive pressure in the abdominal and thoracic cavities, and they contract. The lungs, however, are under negative pressure and expand. The air in the posterior air sacs moves into the lungs and that in the anterior sac passes out the ventral bronchi to mesobronchus and through the trachea to outside. During flight, when more oxygen is required for flight muscles, the respiratory rate is increased, but the rate is usually a whole number of respiratory cycle per wing beat, indicating coordination between the two.

The lung itself is such a small part of this system that at any stage it contains only about 4 per cent of the air that is high in oxygen and low in carbon dioxide, while the anterior air sacs acts as collection and disposal depots for air that is low in oxygen and high in carbon dioxide. This system is most efficient that has evolved among tetrapods. It allows a constant movement of air through the parabronchi and air capillaries.

Mammals: The high metabolic rate of mammals also require large respiratory membranes and rapid, efficient ventilation. The mammalian lung is structurally more similar to that of reptiles than that of birds. It is basically an *alveolar* lung in which each entering bronchus divides into smaller and smaller bronchioles, which finally terminate in microscopic alveoli. The alveoli are very small and are surrounded by fluid having very low surface tension (Fig. 13.17 A).

The various branches of a single bronchiole, together with the bronchiole itself form units, or mass, within the lung which are referred to as lobules (Fig. 13.17 A). The lungs of mammals are lodged in special compartments, the pleural cavities, lodged in the thoracic cavity which is separated from the perivisceral cavity by the diaphragm (Figs. 13.16 A, B).

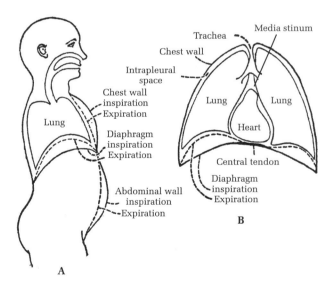

Figures 13.16 A, B
Abdominal and thoracic contours **A.** the relative positions of the chest wall, the diaphragm, and the abdominal wall at the end of maximal inspiration and expiration; **B.** diaphragm at end of normal inspiration and expiration. In quiet breathing there is no marked change in the position of the central tendon; in forced inspiration it may be drawn posteriorly, pulling the heart down with it

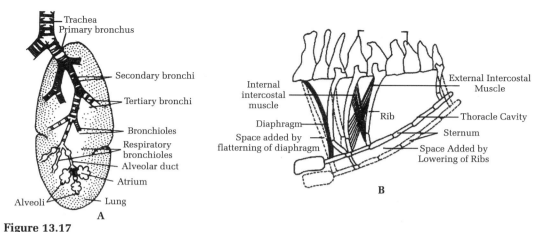

Figure 13.17
A. Tracheal branches and structure of infundibula in rabbit   B. Mechanism of breathing in rabbit. Solid lines showing the position of the ribs, sternum and diaphragm normal position and dotted lines during inspiration.

The thoracic cavity is lined internally by pleura, a membrane derived from the coelomic walls of this region. Between the two lungs lies the pericardium enclosing the heart.

Mechanism of respiration

A musculotendinous **diaphragm** present only in mammals, separates the pleuroperitoneal cavity into two parts. That portion anterior to the diaphragm contains the lungs. Each lung is enclosed in a separate pleural cavity which is lined with a fibro-elastic membrane called the pleura. The part of pleura lining wall of pleural cavity is referred to as parietal pleura. The lung itself is invested by a layer of visceral pleura. Towards the median line the parietal pleurae of the two sides come close together so as to form a two layered septum, the **mediastinum**, separating two pleural cavities (Fig. 13.16 B). The space between the two layers of the mediastinum is called the *mediastinal space*. It widens as it nears the diaphragm and contains the aorta, oesophagus, and posterior venacava. In the region of heart the mediastinal membrane passes over the parietal layer of the pericardium. The pleural cavities are air tight. The mechanism of respiration in mammals consists of alternately increasing and decreasing the size of the pleural cavities. This is accomplished in two ways:

Abdominal respiration   The diaphragm when relaxed is shaped somewhat like an arched vault, and is perforated by the dorsal aorta, oesophagus, azygos vein, thoracic duct, post cava, and vagus nerve. As its rediating fibres shortens by contraction, the vault of the cuplike diaphragm lowers or flattens, thus increasing the space within the thoracic cavity. Causing inspiration, and brings about an increased pressure on the abdominal viscera. The muscular opponents of the diaphragm are the walls of the abdomen. Contraction of the abdominal muscle causes the abdominal visera to push against the diaphragm, bringing about a reduction in size of pleural cavities and resulting expiration is known as abdominal respiration. In big heavy animals, abdominal or diaphragmatic breathing predominates over rib breathing.

Costal breathing   In addition to abdominal diaphragmatic breathing, mammals also utilize the reptilian method of rib muscles to enlarge the thoracic cavity when inspiring air. The method of expanding the pleural cavities is brought about by contraction of external intercostals muscle. These pass form the lower border of one rib to the upper of the rib below and their contraction brings about an elevation of the ribs (Fig. 13.17 B) The ribs are bent, like joined levers at an oblique angle to the vertebral column, and if acted upon by the intercostals muscles the movable sternum, to which they are attached ventrally, move further away form the relatively stationary back bone, thus enlarging the thoracic cavities in which the lungs are located. So it comes about that inspiration is effected not only by the depression of the diaphragm but also by the elevation of the ribs, both efforts calling for muscular activity.

Expiration, on the other hand, is to a large extent automatic through the elasticity of the stretched body walls, the taut cartilaginous ends of the bent ribs and the tensity of the expended lung tissues.

Breathing by means of the ribs is also more pronounced in human females then in males in whom abdominal breathing predominates.

The reason for sexual difference in the respiratory mechanism may be evolutionary adaptation brought about in connection with pregnancy during which period the presence of growing fetus interfere somewhat with freedom of movement of the diaphragm. Jumping animals like kangaroo and monkeys utilize rib muscle rather more than the diaphragm in respiration.

Sound production in vertebrates

In the vertebrate animals the vocal and respiratory organs are intimately associated owing to the fact that the production of sound in caused by the expulsion of air from lungs. Hence in majority of the tetrapods the anterior end of trachea gets modified to form the **larynx**. The larynx is a sound producing organ in tetrapods except birds. Larynx communicates with the pharynx by means of a slit like opening called glottis. Posteriorly it continues into the trachea.

Amphibians: Though the sound may be produced in a few fish but the voice makes its first appearance in the amphibians. In the **urodeles**, the voice in feebly developed or entirely absent. It attains its maximum in the **anura.** The sound producing organs of the frog are located in a type of box called the larynx situated just below the pharyngal cavity at the beginning of the entrance into the lungs. The larynx open into the pharynx through the slit like glottis above and by a pair of openings behind into the lungs. It is held between the stout, bony thyroid processes of the hyoid apparatus to which it is attached by muscle as well as connective tissue. The skeleton of the **larynx** (Figs. 13.18 A, B, C) is composed mainly of the cricoid and arytenoid cartilage, the former consist of a slender ring surrounding the larynx and lying in nearly the same place as the thyroid process of the hyoid to which it is closely attached. At its posterior end it is produced into a spine which extends, backward between the lungs. From near the middle it gives rise to a sort of loop the tracheal process which is bent backward and serve as a means of attachment for the neck or roots of the lungs. The arytenoid cartilages are a pair of semilunar valves, which rest upon the cricoid cartilages, their upper edges form the lateral margins of glottis. They afford attachment of muscle by which the glottis may be opened or closed.

The true sound producing organs consists of a pair of elastic bands the **vocal cords**, extending longitudinally across the larynx (Fig. 13.18 A). They can be seen from above by spreading

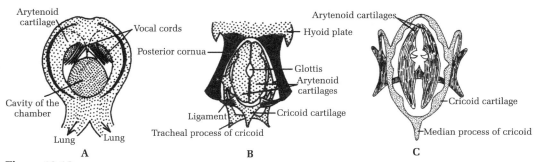

Figure 13.18
Laryngeal skeleton of a frog

apart the two sides of the glottis or from below by removing the membranous floor of the laryngeal cavity. A narrow space (**rima glottidis**) exist between the inner free edges of the vocal cords. It is through this space that the air has to pass on its way to lungs and back. Sound is produced by the expulsion of air from the lungs which sets the free edges of the vocal cords in vibration. Vibrations in the sound are caused by altering the tension on the cords through the action of the laryngeal muscles.

The **vocal apparatus** of the male frog is a pair of vocal sacs situated at the side of the pharynx. These sacs are out pocketings of the pharyngeal wall which extend between the skin and the body. They communicate with the mouth by small openings in the floor, a short distance in front of the angle of the long jaw. The vocal sac are distended during the croaking of the frog through the presence of the air in the buccal cavity. They serve as resonators to re-enforce the sound produced by the vocal cords. They are absent in females.

The simplest type of larynx is found in *Necturus* where only a pair of lateral cartilages encircle the slit-like glottis, caudate amphibians do produce sound but their sound is merely a hiss or slight squeaking.

Reptiles: In reptiles, the larynx in situated at the anterior end of trachea just behind the glottis. It is no better developed than in amphibians. The larynx is made up of a pair of arytenoid cartilages and a median cricoid cartilage (Fig. 13.19). The arytenoid cartilage are located in the anterolateral walls of the larynx and their anterior ends provide support to the glottis The posterior end of arytenoids lie on the anterolateral side of the cricoid cartilages. The two pairs of muscles viz., inner muscularis compressor laryngis and the outer musculus dilator laryngis are found in the walls of larynx the former function as a sphincter for the glottis, whereas the latter acts as a dilator of the glottis. Cricoid ring of larynx in reptiles, remains incomplete. In reptile like crocodilians, the thyroid cartilage is also found in the larynx. **Gular pouch** is met with in reptiles like chameleons. It in in fact, formed by the lining of the larynx between the cricoid cartilages and the outward extension of first tracheal cartilage.

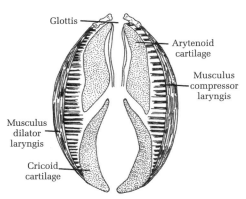

Figure 13.19
L.S. Larynx of *Uromastix*

Most **reptiles** produce just a hissing sound only due to the rapid expulsion of air from the **lungs**. **Guttural** noise, are produced by certain lizards like *geekos* and *chamaeleons* where vocal cords are present. The male alligators are capable of producing **bellowing** sounds during breeding seasons. Such sounds are loud and can be heard over distance of a mile or more.

Birds: The larynx does not function as a sound producing organ in birds and do not produce music. The non-functional larynx is compensated by the presence of **syrinx** located at the junction of trachea and bronchi (Fig. 13.14 B). However, the same usual set of cartilages. i.e., a pair of arytenoids and single cricoid cartilage are still found in reduced larynx in birds. Moreover, the cricoid cartilage is further divided into four pieces called procricoids. Arytenoid cartilages may get ossified in certain birds.

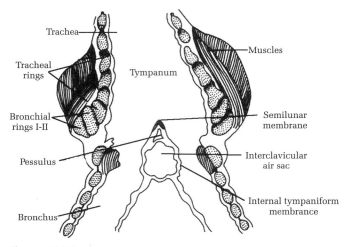

Figure 13.20
L.S. Trachea through syringeal region of Pigeon

The **syrinx** consists of last 3-4 rings of trachea and first – half ring of each bronchus (Fig. 13.20). The cavity of the syrinx is known as **tympanum**. It is lined with a mucous membrane. This mucous membrane forms a cushion like thickening on each side. Internal tympaniformis membrane is formed by the mucous membrane of the median wall of each bronchus. A thin fold of mucous membrane, arising from the angle of bifurcation of trachea, extends forward into the syrinx and is supported by a bar of cartilage called **pessulus**. This fold is known as membrana semilunaris. This fold plays a crucial role in production of sound in birds. The air expelled from the lungs during expiration passes between the tympaniform membrane, whose tension is under delicate muscular (intrinsic syringeal muscle and sternotracheal muscle) control and produces the repertoire of sound of which birds as a class are capable.

Mammals: In mammals the **larynx** or voice box is at the anterior end of the trachea (Fig. 13.21). At the top of this cylindrical box like structure there is a large opening —the glottis, from whose ventral margin projects the cartilaginous flap like epiglottis which serve to prevent food entering the glottis during swallowing (Fig. 13.21) The ventral wall of the larynx is supported by a shield shaped thyroid cartilage. A short distance posterior to this is the cricoid cartilage which forms a ring round the larynx. The dorsal rim of the glottis is supported by a pair of projecting arytenoid cartilage. At the apices of the arytenoids are situated a pair of small nodules, the cartilages of *santorini*. Extending between the thyroid and arytenoid cartilages are two pairs of membranous folds, the **vocal cords** (Fig. 13.21), of these the anterior pair is that of the false vocal cords which may be absent in some. The

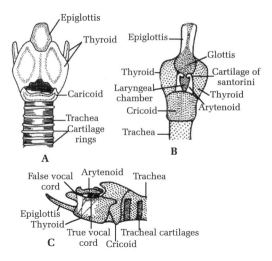

Figure 13.21
Larynx of rabbit
A. Dosal view   B. Ventral view   C. Lateral view

posterior pair is known as the true vocal cords. The vocal cords are folds of the lateral wall of the pharynx, and close the glottis except for a slit between their free edges, and it is through this gap that air enters during the respiratory movements.

The thyroid cartilage in mammals usually consists of two parts which are fused along the mid-ventral line. The prominent protuberance (**Adam's apple**) is an extension of thyroid cartilage and is called the *pomum adami*.

Chapter 14

Circulatory System

The circulatory system is the transportation system in the body which is meant to provide incessant supply of food and oxygen to the tissue and to carry off and remove waste products in order to sustain and maintain life. The chief components of the circulatory system are the blood vessels and the heart; its chief functional unit is the blood itself. With these components, the circulatory system integrates all the body's metabolic activities. The system coordinates the body's chemical actions and its reactions. The following points recapitulate the important function performed by the circulatory system.

- It helps to maintain a constant water and ion content in its interstitial fluids and thus helps to keep the animal osmotically independent of its environment.

- It carries and transmits hormones from the endocrine glands to their target organs.

- The blood carries oxygen which is necessary for the production of energy. Most of the oxygen is attached to hemoglobin molecules in red blood cells, forming oxyhemoglobin. In an area relatively high in carbon dioxide concentration the oxyhemoglobin gives up its oxygen, which diffuses into the tissues. Oxidative metabolism occurs in the tissues and carbon dioxide is carried away in plasma, mostly as bicarbonate ion.

- The circulatory system is of major importance in maintaining a constant body temperature in homeotherms. Body heat, the most of it is produced by muscular contraction, is distributed by way of the circulatory system. When there is an excess of heat, for instance, during muscular exertion, some of the superficial blood vessels dilate and thus bring blood near the surface where heat may be dissipated to the surrounding environment. When the external environment is very cold the superficial blood vessels constrict, little blood then approaches the surface but is circulated instead in the deeper parts of the system, preserving essential heat. In aquatic mammals and birds of the Arctic and Antarctic, the temperature of the extremities is kept just above freezing by heat of arterial blood. Blood before entering the deeper parts of the body from extremities is rewarmed, often by being wrapped about by the outgoing arteries.

- The circulatory system, carries food from the digestive system to all parts of the body and carry waste material to the excretory organs.

- The system can both combat and spread disease. It spreads disease by carrying infective micro-organisms throughout the body, and it can combat disease by both producing and transmitting antibodies.

Hemopoiesis

The blood cells and blood vessels in the developing embryo form from the mesodermal blood islands which lie along the mesoderm bordering the yolk sac. These blood islands give rise to an anastomosing network of blood vessels and is the earliest hemopoietic tissue whose mesenchymal cells differentiate into hemocytoblasts (Fig. 14.1) which later form the blood cells. In the older embryo and in the adult, hemopoiesis is taken over by a variety of organs including liver, kidney, spleen, lymph nodes, intestinal submucosa and bone marrow. Hemopoiesis is continuous throughout life. Worn out blood cells are also continually destroyed and removed, primarily in the spleen, and secondarily in the kidney and liver. In adult vertebrates red blood cells (**erythrocytes**) and granular white blood cells (granular **leukocytes**) are produced in myeloid tissue, and agranular white cells (agranular leukocytes) in lymphoid tissue. In different vertebrate groups, these tissue may be either physically separated from one another or part of the same organ. In adult **teleosts**, erythrocytes form mainly in the kidneys and spleen, granular leukocytes form mainly in the intestinal submucosa and the spleen, and agranular leukocytes form in several mesodermally derived tissue. In *Garpike* and *Bowfins*, which are more generalized **actinopterygians,** hemopoiesis also occurs within skeletal tissues.

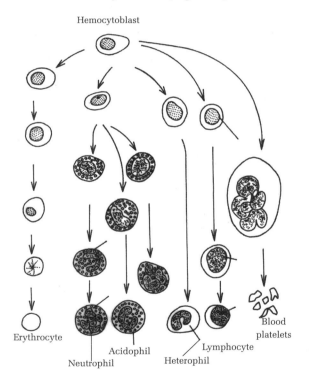

Figure 14.1
The different types of blood cells all differentiated from hemocytoblast

Amphibians: In most amphibians, leukocytes (both granular and agranular) are formed in bone marrow, and most erythrocytes are formed in the spleen.

Reptiles: In reptiles, spleen is the primary site of hemopoiesis in **horned toad**, and the bone marrow is the primary site in other lizards.

In **birds** and **mammals** nearly all agranular leukoctyes are produced in lymph nodes and spleen and nearly all erythrocytes and granular leukocytes are produced in red bone marrow. However under stress, other tissue will produce blood cells. It is apparent that the potential for hemopoiesis is present in a great variety of mesodermally derived tissues but under normal circumstances this potential is expressed by only a few.

Blood Vessels

The blood vessels of all vertebrates are structurally similar (Figs. 14.2 A, B, C, D). The innermost layer called **tunica intima** is made up of squamous endothelial cells lying on a thin basement membrane which may contain elastic fibres as well as reticular fibres. Superficial to the tunica intima is a muscular layer, the **tunica media**, in addition to its smooth muscle cells the tunica media contains connective tissue cells and elastic fibres, whose numbers vary in different vessels. Superficial to the tunica media is the outermost coat, the **tunica adventitia,** it is composed of loose connective tissue and some elastic fibres. The outer portion of the tunica adventitia is continuous with the surrounding connective tissue. Blood vessels are arteries or veins, or diminutives of these, or capillaries. Arteries and veins differ structurally, first in the tunica media which is very thin in veins and very thick in arteries. In both arteries and veins, the tunica media is innervated by autonomic nerves which influence the degree of contraction and thus the diameter of the lumen. When arteries and veins run side by side, however, the vein is larger both in its external diameter and in its lumen diameter (Figs. 14.2 A, B). A second major structural difference is, that veins contain valves at irregular intervals, which allow the blood to flow only towards the heart (Fig. 14.2 C). Arteries have no valves, the blood is propelled by muscular action and cannot flow back. As arteries run toward capillary beds they branch profusely and become smaller in diameter, then called as arterioles (Fig. 14.3). Closer to the capillary bed their smooth musculature is thinned and eventually both the tunica adventitia and the tunica media are lost, only endothelial cells and an incomplete basement membrane form the structure of the capillary bed. The smallest veins into which the blood flows upon leaving the capillaries are called venules, several venules then join together to form a vein. The lumen of most capillaries is just large enough to pass a single blood cell at a time, similar to the situation in fish gills. Although this may seem awkward, but it is, adaptive, as in the gills, it gives a maximum surface area, and it greatly slows the blood and reduces its pressure, both these factors facilitate the diffusion of substances between blood and the interstitial fluids. Because the blood passes much more slowly in capillaries than in arteries, the capillary beds must be capable of handling more volume than the arteries. In liver and certain other organs the capillaries have a much larger diameter and are called **sinusoids,** rather than capillaries. Their walls are formed only of endothelial cells and an incomplete basement membrane.

In general, the capillary network lies between arterioles and venules. However, in gills it usually lies between two sets of arteries, and in portal system such as those of kidney, liver

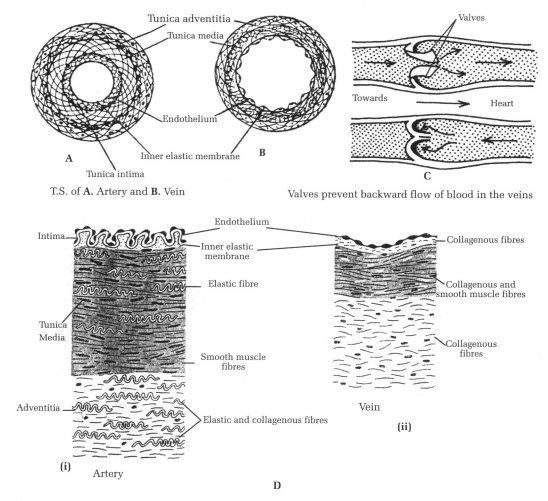

Figures 14.2 A, B, C, D
Structure of Blood Vessels. Structure of a medium-sized artery and its accompanying vein. Note differences in thickness of the muscle layers and the absence of elastic fibers in the vein

and hypothalamus; it lies between two sets of veins. There are frequent short-cut called **shunts**, or anastomoses, between arterioles and venules (Fig. 14.3). If these shunts are open the blood bypasses the capillary network almost entirely; if they are closed, the entire blood flow is forced through the capillary bed. Anastomotic vessels or shunts also frequently occur between adjacent arteries or adjacent veins. Such anastomoses allow for collateral circulation, when for instance, a vessel is blocked or injured.

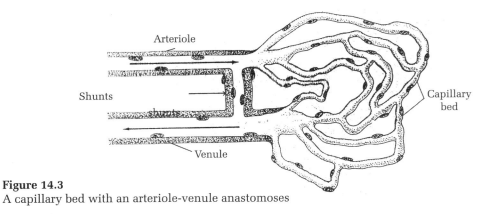

Figure 14.3
A capillary bed with an arteriole-venule anastomoses

Succession of Heart in Various Chordates

The heart is the primary pump for the complex circulatory system. As might be expected the vertebrate heart is developmentally and structurally an extremely modified blood vessel (Figs. 14.4 A, B, C, D). The heart is an unpaired organ, but it is formed from a pair of thin endocardial tubes formed by endocardial cells which are formed from embryonic mesenchymal cells. The two endothelial tube soon fuse to form single endocardial tube which meet above by dorsal and below by ventral mesocardium. Soon mesocardia disappear and endocardial tube thickens to form ventricle. The original endocardial tube becomes endocardium while the surrounding layer develops to form myocardium. Endothelium (here called endocardium) comprising a compact tunica intima, lines its lumen. Its tunica adventitia, which frequently contains adipose tissue, is covered by the visceral pericardium of the pericardial cavity. Its tunica media is greatly thickened and is the most modified of its parts. Instead of smooth muscles the tunica media contains striated cardiac muscles with intercalated discs. These

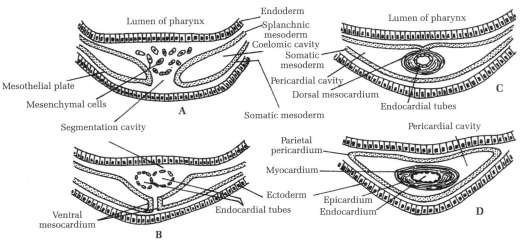

Figures 14.4 A, B, C, D
Section depicting the development stages of the heart **A.** Mesenchymal cells **B.** Mesenchymal cells has formed a pair of endocardial tube **C.** Endocardial tubes unite to form a simple tube which later transform to form heart **D.** Section of formed heart

cardiac muscles, collectively called the **myocardium,** comprise the great bulk of the vertebrate heart (Figs. 14.4 A, B, C, D). The rate of contraction of the myocardium is regulated by the **vagus** nerve. Parasympathetic innervation causes the rate to decrease and sympathetic innervation causes it to increase. Contraction always begins in the sinus venosus or its homologue, where the systemic veins (i.e., veins draining blood from all of the body organs except lungs) come together before entering the heart (Fig. 14.5). This is followed by contraction of the atrium (atria) and then contraction of the ventricle (s). The ventricular contraction begins at the point farthest from where the vessels leave the heart, and continues smoothly toward the openings into the arteries, or in **fish**, to an anterior chamber of the heart called the **conus arteriosus**. These cardiac muscle contractions are controlled and coordinated by specialized, elongated neuromuscular cells called **Purkinje fibres**. There is a group of these cells in the sinus venosus or its homologue, called the sinoatrial node or pace maker, this initiates heart contraction by activating the atrial musculature and conducting the excitation to a second group of Purkinje fibres, the atrio-ventricular node, which lies between the atrium and the ventricle (Fig. 14.5).

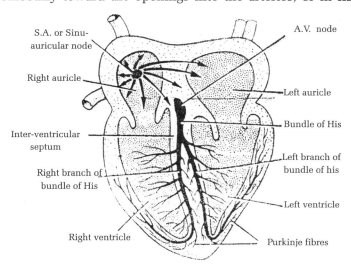

Figure 14.5
Conduction of heart beat

From the atrio-ventricular node other Purkinje fibres spread into the ventricles and causes them to contract.

Cephalochordate: In amphioxus the heart is absent, as such the ventral aorta which is a pulsatile and muscular tubular structure helps in pumping blood. In addition to ventral aorta, at the origin of aortic arches pulsatile bulbils aid in pumping too (Fig. 14.6).

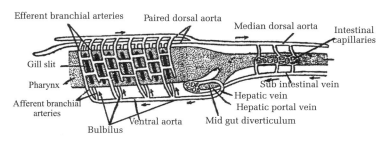

Figure 14.6
Blood vascular system of Amphioxus

Fish: In fish except sarcopterygians (except lungfish) the heart is composed of a folded tube having thick muscular wall and divided into four chambers (Fig. 14.7). At the posterior part of the pericardial cavity is present, the sinus venosus. Venous blood from all parts of the body enters the sinus venosus through paired common cardinal veins (duct of cuvier) and paired hepatic veins. When the sinus venosus contracts, blood is forced into atrium, which

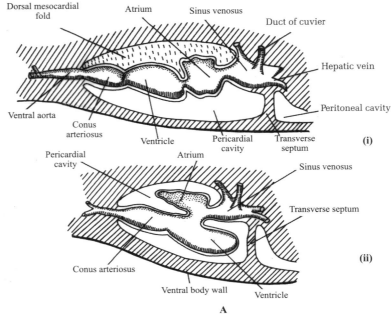

Figure 14.7 A
A highly primitive heart and pericardium **(i)** and **(ii)** a selachian heart

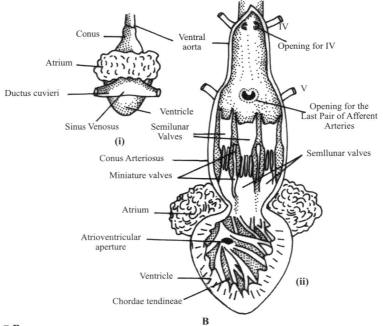

Figure 14.7 B
Scoliodon: The heart **(i)** Dorsal view **(ii)** Dissection from ventral side

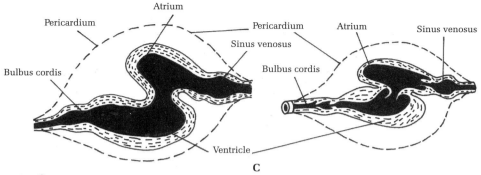

Figure 14.7 C
Heart of *Scoliodon* opened to show chambers and blood flow

is larger and more muscular. When the atrium contracts, the blood, prevented by a valve from passing back to the sinus venosus, is forced into the next and thickest walled portion of the heart, the ventricles. When the ventricle's thick wall contracts it produces the major pressure. The blood, again prevented from flowing backward by a valve, is forced out of the fourth chamber of heart, the conus arteriosus (Fig. 14.7 B). In **elasmobranchs** the conus arteriosus is a long, tubular muscular structure that is continuous anteriorly with the ventral aorta. In **actinopterygians,** the ventral aorta and conus arteriosus together are very short and bulb shaped and are usually called the bulbus arteriosus (Figs. 14.8 A, B, C). In both elasmobranchs and actinopterygians, semilunar valves on the wall of the conus arteriosus prevent the return of blood to the ventricle.

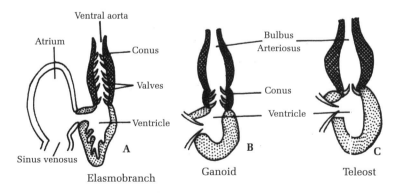

Figures 14.8 A, B, C
Schematic longitudnal section through the hearts of various fish to show relation of the conus and bulbus **A**. Elasmobranch **B**. Ganoid **C**. Teleost

All these four chambers are not laid out linearly but are folded one upon the other in a S-like configuration. This makes the heart more compact, so that it fits into a relatively small pericardial cavity. In **chondrichthyeans,** the pericardial cavity is connected with the pleuro-peritoneal cavity by a pair of pericardial-peritoneal cavity.

Lungfish and amphibians: There are significant differences in the hearts of lungfish and amphibians otherwise in many respects they are similar to those of most fish (Fig. 14.9). The atrium is divided, either partially (in lungfish) or completely (in most amphibians) into right and left chambers. From the sinus venosus right atrium receives systemic venosus blood, which is high in carbon dioxide and low in oxygen concentration.

The left atrium receives blood from lungs via pulmonary vein, with a high oxygen and low carbon dioxide content.

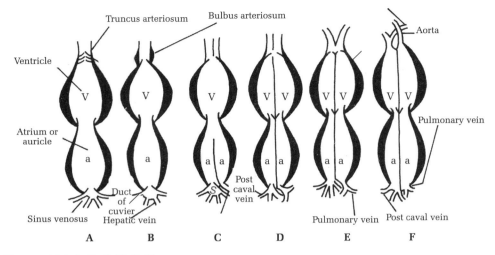

Figures 14.9 A, B, C, D, E, F
Evolution of heart **A.** Elasmobranch **B.** Teleost **C.** Amphibian **D.** Lower reptiles
E. Alligators **F.** Birds and mammals

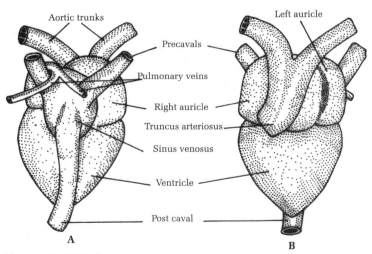

Figures 14.10 A, B
Heart of frog **A.** Dorsal view **B.** Ventral view

The walls of both atria and especially of the ventricle are more heavily muscular than those of most fish, with folds (**trabeculae**) which are more pronounced in amphibian (Figs. 14.10 A, B and 14.11 A). In both lungfish and amphibians a spiral fold in the conus arteriosus (Figs. 14.11 A, B) separates it into two portions along much of its length. In lungfish a partial septum incompletely divide the ventricle into two portions.

Functional studies using X-ray opaque materials in the living heart and studies taking blood samples from various points show that, deoxygenated blood from the sinus venosus, the right atrium, and the right side of the ventricle is directed preferentially into the pulmonary circulation by way of the pulmonary arteries; there it is carried to the lungs, and oxygenated blood is returned to the left atrium. This oxygenated blood passes into the ventricle and preferentially out to the arteries leading to the head and other somatic portions of the body. Although there is no solid septum dividing systemic from pulmonary circulation in lungfish and amphibians, the functional attributes of the heart cause a preferential distribution of oxygenated and deoxygenated blood.

Reptiles: Among reptiles there are two distinct heart types, which differ in the morphology of their ventricles and of the large arteries leading from them. In both types, however, the two atria are completely divided by an interatrial septum (Fig. 14.12 A). The right atrium receives deoxygenated systemic venous blood from the sinus venosus; the left atrium receives oxygenated blood from the pulmonary veins.

From **lepidosauria** to **chelonia** there is a single ventricle (Fig. 14.12 B). A large ventrally placed muscular septum partially divides it into communicating right and left ventricle, left is called the **cavum arteriosum**, the right side which receives blood from the right atrium is called **cavum venosum**, the narrow portion between them is called the interventricular canal (Fig. 14.12 C). The cavum venosum, has a ventral diverticulum called the cavum pulmonale, from which the pulmonary artery goes to the lungs. Exiting directly from the **cavum venosum** are two vessels, the right and left aortae, which carry blood through the systemic arteries to the rest of the body.

When the atria contract, the large, flap like valves between atria and ventricle open and occlude the interventricular canal. This causes all of the oxygenated blood from the left

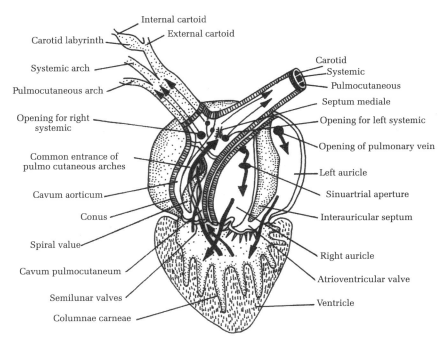

Figure 14.11 A
Internal structure of heart of frog

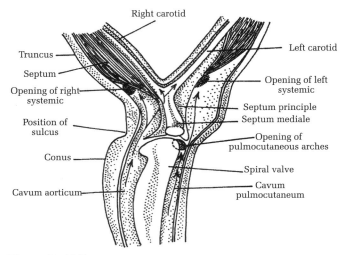

Figure 14.11 B
Conus and truncus of *Rana tigrina* showing course of blood

atrium to pass into the cavum arteriosum, and all of the deoxygenated blood from the right atrium to pass into the cavum venosum and cavum pulmonale. While the atria are contracted the blood of the two atria cannot mix. When the atria relax and the ventricle contracts, however, the flap valves are forced closed and the interventricular canal is opened. Contraction of the ventricle also reduces the volume of its cavities and forces out the blood, which could pass through either of the aortae or the pulmonary artery. However, because there is less resistance to flow in the

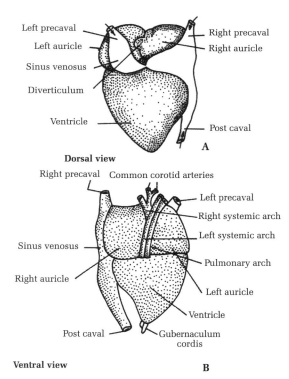

Figures 14.12 A, B
Heart of a lizard **A.** Dorsal view **B.** Ventral view

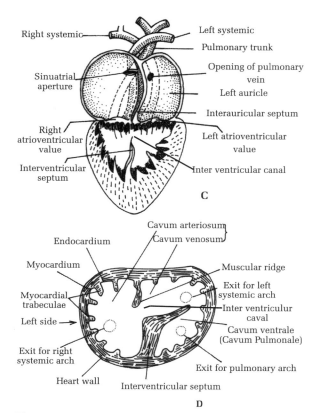

Figures 14.12 C, D
C. Internal structure of heart (after removing its ventral wall) T.S Ventricle of the heart of lizard showing partial partition of the ventricle D. T.S Ventricle of the lizard showing partial partition of the ventricle

pulmonary artery than in the aortae, the first blood to leave the ventricle passes into the pulmonary artery. This is deoxygenated blood which has entered from the right atrium and passed to the cavum venosum and cavum pulmonale. As the ventricle continues to contract, a muscular ridge is brought against the opposite ventricular wall and the cavum pulmonale is thus closed off (Fig. 14.12 D). As the cavum venosum is emptied, more blood enters it through the interventricular canal from the cavum arteriosum, this is oxygenated blood from the left atrium. Because the pulmonary artery is closed off, this blood is forced into the right and left aortae and travels to all parts of the body except the lungs. There is, thus, a highly efficient separation of oxygenated and deoxygenated blood in the systemic and pulmonary circulation of lepidosauria and chelonia.

Other reptiles such as **crocodilians** have a complete interventricular septum, so that blood from the right and left atria remain separate in ventricles (Fig. 14.13 B). Like other reptiles they have no conus arteriosus, but its homologue forms the base of three large blood vessels—the pulmonary artery, coming from the right ventricle and dividing to send an artery to each lung; and the left and right aortae, from the right

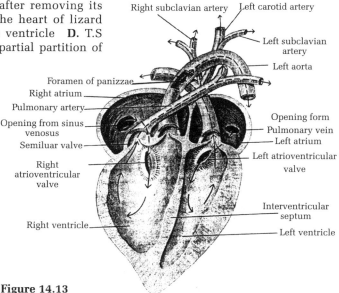

Figure 14.13
A dissected heart of an alligator, showing chambers and major blood vessels

and left ventricles, respectively (Fig. 14.13). After leaving the ventricles, the two aortae cross, and at that point there is a hole, the foramen of **panizzae,** through which their lumens communicate (Fig. 14.13).

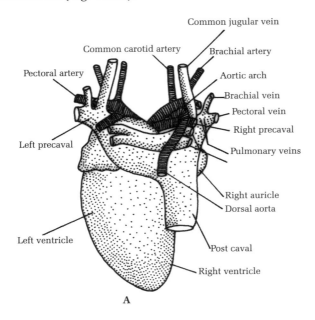

The crocodilian heart functions in the following manner. Atrial contraction forces blood from the right atrium into the right ventricle, and from left atrium into left ventricle. When the atria relax and the ventricles contract, the atrioventricular valves are closed and blood is forced out of the ventricles through the pulmonary artery and right and left aortae.

The right ventricle, containing deoxygenated blood, could conveniently force blood out through both the pulmonary artery and the left aorta. However, under normal circumstances nearly all the blood passes out through the pulmonary artery. The reason for this, again, is that there is less resistance in the pulmonary artery than in the systemic circulation. From the left ventricle blood passes out through the right aorta, is distributed by way of the foramen of Panizzae to the left aorta as well and thus feeds the entire systemic arterial circulation. Thus, crocodilians have the same functional attributes as other **reptiles** but with a different morphological basis. The mechanism by which oxygen-poor blood is shunted to the lungs and oxygen-rich blood is shunted to other parts of the

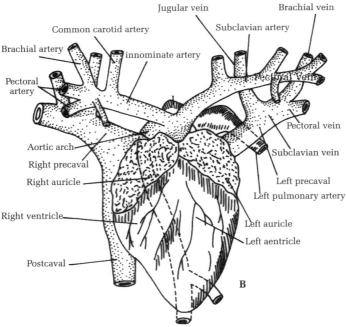

Figures 14.14 A, B
Structure of heart of pigeon   **A.** Dorsal view   **B.** Ventral view

body would seem unduly complicated in reptiles. It would certainly seem much simpler to have a complete instead of a partial interventricular septum, to omit the foramen of Panizzae entirely, and to have only the pulmonary artery coming from the right ventricle and only the systemic arteries from the left ventricle. Such is the situation in birds and mammals, and the reptilian heart, therefore, has often been called transitional. However, recent studies indicate that for most reptiles this complex system, with its potential mixing of oxygen-rich and oxygen-poor blood, has a unique adaptive valve. In all diving tetrapods, the vagus nerve causes the heart rate to slow considerably when the animal dives. The pulmonary arteries constrict, which greatly increases their blood pressure and their resistance to blood entering from the right ventricle. In lepidosauria and chelonia this increased resistance and pressure of the pulmonary arteries cause the preferential flow of blood from the cavum venosum to go to the right and left aortae, similarly in crocodiles the change in pressure and resistance causes most of the blood from the right ventricle to pass into the left aorta and some into the right aorta. Energy is conserved, therefore, by passing the pulmonary circulation at no cost to the animal because sending blood to its lungs while under water would fulfil no purpose. When the diving tetrapod re-emerges into air the heart rate increases, the smooth musculature of the pulmonary vessels relaxes, and normal circulation resumes, the carbon dioxide accumulated during the dive is then dissipated and the blood in the lungs is re-oxygenated. The reptilian circulation, therefore is not a transitional situation, but rather an adaptation to the environmental situation of most reptiles.

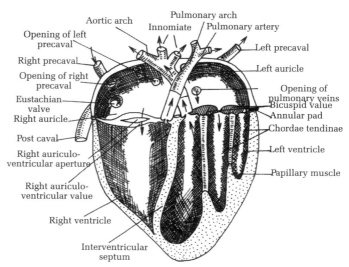

Figure 14.14 C
Dissected heart of pigeon to show internal structure

Birds and **mammals:** In mammals and birds, as already indicated there is a complete separation of both right and left atria and ventricles (Figs. 14.14 A, B, C and Figs. 14.15 A, B, C). The sinus venosus is not a separate entity but has become incorporated into the walls of the right atrium. Systemic blood thus enters directly into the right atrium, passes through the right atrio-ventricular valve into the right ventricle and is then pumped through the pulmonary artery to the lungs. When oxygenated in lungs it is returned through the pulmonary veins to the left atrium, ventricular valve, and pumped out of the left ventricle into the single aorta (left in mammals, right in birds) to all parts of the body except the lungs. This efficient, relatively simple, double circulatory pattern features complete separation of the two circuits. For diving mammals and birds, however, it is less efficient than the reptilian circulatory system.

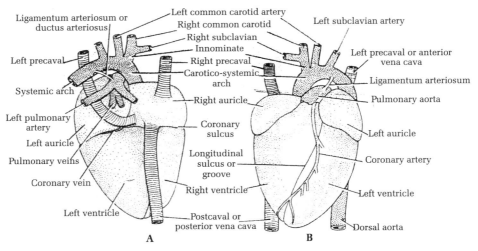

Figures 14.15 A, B
Rabbit. External features of heart   A. Dorsal view   B. Ventral view

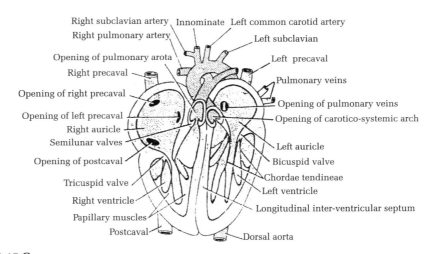

Figure 14.15 C
Rabbit: Internal structure of heart (Dissected heart)

Arterial System

The early development of the arterial system is more or less similar in all vertebrates. Extending anteriorly from the conus arteriosus just ventral to the pharynx, is the **ventral aorta** (Fig. 14.16 A). This vessel gives off paired aortic arches which course dorsally around the pharynx between the developing pharyngeal pouches and then join the paired **dorsal aorta**. Caudal to the pharynx the paired dorsal aorta meet, fuse and form a mid-line dorsal aorta running caudally just ventral to the notochord. Paired segmental arteries come off the dorsal aorta and distribute arterial blood to the somatic musculature paired viscera, and skin, unpaired mid-line arteries from the dorsal aorta travel through developing mesenteries and supply unpaired viscera. The head is supplied by paired anterior extensions of the ventral and dorsal aortae. This basic arterial system is modified in later stages of the development to provide the diversity of arterial patterns as seen in adult vertebrates.

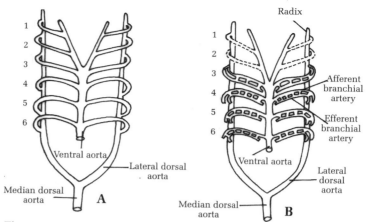

Figures 14.16 A, B
A. Typical plan of aortic arches (in vertebrate embryo)
B. Plan of aortic arches in bony fish

Development of Aortic Arches and Anterior Arteries

In all **gnathostomes** there are six pairs of aortic arches, which develop in relationship to the pharyngeal pouches and skeletal brachial arches, like the latter, these aortic arches are numbered one to six (Fig. 14.16 A). The adult fate of each aortic arch is closely related to the fate of the corresponding pharyngeal pouch. In amniotes the anterior aortic arches (one and two) are frequently lost before the posterior arches (five and six) even develop. In gill breathing gnathostomes the first aortic arch is lost. The more posterior aortic arches are interrupted in the gills by a capillary network. Each aortic arch which supplies gills is divided into dorsal and ventral portions, the ventral portion, which carries deoxygenated blood into the gill, is called the **afferent branchial artery,** the dorsal portion, which drains oxygenated blood from the gills, is called the **efferent branchial artery** (Fig. 14.16).

Chondrichthyeans: In chondrichthyeans the ventral portions of aortic arches two to six form five pairs of different branchial arteries travelling up to their respective pharyngeal arches, each one, except arch two, vascularizes its own holobranch (Fig. 14.17). The second aortic arch, giving rise to the most anterior afferent branchial artery, passes between the spiracle and the first gill slit. Since the spiracle has no respiratory membrane, this afferent branchial artery from the second arch supplies only the anterior demibranch of the first typical gill slit, which is derived from the third embryonic pharyngeal pouch.

Each gill slit has gill lamellae on either side, they are derived from the anterior and posterior demibranchs of two different gill arches. Around the gill slit, collecting from the capillaries of the gill lamellae, is a closed loop of **efferent** arteries. The anterior portion of each loop is called the Pre-trematic efferent artery, and the posterior, the post trematic efferent artery, together they are called the collector loop. Dorsally the entire loop is drained into the efferent branchial artery. The situation is complicated by so called cross-trunks or **anastomotic**

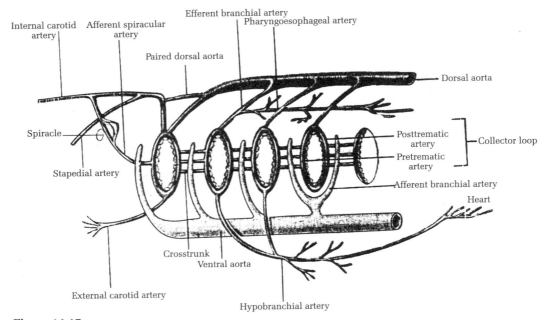

Figure 14.17
Lateral view of the branchial circulation of a dogfish

connections between the pre-trematic efferent branch of one loop and the post trematic efferent branch of the next. These collector loops give rise to four efferent branchial arteries from the dorsal portions of aortic arches three, four, five and six. These four arteries form descending mid-line dorsal aorta. From the middle of the first collector loop arises the afferent spiracular artery, which carries oxygenated blood to the spiracle. The external carotid artery arises from the ventral portion of the first collector loop and carries oxygenated blood to the dorsal portion of the head and to the brain. The hypobranchial artery comes off the ventral portions of the second and third collector loops and carries oxygenated blood to the hypobranchial musculature and to the muscular wall of the heart.

Teleosts: In adult teleosts there are only four functional afferent branchial arteries, which are formed from the ventral portions of aortic arches three to six, these four arteries bring deoxygenated blood to the gill lamellae (Fig. 14.18 A).

The efferent branchial arteries do not form collector loops; each efferent branchial artery collects from an entire, single holobranch and travels up to one of the pair of dorsal aortae as the dorsal portion of the aortic arch 3, 4, 5, or 6. The anterior extensions of the ventral

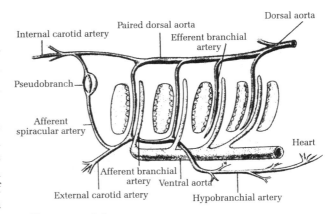

Figure 14.18 A
Lateral view branchial circulation of a teleost

aortae, which would carry deoxygenated blood, become separated from their more posterior parts and gain anastomotic connections with the ventral tips of third and fourth efferent branchial arteries, from which they receive oxygenated blood. They then extend anteriorly as the external carotid arteries, carrying oxygenated blood to the lower jaw. A separate branch of the external carotid, from the second aortic arch, also carrying oxygenated blood, goes to the pseudobranch (where the spiracle would be, if teleosts had a spiracle) and then continues dorsally and joins the anterior portion of the paired dorsal aortae. From this point, like internal carotid, it extends further anteriorly into the head and the brain. The efferent branchial arteries from the 3rd and 4th aortic arches, in addition to contributing to external carotid, also send a ventral branch to the hypobranchial and heart muscles.

Lungfish: The gills of the three genera of lungfish are best developed in the Australian *Neoceratodus*, less developed in the African *Protopterus*, and least developed in south American *Lepidosiren*. In *Protopterus*, aortic arches 2, 5 and 6 form capillary network in the gills. There are no gills for branchial arches 3 and 4, and their aortic arches shunt the blood directly from the ventral aorta to the paired dorsal aortae (Fig. 14.18 B). Anterior extension of the ventral aorta form the external carotid, and extensions of the paired dorsal aortae form the internal carotid arteries. A branch from the sixth efferent branchial artery goes to the lungs as the pulmonary artery (Figs. 14.18 B, C).

Amphibians: In larval amphibians, and in neotenic adults, aortic arches three, four and six are interrupted by capillary networks in the external gills, and aortic arches one, two and five are lost (Figs. 14.19 A, B). During metamorphosis the external gills are lost, and their aortic arches are reconstituted as complete vessels, going from the ventral aorta to the dorsal aorta.

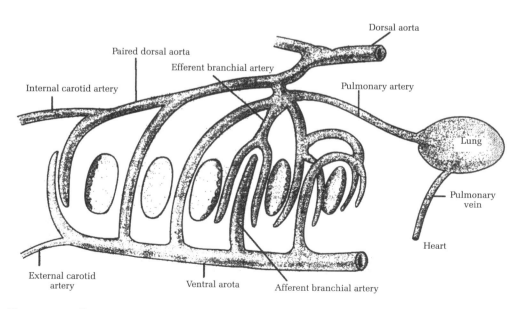

Figure 14.18 B
Lateral view branchial and pulmonary circulation of the lungfish (*Protopterus*)

In adult urodeles and apodans, anterior extensions of the ventral aorta form the external carotid artery, the internal carotid is formed by arch three and the anterior extension of the paired dorsal aorta. On each side, the portion of each paired dorsal aorta between arches three and four is called the **carotid duct**; it remains as an anastomosis between the dorsal aorta and the internal carotid artery (Fig. 14.19 A). Arches four and six persists as the systemic arch and the pulmonary arch, respectively, but both reach dorsally to join paired dorsal aortae which then unite posteriorly to form the descending aorta. The pulmonary artery to the lung comes off the pulmonary arch, as it does in the lungfish.

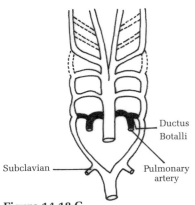

Figure 14.18 C
Modifications of aortic arches in lung fishes

In anurans there is a slightly different condition. The connection between the internal carotid artery and the paired dorsal aortae, formed by the carotid duct in other amphibians, is lost in anurans (Fig. 14.19 B). The portion of the ventral aorta between the third and fourth aortic arches forms the **common carotid** artery; the third aortic arch forms the proximal part of the internal carotid artery. The connection between the dorsal portion of the sixth aortic arch and the paired aortae is also lost; only

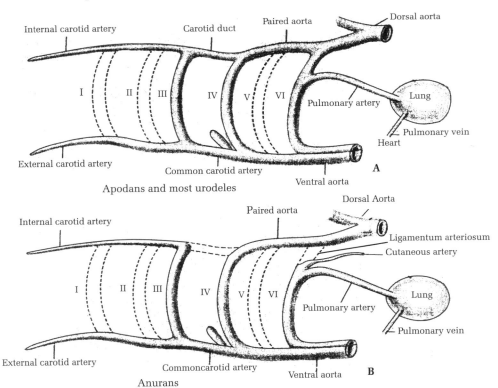

Figures 14.19 A, B
Derivatives of the aortic arches in adult amphibians. Roman numerals indicate aortic arch numbers and dotted lines indicate vessels present during development but lost in the adult in the adult

ventral portion of the sixth aortic arch and its posterior extension to the lung remain, forming the bases of the pulmonary artery.

Reptiles: The aortic arch system of reptiles is very similar to that of anuran amphibians, except in the conus arteriosus and ventral aorta which are divided into three separate vessels coming out of the ventricles (Fig. 14.20 A). One of the vessels derived from the sixth aortic arch, forms the two pulmonary arteries. The fifth aortic arch is lost. The fourth forms the systemic arch, except for the crocodilians, foramen of panizzae there is no connection between the right and left systemic arches. At the base of the right fourth aortic arch arises

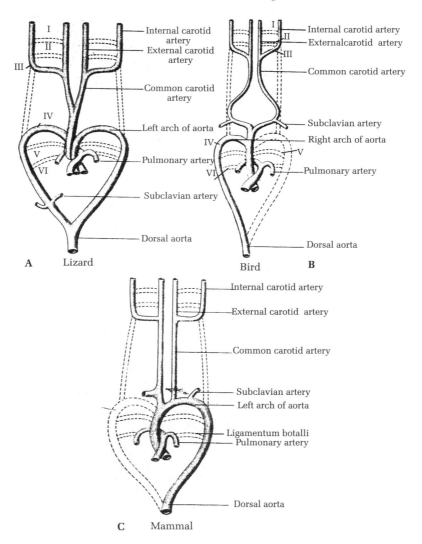

Figures 14.20 A, B, C
Ventral views of amniote derivatives of aortic arches. Dotted lines indicate vessels present during development but lost in the adults. Roman numerals indicate aortic arch numbers. Note connections of the subclavian arteries, which carry blood to the forelimbs. Development of the neck causes a large separation between arches three and four in lizards, birds and mammals

the common carotid artery, developing from the portion of the ventral aorta which runs between aortic arches three and four. The more anterior portions of the ventral aorta form the paired external carotid arteries. The third aortic arch forms the proximal portion of the internal carotid arteries, and extensions of the paired dorsal aortae form its distal portions. Aortic arches one and two, as well as five are lost in adult reptiles.

Mammals and birds: In mammals and birds the ventral aorta and the aortic arches develop in the same manner as those of reptiles, with exceptions only in the systemic arch (4th arch) (Figs. 14.20 B, C). In birds the left systemic arch is lost; all blood to the descending aorta passes through the right systemic arches and back to the descending aorta (Fig. 14.20 B). In mammals just the reverse is true; the right systemic arch is lost, and the left arch persists. Arch three and two when present forms part of the internal carotid circulation to the most of the head. The anterior extensions of the ventral aorta form the external carotid going to the lower jaw. Arch four (the systemic arch) is always retained, at least on one side; in tetrapods it carries most or all of the blood to the descending aorta (Fig. 14.20 C).

In tetrapods and sarcopterygians the ventral part of arch six forms the base of the pulmonary artery; the dorsal part of arch six, which would connect the pulmonary artery and the dorsal aorta, is lost as a blood vessel but remains as a ligament (ligamentum arteriosum) or **ligamentum botalli** (Fig. 14.20 C) in the adult anuran and amniotes.

Venous System

The embryo's first veins develop from blood islands in the yolk sac. These vitelline veins, enter the body and pass through the ventral mesentery of the gut to the sinus venosus. The liver developing within the ventral mesentery and transverse septum, surrounds these vitelline veins (Fig. 14.21). At about the same time the hepatic portal vein courses over from the gut and joins the left vitelline veins, its blood then enters the liver. Within the liver the vitelline veins anastomose with one another and form a complex of sinusoids, which are drained by the large hepatic veins. Blood coming in from either the vitelline veins or the newly developed hepatic portal vein passes through these sinusoids to the hepatic veins and then to the sinus venosus. As these developments occur, paired anterior cardinal veins form in the dorso-lateral head and drain caudally towards the heart, and paired posterior cardinal veins form in the dorsal wall and course anteriorly toward the head. At their posterior ends, the posterior cardinal veins join with one another and receive the caudal vein from the tail. At the level of the heart, the anterior and posterior cardinal veins come together to form the paired common cardinal veins or ducts of cuvier; which enter the sinus venosus laterally. Several subcardinal veins form around the developing kidneys and interconnect with the nearby posterior cardinal veins. Long veins, the lateral abdominal veins, form in the lateral body wall, run the length of the trunk, and empty into common cardinal veins; in addition to draining the lateral body wall these veins also receive blood from the pelvic appendages via the iliac veins and from the pectoral appendages via the subclavian veins. On each side, as the lateral abdominal vein is forming, the portion of the posterior cardinal veins just anterior to the level of the kidney is lost. Blood coming from the caudal region must hence forth pass into the posterior part of the posterior cardinal then into the subcardinal, and then through the capillary beds of the kidney. The subcardinal vein then drains anteriorly into the anterior part of the posterior cardinal vein, and from there blood is carried to the duct of cuvier and to the heart. This is a **renal portal** system, filtering blood from the tail through a kidney capillary system before sending it to the heart.

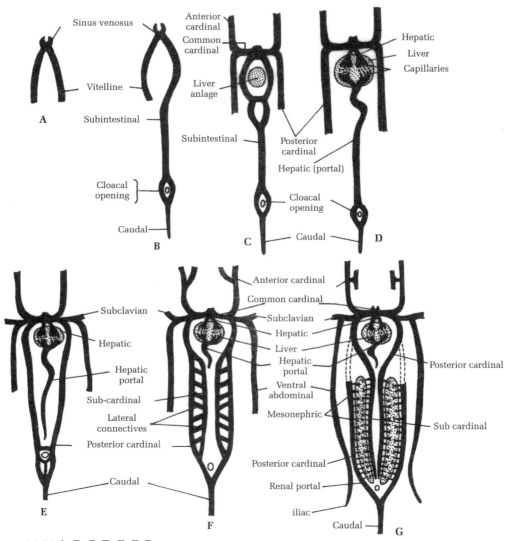

Figures 14.21 A, B, C, D, E, F, G
Development of venous system in elasmobranch (diagrammatic) **A.** Vitelline stage **B.** Subintestinal stage **C.** Comman cardinal stage **D.** Hepatic portal stage **E.** Subclavian stage **F.** Subcardinal stage **G.** Renal portal stage

Fish: In adult fish the venous drainage resembles the general developmental pattern, although with some specialization in various groups, as the yolk sac is reabsorbed, the vitelline veins to the liver are reduced, and their intramesenteric portions remain only as a small ligament in the ventral mesentry. Most teleosts lack lateral abdominal veins. The iliac veins drain

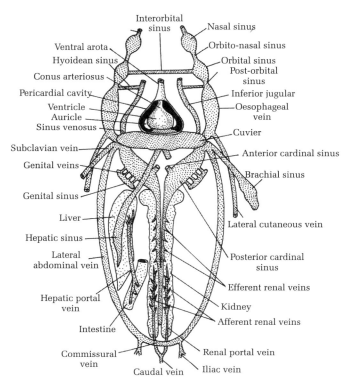

Figure 14.22
Scoliodon Venous system in ventral view (diagrammatic)

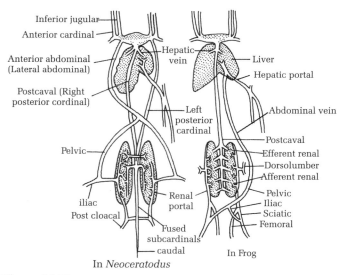

Figure 14.23
Formation of the anterior abdominal and postcaval viens in amphibians

into either the posterior cardinal or the renal portals, and the subclavian veins drain directly into the duct of cuvier.

In **chondrichthyeans** there are large sinuses in the blood vascular system particularly along the anterior and posterior cardinal veins (Fig. 14.22).

Amphibians: The largest vein returning blood to the amphibian heart is the postcaval (Fig. 14.23). It arises between the kidneys and passes anteriorly through the liver to the sinus venosus; it does not itself go through the capillary network, but from the hepatic vein it receives blood which has come through the liver sinusoids. The postcaval vein has two separate embryonic origins. Anteriorly this vein develops from a caudal evagination of the right hepatic vein, posteriorly it is a continuation of the right subcardinal vein.

The posterior and anterior parts of the lateral abdominal veins become separated. The posterior parts receive blood from the iliac veins, the right and left posterior parts then move together ventrally and form the mid-line ventral abdominal vein, which passes through the ventral mesentry of the alimentary canal and empties into the hepatic portal vein (Fig. 14.24). The anterior parts of the lateral abdominal veins remain as cutaneous veins, draining the skin and superficial muscles from the trunk region; and then, joined by the subclavian veins, drain into the ducts of cuvier. In **anurans** the posterior cardinal veins are lost, and the postcaval vein alone drains the kidneys and most of the posterior body; in urodeles although the posterior cardinal veins are retained, the post cardinal vein nevertheless carries most of the blood from the posterior body.

Reptiles: The reptilian venous system is much like that of the **anurans**. The postcaval vein is the primary vein of the posterior body, since the anterior portions of the posterior cardinal veins have completely disappeared. Since no posterior cardinal vein drain into the common and anterior cardinals. These last two vessels cannot be distinguished; the term precaval vein is used to denote them. The anterior cardinal vein distal to the precaval vein is called the jugular vein. Blood from the hind limb may pass to either the renal portal vein or the hepatic portal vein by way of the ventral abdominal veins (Fig. 14.25). The blood in the renal portal vein does not have to pass through the capillaries in the kidney, since there are many anastomoses between the renal portal and the postcaval vein.

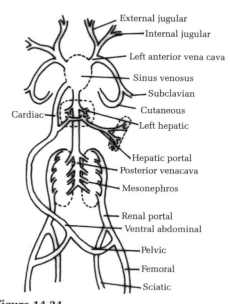

Figure 14.24
Venous system of a frog. (Left auricle and pulmonary vein not shown.) Both renal and hepatic poral system are shown

Birds and mammals: The venous system of a bird (Fig. 14.26) is nearly identical to that of the **reptiles** except that in birds the renal portal circulation is almost completely lost, and the iliac veins drain directly into the postcavals (although some branches pass through the kidney, perhaps with slight renal portal circulation). Because of this large connection between the iliac and postcaval veins, the ventral abdominal vein, here called the inferior mesenteric, is quite small and carries little blood.

The mammalian system (Fig. 14.27) is also similar to that of reptiles, the hepatic portal is basically the same, but with increased drainage by the postcaval. The abdominal vein is not present in adult mammals. The postcaval vein develops ontogenetically from several different vessels, including the right hepatic vein, the right subcardinal vein, the sub-supra cardinal anastomosis, the posterior part of the right supracardinal vein, and the posterior part of the posterior cardinal vein.

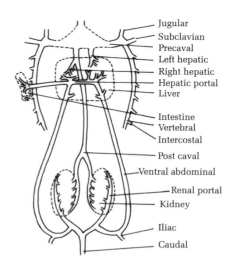

Figure 14.25
Venous system of a reptile

Portal System

Hepatic portal system Hepatic portal system can be noticed in vertebrates beginning with the **cyclostomes**. Blood from different parts of the alimentary canal and associated glands is drained by several branches which unite to form a single large **hepatic portal vein**. It runs to the liver lobes into which it breaks up into capillaries. From each liver lobe blood is collected by a large hepatic sinus which opens into the sinus venous.

Renal portal system The renal portal system is found for the first time in the chondrichthyes (Fig. 14.22). Blood from tail is collected by a median caudal vein. It runs forward ventral to caudal artery in the haemal canal of tail vertebrae. Dorsal to cloaca it bifurcates into right and left renal portal veins. Each of these continues forward dorsal to the kidney of its side giving off capillaries.

Significance of the portal system Portal system is a part of the venous system in which blood is collected from one set of organs and is conveyed to the capillaries of another organ, before it reaches the heart. This type of intervention of a portal organ plays a significant role in removing useful (food) and harmful (waste matter) metabolites from the blood circulation. There are three types of portal systems operating in a vertebrates body:

(a) **Hepatic-portal system** Here a hepatic portal vein returns blood from the intestine and breaks into a portal system of capillaries in the liver.

(b) **Hypophysial portal system** The venous blood collected from near the hypothalamus of the brain by a hypophysial portal vein which form a portal capillary system in the anterior part of the pituitary gland.

(c) **Renal portal system** The venous blood is collected from the legs and the hinder part of the body by renal portal vessels and passed on to the capillaries of the kidneys.

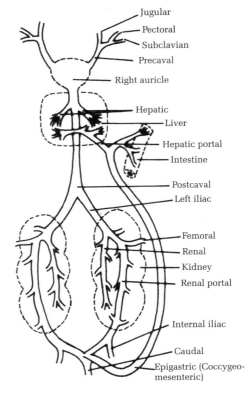

Figure 14.26
Venous System of bird, left auricle and pulmonary vein not shown

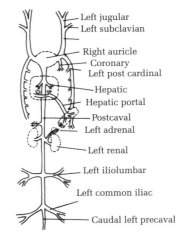

Figure 14.27
Venous system of lower mammals

Amphibians: In frog the veins which carry blood to a capillary system in kidneys constitute the **renal portal system**. Blood of each hind leg is collected by two veins, an outer femoral and an inner sciatic (Fig. 14.28). On entering the abdominal cavity the femoral divides into a dorsal renal portal and a ventral pelvic vein. The renal portal unites with the sciatic and while running along the outer border of the kidney of its side, it receives blood from the lumbar region by dorso-lumbar vein. Renal portal vein enters the kidney by several branches which break up into capillaries. The pelvic veins of both sides unite to form a median ventral or anterior abdominal vein. It receives blood from urinary bladder and ventral abdominal wall and runs forward to enter liver into which it breaks up into capillaries. Before entering liver the anterior abdominal and hepatic portal veins are connected by a small loop.

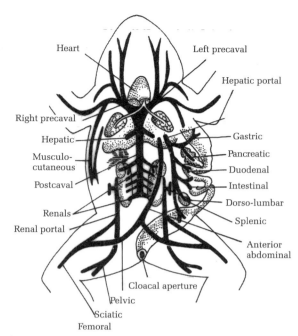

Figure 14.28
Frog: Venous system in ventral view and hepatic and renal portal vessels

A large **hepatic portal vein** is formed by the fusion of several branches from stomach, intestine, spleen and pancreas. It carries blood of alimentary canal, laden with digested foodstuffs, to the liver into which it breaks up into capillaries. Connected with hepatic portal vein, in the region of liver, is the anterior abdominal vein (Fig. 14.28).

Reptiles: In reptiles the blood from the tail is gathered by a median caudal vein. It runs ventral to the caudal artery through the haemal canal of the vertebrae. Reaching the kidneys, it divides into two afferent **renal** or **renal portal veins**, running forward over the ventral surface of the kidneys and partly buried in their matrix. They receive cloacal and rectal veins from the cloaca and rectum respectively (Fig. 14.29). The two renal portal veins break up into minute capillaries inside the kidneys.

The blood from each hind limb is collected by an internal iliac (or sciatic) and an external iliac (or femoral) vein. While entering the abdominal cavity, they unite to form a common iliac vein. It is connected to the renal portal vein of its side by a small transverse connection (the iliac – afferent union), near the middle of the kidney. Each common iliac is further continued antero-laterally as the pelvic vein. On the way, it receives parietal veins from the posterior muscles, vesicle vein from urinary bladder, epigastric and ischial veins from dorsal body wall and adipose veins from fat body. A fine median pubic vein joins both the pelvis. The two pelvics meet each other mid-ventrally in front of the kidneys to form an

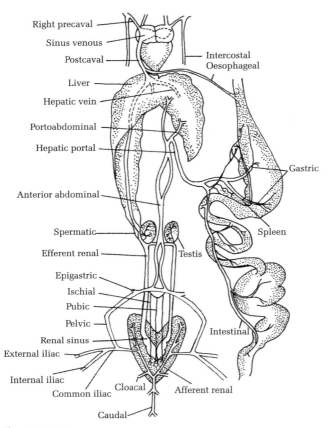

Figure 14.29
Uromastix venous system and hepatic portal and renal portal system

anterior abdominal vein. It becomes double at places which is a primitive character indicating imperfect union of two lateral abdominal veins found in fish. At the anterior end, the two components again become separated, one entering the left liver lobe directly, while the other joining the hepatic portal vein to form the porto-abdominal vein before entering the left liver lobe.

The blood from various regions of the digestive tract is carried to the liver by a large **hepatic portal vein** (Fig. 14.29). It is formed by the union of several branches such as anterior and posterior gastric from stomach, lienogastric from spleen and a part of stomach, duodenal from pylorus, anterior intestinal from ileum, and posterior intestinal from caecum, colon and rectum. Anteriorly, the hepatic portal vein unites with one branch of anterior abdominal vein to form the porto-abdominal vein which enters the left lobe of the liver to branch into capillaries. As already stated, the blood from the liver is collected by the hepatic veins which join the postcaval.

Functions of the portal system

a) The hepatic portal system plays an important role in following metabolic activities of the body.

   (i) As impure blood contains dissolved food material, the liver absorbs carbohydrates and convert it into glycogen and store it in the liver to be consumed later.

   (ii) In the liver, the carbolic acid, cresol, indol etc., are changed to harmless products.

   (iii) The amino acids get converted to urea.

   (iv) Ammonia is changed to Non-toxic product.

   (v) The hepatic cells absorbs – sugar, protein etc., which is later on used for providing energy.

b) The renal portal system plays a significant role in removal of urea, CO_2 etc., from blood as these harmful products get absorbed and eliminated in kidney and are passed to out side with the help of water.

c) The hypophysial portal system enables the hormones of hypothalamus to reach the anterior pituitary.

Lymphatic System

Development and Origin

Lymphatic vessels develop quite independently, in the later stages of the embryo, in comparison to the development of the vessels of blood vascular system. The connections of lymphatic vessels with the blood vessels are acquired later. Appearance of fluid filled spaces in the mesenchyme, is the first sign of formation of lymphatic vessel during development. These mesenchymal cells differentiate into flattened endothelial cells which later line the lymphatic vessels. Various lymph spaces fuse and branch in a complex manner, so that ultimately an elaborate system of anastomosing vessels is formed throughout the body.

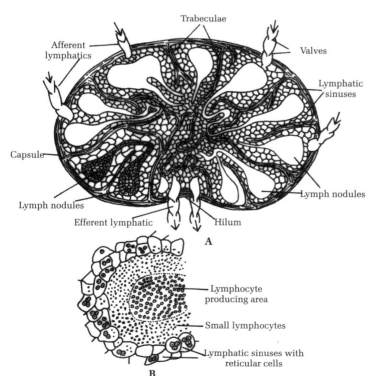

Figures 14.30 A, B
Diagrammatic section **A.** Lymph node in detail
B. A portion of nodule magnified

Initial lymphatic spaces appear near larger veins. Enlargement of these lymphatic network results in lymphatic sacs or sinuses in several regions. Larger lymphatic vessels usually course along with a vein and its corresponding artery. Smaller lymphatic vessels, however, do not have the tendency to unite to form a single vessel and several may be found close to a vein and its companion artery. Secondary connection of lymphatic vessels and veins takes place with some of the large veins near heart where blood pressure is at the lowest. The valves of the lymphatic system make their appearance considerably earlier than those of venous system. The first one to form are those in the larger lymphatic vessels in the vicinity of heart.

Lymphatic nodes (Figs. 14.30 A, B) and nodules are formed when connective tissue elements condense around lymphatic plexuses associated with strands of mesenchymal tissue. Their development begins only after the primary vessels of the lymphatic system have been formed. Lymphatic nodules in man usually develop after birth.

When the small lymphatic clefts and spaces first make their appearance, they are filled with tissue fluid. Some blood corpuscles may be present in these early vessels, probably coming from adjacent mesenchymal cells which are in the process of differentiating into blood forming tissue.

Structure The smallest components of a lymphatic system are **lymph capillaries**. Most of the tissue fluid called lymph diffuses through the endothelial walls into the lymphatic system. Although the lymphatic system is a closed system, it does not in itself form a complete circuit and in this respect differs from the blood vascular system. Lymph capillaries are tubules of somewhat greater diameter than blood capillaries. Also their diameter is not uniform throughout, no valves are present in the lymph capillaries and their walls are extremely thin, consisting of a single layer of flat, squamous endothelial cells. The blind ends of the tubules appear as small, distended knobs (**lacteals**).

Larger vessels formed by the union of lymph capillaries have thicker walls and contain valves. In still larger vessels, three layers comparable to those of small arteries and veins make up the walls. So in blood vessels, they are called **tunica intima, tunica media** and **tunica externa** or **adventitia.** These layers are generally not so distinct as those in the blood vessels. The valves which prevent back flow, are more numerous than venous valves and occur at close intervals. The walls of larger lymphatic vessels are supplied with tiny blood vessels similar to the vasa vasorum supplying the larger vessels of the blood vascular system. The nerve supply to lymphatic system is abundant. Before the lymph, in the lymphatic vessels of mammals, enters the blood vascular system, most of it filters through small, bean shaped structure known as lymph nodes. Phagocytic cells in the nodes remove and even destroy particles of various kinds as well as bacteria which may be carried via the lymph to the nodes. Two main types of cells are found within lymph nodes:

1. **Plasma cells** which help in production of antibodies.
2. **Lymphocytes** which enter the blood stream.

Lymphatic nodules are small, usually spherical, composed largely of lymphocytes and plasma cells, they may appear and disappear. Numerous lymphatic nodules are located within lymph nodes (Figs. 14.30 A, B). Small lymph capillaries of intestine are called lacteals. They are concerned with fat absorption. Several factors are instrumental in propelling the slowly moving lymph through lymphatic vessels and nodes. These include the muscular activity of various parts of the body tending to squeeze the fluid and pulsations of neighbouring arteries. Pressure builts up in the smaller vessels by osmosis and absorption of tissue fluid and, action of pulsating lymph hearts.

Lymph hearts (Fig. 14.31) consist of enlargement in lymphatic vessels which have contractile walls. They are generally situated near the point where lymph enters the venous system. The rhythm of the beating lymph hearts bear no relation to the beat of the heart itself.

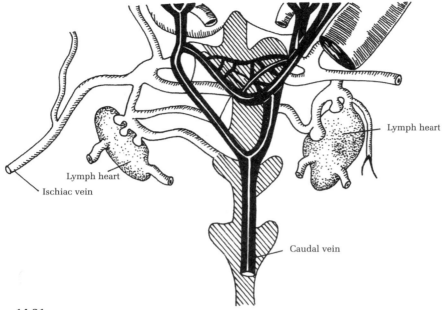

Figure 14.31
Position of lymph hearts in an alligator

Lymphatic system in various vertebrates

Fish: Cyclostomes and elasmobranchs do not have a defined lymphatic system. The lymphatic vessels of other fish are well developed (Fig. 14.32 A). The *eel* has lymph heart in tail. *European catfish* has two caudal lymph hearts. Lymph node seems to be lacking in fish.

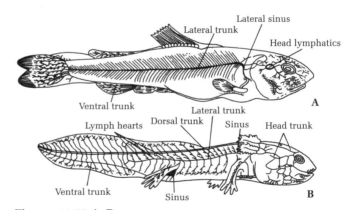

Figures 14.32 A, B
A. Lymphatics of a trout embyo B. Lymphatics of larva of salamandar

Amphibians: In urodele amphibians (Fig. 14.32 B), two main sets of lymphatic vessels are present and about fourteen to twenty hearts are present. Anurans are characterized by missing of large lymph sacs or spaces beneath the skin. Two pairs of lymph hearts are usually present in adult forms (Figs. 14.33 A, B). Lymph hearts are more numerous in larval and tadpole stages. Over 200 lymph hearts are present in caecilians.

Reptiles: Well developed lymphatic system is present in reptiles. In snakes the lymphatic vessels and sinuses are exceptionally large and numerous. A posterior pair of lymph heart is found in many reptiles (Fig. 14.31).

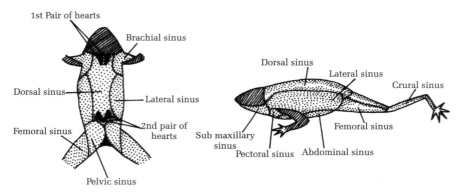

Figures 14.33 A, B
Frog   **A.** Lymph sinuses and Lymph heart (dorsal view)   **B.** Lymph sinus (Lateral view)

Birds: Transitory lymph heart have been observed during embryonic development. Bursa fabricii (Fig. 14.34) a lymphoid organ present in young birds is important in the production of lymphocytes.

Mammals: Lymph hearts are altogether lacking. In mammals lymph nodes are particularly abundant in superficial regions of head, neck, axillae and groin. **Peyers patches** are masses composed of numerous lymphatic nodules in the wall of small intestine. They are numerous especially in the region of ileum. **Lacteals** are small lymph capillaries in the villi of small intestine (Fig. 5.32). The food carried by the lymph, gives the latter a milky white appearance. The term **chyle** is used to refer to this milky fluid. Other lymphatic organs are tonsils, adenoids and thymus gland.

Spleen is considered to be a **hemolymphatic** organ, because it is interspersed in the blood stream rather than in lymphatic vessels. It means the blood rather than lymph, filters through spleen.

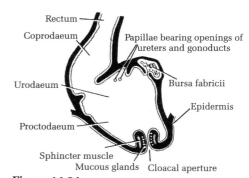

Figure 14.34
Lymphatics of birds

Chapter 15

Excretory System

Metabolic activities occurring in the animal body produce nitrogenous wastes, carbon dioxide, salt, water and pigments. These waste material become toxic if allowed to accumulate beyond a certain concentration. Hence all such organs which assist in removal of these waste products of metabolism from the body are called excretory organs. These organ help in regulating the volume, composition, pH and osmotic pressure of the body fluids by removing these harmful substances and conserving materials necessary for normal functioning by *osmoregulation*. Thus osmoregulation is imperative for a vertebrates survival because it maintains **homeostasis** by keeping composition of body's internal fluid environment constant and independent of external environment.

If extracellular fluid is too concentrated, osmotic pressure causes water to move out of the cells into the extracellular fluid and permeable ions to pass into the cells. This changes the cell's internal environment so drastically that they can no longer function and they die. On the other hand, if the extracellular fluid is too dilute, the osmotic pressure causes water to enter the cells until they literally blow up. The complex process of maintaining an optimum extracellular fluid composition involves the continual activity of lungs, gills, skin, some times salt-secreting glands, and most notably, the kidney to balance the input and output required by the animal's metabolic functions.

The primary nitrogenous waste produced by cell metabolism is ammonia, formed by the deamination of amino acids. Because ammonia is toxic, it must either be eliminated very rapidly, which requires large amount of water or be chemically transformed into something less toxic. Tetrapods, therefore, whose osmotic balance depends upon water conservation, chemically transform ammonia into less toxic substances, most frequently urea and uric acid. Animals which excrete ammonia are called ammonotelic and excretion of ammonia is termed as the ammonotelism where as excretion of urea is known as ureotelism and the animal termed as ureotelic rest all animals that excrete uric acid are called uricotelic and the excretion of uric acid is known as uricotelism.

The mechanism of osmoregulation varies with different environments. The body fluids of freshwater vertebrates, for instance, are more concentrated than their environment, and those of marine vertebrates are usually less concentrated. Osmotic pressure would tend to dilute the former and dessicate the latter, and active mechanisms are required to maintain proper balance. Terrestrial vertebrates, particularly those in very dry climates, must conserve water and replace the water that is lost through urine and faeces, during respiration, and through the surface of the skin, this is done by intake through the alimentary canal and by synthesising metabolic water (water synthesized as a result of metabolic reactions in the body).

Succession of Kidneys and Ducts

In all vertebrates the kidney develops embryologically from mesomeric tissue (of intermediate mesoderm), which lie in a retroperitoneal position along the dorsal edge of the developing coelom (Fig. 15.1). Like the vertebral column and somatic muscles, the developing kidneys are segmented, like many other structures, they develop in a cephalocaudal sequence. The most anterior nephrogenic tissue is the first to differentiate into kidney tubules, this portion is called the **pronephros**. The intermediate portion develops next as the **mesonephros**. The most posterior portion differentiate last, and in animals where only this portion persists in the adult it is called the **metanephros** (Fig. 15.2).

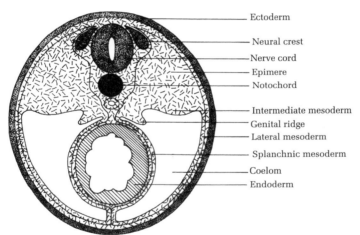

Figure 15.1
Cross-section through a developing vertebrate to show the relationship of the nephrogenic intermediate mesoderm to other tissues

It is in the kidney tubules that the urine is formed. In the pronephros of most vertebrate one kidney tubule develops per each body segment. These tubules join distally and form a large duct which grows caudally, adjacent to the undifferentiated posterior nephrogenic tissue, and then all the way to the cloaca where it terminates. Unfortunately, this primary urinary duct has been given many names: pronephric, holonephric, archinephric, opisthonephric and wolffian duct (Fig. 15.3).

In mesonephros and metanephros more tubules usually develop per segment even as many as several hundred for most vertebrate groups these tubules drain into the wolffian duct or, in amniotes, the separate, independently evolved urinary duct called **ureter** (Fig. 15.3).

Holonephros The kidney of the larval hagfish is structurally and functionally so generalized that it is thought to be very similar to a phylogenetically early kidney (Fig. 15.4). It is called the holonephros. It lies in a retroperitoneal position, but extends the entire length of the body. Not surprisingly it develops in a cephalocaudal sequence and has one kidney tubule per segment throughout its length. The tubules from the pronephros portion join and form the wolffian duct, which runs the entire length of the nephrogenic tissue and terminates at the cloaca. As the mesonephros and metanephros develop, their tubules also hook on to the wolffian duct.

Each holonephric tubules originate at the body cavity; its lumen is continuous with the coelomic fluid through an opening at its neck, called the **nephrostome**. The tubule cells in the nephrostome are *ciliated* and pump coelomic fluid into the tubule (Fig. 15.4). Blood comes to each holonephric kidney tubules in one of the pair of renal arteries from the dorsal aorta. This renal artery terminates in a capillary knot, called an **external glomerulus** against the serosa of the coelom. From the glomerulus, small vessels carry blood to the walls of the tubule where another capillary network forms. A renal vein drains the tubules capillary blood into the posterior cardinal veins, there is no renal portal system.

Blood pressure is very high in the external glomerulus, and its walls are thin. Much of the blood, with its burden of nitrogenous wastes is filtered out of the capillaries into the coelom, remaining in the blood capillaries are some plasma and the blood cells and larger proteins which cannot pass through this filter. Coelomic fluid is drawn through the nephrostome by the cilia in the neck of the tubule; as the coelomic fluid passes through the tubule some essential materials, such as glucose, are resorbed into the blood. The

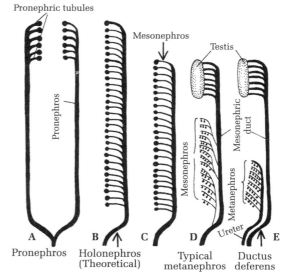

Figure 15.2
Development of kidney in vertebrates
A. Pronephros **B.** Holonephros **C.** Mesonephros
D. and **E.** Metanephros

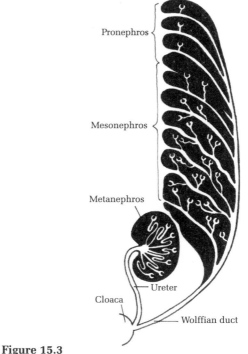

Figure 15.3
Diagrammatic presentation of pronephros, mesonephros and metanephros in vertebrates

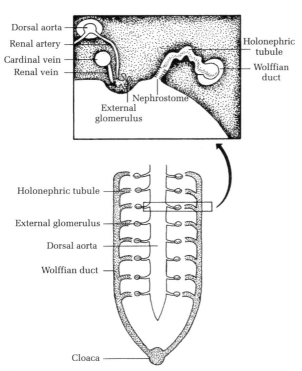

Figure 15.4
A holonephric kidney (hagfish)

urine then passes out of the holonephric tubules into the wolffian duct and is carried down to the cloaca for excretion.

Because the tubules are short, little water can be resorbed through them and the urine is copious. There is some active transport of salts; the urine is, therefore, slightly hypotonic to the body tissues, which is necessary in order to keep the animal body fluids slightly hypertonic to sea water. Hagfish retains higher salt concentration in their tissues than do other vertebrates, the adult does it by an opisthonephric kidney.

Opisthonephros In anamniote embryos (other than hagfish) the most anterior mesomeric tissue forms a pronephros, which lays down the wolffian duct as described and which may also function as the embryonic kidney. In the anamniote adult the pronephric tubules may or may not be resorbed, but in any case these tubules have no excretory function. The adult anamniote kidney is the opisthonephros, which develops from the intermediate and posterior portions of the mesomeric tissue (Fig. 15.5). Unlike the pronephros it has several tubules per segment, the number increasing caudally and each tubule is longer and more convoluted than the relatively short straight pronephric tubules.

In the opisthonephric kidney each renal artery (or rather, each small branch of the renal artery) terminates in a tuft of capillaries which is called an **internal glomerulus** because it lies within a double walled cup shaped termination of the kidney tubule called Bowman's capsule (Fig. 15.6).

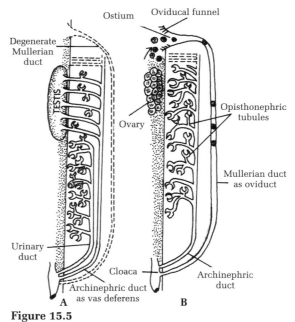

Figure 15.5
A. Kidney and genital duct of male anamniota
B. Kidney and genital duct of female anamniota

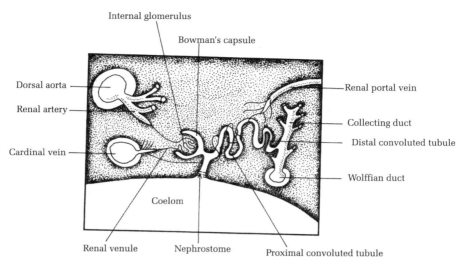

Figure 15.6
A single opisthonephric tubule and its vascularization (Many opisthonephric kidneys lack nephrostomes)

The **internal glomerulus** plus Bowman's capsule comprise the *renal corpuscle*. Electron microscopy reveals that the cells of the inner wall of Bowman's capsule called *podocytes*, contain small spaces through which the filtrate from the glomerulus can pass into the kidney tubule sometimes but not always there is a nephrostome connecting the kidney tubules with the coelomic cavity (Fig. 15.6). Each opisthonephric tubule is composed of a Bowman's capsule, a short and relatively straight neck region, a convoluted region called – the proximal convoluted tubules. Another short, straight segment and a distal convoluted tubule. Several distal convoluted tubules join in a collecting duct, which in most anamniotes empties into the wolffian duct but in some empties into a different duct, formed by the coalescence of collecting ducts by one or other of these ducts the urine is emptied into the cloaca.

The internal glomerulus receives blood from a small branch of the renal artery called the afferent arteriole and is drained by the efferent arteriole, which then forms a capillary network wrapping around the proximal and distal convoluted tubules and the intervening small, straight segment. This capillary network around the tubules also receives blood from the renal portal vein, which drains the hind limbs and tail. However, the renal portal vein supplies no blood to the glomerulus. The capillaries from both the efferent arteriole and the renal portal system are drained by renal venules into renal veins which then drain into either the postcardinal or postcaval veins, depending on the anamniote.

At the glomerulus the blood is filtered into the kidney tubule, except for some plasma, the blood cells, and the largest protein molecules. The filtrate passes down the kidney tubule, and throughout this passage there is selective resorption of salts and sugars and selective secretion of additional waste products, particularly of urea and ammonia, there is very little resorption of water. Thus the final product, urine, has a distinctly different composition than the filtrate at Bowman's capsule.

Metanephros In amniotes a pronephros develops very early and lays down a wolffian duct; it is never a functional kidney, however. The mesonephros develops next from the intermediate mesomeric tissue; the mesonephros has a tubule structure similar to that of the anamniote opisthonephros and is drained by the wolffian duct. This is the functional kidney of the embryo and early foetus.

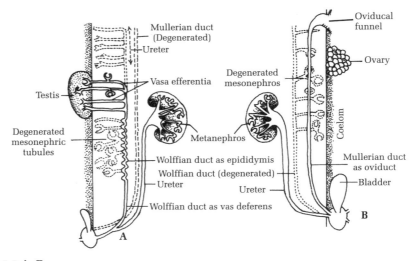

Figures 15.7 A, B
A. Kidney and genital duct of male ammniota   B. Kidney and genital duct of female ammniota

The kidney of the adult amniote, however, is the metanephros (Fig. 15.7). It develops from the most posterior portion of the mesomere; with many more kidney tubules than in any other kidney type. Each metanephric kidney tubule like each opisthonephric tubule, is made up of Bowman's capsule, a very short neck region, a proximal convoluted tubule, an intermediate region, and a distal convoluted tubule; many distal convoluted tubule empty into a single collecting duct.

However, the collecting ducts of the metanephros do not enter the wolffian duct. Instead, at the junction of the wolffian duct and the cloaca an evagination grows anteriorly towards the metanephros, forming the **ureter**; it enters the metanephric tissues and divides into large funnel like structures called calyces (Fig. 15.8). It is into these calyces that the

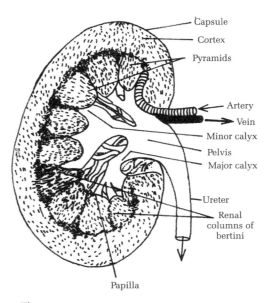

Figure 15.8
Frontal section of human kidney

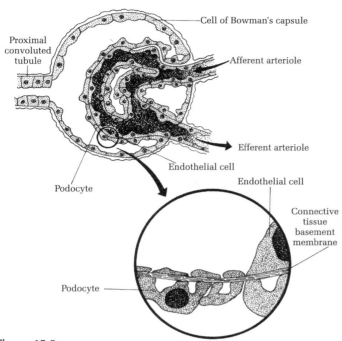

Figure 15.9
Detailed structure of Bowman's capsule and glomerulus

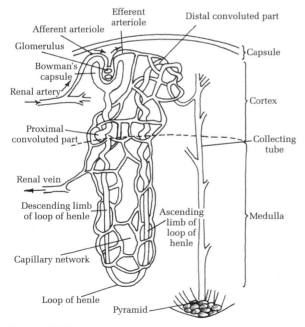

Figure 15.10 A
Details of Vascularization of a single opisthonephric tubule

collecting ducts of the kidney empty. Therefore the metanephric kidney has a dual origin- partially from the posterior nephrogenic tissue and partially from the budding of the cloaca.

There are also functional similarities between the metanephros and the opisthonephros. In the renal corpuscle, only a thin connective tissue membrane separates the glomerular capillaries from Bowman's capsule; here blood plasma is filtered through fenestrations in the capillary endothelium and spaces between the podocytes of Bowman's capsule. This efficient mechanism filters about 90 per cent of the blood plasma into the cavity of Bowman's capsule (Fig. 15.9), from which it passes through the neck region into the proximal convoluted tubule. The proximal convoluted tubule is made up of truncated, pyramid shaped cells (Figs. 15.10 A, B) with strong brush border of microvilli adjacent to the lumen and large interdigitations in the plasma membrane adjacent to the thin basement membrane. These cells are metabolically very active and are responsible for selective reabsorption of glucose, amino acids, and various salts from the filtrate. At the same time additional nitrogenous wastes are secreted from the blood stream into the lumen of the proximal convoluted tubule. The intermediate segment, of variable length, has a low cuboidal to squamous epithelium and a primary site of water resorption (Fig. 15.10 B). In the distal convoluted tubules, where the cells are somewhat flattened and their plasma membranes have fewer convolutions, the final formation of urine takes place, some more water is resorbed, the pH is adjusted and there is selective sodium resorption. The kidney activity is adjustable over a wide range. If, for instance, the body is flooded with sodium, very little is excreted. Furthermore, within large tolerances, the amounts of wastes and water that are excreted are independent. When there has been a large water intake the urine simply becomes more dilute, and when there has been a little water intake, it becomes much more concentrated.

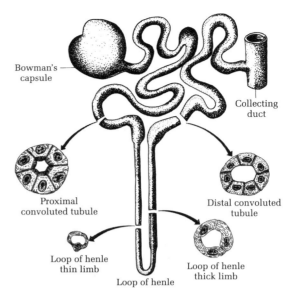

Figure 15.10 B
Details of mammalian metanephric uriniferous tubule

The mechanism of this control are not fully understood, but some elements are clear. Myoepithelial cells form a **juxta-glomerular apparatus** which enfolds the afferent arteriole just before it enters the glomerulus. Contraction of this apparatus greatly diminishes the amount of blood flowing through the glomerulus and thus the amount of filtration that can occur. It is also thought that these myoepithelial cells secrete the enzyme renin which, it will be recalled, can transform the angio-tension I in the blood to angio-tension II, which in turn stimulates further vasoconstriction.

One portion of the distal convoluted tubule, the **muscula densa**, has specialized slender, columnar cells which apparently respond to the osmolarity of the urine and pass this information to the juxtaglomerular cells, thus influencing whether or not renin is secreted. The mechanism of such a feedback system are unclear, and in fact the whole idea, although consistent with current experimental evidence, is still speculative.

The intermediate segment of the amniote kidney tubule is relatively short in the most reptiles, long and somewhat looped in mammals where it is called the loop of Henle (Fig. 15.10 B). Its function is best understood in mammals. The mammalian renal corpuscles, proximal convoluted tubules, and distal convoluted tubules are in the outer portion, or cortex, of the kidney, the collecting ducts, calyces, and loops of Henle are in the deeper part or medulla. The loop of Henle pass from the proximal convoluted tubule down parallel with the collecting ducts and finally loop back, travel up, and join the distal convoluted tubules. Because the blood in the collecting ducts and the urine formed in the loops of Henle flow in opposite directions there is a counter current distribution which facilitates and accelerates the absorption of water from the urine further influenced by the fact that the interstitial fluid in the medulla is hypertonic to that of other parts of the body. The longer the loop of Henle, the more opportunity there is for water resorption to occur. They reach their extreme size in some desert mammals where they pass far into the ureter before looping back up. Thus the loop of Henle plays an extremely important role in the selective resorption of water, in mammals at least, this occurs by osmosis rather than by active transport; so far as is known there is no means for the active transport of water in vertebrates.

The intermediate segments also tend to be longer in reptiles and birds inhabiting drier climate than in those for whom water is abundant; thus suggesting that the water conserving role of this part of the kidney is not limited to mammals. However, since water is resorbed in the avian and reptilian cloaca, they had less selective pressure for the development of long loops of Henle than mammals.

Maintenance of water balance: In vertebrates certain organs other than the kidney are involved in osmoregulation. It is therefore appropriate to look at the functional anatomy of the entire water balance system, particularly as it relates to the environmental stresses.

Fresh water fish Fresh water fish have an internal osmotic pressure greater than that of their environment, that is, they are hypertonic. This being so, there would be a tendency for water to diffuse into them, and in absence of some control, their water balance could not be maintained. Though the mucous layer over the skin renders it nearly impermeable to water. However, the alimentary canal and the gills must remain permeable to water, and through these structures water constantly diffuses into the animal. With so much water coming in, it is necessary to produce a copious dilute urine. The opisthonephric kidneys of freshwater fish have large glomeruli for this purpose and short tubules which resorb little water. With so much water being excreted so rapidly, ammonia does not need to be transformed but can be excreted before it can build up to a toxic level. In most fresh water fish, the chloride cells of the gills actively secrete sodium, and thus play a minor role in osmoregulation.

Marine vertebrates All vertebrates in a salt water environment (except hagfish) have the opposite metabolic problem. Their body fluids are hypotonic to sea water, and there is a tendency for water to diffuse out, thus drying up the animal. Moreover, sodium diffuses in through the gills and alimentary canal, thus potentially increasing the toxicity of the body fluids and putting an undue osmotic strain on the cells. It is interesting that every vertebrate class has at least a few marine members, which have evolved somewhat different mechanisms to maintain metabolic balance. All these mechanisms have two characteristics in common.

Firstly, all involve relatively impermeable body surface which reduce the amount of water diffusing out of the body; in salt water fish it is most frequently the same sort of mucous layer that protects freshwater fish, and in marine tetrapods it is a very dense dermis.

Secondly, no marine form excretes ammonia, the necessary large volume of water would further deplete the animals internal water reserves.

The hagfish, alone among marine vertebrates, has salt concentrations greater than sea water and an osmotic gradient similar to that of fresh water fish. Thus it also has similar water balance problems, and it is not surprising that it has large glomeruli and a copious urine excretion. Precisely how the hagfish maintain other metabolic activities is not understood, in most vertebrates, for instance, nerves could not conduct action potentials in the presence of so much sodium.

All but a few species of chondrichthyeans live in salt water. They maintain osmotic equilibrium in a complex way. Their high protein diet yields a large amount of ammonia, which is rapidly converted to urea in the liver. The urea is then maintained in the body tissues and blood at concentrations which in most vertebrates would cause uremia and death, making the body quite hypertonic and causing water to diffuse into the body tissues through the gills and alimentary canal. Not surprisingly, chondrichthyeans have a very large renal corpuscle and produce a copious and dilute urine; water is thus excreted as rapidly as it enters. Excess salts is actively excreted by the *rectal* or *digitiform gland*, which produces a concentrated sodium chloride solution and excrete it into the rectum from which it passes out through the cloaca. It is interesting that the few fresh water chondrichthyeans also retain high concentrations of urea compared to other freshwater forms. This increase their osmotic pressure, causing more water to come into the animals, therefore, puts a greater burden on their kidney. Marine coelocanth (crossopterygian) *Latimeria* indicate that it also retains a high urea content; evidently its osmoregularity mechanisms are similar to those of chondrichthyeans. Marine teleosts also convert most of their ammonia into urea, but the urea is then rapidly excreted by the kidneys and perhaps even by the gills. Their kidneys are smaller than those of chondrichthyeans with very small glomeruli; or in some forms, none at all. This minimizes filtration. The urine is formed by the direct secretions from the capillaries of the renal arteries and renal portal veins which surround the kidney tubules, and is very concentrated, being primarily urea with a small amount of water. Like all marine forms, marine teleosts must eliminate excess salt. Their chloride cells are more profuse than those of freshwater teleosts and actively excrete a hypertonic sodium chloride solution.

Amphibians: When not on land, most amphibians inhabit fresh water, and unlike fish their skin is highly permeable to water. Their tissue is hypertonic, and thus water tends to flood into the body; however, on land they tend to lose a great deal of water by evaporation from the surface. Their kidneys have large glomeruli and produce a copious, dilute urine, which is not passed directly to the outside; but after entering the cloaca it backs up into a large sac, the urinary bladder, where it may be stored for a considerable period before being removed (Figs. 15.13 A, B). Thus when the animals are on land they can offset the effects of evaporation by resorbing water from their urinary bladder. Toads, which are more terrestrial than most amphibians, have a skin which is denser, less-moist, and much less permeable to water; their kidney tubules are larger and their glomeruli smaller, producing a more concentrated urine with less water loss. One of the very few marine amphibians is a crab-eating frog of South eastern Asia whose internal milieu is slightly hypertonic to the sea. This hypertonicity is maintained by retaining slightly higher sodium chloride concentration than do most vertebrates rather than by retaining urea as chondrichthyeans and the coelocanth (*Latimeria*) do.

Reptiles and birds: The more completely terrestrial vertebrates-reptiles, birds and mammals have metanephric kidneys, which have a greater capacity for concentrating urine and thus for reducing water loss. Reptiles (except fresh water turtles) and all birds excrete their nitrogenous wastes primarily as uric acid. Because uric acid is almost insoluble in water it does not add to the osmotic pressure. And thus it helps in water conservation. Reptiles have reduced number and sizes of glomeruli (not unlike marine teleosts) and most of the urine is formed by direct secretion of wastes but very little water from the blood into the kidney tubules. The reptilian intermediate segment is very short, of the small amount of water in the wastes very little is resorbed, for there must be some water to carry the uric acid down the ureter to the cloaca or else it would "Stop up the excretory plumbing". Thus cloaca itself is the major water-resorption organ in the reptiles. Their urine and faecal material are mixed, most of the water is reabsorbed, and a nearly dry, pasty mixture is extruded. Birds have larger glomeruli, so that more filtration occurs, but also a larger intermediate segment which resorbs much of the water; some water is also resorbed in the distal convoluted tubule. As in reptiles, the remaining water and metabolic wastes, primarily uric acid, are passed down, to the cloaca and mixed with faecal material, water is resorbed and the pasty residue is extruded from the cloaca.

Mammals: The metanephric kidney is not only an excretory organ. It acts as a homeostatic organ by regulating body fluids and thus stabilizing the internal milieu of the animal body. It is quite common, if the animal drinks a great deal of water, the excess water escape the body through urine which becomes copious or dilute. On the other and if body is deficient of water, the water is reclaimed from urine there by making it concentrated. The kidneys play an important role in maintaining osmotic pressure of the blood and tissue fluids.

Studies on quite a few mammals have shown that 1% glucose, amino acids and urea are reabsorbed in the proximal convoluted tubule. About 99% water get reabsorbed in the nephron including loop of Henle. The highly concentrated urine contain water, urea, uric acid creatinine and ammonia etc. The yellow colour of the urine is due to pigment *urochrome* which is formed from the break down of RBC.

Urinary bladder

The bladder is especially important in tetrapods living in an arid habitat, where water conservation has survival value. Hence adaptive value of the tetrapod urinary bladder seems to be its capacity to store water that may be required later. The bladder develops from the extra embryonic membrane and from the allantois also. However most vertebrates have a urinary bladder. The bladder is absent in cyclostomes and elasmobranchs among fish, snakes, crocodile, some lizard and birds other than ostrich. There are three general types of urinary sacs or bladders, namely tubal, cloacal and allantoic.

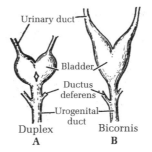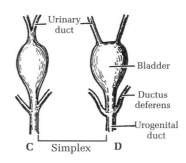

Figures 15.11 A, B, C, D
Types of urinary bladders in fish **A.** Duplex in *Gadus*; **B.** Bicornis in *Lepidosteus*; **C** and **D.** simplex with united urinary ducts

(i) The **tubal bladders**, which are present in most ganoid fish, are formed by the widening or enlargement of urinary ducts (Fig. 15.12). In many fish two independent bladders may form **vesica duplex** (Fig. 15.11 A) one near the end of each urinary duct, with the two ducts afterwards uniting into common passage way of exit; or the two may run together into a common bilobed bladder **vesica bicornis** (Fig. 15.11 B), as in *Lepidosteus* and some other ganoids, or finally the two excretory duct may first fuse and then expand into a single bladder, **vesica simplex** (Fig. 15.11 C) as for example in *Pike*, Esox.

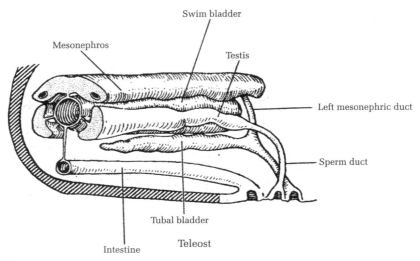

Figure 15.12
The unpaired urinary bladder arises as a bud off the conjoined caudal ends of the two mesonephric ducts (male teleost *Pike*) lateral view

(ii) The **cloacal bladder** occurs in dipnoans, amphibians (Fig. 15.13 B) and monotremes. It is a diverticulum of the cloacal wall opposite the point where the urinary ducts, with which it has no direct connection, enter. It is located dorsally in lungfish and ventrally in amphibians. In perennibranchiate amphibians it is considered elongated, but rounded and broadened in frogs and toads. Frequently, it is bilobed while in some urodeles, for example *Salamandra*, *Triton* and *Eurycea*, the lobes are prolonged into horn like process.

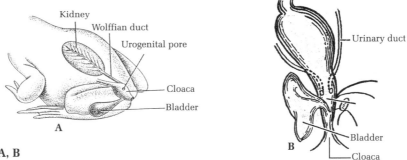

Figures 15.13 A, B
A. Urinary bladder of frog   B. Cloacal bladder of an amphibian

(iii) The **allantoic bladder** arises form the enlargement of the proximal or basal end of the embryonic allantoic stalk. It is characteristic of mammals and of such reptiles as turtles and certain lizard, that have a bladder. In case of other amniotes, like snakes, crocodiles, some lizards and birds the whole allantois degenerate, without developing into a bladder.

In mammals, at the time of birth when umbilical cord connection with placenta is severed, a portion of allantoic stalk is left within the body. Probably, it is this part of the allanotoic stalk which later enlarges to form bladder, and also the **urachus**, or vesico umbilical ligament which anchors the bladder to the inner body wall at the umbilicus. Thus the proximal end of the allantois stalk enlarges into hollow sac of the bladder while the distal part within the body wall undergoes quite a different fate in being transformed into a solid ligament utilized as a rope to support the bladder.

According to Arey, allantois contributes nothing to the bladder or urachus. The mammalian bladder is a derivative of the embryonic cloaca. Both allantoic and cloacal bladder may be attached to the ventral body wall by a remnant of the ventral mesentery, the ventral ligament of the bladder.

However most vertebrates have a urinary bladder. The bladder is absent in cyclostomes and elasmobranchs among fish, snakes, crocodilian, some lizards and birds other then ostriches.

Fish: Most of the fish have the terminal enlargements or evaginations of the mesonephric duct which are known as tubal bladders. The urinary sinus of male sharks and urinary sinus of female both of which have neither urogenital nor urinary papilla are the closest structure to a urinary bladder in shark. In dipnoans a small diverticulum of the dorsal wall of the cloaca is called urinary bladder (Fig. 15.12).

Amphibians: In amphibians and all remaining vertebrates the bladders develop as evaginations from the ventral wall of the cloaca (Fig. 15.13 A). The urine gets collected in cloaca and then backs up into the bladder, where it is stored.

Reptiles: Turtles and some lizards have large bladders and some fresh water turtles have two accessory bladders. The urine back up into the bladder from the cloaca. In female turtles the water stored in bladder is used by female for moistening the soil when building nest for the eggs.

Birds: There is no urinary bladder in bird. The lack of bladder reduces body weight and help in flight. However *ostrich* has a urinary bladder.

Mammals: In mammals the kidney ducts empty directly into the bladder, and the bladder is drained by urethra (Fig. 15.14). The adaptive value of the tetropod bladder seems to be its capacity to store water that may be needed later. This water may be reabsorbed from the bladder. An endocrine gland called pituitary play an important role in controlling the reabsorption of water form the bladder by releasing a hormone called ADH (Antidiuretic hormones), it causes active resorption of water from bladder and the animal ceases voiding of urine.

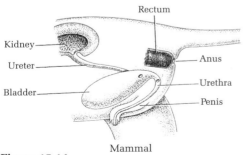

Figure 15.14
Urinary bladder of mammal

In man urinary bladder lies in the pelvic cavity and is bean shaped, muscular saclike structure. Its inner lining is composed of transitional epithelium. The muscular layers of urinary bladder is called *detrusor* muscle. The urinary bladder has a triangular area the *trigone* between the three openings. Two openings through which the ureter enters the bladder and one opening through which the urethra leaves the bladder. The urinary bladder stores urine temporarily.

Chapter 16

Reproductive System

While adaptations that evolved in other systems ensured the survival of the individual, those of reproductive system ensured the survival of the species. The ability of an animal to produce viable offspring is linked to the evolutionary development of an animal. The morphological units of reproduction are the sperm and egg cell, collectively called gametes. They require an organ the gonad, in which they are developed and stored until released. Male and female gametes must be effectively brought together for fertilization. Finally enough fertilized eggs must develop into adults and reproduce themselves for the perpetuation of the species.

Gonads

Gonads are compound glands which in addition to developing and storing the germ cells, secrete reproductive hormones. They develop from paired masses of mesodermal tissues on either side of the developing dorsal mesentery. However, unlike the developing kidney, the gonads show no segmentation.

Proliferation of this tissue forms the genital ridge which bulges into the developing coelom (Fig. 16.1). The genital ridge is lined with well-defined germinal epithelium which eventually will form the germ cells, however, the germinal

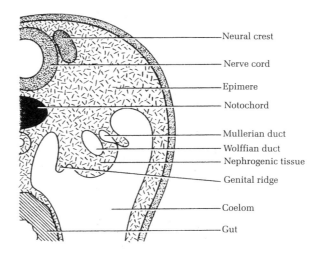

Figure 16.1
Formation of mullerian duct from the lining of the coelom adjacent (except chondrichthyes and some amphibians)

epithelium proliferates as cordlike buds that grow out into the rest of the genital ridge. These cordlike masses are called the sex-cords of the embryonic gonad (Fig. 16.2).

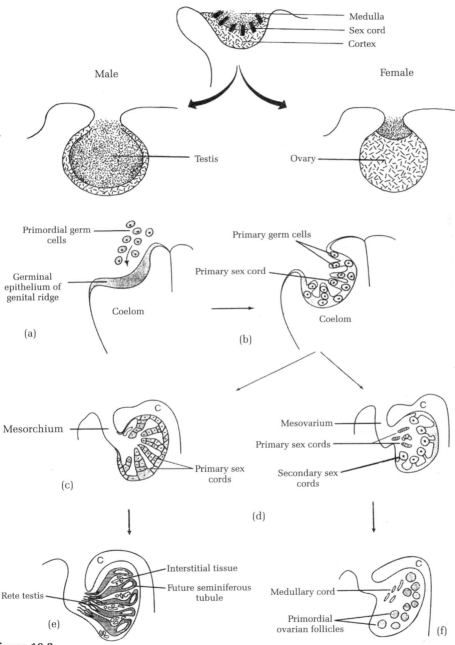

Figure 16.2
Development of mammalian gonads. Different stages of formation of male gonads (Testis) and female gonads (Ovary)

Indifferent or sexless gonad

The indifferent (sexless) gonad gradually takes on a more definitive shape developing an outer cortex and an inner medulla. From there on the presumptive ovary and presumptive testis develop differently. In ovary the outer cortical portion differentiate and in the testis the inner medullary tissue develops and the outer cortical tissue is resorbed (Fig. 16.2). In most vertebrates gonads are symmetrical and paired.

Testis

Although the gross-shape of the testis varies considerably among vertebrates, its internal structure is remarkably constant (Figs. 16.3 A, B). There is an outer connective tissue capsule, tunica albuginea which also contains smooth muscles. In most vertebrates extension of this layer deep into testis, divides it into a number of compartments. Within each compartment are seminiferous tubule containing sperm producing cells, and between the tubules are blood vessels, nerves, loose-connective tissue, and the interstitial cells of Leydig—which produce male hormone—testosterone. A seminiferous tubule is a long and coiled and has narrow lumen. Sperms are produced along its entire length. Its outer portion is formed by a basement membrane, upon which rests epithelial cells and large nurse cells or sertoli cells. Sertoli cells do not themselves produce sperms, but facilitate sperm production from germinal epithelial cells which undergo spermatogenesis. Spermatogenic cells called spermatogonia lie on the basement membrane between sertoli cells. During wave of spermatogenesis the spermatogonia undergo mitotic divisions into primary spermatocyte through meiosis which is a two step process; the first cells produced by meiosis are the secondary spermatocytes; the second cells, which are haploid, are the spermatids. Spermatids mature in a process called spermiogenesis, which involves no further cellular division but their differentiation into spermatozoans.

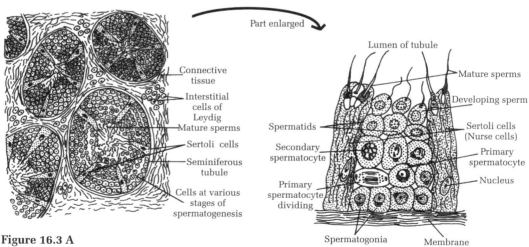

Figure 16.3 A
T.S of mammalian tests showing several cross sections of seminiferous tubules, leydig cells, and connective tissues septum dividing compartments

Figure 16.3 B
Single seminiferous tubule (magnified)

In many anamniotes-seminiferous tubules are shaped like cysts and are empty except for the period just before and during spawning. At the breeding time, the spermatogonia just outside these cysts become mitotically active and form many pre-spermatocytes which migrate into the cysts. There they undergo meiotic division into secondary spermatocytes and spermatids, which later differentiates into spermatozoans. Sertoli cells in the walls of the cysts function similar to those of amniotes. In most vertebrate groups the seminiferous tubule terminate in modified mesonephric tubule called efferent ductules, with one ductule draining many seminiferous tubules. These efferent ductules carry sperms out of the testis.

Dispersal of Sperms

In **cyclostomes** and **salmonidae** sperms are released from testis directly into the body cavity. They travel in the coelomic fluid to the posterior part of the cavity and exit the body by a small genital pore leading to the urogenital sinus. In teleosts except salmonidae the sperms travel from the testis to the urogenital sinus in a sperm duct which develops from the testis and its mesentery, the mesorchium.

Position In amniotes and in some anamniotes testis do not remain in their original embryonic position in the anterior dorsal wall of the body cavity but migrate caudally. In monotremes whales and sirenia (e.g. Dugong) they are retained within the body cavity throughout their life, but in other mammals testes lie, for at least some of the time in an extra abdominal extension of the body cavity—the scrotum (Fig. 16.4). Testes may descend into the scrotum only during periods of sexual activity (as in some marsupials most rodents, insectivores bats and some ungulates such as camels and some primates. Testes may also descend into the scrotum during late foetal life and remain there permanently (e.g. some marsupials

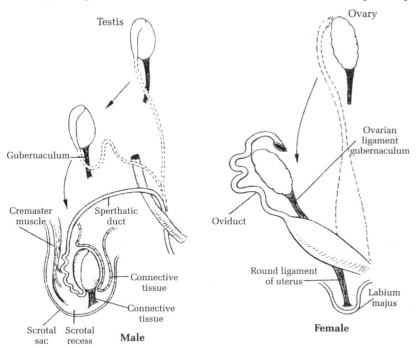

Figure 16.4
Displacement of mammalian gonads, ventral view. The ovarian ligament and round ligament of the uterus collectively are homologous with the male gubernaculum. Arrows indicate the displacement of gonads. Showing spermatic duct arching over the ureter. Testis are displaced to scrotal sacs,–in males. In females–ovary are displaced from original position.

and most carnivores ungulates and primates). Regardless of when it occurs the migration and descent of the testes are complex matters controlled primarily by hormones (Fig. 16.4). Part of the process is accounted for by a ligament the *gubernaculum*, which forms just behind the early testis and holds it to the peritoneum. It seems strange to find these vital organs outside the body cavity where they are vulnerable to many accidents. However, the temperature in the scrotum is lower than that in the body cavity and it has been found that in mammals whose testes are normally descended, spermatogenesis will not occur if the testes do not descend or if scrotum is artificially kept warm. On the other hand, the testes are kept warm in cetaceans, sirinians, monotremes and all birds. The **cooling mechanism** is clearly understood. The scrotum is relatively thin and has a large surface area, so heat is readily dissipated. Moreover, the arteries going to the scrotum are entwined with the veins returning from it, through a plexus of vessels called the **pampiniform plexus**. This provides a counter current heat distribution system which further cools the scrotum (and also helps rewarm the blood returning to the body cavity). What is not understood is why the testes should descend in some animals for spermatogenesis to occur. It is clear that it is not necessary in most animals, and one wonders how and for what adaptive reason, such a mechanism could have evolved, in the first place.

Ovary

Ovaries differentiate from the cortical portion of the sexually indifferent gonad, they produce germ cells (eggs or ova) and female hormones (Figs. 16.2 A, B).

There are three types of ovaries (Fig. 16.5).

(i) In most vertebrates the ovary is a solid structure in which the developing eggs are embedded. When mature, the eggs are expelled from ovary into body cavity in a process called ovulation in chondrichthyes, aves, mammals, crocodiles and also chelonia of reptiles.

(ii) In amphibia and the lepidosauria the ovary is hollow or saclike, eggs develop within its walls and when mature rupture through the walls into the coelom.

(iii) In teleosts, ovary is saclike but mature eggs rupture not into the coelom but into the cavity of the ovary.

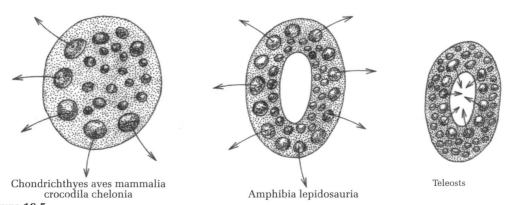

Chondrichthyes aves mammalia crocodila chelonia   Amphibia lepidosauria   Teleosts

Figure 16.5
Types of ovaries found in vertebrates. The arrows indicate the direction of ovulation

Microscopically all ovaries are composed of a loose connective tissue—stroma in which are embedded a number of ovarian follicles in various developmental stages (Fig. 16.6). Each ovary follicle is made up of a large number of follicle cells surrounding a single diploid oogonium. Only oogonia will produce egg but follicular cells must be present for their development and maturation. In most vertebrates, the maturation of egg involves mainly the formation of nutrient substances—yolk. Protein component of the yolk are synthesized primarily in liver and carried by blood to ovaries, and then picked up by follicle cells which deposit yolk in the eggs. Great amount of yolk is seen in eggs of reptiles, aves, monotremes and chondrichthyeans—hence they are **telolecithal**. Small amount of yolk is present in teleost, amphibian egg-**mesolecithal**. Yolk is scanty in therian mammals—**isolecithal**.

As more yolk is added to the egg its overall size (but not its cytoplasm) increases in some animals by as much as a million times. As this happens the ovary itself become larger, the maturing follicles bulge out and outline of developing egg can be seen on the surface of the ovary. Cell divisions, particularly meiotic, are also important in egg maturation. These do not begin until the egg is almost fully formed in follicle and are not complete until after the egg has been fertilized.

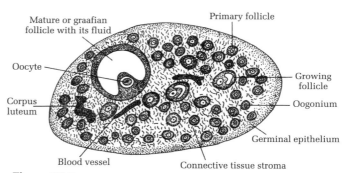

Figure 16.6
A section of mammalian ovary

The divisions are asymmetrical, and only one mature egg is formed from each two step meiotic division. The other cells which result from the division are called polar bodies. They contain the same amount of chromatin but are very much smaller, they do not have the capacity to be fertilized, and are eventually resorbed.

Before ovulation, follicle moves to surface of ovary minimising the amount of ovarian tissue between the egg and the surface. Then the follicle ruptures, casting the egg out into the coelom or into cavity of ovary. The cells of ruptured follicle remain in ovary and undergo rapid cytological changes, becoming corpus luteum. This structure in mammal produces important hormone for maintenance of pregnancy.

Genital ducts

No genital ducts exist in the cyclostomes and the genital products in both sexes are discharged into the body cavity, from where they escape to the exterior by a pair of slit like opening called the genital pores. In all other anamniota, there are definite genital ducts, however, there are marked differences in the two sexes (Fig. 16.7). As a rule sperms are never shed in coelom while ova are first shed in coelom, then picked up by genital ducts, which carry them to the exterior. In all other vertebrates the sperm are transported by modified portions of the embryonic kidney. The modified posterior pronephric tubules and anterior mesonephric tubules become the efferent ductules, carrying sperm from the seminiferous tubules out of the testes (Fig. 16.7 A). In male fish and amphibia, some of the anterior nephric tubules of opisthonephros connect with the testis of that side and collect sperms, these tubules are known as vasa efferentia. Vasa efferentia at the other end are connected with the archinephric duct (wolffian duct of amniota), which thus serves as the urogenital duct (Fig. 16.7) In *Scoliodon* vasa efferentia pass through the anterior part of the opisthonephros and connect with archinephric duct, which leads into cloaca. Male elasmobranchs have another pair of ducts in the embryo. These are mullerian ducts. Remnants of mullerian ducts persists in male as **sperm sac** (Fig. 16.12 B).

In teleosts no part of kidney serves genital purpose. An altogether new duct, sperm duct, runs along the length of the testis and is connected with it by a number of branching canals.

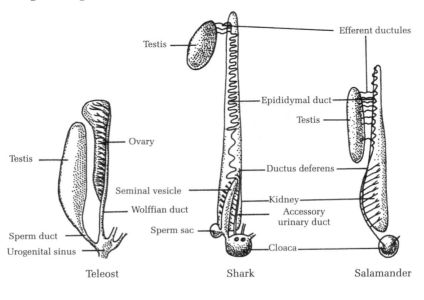

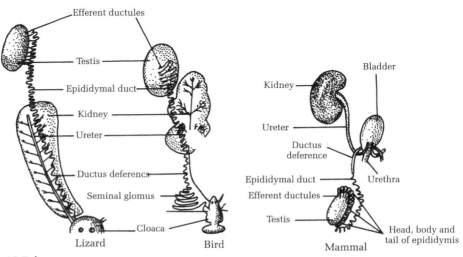

Figure 16.7 A
Ventral view of male gonads and genital ducts in diverse vertebrates

The sperm duct either connects with the archinephric duct or opens out by an independent aperture (Fig. 16.7). The sperm duct of such teleostean fish is not the vas deferens. In male amphibia, vasa efferentia either connect directly with the archinepheric duct or first pass into a longitudinal **bidder's** canal, which in turn leads into archinephric duct. Bidder's canal is present in opisthonephros of female also but has no significance.

In female anamniota the oviducts, in majority of cases are modification of mullerian ducts which in male are embryonic or left as vestiges (Fig. 16.7 A).

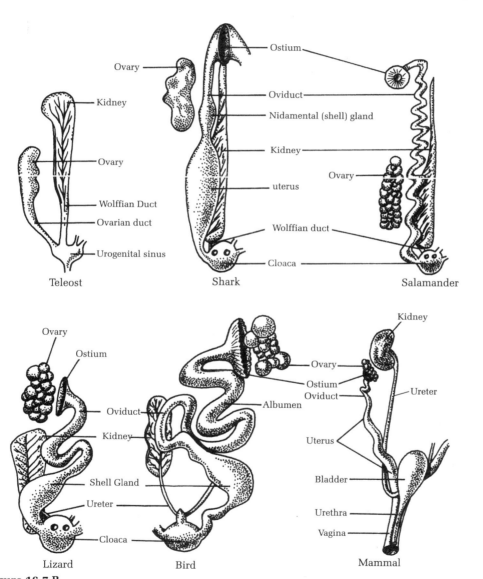

Figure 16.7 B
Ventral view of the female gonads and genital tracts in diverse vertebrates. The right ovary and oviduct are shown except in birds, where the right ovary is degenerate

Origin of mullerian duct: There is a considerable evidence that in elasmobranchs and amphibians the ostium is formed by an enlargement of a pronephric nephrostome, but not, apparently, in other vertebrates. The remainder of the mullerian duct is formed in either of the two ways (Fig. 16.8).

In elasmobranchs and tailed amphibia the archinephric duct *splits* longitudinally into two, one of these remain in service of the kidney as **archinephric duct** proper, the other becomes the **mullerian duct**. Anterior end of the mullerian duct soon after its formation becomes connected with one or more pronephric tubules, the funnels of the tubule persist in the

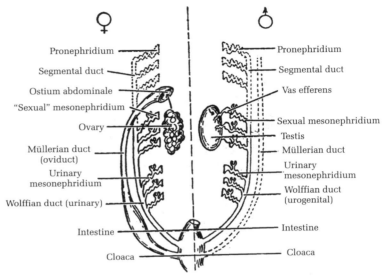

Figure 16.8 A
Mesonephridial stage of fish and amphibians (Anamniota)

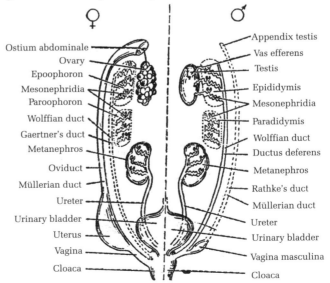

Figure 16.8 B
Metanephridial stage of reptiles, birds and mammals (Amniota)

adult to form oviducal funnel and the ostium. The mullerian duct never connects directly with the ovary. Mullerian duct either open into urogenital sinus of cloaca or to the exterior.

In all other **anamniota**, and also in amniota mullerian ducts arise *independently* from groove or invagination of the peritoneum covering the ventro-lateral part of the opisthonephros in anamniota or the mesonephros in amniota near its anterior end

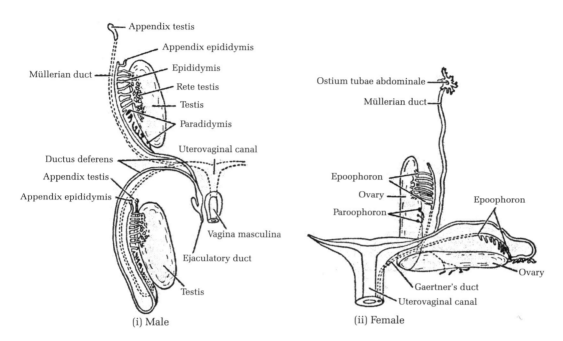

Figure 16.8 C
Position of the urogenital organs before and after descent (i) Male (ii) Female

The edges of the grooves or invaginations fold and meet converting the grooves into tubes (Fig. 16.1). These are the rudiments of the mullerian ducts, which grow back to open into cloaca, anterior and of tubes remain widely open as ciliated funnel. Ova are discharged in the body cavity and caught in the oviducal funnel and passed down the oviduct. In female teleosts oviducts are short (mullerian duct) and usually continuous with the gonads. The gonad is surrounded by the fold of peritoneum forming a sac continuous with the oviduct. The ovaries are thus completely shut off from abdominal cavity and ova fall into the oviducts which leads to the outside since cloaca is absent.

Genital ducts of amniota

Male amniota The wolffian duct are retained, while mesonephric tubules are non-excretory and most of them degenerate. Some of the mesonephric tubules, however, always persists and establish connections with the testis on each side to become vasa efferentia leading into the wolffian duct or the vas deferens. Anterior part of wolffian duct generally, becomes convoluted to form epididymis. Mullerian ducts may persists as functionless vestiges in the adult (Fig. 16.8 B).

Female amniota Wolffian duct disappear but some part of mesonephric kidney on each side may be left as vestiges. Mullerian ducts, developed independently from the mullerian grooves or invagination persists on oviduct, communicating at one end with body cavity by means of oviducal funnel, while at the other, leading to the outside through the cloaca in reptiles, birds and egg laying mammals, or directly to outside in other mammals (Fig. 16.8 B).

Degenerate and Rudimentary organs

The young embryo presents a condition with respect to the reproductive apparatus, that suggests a hermaphrodite with the rudiments of both sexes present. As development proceeds one sex becomes dominant, and the structures which characterize the other sex fade into the background as degenerate or rudimentary remains. There results a homology or equivalence in the anatomical details of the two sexes.

The organs that are typed in boldface are functional while the others are either degenerate useless structures or of doubtful function (Table 16.1).

Table 16.1: Table of homologies of structures derived from the mesonephros and its associated ducts

		Male	Female
Mesonephros	Sexual part	Appendix epididymis **Epididymis** (in part) **Rete testis**	Epoophoron (in part)
	Urinary part	Paradidymis Ductus aberrantes	Paroophoron
Wolffian duct	Proximal part	**Epididymis** (in part) **Ductus epididymis**	Epoöphoron (in part)
	Distal part	Ductus deferens	Gaertner's duct
Müllerian	Proximal part	Appendix testis	**Fallopian tube**, **uterus**
	Distal part	Vagina masculina Colliculus seminalis	**Vagina** **Hymen**

The **epoophoron** of human female is composed of eighteen or twenty anastomosing mesonephridia that are closed at both ends. In ruminants perrisodactyl and pigs, the mesonephridia forming the epoophoron are connected with a fragment of Gaertner's duct which corresponds to the wolffian duct in male (Fig. 16.8 C)

The **paradidymis** and the **paroophoron** in the two sexes respectively are all that remain of the posterior mesonephridia. The paradidymis lies within the spermatic cord near the globus major of the epididymis. Both the paradidymis and its homologue in the female are found in older embryos and young children.

The **ductuli aberrantes** are also tubules, originally nephridia, blind at one end and opening into the duct of the epididymis. There may be one, two or several of them, although the number is usually two. They lie between the testis and the epididymis. The inferior ductule which is the more constant of the two, may attain the length of two inches in man.

The **appendix epididymis** which is a degenerate tip of the mesonephros, lies upon the globus major of the epididymis.

The **appendix testis**, a small spherical sac attached to the testis represents the tip of the mullerian duct. The other end of the embryonic mullerian duct remains in the male in the form of the vagina masculina, a small sac homologous with the vagina. It is embedded in the prostate gland along with the base of the urethra and in usually distally bifid (Fig. 16.8) which is additional evidence that it represents the remains of coalescing oviducts.

Around the opening of the vagina masculina is a small fold of tissue, the colliculus seminalis, that marks the end of the mullerian ducts, and is homologous with the hymen of the female, which partially separate the vagina from the vestibule, and like wise locates the true termination of the mullerian ducts.

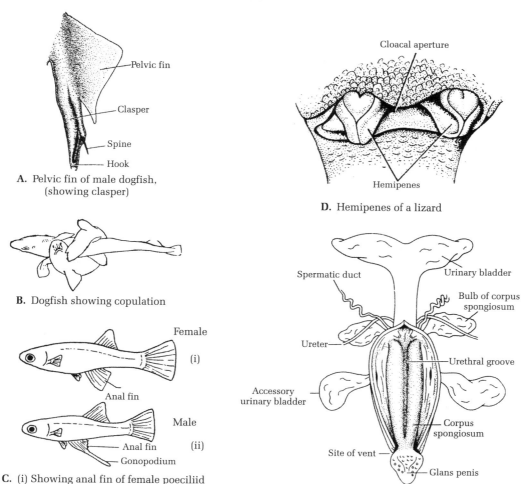

A. Pelvic fin of male dogfish, (showing clasper)

D. Hemipenes of a lizard

B. Dogfish showing copulation

C. (i) Showing anal fin of female poeciliid fish, (ii) anterior portion of anal fin of male is modified to form the gonopodium

E. Urogenital system of a male turtle (Ventral view)

Figures 16.9 A, B, C, D, E

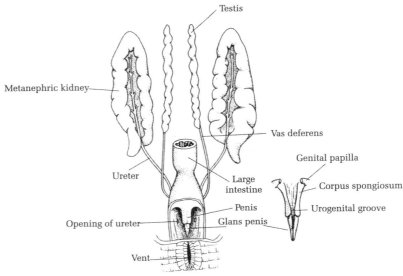

Figure 16.9 F
Urogenital system of a male alligator. Ventral view

Fertilization

There is a close inverse relationship between the mechanisms ensuring fertilization in any taxon and the number of germ cells that must be produced.

External fertilization

Fertilization can take place outside if both eggs and sperms are released into the water, and this system, called external fertilization, is common among aquatic vertebrates. In many osteichthyes, fertilization is ensured simply by the enormous number of germ cells released. Although this system is wasteful, enough eggs will be fertilized purely by chance to continue the species. In osteichthyes and cyclostomes, occurs more elaborate sexual behaviour which decreases the odds against fertilization. Hence in these fish fewer eggs and sperm are produced. Stickle back, a teleost goes through a precise courtship ritual which includes building a nest, the female discharges its eggs into the nest, and the male releases its sperms while swimming just above the nest. The female lamprey builds a nest and then the male and female, warp around each other, release their germ cells simultaneously into the nest, where the young develops. In anuran amphibians the male mounts the female, clasping her trunk in an amplexus, which aids the release of eggs through the cloaca. As the eggs are released the

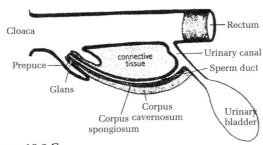

Figure 16.9 G
Longitudinal section through cloacal region of male monotreme, showing retracted penis

male releases sperms, and many eggs are fertilized. In urodele amphibians there are courtship rituals culminating in a walk; the male goes first laying down spermatophores, the female follows, dragging her cloaca over the spermatophores and thus picking up the sperm. When the eggs are laid, while on their way to the outside they are fertilized in cloaca by the sperm.

Internal fertilization

A higher percentage of eggs will be fertilized if the male inserts the sperm directly into the female. This is done by an intromittent organ, several types of which have evolved in different vertebrates (Fig. 16.9). In chondrichthyes, the male's pelvic fin is modified as the **clasper organ** (Fig. 16.9 A), which includes a skeletal support, a siphon gland, and a siphon sac; a groove runs from the cloaca through the clasper organ to its tip. The siphon gland produces a mucopolysaccharide secretion, which is stored in the siphon sac and then added to the sperm as it passes along the groove. During copulation the male elasmobranch wraps himself around the female and inserts a single clasper into her cloaca and up into a mullerian duct (Fig. 16.9 B). The muscular walls of his urogenital tract contract, forcing the sperm out through the clasper groove and up into the mullerian duct. The sperm by their own mobility swim up the mullerian duct to the ostium where fertilization occurs.

There are many teleosts with internal fertilization. The male's sperm ducts terminate at the base of the anal fin, which is modified as an intromittent organ called the gonopodium (Fig. 16.9 C). This is inserted into the female's ovarian duct; the sperm, passes directly into the duct, swim up into the saclike ovary where fertilization occurs. **Ascaphus**, an amphibian has a functional intromittent organ, it can evert its cloaca and place it into the female's cloaca. It is this everted cloaca which gives the animal its common name, the tailed frog.

Two types of intromittent organs have evolved in reptiles. In **snakes** and **lizards** there are paired, vascularized, eversible sacs, the hemipenes (Fig. 16.9 D), opening into the posterior part of the cloaca, when everted, each hemipenes has a spiral groove along its medial surface through which the sperm pass when the two hemipenes are brought together and inserted into the female cloaca. In **turtles** and **crocodilians** (Fig. 16.9 E) there is a single penis, formed in the ventral wall of the cloaca by large, paired ridges of vascularized erectile tissue called the corpora cavarnosa. With sexual excitation the erectile tissue becomes turgid, closing a groove which lies between the two corpora and converting these structures to a tube. When this penis is inserted into the female's cloaca the sperm pass along the tube into the female. Few birds have a true copulatory organ, and for those that do not (ducks, swans, geese and flightless birds such as ostrich) it is morphologically and functionally almost identical with that of turtles and crocodilians. Internal fertilization is necessary for all birds, however, so that shell can be laid down around the fertilized egg before it is passed to the outside. Copulation in birds without a penis involves what is called the **cloacal kiss**. In some birds this is a graceful and beautiful maneuveur performed in flight in which the male's cloaca is for an instant placed against the females cloaca. In order to pass sperm successfully from male to female, this brief event must be precisely timed. To ensure this, it is preceded by a ritualized **courtship dance** in which the male and female fly parallel to one another, the male below and the female above until one dips down and the other tips up for the climactic event.

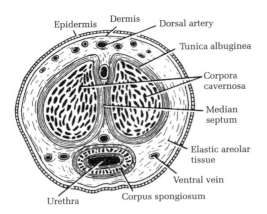

Figure 16.10
Rabbit. T.S. of penis

The **mammalian** penis has two masses of erectile tissue called the corpora cavernosa, homologous to the corpora cavernosa of reptiles, and in addition a third, mid line mass, the corpus spongiosum. The corpus spongiosum surrounds the urethra, a closed tube through which the sperm pass (Fig. 16.10). Both the corpora cavernosa and the corpus spongiosum contain erectile tissue. In many mammals (except humans) there is an ossification in the connective tissue between the two corpora cavernosa this is called the **os-penis** or **baculum**. The corpus spongiosum is expanded at its distal end to form the glans of the penis, over which is a fold of skin, the prepuce. Erection occurs by dilation of the arteries supplying the erectile tissues and a simultaneous constriction of their veins, causing a build up of pressure. Almost any stimulation of the penis will cause a reflex erection if the appropriate hormonal balance is present. Ejaculation occurs following a erection and further stimulation. It involves violent contractions of the muscles of the ductus deferens and accessory male glands, which force the sperm and seminal fluid out of the urethra.

An intromittent organ to be successful, requires complementary structures in the female. In Most nonmammalian vertebrates the intromittent organ is received by the cloaca only; however, the chondrichthyean clasper organ is pushed up into the distal portion of the mullerian duct, and in teleosts, where there is no cloaca, the gonopodium is inserted directly into the ovarian duct. Among mammals, only monotremes have a cloaca to receive the penis. In all other mammals the distal portions of the mullerian duct are modified to form the genital orifice, which receives the penis. The outer part of the females genital opening, the urogenital sinus, is formed by an infolded portion of surface skin. Within the wall of the urogenital sinus lies the clitoris, a small homologue of the male's penis containing erectile tissue homologous to the corpora cavernosa (not corpus spongiosum). Upon sexual excitation the clitoris is engorged with blood and becomes erect. Internal to the urogenital sinus is the most distal portion of the Mullerian duct, the vagina. Between urogenital sinus and vagina is a fold of tissue, the hymen, which ruptures and bleeds at the first intercourse (Table 16.1). In marsupials the distal ends of the right and left mullerian ducts have not joined, and there are paired vaginas. It is not surprising, therefore, that the distal end of the marsupial penis is also paired, having what is called a **bifid glans**. In placental mammals the distal ends of the two mullerian ducts fuse and form a single, midline vagina, which receives the penis during intercourse.

In **mammals** mullerian ducts give rise to oviducts, uteri and vaginas. The embryonic mullerian ducts always fuse at their caudal ends. As a result, the uteri are typically paired and there is a single vagina. The oviducts, or fallopian tubes as they are sometimes called, are relatively short, of small diameter and convoluted. The ostium is surrounded by a fimbriated infundibulum.

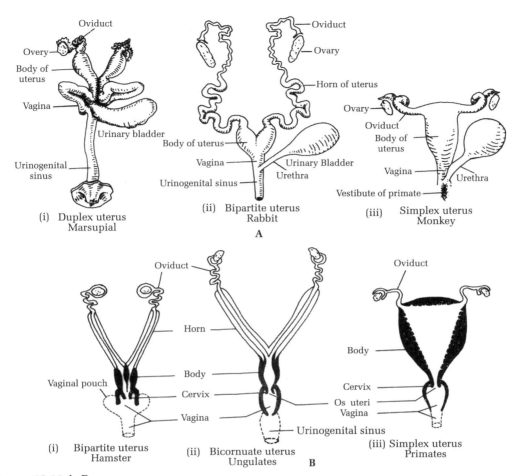

Figures 16.11 A, B
A. Reproductive tracts of female mammals, (i) Marsupial (ii) Rabbit (iii) Monkey
B. Uterine types among mammals. Black regions represent fused caudal ends of the mullerian duct

Uterus

In **monotremes** and many **marsupials** there is no fusion of the mullerian ducts. Therefore the uteri are double throughout **duplex uterus** (Fig. 16.11).

In placental mammals there are varying degrees of fusion of the caudal ends of the mullerian ducts, which results in two uterine horns and single uterine body. When there are two complete lumens within the body of the uterus, it is said to be a **bipartite uterus.**

When there is a single lumen within the body and two horns, the uterus is said to be bicornuate. There are, however, many species with uteri transitional between the bipartite and bicornuate condition (Fig. 16.11). When there are uteri horns, the blastocysts implants in horns. In some mammals one horn is much larger, and the blastocysts always implant in the enlarged horn, the right in Impala, even though both ovaries produce eggs.

In apes, monkeys, man, some bats, and armadillos no uterine horns develop and the oviducts are only the remaining paired portion of the mullerian duct. They open into the body of the **simplex uterus.** Except in ectopic pregnancies in which blastocyst implant in abnormal locations such as the oviduct (tubal pregnancies) or the twins, triplets, quadruplets or quintuplets all implant in the body of the uterus. The body of the uterus narrows to form a cervix (neck), the lower end of which projects into the vagina as the lips of the cervix. The lips surround the opening (os-uteri, mouth of the uterus) leading from the uterus into the vagina. The cervix must dilate for the young to be delivered. The fertilization takes place in upper part of the oviduct. The uterine lining, or endometrium becomes highly vascular under the stimulus of hormones prior to implantation of the blastocyst. The thick muscular layer of the uterine wall, the myometrium assists in ejection of the young at birth provided it too, has been hormonally prepared for this action.

Vagina

Typically, the vagina is the fused terminal portion of the **mullerian ducts** opening into the urogenital sinus (ungulates).

In many rodents and primates, however, the vagina is extended so as to open directly to the exterior. Like other mullerian duct derivatives, the vagina has a muscular wall. The lining is specialized for reception of male intromittent organ. In monotremes, the uteri open directly into the cloaca so that there is no vagina. The vagina in marsupials is unusual. Just beyond the uteri the two mullerian ducts meet to form a median vagina, which may or may not be paired internally.

Beyond the median vagina the two mullerian ducts continue to the urogenital sinus as paired (lateral) vaginas. The pouch like median vagina projects posteriorly and lies against the urogenital sinus. At birth the foetus is forced through the partition directly into the urogenital sinus and thus bypass the paired vaginas. In kangaroos, both routes may be used as a birth canal. The opening between the median vagina and the urogenital sinus may remain throughout life, once perforated.

In **Opossums**, therefore, the chief function of lateral vaginas is to serve as a path for sperm in their journey to the egg. As an adaptation to dual vaginas, the penis of male is forked at the tip, and one tip enters each lateral vaginal canal, where the semen is discharged.

Cloaca

In many vertebrates the cloaca is divided into two partial chambers, the coprodaeum and urodeaum. The urodeaum is meant to receive the urinary waste, where a coprodeaum receives the faecal matter. Both jointly open into the proctodeum. In placental animals instead of single opening, cloaca is completely divided because of urorectal fold which separate the two chambers. Further separation of urogenital sinus takes place in the females of majority of primates, including man, and in rats, mice, where genital ducts and urinary passage are independent. In absence of cloaca in adult, the ureter open into the urinary bladder. The bladder narrow down to a channel the urethra, that leads in male, through the penis and open at its tip serving as a urogenital passage. In female, it leads into the urogenital sinus or directly to the outside through clitoris.

Accessory Reproductive Glands

The ovaries and testes are not the only glands in the genital system. Accessory genital glands along both the male and female reproductive tracts play important roles in the reproductive process.

Male In the human male the seminal fluid is produced by a single pair of glands, the seminal vesicles, and a midline gland, the prostate. There are seminal vesicles in most eutherians except cetaceans and many carnivores. There are none in monotremes or marsupials. The prostate is a large mid line gland lying at the point where the two ducts or deferens join the urethra. It is a compound tubular alveolar gland with several ducts opening into the urethra. Its slightly alkaline secretion forms the bulk of the seminal fluid. A pair of smaller glands, the bulbourethral or cowper's glands lies distal to the prostate gland and opens into the urethra. During erection but before ejaculation these glands produce an alkaline mucous secretion, which neutralizes the slight acidity of both the male urethra and the female's vagina and thus prepares the genital tracts to receive the sperm.

Accessory male glands producing seminal fluid are not peculiar to mammals, but in one form or another are found in most male vertebrates. For instance, a mucopolysaccharide is produced by the male elasmobranch's siphon gland. Chondrichthyeans also have a **sexual kidney** or **Leydig's gland.** This is the modified anterior portion of the opisthonephric kidney. Sharks have two distal modifications of the wolffian duct; one is a simple increase in the duct's diameter, forming a seminal vesicle, and the other is a saclike dilation of the seminal vesicle, called the sperm-sac. Both these structures may store sperm prior to ejaculation. Such modifications of both the wolffian duct and anterior opisthonephros have evolved in similar but not always identical ways in different groups of vertebrates.

Female There are also accessory glands in the female's genital tract, which develop as modifications of the mullerian duct. These are most prominent in non-mammalian vertebrates whose eggs are shelled or have a heavy albumen layer around the yolk. Even mammals, however, have vaginal and cervical glands which help to lubricate the genital tracts. They also have important uterine glands whose albuminous secretion helps repair the epithelial uterine wall following menstruation in primates, or estrous in non-primates. In addition, there are some accessory sex-glands which are not directly connected to the genital tracts. Most notable are scent-glands often located just deep to the anus which are attractive to the opposite sex.

Comparative Anatomy of Urogenital System

Cyclostomes: The hagfish are **hermaphrodite**, the anterior part of the unpaired gonad produces ova and posterior part produces sperms. Hagfish possess the most primitive type of urogenital system. Self-fertilization is not possible, because of the time interval in the maturity of ovarian and testicular parts of the gonad. Sperms and ova are thrown out from the body cavity through the genital pores.

Fish: In *Scoliodon* in the **male urogenital system**: opisthonephros is the functional kidney of the adult. Two ribbon-shaped kidneys extend from one end of the body cavity to the other (Fig. 16.12). The anterior half of each kidney is set aside for transporting sperms and is called genital part or leydig gland, remaining half is the functional kidney and is renal

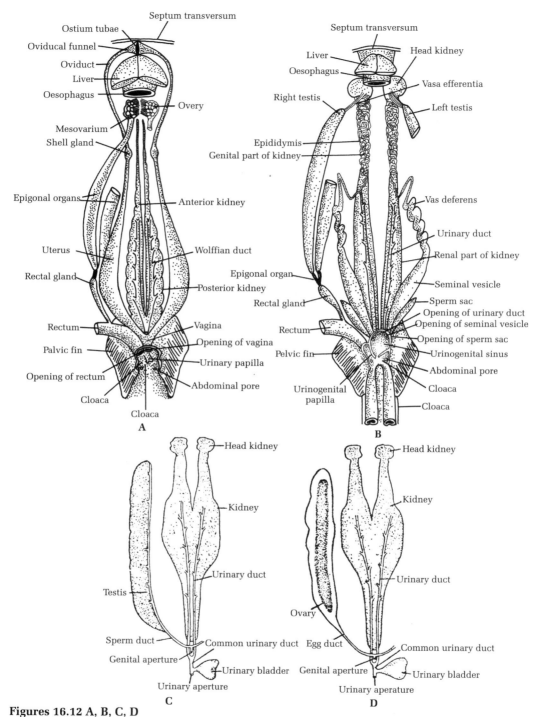

Figures 16.12 A, B, C, D
A. Female urinogenital system (*Scoliodon*) **B.** Male urino-genital system (*Scoliodon*) **C.** Urinogenital system (Testis of one side) (*Labeo*) **D.** Urinogenital system (Ovary of one side) (female *Labeo*)

part. Renal part or the opisthonephric kidney has well developed nephric tubules, some of which have coelomic funnels. Tubules have internal glomeruli and communicate with collecting tubules, which in turn unite to form a common urinary duct. Urinary ducts of the two sides open separately into the urogenital sinus.

Testes, one on each side, are elongated glandular organs from the base of the liver to the rectal gland and suspend into the body cavity by mesorchia. Testis is composed of numerous lobules lined with germinal epithelium. Lobules lead into central duct, which passes into vas deferens which in adult is highly convoluted to form ductus epididymis. Vas deferens dilate to form seminal vesicle. Seminal vesicles of two side, open behind into a common urogenital sinus opening at urogenital papilla. A pair of sperm sac representing vestigial mullerian duct project into body cavity from the ventral wall of urogenital sinus.

Internal fertilization is a rule for which males are provided with erectile copulatory organs in the form of a pair of claspers. Claspers have anterior opening apopyle, communicating, with cloaca and receives sperms, posterior opening called hypopyle, serves as exit to pour sperm in the females cloaca. A pair of sacs, the siphons with heavy muscular walls lie beneath the skin. Before copulation, the siphon are filled with water and the apopyle is kept open. Muscular walls of the siphons contract at the same time forcing the contained water into the grooves of the claspers that flush the spermatozoa into the cloaca of the female. Thus siphons are meant to eject sperms.

In female urogenital system there is no direct connection between the kidneys and the genital organs in female (Fig. 16.12). The anterior parts of kidneys are extremely reduced. A pair of flattened kidneys occupy the same position as they did in the male. Anterior part of the kidney corresponding to the genital part of the male kidney is thin and not well developed as the renal part. The archinephric duct, receives nephric tubules all the way. The two archinephric ducts join to form urinary sinus which communicate with cloaca through urinary papilla. A long tubular non-glandular lymphoid tissue, the epigonal organ connect each ovary with rectal gland. Ovaries lie near the base of liver and are suspended by mesovaria. Mature ova are enclosed into ovarian or graafian follicles, which bursts to liberate them into body cavity. Mullerian ducts form the oviducts. Their anterior ends reach the mid-dorsal line and two oviducts communicate with coelom by a single large median longitudinal slit, the ostium tube. The oviduct proceeds backwards and enlarges to form a shell or oviducal gland. Further backwards, each oviduct enlarges to form uterus. The two uteri unite to form a vagina which opens into the cloaca.

Male urogenital system: In the *Labeo rohita* the opisthonephric kidneys are elongated, brown masses above the air bladder, with middle and posterior parts fused. The head kidney have generated nephric tubules. The archinephric duct runs the whole length of each kidney posteriorly the ducts of two sides unite to form common duct which is swollen to form bladder. Since cloaca is absent in teleosts, the bladder communicate directly with the outside by urinary opening behind anus. Testis are a pair of glandular structures, about as long as the kidneys. The archinephric duct have nothing to do with genital organs. Sperm duct which is formed by folding of peritoneum in form of a tube receives sperms through finer tubes connected with testis, sperm duct opens out independently. Copulatory organs are absent.

In the **female genital system** the kidneys and their ducts have the same plan as in male, the ovaries are paired structures of large size, a fold of peritoneum grows from the posterior part of the abdominal cavity and become continuous with the ovary of each side. Ova are discharged on maturity into central coelomic space of the ovary from where they pass on into the peritoneal tubes which thus function as oviducts and open to the exterior by a separate apertures. Fertilization is external. Variation in the urogenital system of fish are numerous — a few important ones such as (i) kidney as well as gonads show various degrees of fusion (ii) in some elasmobranch, only the right side ovary is developed, the other atrophies (iii) elasmobranchs may be oviparous, ovoviviparous or viviparous (iv) male elasmobranchs have erectile copulatory organ (v) most teleosts are oviparous, a few are ovoviviparous.

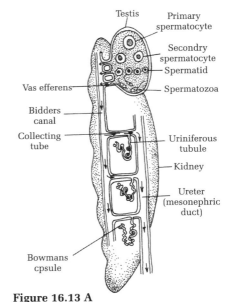

Figure 16.13 A
L.S. kidney showing Bidder's canal

Amphibia: In *Rana tigrina*, the kidneys in both sexes are shifted to more posterior positions and there is no external differentiation of genital and renal parts (Fig. 16.13). The nephric tubules have lost connections with their funnels, some of which, however, stay on the ventral surface.

In **male frog** some anterior tubules have become vasa efferentia taking the sperm into the bidders canal lying along the medial edge and inside the kidney (Fig. 16.13). This canal then

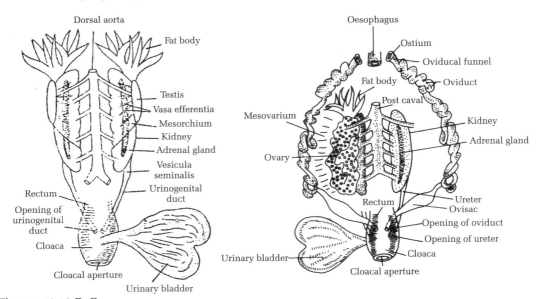

Figures 16.13 B, C
 B. Urinogenital system of male frog   **C.** Urinogenital system of female frog. (Ovary of one side)

connects with the anterior part of the archinephric duct. The archinephric duct are situated within the kidneys along their lateral margins. The archinephric ducts of the two sides dilate soon after leaving the kidney to form ampullae or seminal vesicle, where sperms are stored temporarily. Vasa deferentia open into the cloaca separately. Urine is temporarily stored in the urinary bladder. Testes are ovoid and compact, situated near the anterior ends of the kidneys and suspended by mesorchia. Fat bodies or corpora adiposa are associated with the gonad. Adrenal or suprarenal glands are located on the ventral sides of the kidneys. Copulatory organs are not found, false copulation or **amplexus** takes place. Fertilization is external.

In the **female frog** kidneys have no connection with the ovaries. The archinephric ducts are entirely in service of kidneys. Ovaries shed mature eggs into the body cavity. The mullerian ducts acts as oviducts. Oviducts forms ovisacs or uteri for storing the eggs temporarily, uteri open into the cloaca separately. Fat bodies are present in the female. The walls of oviduct are glandular and secrete albumen. Reduced bidder's canal are present in females also, but have no significance. It may be an endocrine gland.

Reptiles: In male *Uromastix* metanephric kidneys are placed in the pelvic region of the abdominal cavity (Fig. 16.14). The fused posterior part forms a V-shaped kidney, a connective tissue matrix separates the two kidney. Ureters arise from the anterior one-third of the ventro-lateral regions of kidneys, each passes backwards, being embedded in the central connective tissue inside the peritoneum, covering the ventral surface of the kidney and opens in the middle compartment of the cloaca. Ureters and genital ducts open separately in the same compartment. Ureters in male *Calotes* open into vasa deferentia at their distal ends. Urinary bladder arises from the postero-ventral border of the compartment, it is filled

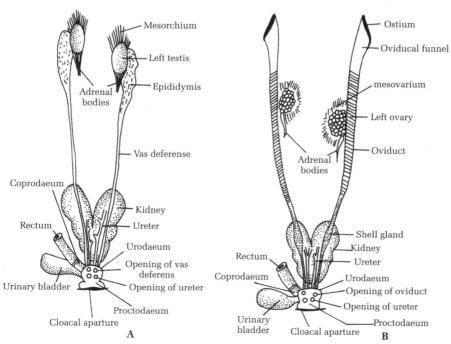

Figures 16.14 A, B
A. Urinogenital system of male *Uromastix* B. Urinogenital system of female *Uromastix*

with urine. Testes are ovoid bodies placed asymmetrically much anterior to the kidney. Generally, the testis of the right side is slight ahead of the left. Testes are suspended in body cavity by mesorchium. Some anterior tubules of the mesonephros are connected to the seminiferous tubules of testis of each side to become the vasa efferentia, which in turn connect with the wolffian duct. Posterior part of the mesonephric kidney is left as the wolffian vestige lying obliquely at the posterior margin of the testis. Anterior part of the wolffian duct, now the vas deferens, is thickened and thrown into convolutions to form the epididymis. Vas deferens then open into urodaeum. Vestigial mullerian ducts persists as a pair of peritoneal thickened strands. Adrenal glands are posterolateral to the testes.

Copulatory organs of *Uromastix* and *Calotes* are a pair of structures called hemipenis or Hemipenial sacs. The shape and size of hemipenis is dependent upon the degree of eversion, but it has a distinct hardened pedicel (Fig. 16.9), a soft anterior glans, and a furrow or sulcus along the outer wall, through which the spermatic fluid passes out during copulation.

In the female urinogenital system of *Uromastix* the position and arrangement of kidneys and ureter is the same as in the male. Mesonephros is left as wolffian vestige or epoophoron (Fig. 16.14). Ureters and genital ducts open independently in cloaca. The ovaries are a pair of white oval bodies and nearer the kidneys then the testes. Ovary is attached to abdominal wall by mesovarium. The oviducts of each side originates in the anterior part of the abdominal cavity by a wide mouthed oviducal funnel with its ostium facing outwards. It runs backwards passing from outside the ovary, presenting striated appearance throughout due to special elastic fibre bands in its wall. A broad ligament of coelomic epithelium serves to hold the oviducal funnel and some part of the oviduct in place while passing along the ventral surface of the kidney. The oviduct is enlarged into a shell gland or ovisac which besides lodging the mature fertilized eggs, secrete a part of the egg shell. The terminal part of the oviduct forms so called vagina, that opens into urodaeum anterior to urinary aperture. Hemipenes are also present in female but are very small and insignificant. Secretions of the anal or cloacal glands in both sexes help in bringing the sexes together. These glands lie on either side of the base of the tail and open on the sides of the penial openings. The tails of the copulants are turned out laterally so that their cloacal openings come together and opposed to each other. The hemipenis of the side nearest to the female is everted out taking the form of a small finger and penetrated into the proctodaeum of the female. The glans of the hemipenis extends further into the vagina. The spermatic fluid is discharged into the urodaeum from where it is squeezed into the vagina by the contraction of urodaeum. Fertilization takes place high up in the oviducts. Egg shell is secreted in the ovisacs and shelled or **cleidoic eggs** are deposited outside the body.

Spermatic fluid is not conveyed directly into oviduct, hence **hemipenis** is not an intromittent organ, but an organ of prehension to make a firm hold on the cloaca of female.

Birds: In male genital system of *Columba livia*, the kidneys lie in the pelvic region (Fig. 16.15). Each is divided into three lobes by deep fissures. Ureters arise from the first lobe of each kidney and pass straight on to the urodaeum into which they open separately. Testes are large ovoid bodies, often the left is somewhat larger than the right and attached by mesorchia to the ventral surface of the anterior ends of the kidneys. Vasa efferentia from the testis are connected to the small epididymis formed by the wolffian duct. Wolffian duct parallels the ureter as a highly convoluted vas deferens and enters the urodaeum independently after having formed a swollen seminal vesicle. At the end the vas deferens is intensely coiled to form an erectile nodule or papilla projecting into cloaca. During copulation, the proctodea of male and female are everted and pressed together so that sperms are ejected directly into female urodaeum, from where they are squeezed into the oviduct by the contraction of the urodaeum. A temperature, lower than that of the body at the nodule, probably hastens maturity of the sperms.

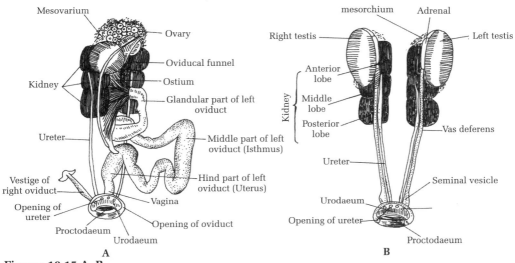

Figures 16.15 A, B
A. Pigeon: Female urinogental system   B. Pigeon: Male urinogental system

In **Female genital system** the kidney and ureters occupy the same place in the female (Fig. 16.15). Both the ovaries are present in the embryo but later the right side ovary and the right oviduct degenerate. The functional oviduct is a long specialized muscular and convoluted tube divisible into several regions. The first part, oviducal funnel, has a wide *ostium* with a fimbriated margin, next is the **magnum** or glandular part, whose numerous tubular glands pour albumen around the egg. This is followed by a short tubular, **isthmus**, leading into the swollen part called uterus, glandular epithelium of which secretes the shell over the egg. Last part of the oviduct is short vagina; opening into the urodaeum. This part secretes mucus to help egg-laying. Fertilization of the ova takes place in the upper part of the left oviduct. There is a band of cilia in the pigeon's oviduct that beats towards the ovary and guides the spermatozoa to the place of fertilization. Atrophy of one ovary, and putting the oviduct of the same side out of use, is related to some facts; (i) that the birds lay large sized eggs, and development of such big eggs in the two oviducts simultaneously was practically not possible without adversely affecting the egg and (ii) their functional presence would have been against the weight reducing tendency. (iii) A reasonable guess for the loss of one ovary is the problem of providing enough calcium for the shells of two eggs at a time. The ultimate source of calcium for the eggs is bone tissue of the mother bird. The size of the ovary fluctuates during the year, it is very extensive during the mating period. Large sized yolky eggs (polylecithal) pass on to the surface of the ovary forming as many egg follicles. A mature follicle is quite big and finally bursts through a preformed, non-vascular band, the **stigma or cicatrix**, situated on the surface of the follicle. After each follicle has burst, it shrinks very quickly; there is no corpus luteum. Oocyte thus released in the body cavity is quickly swallowed by the oviducal funnel which lies very close to it and is passed back down the oviduct in a spiral manner, with the help of peristaltic contractions of the longitudinal and circular muscles of the wall of the oviduct. Two shelled eggs are laid in a clutch, the second egg is laid two days after the first one. The eggs are incubated by the parents for about a week in tropical regions.

Mammals: Rabbit shows distinct sexual dimorphism. Males and females differ in body size and external genital organs.

Male genital system   The male genital system consists of a pair of testes, a pair of scrotal sacs, a pair of epididymes, a pair of vasa deferentia, uterus masculinus, urethra, penis, uringenital aperture, and accessory sex glands.

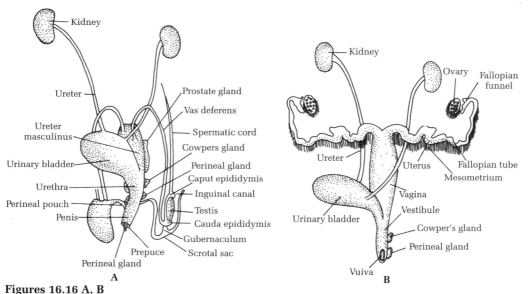

Figures 16.16 A, B
A. Urinogenital system of male rabbit   B. Urinogenital system of female rabbit

(i) **Testes**   The tests are pink, oval bodies that develop near the kidneys, but later shift backward and descend into the scrotal sacs. Temperature of the scrotal sacs is relatively low and suitable for the maturation of the sperms. A testis is enclosed in a fibrous coat, the tunica albuginea. Outside this coat is a serous membrane, the tunica vaginalis, which continues over the gubernaculum and then lines the scrotal sac. Around the testis between the two serous membranes is a narrow coelomic cavity containing coelomic fluid. Ingrowths of the albuginea, called septa, divide the testis into lobules, which contain convoluted seminiferous tubules. The seminiferous tubules are coiled and blind at one end, but on the inner side of the testis they become straight and join into a network known as the rete testis. From the rete testis fine tubules, the vasa efferentia, lead into a long, narrow, closely convoluted tube, the epididymis that finally joins a wider muscular tube, the vas deferens. The latter leaves the testis. Each seminiferous tubule is walled by a single layered germinal epithelium. The germinal epithelium, by mitosis and then grow into primary spermatocytes. Each primary spermatocytes, by first meiotic division, gives rise to haploid secondary spermatocytes. Each of which by second meiotic division produces two haploid spermatids. The latter get attached to a sertoli cell and change into spermatozoa by a process called spermiogenesis. Mature spermatozoa become free in the seminiferous tubule.

(ii) **Scrotal sacs** The scrotal sacs lie on the sides of the penis. Each scrotal sac encloses a coelomic cavity lined by a serous membrane called the tunica vaginalis the cavity contains coelomic fluid which permits sliding of the testis. Each scrotal sac retains its connection with the abdominal cavity through a narrow passage, the inguinal canal.

(iii) **Epididymes** A mass of long, narrow, closely convoluted genital duct lies along the inner side of each testis. This is called the epididymes. It extends to the top and base of the testis as the caput epididymis and the cauda epididymis respectively. The cauda epididymis is connected to the bottom of the scrotal sac by a short, thick, elastic cord, the gubernaculum. An elastic cord connects the caput epididymis with the dorsal abdominal wall through the inguinal canal. It is called **spermatic cord**. It consists of spermatic artery, vein and nerve held by connective tissue. The testis communicates with its epididymes by fine, ciliated ductules, the vasa efferentia.

The epididymis has a thin slightly muscular wall with glandular lining. While passing through the epididymis, the spermatozoa undergoes a physiological maturation, probably by getting nourishment from the secretion of its glandular lining.

(iv) **Vasa deferentia** From the cauda epididymis, the genital duct becomes straight which is now called the vas deferens. It leaves the scrotal sac and enters the abdominal cavity through the inguinal canal. Here, it curves round the ureter of its side and runs backwards to open into the uterus masculinus through the anterior part of its ventral wall.

(v) **Uterus masculinus** The uterus masculinus is a large median sac with a bifid anterior end. Posteriorly, the neck of the urinary bladder and the uterus masculinus unite to form a common passage, the urethra.

(vi) **Urethra** The urethra is a straight tube with thin, highly vascular wall. It passes into the penis.

(vii) **Penis** The penis is a cylindrical, erectile organ. It encloses the urethra. Its posterior wall is formed of a highly vascular, spongy tissue, the corpus spongiosum. It surrounds the urethra. Its anterior wall is strengthened by a pair of hard ligamentous bodies the corpora cavernosa. The tip of the penis is called the glans penis. It is covered by the loose, retractable fold of skin the prepuce or foreskin. The urinogenital aperture of the male lies at the tip of the penis. This is a common outlet for urine and spermatic fluid.

(viii) **Accessory sex glands** The accessory sex glands include the prostate gland, cowper's glands, perineal glands and rectal glands.

(ix) **Prostate gland** It lies on the dorsal and lateral sides of the uterus masculinus which opens into the urethra by fine ducts. It secretes a whitish fluid. The latter nourishes and, activates the spermatozoa that have been lying passively in the epididymis during the storage period.

(x) **Cowper's glands** These are a pair of small ovoid bodies situated posterior to the prostate glands. They secrete a clear, viscid, slightly alkaline fluid. The latter neutralizes the acidity of the urethra and vagina to protect the sperms from the action of the acid.

(xi) **Perineal glands** These are a pair of dark, elongate structures lying next to the Cowper's glands. They open to the exterior in the perineal pouches on small papillae. Their secretion gives characteristic odour to the animal. This odour probably helps in attracting the mate.

(xii) **Rectal glands** These are a pair of yellowish glands. They lie on the sides of the terminal part of the rectum. Their function is not known.

Female genital system The female genital system comprises a pair of ovaries, a pair of oviduct, vagina, vestibule, vulva, certain accessory sex glands, and a few pairs of teats.

Ovaries The ovaries are whitish, ovoid, solid, somewhat flattened organ about 18 mm long. They are attached to the dorsal body wall by double folds of peritoneum the mesovaria. The surface of each ovary shows several small, rounded, semi-transparent projections, the ovarian or graafian follicles. When ripe, the graafian follicles burst to shed the ova.

After discharge of the oocyte, the follicle cells encroach the antrum by proliferation and form a yellowish endocrine body, the corpus luteum. The latter soon disappears if fertilization does not occur. If, however, fertilization takes place, the corpus luteum grows during pregnancy and persists for some time after the birth of the young one, functioning as a temporary endocrine gland.

Oviducts The oviducts are formed from the mullerain ducts of the embryo. Each oviduct consists of three regions; fallopian funnel, fallopian tube and uterus.

Fallopian funnel It is the expanded funnel-like anterior end of the oviduct. It lies near the outer border of the ovary. Its margin bears small projections called the fimbriae. The latter are covered with cilia. Opening of the funnel is called the ostium. It receives the ova shed by the ovary. The fallopian funnel leads to the fallopian tube.

Fallopian tube It is a narrow, thin-walled, muscular, convoluted tube. It is lined by ciliated epithelium having gland cells as well. It is suspended from the dorsal body wall by a double fold of peritoneum termed the *mesosalpinx*. The fallopian tube insensibly leads into the uterus.

Uterus It is broad, thick-walled, relatively less convoluted, muscular organ with glandular lining. It is highly vascular and distensible. The uterus is suspended from the dorsal body wall by double fold of peritoneum called *mesometrium*. Development of the foetus (embryo) takes place in the uterus. The two uteri extend towards the median line, cross the ureters and unite to form the vagina.

Vagina The vagina is a wide median tube with muscular wall. It runs straight backwards dorsal to the urinary bladder. It is suspended from the dorsal body wall by a double fold of peritoneum. It joins behind with the neck of the bladder to form a common passage, the urinogenital canal or vestibule.

Vestibule The vestibule is a wide median tube that extends backwards through the pelvic cavity ventral to the rectum. Its wall is very vascular. It opens to the exterior by female genital aperture called vulva.

Vulva It is a slit-like aperture situated in front of the anus. A small, rodlike organ lies on the anterior or ventral wall of the vestibule just inside the vulva. It is called the clitoris. It is erectile and highly sensitive. It corresponds to the penis of the male.

Accessory sex glands The glands associated with the female genital system includes the Bartholin glands, perineal glands and rectal glands. There is no prostate gland.

Bartholin's glands These are a pair of ovoid bodies lying on the roof of the vestibule. They correspond to the Cowper's glands of the male. They are greatly reduced and may even be absent. They secrete an alkaline fluid. It neutralizes the traces of acid of the urine present in the vagina. It also lubricates the vagina to facilitate copulation.

Perineal glands These resemble those of the male in structure as well as function.

Rectal glands These are also similar to those of the male.

Teats There are four or five pair of the teats or nipples or mammae, on the ventral side of the trunk. The mammary glands embedded in the skin open on these teats. The teats become enlarged after birth of the young ones, and the glands start secreting milk for their nourishment. The process of giving birth to the young ones is known as **parturition**. The duration between fertilization of the ovum and birth of the young ones is called period of gestation. It is only one month in rabbit. The female rabbit starts bearing young ones at the age of 6 months, produces 2-8 litters in a year, each litter comprising 3-8 young ones.

On the basis of the site of embryonic development all vertebrates can be divided into three major categories.

Oviparous In oviparous forms which are quite common in every class of vertebrates among anamniotes, the egg is usually laid in a protective gelatin coat of mucus and develops outside the mother. However, in several groups additional protective devices have evolved. The nests built by the *Lamprey* and many teleosts keep the eggs in a relatively secluded area during development. In the oviparous chondrichthyeans the nidamental (shell) gland, located in the anterior portion of the mullerian duct, secrete a gelatinous shell around the eggs as they pass down the duct, the young develop within this shell and then hatch out of it. Other anamniotes have pockets in the skin, called brood pouches, where the eggs are retained during development. There are many oviparous amniotes such as all birds, most reptiles, and the monotreme mammals. In these, the fertilized egg is protected by layers of albumen, shell membranes, and a calcareous shell, all secreted by specialized portions of mullerian duct as the egg passes to the outside. The calcareous shell prevents injury and dessication and yet permits respiration because it is permeable to air.

Ovoviviparous This include some chondrichthyeans, some teleosts, and many reptiles, the egg is retained in the mullerian duct but gets all of its nourishment from the yolk. In ovoviviparous teleosts the brood chamber is the ovarian sac, in chondrichthyeans and reptiles it is a dilation of the mullerian duct, called the uterus. Young in these forms do not emerge from the mothers genital tract until they are well developed and can feed themselves.

Viviparous Viviparity seems a logical extension of ovoviviparity, providing the young both protection and nourishment by the mother. The placenta is responsible for intimate relationship between the maternal circulatory system and that of the developing young, allowing exchange of gasses, nutrients, and waste products. The precise form of this relationship between blood streams varies considerably, particularly in regard to how large and thick the placenta is, and has how many membranes. However, the maternal and foetal blood streams always remain separate except in abnormal situations. Metabolic products pass by diffusion from one to the other, through the placenta.

Viviparity occurs in many taxa, including teleosts whose placenta is formed in the ovarian sac, or, in some, in the ovarian follicle. In some sharks and reptiles e.g., rattle snakes and in all mammals except monotremes there is a uterus. This modification of the mullerian duct acts as the receptacle for the developing young and the placenta is located there.

Parental Care in Different Classes of Vertebrates

Behaviour concerned with the care of the young including the preparations made before they are hatched or born is called parental care. It is among vertebrates that parental care is most intense culminating in the long period of infancy in primates and especially in man.

Fish: The eggs of some fish and of birds are kept in nests. A male stickle back makes a nest in which the female lays its eggs and, after fertilizing them, the male guards the nest and even chases off the female. By vigorous movements of its fins and tail the male drive freshwater through the tunnel of the nest and over the eggs, ensuring that they have a good supply of oxygen. Some species of the cichlid (*Tilapia*) are mouth-brooders, the female takes the eggs and sperm into her mouth and she subsequently holds the fertilized eggs in her mouth until they hatch. On hatching the young are allowed to leave, but return to the parents mouth rapidly at the approach of danger. Fish unlike birds and mammals, do not feed their young.

One of the most curious is the habit among the sea horses and the pipefish, in which the fertilized eggs are carried in a pouch in the male.

Nest building fish are geographically wide spread and the habit appears in many orders such as the Bowfin of northern America, the lungfish of Africa, and the well known stickle backs of North America and Europe. Some nests are merely hollowed out depressions in bottom, as in lungfish, but the stickle back collects a mass of vegetation in which the nest in made. *Salmon* and *Trout* excavate a depression in a riverbed among gravel.

Floating nest are made by a number of Anabantids (the gouramis and the siamese fighting fish) and south American catfish, in which the eggs are suspended in a mass of bubbles produced by the fish. Nest building involves the owners of the nest in territorial behaviour.

Some catfish and cichlids have adopted the habit of mouth brooding. One of the pair (in catfish the male, in cichlids, the female) will gather the fertilized eggs into its mouth and carry them around until they hatch.

Amphibians: The *surinam toad* has an unusual method of breeding; the egg develops in small pits in the female back. Cases of parental care are known from urodela and apoda as well. *Amphiuma* keeps a guard by coiling around the eggs, while the eggs are attached to the neck of the salamanders, Desmognathus. Limbless *Ichtypophis* (female) lies coiled round the eggs in a hole underground.

Curious instances of parental care are known from each group of amphibians. A number of different species of frogs and toads construct nests or shelters in which the eggs are laid, in others, parents carry the egg or progeny over their own bodies.

All such instances can be briefed into three categories:
 a) those laying egg in water
 b) those laying egg outside water
 c) those carrying egg over the body

a) **Egg laying in water** *Hyla faber* (Tree frog) come down from their natural home on trees, the female dig up mud from bottom of shallow pond, make small enclosures, the nurseries, each about 30 cm in diameter. Eggs are laid in them and hatched out larvae remain safe from predatory fish and insects.

b) **Egg laying outside water** *Leptodactylus mystacinus* of South America has left water completely and lives on land. Its nest is a hole in the mud quite near a pond. A gelatinous material from oviduct form a foam or froth that fills the hole and eggs are laid in it. In a few days it starts raining and by that time the larvae are hatched out. Soon the pond swells and include all such nest around it.

Chiromantis lays eggs in similar manner but the froth is beaten by hind limbs and raised into huge mass of jelly.

Rhacophorus schlegelii of Japan has evolved a novel technique to see that its eggs are well cared for. The copulating males and females burrow deep into the ground on the bank of a standing pond. Eggs are laid in a froth and parents retire through a tunnel they dig, the tunnel slopes down straight into pond water. Froth settles down and turns into a fluid shortly before the tadpoles are readyz to hatch out.

c) **Carrying eggs over body** Direct care by the father is best shown by the famous mid wife toad *Alytes obstetricans*. It accepts the entire mass of eggs and ties them around the waist and thighs, then retires to some shady and moist place, occasionally coming out at night to moisten the eggs with dew or pond water and quietly retreats back.

In *Pipa americana*, the female takes charge of the eggs over its back. Marsupial frog *Nototrema* where, a pouch develops on back of female for storing the string of eggs.

Arthroleptis of Central America lays eggs over moist soil; the parents keep waiting for the tadpoles to come out. Tadpoles are lifted on to the back and carried about till changed into tiny frogs.

Reptiles: The reptiles had left the aquatic environment even for laying eggs and had developed special means to provide everything necessary within the egg. As a food supply for the embryo, the egg of a reptile or a bird contains a large amount of nourishing yolk. A fluid filled sac, amnion is a substitute for the pond of an amphibian. It is because of cleidoic eggs that the reptiles and later vertebrates have been able to eliminate the aquatic stage in their development, and this is the main reason of their success upon land. The newly hatched individual immediately starts an existence of its own. *Hemidactylus* – female have the habit of brooding the eggs; it is something unusual for lizards. In some lizards and snakes, eggs though usually laid (oviparous), but retained in female for development.

In turtles, female swims ashore and selects a suitable nesting site on sand. Once this is done, it scoops out a hole with its flippers for its body to rest in. It then deepens the hole, using flippers alternately to dig and prevent the sand falling back into the excavation; then the eggs are laid, in hole which is about 10″ wide and 8″ deep. Finally the animal leaves out and throws back showers of sand until the sight of egg laying is quite concealed.

In snakes i.e. *Vipera berus* which is viviparous species the yolk sac is still large at the time of birth and it provides the young with nourishment during their approximately four week period of activity in the autumn periods to hibernation as well as during the subsequent winter months.

Phrynosoma, horned toad, found in Western American deserts, has the habits of burrowing in sand. This is the only **iguanid** in which the egg develops within the mother.

Aves: a) Birds can be classified into two types as far as parental care is concerned.

(i) Altricial birds or animals which are helpless at birth or hatching and require parental care for some time.

(ii) Precocial bird or animals which are able to move about and care for themselves to a considerable extent immediately following birth or hatching.

Birds may be either altricial or precocial. These, however, are relative terms and the degree of dependence of the young on the adults may range from total to none.

b) Young birds have evolved many special behaviour patterns in connection with feeding habits, the most important being the inducement to the parent bird to regurgitate food on seeing the coloured inside of the throat and mouth of the young. Later when the young leaves the nest, but are still being fed by the parents, they develop a begging call which has a similar effect.

Physiological changes take place in the breeding bird which not only affect its reproductive organs. For example a patch on the breast—the **brood patch**, becomes abundantly supplied with blood vessels and the feathers in that area may moult. The result is that a bird can settle itself down on to its eggs bringing the warm patch into contact with them. The need to incubate involves remarkable feats of endurance. An emperor penguin parent will hold its single egg on its feet, covered by a fold of skin, for seven or eight weeks while its partner is away at sea, feeding. The parents exchange places, taking their turn in feeding and incubating. Nest sanitation is of the great importance, not only for reasons of hygiene but also because the white droppings may make the nest conspicuous. So the parent birds remove the droppings of the very young which, when they are older, move to the edge of the nest to defecate. Pieces of egg shells are also removed from the nest. Penguins nest on the ground, they are usually faithful to mate and nest site throughout life. The more southerly species undergo a long fast during incubation period. Such as little blue penguin nest in burrows or crevices. Emperor and king penguin make no nest but carry their single eggs on their feet, covered with a flap of skin. Most species nest in colonies or "rookeries".

Mammals: Some mammals also make nests for their young. Rabbits, for example, build nests lined with hair, which the female plucks from her body. Toward the time of birth the hair on the parts of the mother's body that she can most easily reach become loose and can easily be plucked out.

A pregnant *Funambulus* female builds up a nest in a hollow of the tree by means of twigs, leaves, grass, cotton, wool and its own fur. Two to four live and blind youngs are borne at a time after about 28 days of gestation. Some squirrels hibernate in winters during which the body is curled up and the bushy tail covers the body like an umbrella, thus serving as blanket. Young mammals must suck the mother's nipples to obtain the milk they need and yet the ability to locate the milk source seems to be a matter of experience. The parent often helps the young animal in its search by pushing it into the right position.

A mother cat will leave its kittens to feed itself for an hour or so. On return it wakes the kittens by nudging them, circles around them and settles down, enclosing them in an oval formed by the body and legs. This restricts their search to the underside of the body and so aids their learned feeding behaviour.

Care of the new born extends also to their bodily functions. Kittens after birth, are unable to urinate or defecate without the mother's help. It is entirely due to licking and stimulation that they are able to rid their bodies of waste products in the first few days of their life without the mother's attention they would die.

This behaviour probably arise from the fact that in the wild state it is certainly safer for the mother to collect the excrement and keep the later clean much in the same way as the parent bird keeps the nest clean.

There is, of course, much evidence that mammals teach their young some of the activities which are essential for life, for obtaining food often the young are encouraged to imitate a parent or are punished if they go to a place which is unsafe.

Chapter 17

Nervous System

All organisms live in an external environment that is some time friendly, some time inimical and seldom neutral. Under such conditions the responsibility of nervous system is to acquaint the organism with its external environment by sensing environmental changes and prepare the organism to adopt according to environment or to respond on the basis of its previous experiences. Obviously, the receptivity of stimuli in an animal, as well as the range and intensity of its response, and hence its degree of fitness for survival in a particular environment are all determined by its genetic memory which every animal has obtained through countless generations which is inherited to every individual from its progenitors.

Hence as experiences multiply, information accumulate and the rewards of the success and penalty of past mistakes, all these assessment are then transferred to memory site where they get stored in the brain which is the principle site of information storage. Besides storing, it analyses the sensory information, correlates them and then issue appropriate command. In this manner, the memory store is continually modified, readjusted and referred during the life time of the animal, with time, further experiences are gained. This learning ability is found exclusively in the higher forms of animals. In higher forms the nervous system along with endocrine system, serves to coordinate and integrate the activities of various parts of the body so that they act harmoniously as a unit.

The nervous system comprises

  a) The **central nervous system** which includes, the brain and the spinal cord.

  b) The **peripheral nervous system** which comprises the nerves arising from the brain and spinal cord. It is further divided into:

   (i) **peripheral central system** which includes nerves (cranial and spinal) in connection with parts of the body that are under voluntary control, and

   (ii) **peripheral autonomic system** which consists of nerve fibres associated with spinal nerves and certain cranial nerves that supply the organs under involuntary control.

Meninges

The parts of the central nervous system lie well protected within the head and vertebral column. Skeleton of head is surrounded by a tough collagenous membrane called perichondrium, in case when the former is composed of cartilage, periosteum when it is made up of bone. These membranes are commonly called endorachis (Fig. 17.1). In addition to endorachis, the brain and spinal cord are protected by membranes known as meninges, minimum number of meninges is one and the maximum is three.

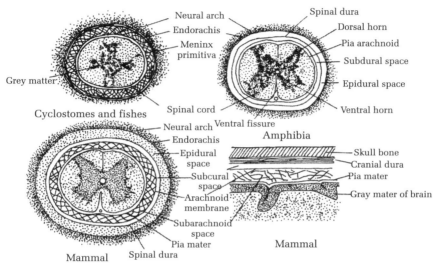

Figure 17.1
T.S of spinal cord of different vertebrates to show meningial membranes

In **cyclostomes and fish**, there is one meninx, the **meninx primitiva** (Fig. 17.1) fitting closely over the brain and spinal cord. The space between the meninx primitiva and the endorachis, called perimeningeal space filled in with lymph like, cerebrospinal fluid traversed by strands of connective tissue that connect the meninx primitiva with the endorachis.

In **amphibia, reptilia and birds,** (Fig. 17.1) the meninx primitiva is replaced by two more or less distinct meninges below the endorachis. These are an outer spinal dura or dura mater formed mostly of connective tissue, and an inner vascular pia-arachnoid membrane derived from the cells of neural crest. Dura mater covering the brain is called **cranial dura** and is continuous with the spinal dura at the foramen magnum. The spaces both above and below the dura mater, are filled with cerebrospinal fluid. Space between the Dura mater and pia-arachnoid membrane is the subdural space. Epidural space contains, besides cerebrospinal fluid fatty and connective tissue and a number of veins.

The **mammals** have three distinct meninges (Fig. 17.1) the pia-arachnoid membrane having split up into an inner pia mater and an outer arachnoid membrane crossed by a network of fibres; the latter having derived its name from *arachne* meaning spider's web, with which it resembles in appearance. Arachnoid membrane and pia mater, though separated are collectively known as **lepto-meninges**. Pia mater closely fits over the surface of the brain and spinal cord, and at places is even carried within them. A new space is added between the arachnoid membrane and the pia mater, known as sub-arachnoid space. This space with subdural and epidural spaces are filled with the cerebrospinal fluid. Cerebrospinal fluid filling the cavities of brain and spinal cord is mostly secreted in the parts of brain. It flows from front to back and through a median **foramen of magendie** and two lateral foramina of **Luschka** in the roof of medulla oblongata, it communicates with the fluid in the sub-arachnoid space. There are special sites in the arachnoid layer containing arachnoid villi, in which the cerebrospinal fluid is absorbed back into the venous system.

Central Nervous System

Development of neural tube

The rudiment of brain and spinal cord first appears as an ectodermal thickening on the dorsal side of the embryo, the thickening is known as **neural** or **medullary** plate (Fig. 17.2). By a process of differential growth and cell migration, longitudinal ridges, called neural folds, arise from both side of the plate. Folds soon begin to approach one another making,

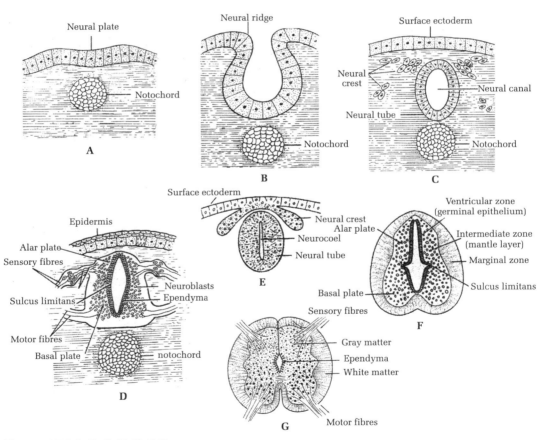

Figures 17.2 A, B, C, D, E, F, G
A, B, C. Formation of spinal cord from neural plate   **D.** Neuroblast differentiation into neurons
E. Differentiation of neural tube in region of spinal cord   **F.** Later developmental   **G.** Sensory fibres from a dorsal root ganglion entering alar plate to synapse with secondary sensory neurons, and motor fibres are emerging from basal plate neuroblasts to innervate striated muscles

firstly a shallow neural groove between them and the neural plate, and finally converting the neural plate into the neural tube. The neural folds fuse and the neural tube sinks down to take a position immediately above the notochord; the epidermis over the tube is reformed. Neural groove is wider in front than behind, in the region of spinal cord. Meeting and fusion of neural folds are not simultaneous at all places and, for some time, the neural tube communicates in front with the outside by the neuropore, and behind with the alimentary canal through a passage called **neurenteric** canal. Lateral parts of the thickening, which give rise to the neural plate, do not get folded into the nerve-tube, when the folds meet. They lie just to the side of the points of fusion of the neural folds and form the neural crest. The neural crest cells migrate from the neural crest to all parts of the body and give rise to many structures, such as sensory neurons, whose cell bodies form the dorsal root ganglia, **Schwann cells** on the nerves, pigment cells, mesenchyme cells in the gills, odontoblast, for the teeth, chondroblasts for the cartilages, osteoblasts for bones, and certain meninges. Soon after the formation of neural tube, its anterior part enlarges and constricts at two places to form the three primary brain vesicles (Fig. 17.3) the remaining part of the tube gives rise to the spinal cord. The central canal of the neural tube in the brain vesicles will give rise to the ventricles of brain which remain continuous with the central canal of the spinal cord. In the region of spinal cord a fibrous dorsal septum grows up from the roof of the central canal and joins the pia mater. The junction is in the form of a groove called dorsal fissure, which also marks the meeting point of the neural folds in the early development. Ventro-lateral parts of the spinal cord outgrow their dorsal counter parts and meet in the mid-ventral line to form the ventral fissure of the spinal cord.

i) Differentiation of Brain parts

Two constrictions in the brain vesicle divide it into three primary brain vesicles called **forebrain** or **prosencephalon**, **midbrain** or **mesencephalon** and hind **brain** or **rhombencephalon**. (Fig. 17.3) The three vesicles soon become five by a paired out pocketing of the anterior end of the forebrain, called **telencephalon**, and a median outpocketing of the hind brain forming the **metencephalon**. The telencephalon is designed to form the cerebral hemispheres and the remaining part of the forebrain to become the **diencephalon** or **thalamencephalon**, also called the "**tween brain**". Antero-ventral part of the telencephalon grows forward to differentiate into olfactory lobe, constituting the rhinencephalon. From the mid brain grow out paired dorsal outgrowths called optic lobes. The

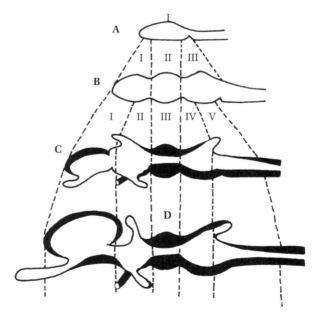

Figure 17.3
Differentiation of the encephalon **A.** Primitive encephalon **B.** Division into anterior presencephalon, a mesencephalon and posterior rhombencephalon **C** and **D.** Further division into the five regions of the brain

metencephalon gives rise to the cerebellum, the remaining hind brain called myelencephalon, develops into the medulla oblongata, which remains continuous with the spinal cord behind. With the fate of various parts of the brain thus established, further differentiation gives rise to the brain of an adult vertebrate. The **roof** and side walls of the telencephalon bulge outwards and forward in front of the original anterior wall of the telencephalon (lamina terminalis) to differentiate into cerebral hemisphere. The central canal of the original neural tube, called neurocoel consequently, becomes remodelled to form the ventricles of the brain. Paired cavities of the cerebral hemispheres are known as paracoel or the lateral ventricles, and their extension, if any, change into the olfactory lobes, as olfactory ventricles, or rhinocoel (Fig. 17.5) The roof and dorsolateral walls of the hemispheres are called the **pallium**; while the floor of the ventrolateral sides becomes thickened into **corpora striata** (Fig. 17.4)

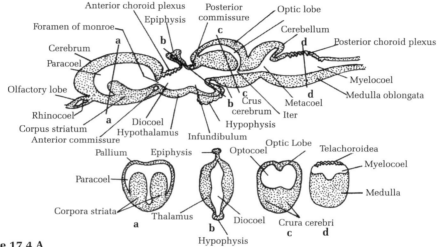

Figure 17.4 A
Development and differentiation of brain. V. L. section of brain. Below a, b, c, d sections taken at different levels as indicated

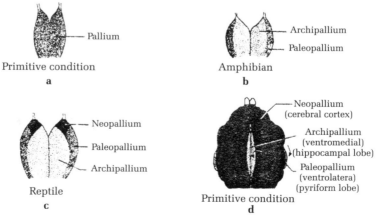

Figure 17.4 B
Evolutionary progress in development of pallium

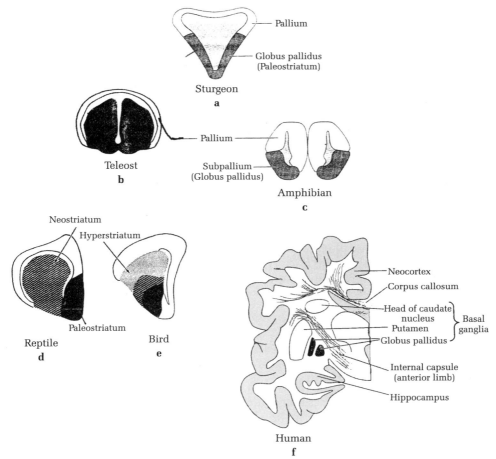

Figure 17.4 C
Evolutionary progress in development of striatum. Evolution of the cerebral hemispheres as seen in cross sections. Reptiles have added a neostriatum to the old paleostriatum. Birds added a hyperstriatum. The striated complex (now called basal ganglia) present in the mammal, and the addition of a cortex on the surface of the mammalian hemisphere.

Evolutionary significance of cerebral hemispheres: As we ascended the vertebrate scale we see the cerebral hemispheres enlarge in size and undergo the greatest change than any other part of the brain. The roof of the cerebral hemisphere is called pallium which covers the lateral ventricles. It is the pallium which has become highly developed and modified and has contributed to the evolution of the cerebral hemispheres of higher group of vertebrates.

Initially the pallium is relatively thin walled, and the gray matter is present only on its inner wall adjacent to the ventricle. In teleosts the gray matter of the pallium has pushed laterally and downward towards the corpus striatum, which is very large and well developed hence in them the pallium is thin and nonnervous layer of the roof.

In **amphibians** the arrangement differs but little from that of fishes except that pallium is thicker and more gray matter has moved to the peripheral position. As a result of this the pallium can now be divided into two regions, a dorsal medial archipallium and a more lateral paleopallium (Fig. 17.4 B) As a result the cerebral hemisphere of amphibians are relatively larger than those of fishes.

In **reptiles,** the changes in telencephalon are very conspicuous and quite prominent, as a result of this cerebral hemisphere become very large in size. The corpus striatum is well developed. In certain reptiles, a new area, the neopallium (Fig. 17.4 B) has appeared on the outer portion of each cerebral hemisphere at its anterodorsal end between archipallalial and paleopallial areas.

In **birds** the cerebral hemisphere are large but because of great size of corpus striatum and not due to pallium.

In **mammals,** it is the growth and development of the neopallium that accounts for the large size of the cerebral hemisphere (Fig. 17.4 B). In mammals, especially man, the cerebral cortex reaches the climax of its development. The neopallium has grown to a relatively enormous degree (Fig. 17.4 B) pushing the archipallial area medially and ventrally so that it becomes folded and lies in ventromedial portion of cerebral hemisphere where it is known as hippocampal lobe (Fig. 17.4 B) likewise the development of neopallium, and the paleopallium, originally in a latter position is pushed ventrally where it becomes pyriform lobe.

In almost all vertebrates the floor of each cerebral hemisphere differentiate into a thickened corpus striatum. The corpus striatum is thick ventrolaterally in fish and each cerebral hemisphere consists chiefly of a paleostriatum (Fig. 17.4 C). The striatum, lateral ventricles and thin roof (pallium) constitute the entire cerebral hemisphere (Fig. 17.4 C). Sturgeon and teleosts.

The striatum of amphibian receives a large number of sensory fibres and develop additional nuclei that have evolved in the formation of subpallium (Fig. 17.4 C) which enable the cerebrum to participate more extensively in coordinating sensory information and in directing responses. In specialized reptiles additional nuclei constituting a neostriatum are added to the striatum and these nuclei receive many more sensory fibres from the thalamus (Fig. 17.4 C).

Birds cerebral hemisphere are essentially reptilian. The striatum reaches peak development and additional strata of nuclei the hyperstriatum are superimposed on the neostriatum (Fig. 17.4 C). The corpus striatum is less developed in mammals as compared to lower forms.

In mammals the striatum and other nuclear mass, now collectively called the basal ganglia, continue to play an important role in nervous system. The globus pallidus is one of a group of striated nuclear masses known as striatum in amniotes. The nuclei of the pallium and the globus pallidus persists in tetrapods, although eventually becoming subsidiary to more recently evolved nuclei. But the cerebral cortex on the greatly expanded roof i.e. pallium has become the most conspicuous part of the mammalian brain, Fig. 17.4 C, (vi).

The development of the many additional component of the cerebrum of mammals has displaced the globus pallidus to a position deep within the hemisphere. It has been joined by adjacent striated nuclei, a caudate nucleus, a putamen and an amygdaloid nucleus. Early homologue of the associated nuclei are present in nonmammalian amniotes. Collectively these nuclei constitute what has been known as basal ganglia (Fig. 17.4 C).

Thickenings and outgrowths appear in the diencephalon in which lies the 3rd ventricle or diocoel separated from the lateral ventricles by the lamina terminalis, but communicating with them by paired interventricular foramina of monro. The more dorsal part of the diencephalon is called epithalamus. It is made up of non-nervous tissues, which together with vascular pia mater covering forms the tela-choroidea. Villi-like processes of the tela-choroidea (Fig. 17.4) extend into the diocoel, the **tela choroidea,** with its processes forms the anterior choroid plexus, which produces the cerebrospinal fluid filling all the ventricles of brain and other spaces. The plexus often extends in the roofing wall of lateral ventricles through the foramina of monro. (Fig. 17.4). Behind the anterior choroid plexus, the roof of the diencephalon gives off two median projections, an anterior **parietal body**, and a posterior **pineal body** or epiphysis, the two collectively are known as **epiphyseal apparatus**. Lateral thickenings of the diencephalon form the thalami, the ventral part forms the hypothalamus (Fig. 17.4). The later consists of several parts: (i) optic chiasma is the anterior most part, where the two optic nerves cross as they pass from the eyes to the brain, (ii) the tuber cinerarium behind the optic chiasma is a median outgrowth, followed by (iii) the infundibulum, to which is applied (iv) the **hypophysis** (Rathke's pocket)

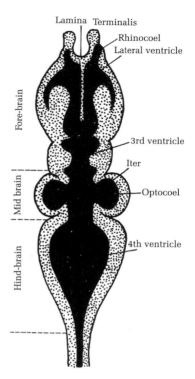

Figure 17.5
Horizontal section of a generalized brain showing the ventricles of brain

developed as an evagination from the mouth cavity of the embryo. The infundibulum and hypophysis together constitute the pituitary body. The mid brain undergoes relatively less modifications. The dorsolateral walls become thickened as tecti opticum, which bulge out into a pair of optic lobes or corpora bigmina. Cavity of mid brain, the mesocoel (Fig. 17.5), is carried into the optic lobes as optocoels. The floor and side of the **mid brain** are thickened into tracts of grey matter, the **crura cerebri** or tegmentum, connecting, the diocoel with the metacoel (IV ventricle in the hind brain) is known as cerebral aqueduct or iter or **aqueduct of sylvius.**

A median dorsal outpocketing of the hind brain gives rise to the cerebellum (Fig. 17.4) the part of the neurocoel it carries with it is called the metacoel or cerebellar ventricle. The remaining part of the hind brain differentiates into the medulla oblongata having thick ventral and lateral walls and enclosing the IV-ventricle or myelocoel. The roof of medulla is non-nervous and gives rise to posterior choroid plexus.

Commissures The two sides of the brain are joined at definite places by fibrous tracts called commissures (Fig. 17.4). An anterior commissure, connecting olfactory region of the brain is located in the lamina terminalis. A posterior commissure lies in the roof of the diencephalon just in front of the optic lobes. In higher vertebrates, a dorsal or superior or habenular commissure, similarly, lies in front of the pineal body and a soft or middle commissure connects the two thalami of the diencephalon. Soft commissure is not fibrous and is not actually a commissure, since it is formed by the meeting of the inpushings of the thalami. Some more commissures are present in the brain of the mammals.

Grey matter and white matter Distribution of grey matter and white matter, and of the four functional components viz; the somatic sensory SS, somatic motor SM, visceral sensory VS and viseral motor VM, in the posterior part of the medulla oblongata, is the same as in the spinal cord (Fig. 17.2). In the anterior part of the medulla, grey and white matters are irregularly intermingled, while in the cerebellum, midbrain and cerebral hemisphere this simple arrangement is greatly distorted. In these parts the grey matter comes to lie outside and surrounding the white matter.

Functions of the Various Components of Brain

The cerebral hemispheres represent an immense development of somatic sensory components. In lower vertebrates, such as **cyclostomes, fish** and **amphibia**, these are concerned mainly with receiving olfactory impulses from the olfactory organs and relaying them to the diencephalon. In higher vertebrates, the function is shadowed largely by the more important function of receiving impulses from many other important receptors, such as **eyes, ears** and **skin** and directing appropriate responses in the light of past experiences stored there in. The sum of these activities characterizes the psychic life, such as intelligence, thought and sensation (Fig. 17.4). By means of the ascending and descending fibres, and connections with other parts of brain, the hemispheres have a direct control of the muscles and other effectors.

The thalamus of diencephalon contains a number of cell bodies of neurons, which serve as **relay centres** for the sensory fibres coming from various receptors and having their destination in the cerebral hemispheres. Some of the ventrally located cell bodies in the thalamus relay impulses originating in the hemispheres and directed to effector organs. The floor of diencephalon (hypothalamus) is composed of visceral sensory and visceral motor components and serves to regulate many of the activities directed through autonomic peripheral system. Hypothalamus controls the metabolism of fat, water and carbohydrates, integrates the activities of detector, intake and digestion of food and also controls the amount taken, regulates the body temperature and blood pressure, genital functions and sleep. Hypothalamus also produces a hormonelike substance called hypothalamic D. Hypothalamus and pituitary are intimately connected and therefore influence and control a number of activities.

The optic lobes are composed of somatic sensory components. In lower vertebrates, they serve as centres for the visual senses and receive impulses from the eyes. In higher vertebrates, such as mammals where there are four optic lobes, the anterior pair receives impulses from the eyes, while posterior pair has centres for receiving impulses from the ears that help in maintaining the correct posture. The crura cerebri have immensely developed somatic motor component serving to communicate impulses from the fore brain to the hind brain.

The **cerebellum** represents an immense development of the somatic sensory component. It receives fibres from the ear and send impulses to the motor centres of the crura cerebri. It is concerned chiefly with the balance or equilibration of posture of the body and voluntary movements initiated in the cerebral hemispheres and, thus, co-ordinates the neuromuscular mechanism of the body.

Lastly, all the four functional components are represented in the medulla oblongata in its thickened ventral and lateral walls. In its visceral motor area are located centres for the regulation of heart beat, respiration and metabolism in general. The visceral sensory area has centres for the sense of taste (only in **fish** and **amphibian** larvae). The somatic sensory area has centres for receiving impulses from the lateral line system (only in **fish** and **amphibian** larvae) and the ear.

Differentiation of spinal cord Spinal cord is usually a slightly dorso-ventrally flattened tube. It is very much flattened in **cyclostomes** and **fish** and is of fairly uniform diameter throughout its length. In the tetrapoda, it presents cervical and lumbar enlargement associated with the concentration of nerves supplying the limbs (Fig. 17.15). Such enlargements are absent in limbless tetrapods such as snakes. The two enlargements vary in proportionate size with the difference in the degree of development of the two pairs of limbs. The length of spinal cord fails to keep pace with that of vertebral column, and is therefore usually shorter than the latter, however, a non-nervous portion of spinal cord, the **filum terminale**, runs through some more vertebrae. In mammals, the nervous part terminates in the lumbar region even if there is a tail. A pair of spinal nerves arises; one from each side of spinal cord, and correspond to a body or vertebral segment. Each nerve originates by a dorsal nerve root and ventral nerve root (Fig. 17.14 A).

Spinal cord itself is composed of two kinds of matter: (1) The grey matter, surrounding the central canal and consisting of (i) axonic fibres of the afferent or sensory neurons passing into it through the dorsal nerve root and (ii) the cell bodies of association or internuncial and efferent or motor neurons. (2) The white matter, almost surrounding the grey matter and consisting chiefly of axons with white medullary sheaths. The cell bodies of afferent or sensory neurons are situated outside the spinal cord over the dorsal root, where they form the ganglionic swelling, their dendrites lie in the receptor surface. The cell bodies of motor neurons together with their dendrite lie in the ventro-lateral parts of the grey matter, their axonic fibres (medullated and non-medullated) pass through the ventral nerve root to contact the effector organs. The shape of the grey matter varies with the degree of development of ventro-lateral parts of the spinal cord where motor neurons are concentrated. In the amniota, it usually assumes H-shape or a butterfly shape, (Fig. 17.14 A) the limbs of the H or wing-tips of the **butterfly** form the dorsal and ventral horns or columns. The connecting bars above and below the central canal are, respectively, known as dorsal and ventral grey commissures. The characteristic shape, that the grey matter assumes, leaves the white matter divided into longitudinal columns called **funiculi**. Thus, between the dorsal columns of grey matter, on either side of the ventral fissure. A lateral funiculus lies on either side between the dorsal and ventral columns. A funiculus is composed of white medullated fibres.

Functions of the spinal cord

Familiar experiments showing reflexes with frog, whose brain has been damaged, go to prove that the spinal cord is responsible for local and spontaneous reflexes arising out of the stimuli operating in the external environment. In such simple reflexes, signals are picked up at the receptor surface by the dendrites of the afferent or sensory neurons, and after passing through their cell bodies in the dorsal ganglionic swelling, signals are transmitted to the grey matter through the sensory axons. Fresh signals are then transmitted to the cell-bodies of motor neurons and then by their axons, pass via the ventral nerve root, to the effector organs. A sensory neuron may synapse directly with the motor neurons, the arc thus completed is called a **monosynaptic reflex arc**. But more commonly, a sensory neuron synapses with one or more motor neurons, such complex arcs are called **polysynaptic**. Internuncial neurons on the same side of the spinal cord or brain are called association neurons while those crossing on the opposite side are called commissural neurons.

Besides carrying out local reflexes, the spinal cord is conducting pathway of signals. Some of the signals arriving in the grey matter are communicated to the brain through internuncial neurons whose axons synapse either directly with the efferent neurons of the brain or are relayed to the brain by the internuncial neurons of the brain. Similarly, signals from the brain are communicated to the effector organs by internuncial neurons of the brain which make synapses either directly with the spinal motor neurons or through one or more internunical neurons in the way (Fig. 17.14 A). Thus, the spinal cord also contains both ascending and descending fibres for the brain, through these cerebrospinal fibres, brain controls some of the actions of the spinal cord. The spinal cord, with its two important functions just described, sends out throughout life continual streams of impulses to the muscles and other effectors, which help it to regulate the body posture, and initiate movements which are, of course, guided by the stored information (memory and experience) in the brain and transmitted to spinal cord through descending fibres.

The functional nerve components

These are different receptors and effectors for dealing with external and internal stimuli, respectively (Fig. 17.14 A). Receptor receiving external stimuli and located in the skin and its derivatives are called **somatic receptors**, and the effectors, which respond to the stimuli coming from the external world, are called somatic effectors, such as the striated muscles below the skin. Similarly, there are **visceral receptors**, such as the mucous membrane of the alimentary canal and visceral effectors, such as the smooth muscles and glands in the wall of alimentary canal. Receptors, whether somatic or visceral, are always innervated by sensory neurons, which must have their cell-bodies situated in the dorsal horns of grey matter. Similarly, all effectors are innervated by motor neurons, which have their cell-bodies in the ventral horns of grey matter of central nervous system and their fibres always leaving via the ventral nerve root. This leads to the conclusion that the spinal nerve formed by the union of the two nerve roots contains four kinds of fibres—somatic sensory, somatic motor, visceral sensory, and visceral motor dealing respectively with internal and external reflexes. Anatomically these four kinds of component fibres (Fig. 17.14 A) have relations with similarly named definite areas or columns of the grey matter. From the ventral side of the spinal cord (also medulla of brain) upwards these are (1) somatic motor (SM), located in the ventral horn and containing large-sized cell bodies of somatic motor neurons, (2) visceral motor (VM) above the somatic motor area, composed of smaller motor cell bodies and forming the lateral horn of the grey matter, (3) visceral sensory (VS) is the lower part of dorsal horn and (4) somatic sensory (SS) is the upper part of the dorsal horn.

Comparative Account of Central Nervous System

In *Branchiostoma,* the spinal cord is least modified and its anterior end terminates in a brain like structure of two vesicles. Spinal cord is some what triangular in a cross-section with a slit-like central canal. There are no medullated fibres, hence no such differentiation as grey and white matter exists. The dorsal and ventral nerve-roots do not join. Nerve cells near the central canal are exceptionally large sized.

While doing comparative study of brain from fish to mammals a few important points are to be noted which are as follows.

(i) There is a tendency to increase the size of cerebral hemispheres in proportion to other parts of the brain.

(ii) The grey matter which at first is concentrated on the roof of the paracoels, tends to move towards the surface and thickens to form the cerebral cortex, its thickness is directly proportional to the intelligence.

(iii) The cerebral hemispheres become more and more connected with each other by means of commissures to co-ordinate the actions of the two hemispheres.

(iv) The surface area of the hemispheres tends to increase by folding.

(v) Olfactory lobes and other olfactory centres are better developed in lower vertebrates than in higher vertebrates.

(vi) The large size of cerebral hemisphere in lower vertebrates is not because of cerebral cortex, but because of large corpora striata.

(vii) Cerebellum is more extensive and advanced in animals which move in more than one plane and whose muscular responses are far more numerous than in animals leading quiet life.

Fish: The spinal cord and brain are covered with a single meningeal membrane. The spinal cord varies in the length in *Scoliodon*, it extends almost the whole length of the vertebral column, in many others, it falls short of the back bone. In a transverse section of the spinal cord of *Scoliodon*, it is almost rounded enclosing a narrow central canal, the dorsal fissure is shallow and a ventral fissure is well marked, however, in most fish, a ventral fissure is insignificant. There is a clear differentiation of grey and white matters, the former arranged in the form of a triangle with a dorsal apex.

The brain of *Scoliodon* (Fig. 17.6) is large and the usual three divisions are well represented. The forebrain consists of a large and smooth cerebrum or telencephalon, a median longitudinal fissure, dividing the cerebrum into a pair of cerebral hemisphere in **fish**, such as **European Dogfish**, is not marked in Indian **shark**. At the anterior end of the cerebrum are the olfactory lobes (rhinencephalon) not quite distinct from the cerebrum. From each olfactory lobe arises a stout stalk, the olfactory tract or peduncle, which becomes expanded at its distal end into a bilobed olfactory bulb. Olfactory bulb of each side is applied to the outer surface of the olfactory or nasal sac. Large size of the olfactory sacs and their relations with the brain suggest that cerebrum receives sensory fibres from the nasal organ and serves mainly as an apparatus for analysing the olfactory impulses. Sharks discover their food largely by smell. Cerebrum also receive sensory fibres from other receptors. Roof of the cerebrum is quite thick but the grey matter is concentrated only around the lateral ventricles.

These cavities are extended into the olfactory bulb through the olfactory tracts to form rhinocoel of olfactory ventricles. On the ventral side of cerebrum are a pair of small or terminal nerves, arising from the olfactory sacs and passing into the cerebrum through mid-ventral neuropore. A ganglionic swelling lies half-way down each nerve. Corpora striata on

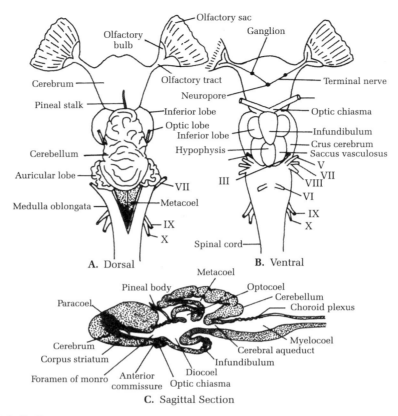

Figures 17.6 A, B, C
Brain of *Scoliodon* **A.** Dorsal view **B.** Ventral view **C.** Sagittal section

the lateral and ventral walls of the cerebrum are present. Their function is not yet well understood.

Diencephalon is almost completely shadowed by the large sized cerebellum. The cavity of the diencephalon is called diocoel, its non-nervous roof forms the anterior choroid plexus. From the posterior dorsal portion of the diencephalon arises a slender pineal body directed forward and upward. Lower part of the diencephalon (hypothalamus) presents:

(i) Optic chiasma formed by the crossing of the optic nerves in front of the infundibulum. The infundibulum, as a median outgrowth is flanked on either side by a thick-walled sac called inferior lobe.

(ii) The saccus vasculosus, as a thin walled dorsal expansion of the distal end of the infundibulum this structure is a pressure receptor, since it is well-developed in deep sea-fish. In aquatic amphibians it serves as a stato-receptor having connections with the cerebellum and responding to stimuli due to gravity changes.

(iii) The hypophysis, which together with the distal part of infundibulum, forms the pituitary gland. Part of the diocoel extends with the infundibulum. Diocoel communicates with lateral ventricles by the interventricular foramina of monro, and the iter connects it with the myelocoel.

Mid brain is large and its roof is enlarged, into a pair of rounded, optic lobes or corpora bigemina, each having optocoel. A considerable part of the optic lobe is concealed by the cerebellum. Optic lobes are visual centre and receive sensory fibres from the retina via optic nerves. The roof of midbrain also receives fibres from other centres, such as olfactory, gustatory and from skin. The floor of mid brain is thickened to form crura cerebri, which serves as a pathway along with efferent fibres from the mid brain, which first, pass down, then proceed back into the spinal cord controlling some of the activities of the latter. In the hind brain, cerebellum is very large and hangs over a considerable part of the mid brain and diencephalon in front and covers some part of the medulla behind. Its surface is further increased by numerous convolutions. The deep transverse furrows divide the cerebellar surface into three lobes. The part of the ventricle it carriers into it, is called the metacoel or IV ventricle situated at the anterolateral angles of the medulla on its dorsal side but belonging to the cerebellum are prominent, irregular and hollow projections called auricular lobes or corpora restiformia or restiform bodies corresponding to the floccular lobes of crocodilians, birds and mammals. Each of them contains a continuation of metacoel, called auricular recess. Cerebellum receives sensory fibres from the ear and the organ of lateral line system and through the auricular lobes sends out impulses to the voluntary muscles. The auricular lobes correlate muscular movements that keep the equilibrium of the body. Medulla oblongata is triangular in outline; its sides and floor are thick, but roof is thin and non-nervous, forming the posterior choroid plexus. Cavity of the medulla, the myocoel, continues backwards into the spinal cord. Medulla has centre for regulating and controlling many visceral activities, such as respiration and heart-beats.

The brain of **bony fish** is built on the same plan, with certain characteristics of its own, which are as follows.

(i) The large size of the cerebrum is mainly due to the great development of corpora striata while the roof of the cerebrum remains thin and membranous, and the cerebrum is essentially concerned with the impulses arising from the olfactory receptors;

(ii) Diencephalon is poorly developed because most of the optic fibres do not end here, but pass into the mid-brain, a saccus vasculosus is present;

(iii) The mid brain is sometimes the largest part of brain and has the same functional importance as in the shark.

(iv) The large sized cerebellum sends out a special outgrowth, the valvula cerebelli, under the mid brain.

Amphibians: Brain of tailed amphibians, such as *Triturus*, is similar to that of fish. The cerebral hemispheres are elongated a saccus vasculosus is absent. The roof of midbrain forms optic lobes but the latter are not well defined.

In the brain of frog (Fig. 17.7) the fore brain consists of large, swollen and somewhat elongated cerebral hemispheres, divided in the middle by a longitudinal fissure. In front of the hemispheres are the large-sized olfactory lobes fused in the median plane. Through them olfactory fibres from the nasal sacs pass into the cerebral hemispheres. The lateral walls and the floor of hemispheres are thickened to form the corpora striata, which are joined with each other by the anterior commissure. Another commissure, the hippocampal, lies above the anterior commissure. The roof or pallium is also thickened, grey matter is concentrated around the lateral ventricles, but some cells of the grey matter have migrated to peripheral positions (Fig. 17.4 B). Besides analysing olfactory impulses, the cerebral hemispheres of frog receive tactile and optic impulses, respectively from skin-receptors and the eyes. Efferent fibres from the hemispheres reach only to some part of diencephalon and midbrain.

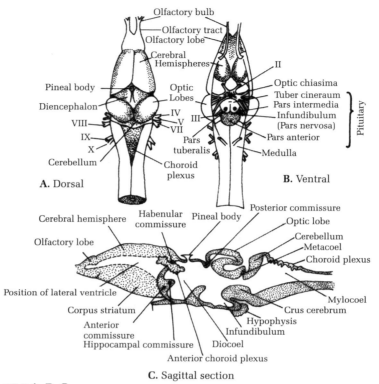

Figures 17.7 A, B, C
Brain of frog **A.** Dorsal view **B.** Ventral view **C.** Sagittal section

Diencephalon appears as a rhomboidal area on the dorsal side, its roof forms the anterior choroid plexus. The pineal body is in the form of a simple sac. Its position over the head is marked by a patch known as **brow spot**, which is apparently a vestige of what might at one time have been a third eye. Immediately behind the choroid plexus lies the habenular commissure. A posterior commissure limits the diencephalon from the mid brain. The constituent parts of the hypothalamus are well represented. Its main part is a bilobed prominence, the tuber cinereum, lying immediately behind the optic chiasma. The infundibulum forms the posterior lobe of pituitary. Other lobes of pituitary are distinctly marked. Diocoel communicates with the lateral ventricles by foramina of monro and is also connected with the myelocoel by iter. Diencephalon receives a number of afferent optic fibres from the eyes, and some from the skin receptors.

Mid brain has large optic lobes and thickened crura-cerebri. Considerable amount of grey matter has moved away from the optocoels to peripheral positions. The optic lobes receive fibres of the optic and also of olfactory and auditory nerves. Efferent fibres from here leave for crura cerebri on their way to medulla and, perhaps, to spinal cord as well.

Cerebellum is small, represented only by a transverse band on the dorsal side of the hind brain, and contains only a very small metacoel. Small size of cerebellum of tail-less amphibians is correlated with the rather quiet and less active life than that of the tailed amphibians and fish. Medulla oblongata is well-developed, its roof forms the posterior choroid plexus, myelocoel continues back into the spinal cord.

Reptilia: The brain, (Fig. 17.8) such as that of *Uromastix* shows some advancement over that of the frog. The smooth cerebral hemisphere are comparatively large, being roughly oval in outline and separated from each other by a deep dorsal median longitudinal fissure. Anterior part of each hemisphere, representing the olfactory lobe is drawn forward to form the narrow and delicate olfactory tract or peduncle, which terminates into swollen olfactory bulb. Cerebral ventricles are continued into the olfactory bulbs forming rhinocoels. The large size of cerebral hemispheres is not because of thickened roof, which remains comparatively thin, but because of the large fibrous corpora striata. The anterior commissure connects the two corpora striata which receive many efferent fibres from the sidewalls of the diencephalon, indicating a tendency of transferring more functions to the cerebrum. Immediately above the anterior commissure lies the hippocampal commissure, connecting the hippocampal areas of the cerebral hemispheres, which are really the olfactory centres receiving afferent fibres from the olfactory organs. Formation of cerebral cortex started in the amphibia has been carried a little further; there are more cells in the peripheral parts of the pallium, but an elaborate cortex has yet to appear. However, in crocodilians, cells of the grey matter have migrated to more peripheral positions in the outer portions of cerebral hemispheres at their antero-dorsal ends. These new areas of concentration are called neopallium and constitute the cerebral cortex (Fig. 17.4 B).

The diencephalon is small and pressed between the cerebrum and mid brain. Its roof, forming the anterior choroid plexus, is extended into the lateral ventricles through the foramina of monro. Thalami are thick and many of the fibres of the optic nerves end here. The posterior and habenular commissure lie behind the choroid plexus. The epithalamus has a well developed epiphysial or pineal apparatus consisting of an anterior eye-like parietal body and a posterior pineal body. The former, it is believed, used to function as an eye, opening to the outside through the parietal foramen in the skull. The parietal eye still continues to discriminate light from dark in many lizards, such as *Varanus*, *Anugis* and *Lacerta*, and in the tuatara (*Sphenodon*), lizard-like reptile of New Zealand. The hypothalamus has all its constituents well represented. Large optic tracts of the optic chiasma decussate and end in the thalami and the mid brain.

Optic lobes with their optocoels are oval bodies but relatively smaller than the cerebral hemispheres and the floor of the mid brain is thickened into crura-cerebri.

Cerebellum is not bigger than its relative size in frog, probably because these animals move mainly in one plane. Crocodiles, on the other hand, moving in more than one plane like fish, have a large sized, cerebellum, consisting of one median and two lateral lobes, respectively known as vermis and floculi, the later corresponding to corpora restiformia of fishes. Medulla oblongata is broad and thick, with the posterior choroid plexus and a strong ventral flexure. Where it passes into the spinal cord.

Birds: The brain (Figs. 17.9 A, B) as that of the pigeon shows considerable advancements over that of reptiles. Brain, as a whole, is of large size relative to the body of bird. Cerebral hemispheres are large and pushing back against cerebellum and there is a corresponding large cerebellum. The immense size of these parts of the brain causes the optic lobes to hang out laterally and the diencephalon to cave in and be represented on the dorsal side very slightly.

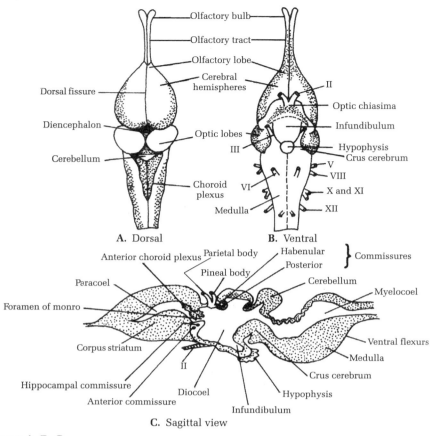

Figures 17.8 A, B, C
Brain of *Uromastix* Dorsal and Ventral views and sagittal section

Smooth cerebral hemispheres are larger than any other part of the brain. Their large size is mainly due to the immensely developed corpora striata, occupying the ventro-lateral positions. The roof is thin and there is little development of cerebral cortex. Olfactory lobes are very much reduced, and the olfactory bulbs are correspondingly small, and mostly represented on the ventral side. This is because of the poorly developed olfactory organs. Birds flying high above the ground, cannot depend much upon their sense of smell. Their finer senses are restricted to the eyes, ears and beaks. The hippocampal areas concerned with the analysis of olfactory impulses in reptiles are, therefore, very much reduced in Birds. Corpora striata do not have direct control over spinal activities, but fibres pass back from here to thalami and parts of mid-brain from where impulses are relayed through other fibres contacting the spinal cord. Corpora striata are joined by the anterior commissure.

Diencephalon is mostly visible on the ventral side, its epiphysial apparatus consists of the pineal body alone. The side walls of thalami are large and receive, besides optic fibres, afferent fibres from the tactile, pain, temperature and perhaps auditory sources. In the hypothalamus, the optic chiasma consists of optic fibres or tract that decussate completely and end mainly in mid brain. The remaining hypothalamus is small and there is no intermediate lobe is the pituitary. The diocoel communicates with the lateral ventricles

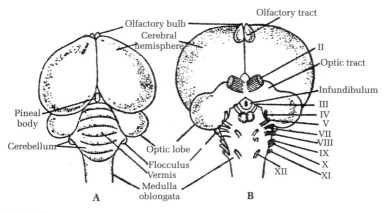

Figures 17.9 A, B
Pigeon brain in A. Dorsal view B. Ventral view

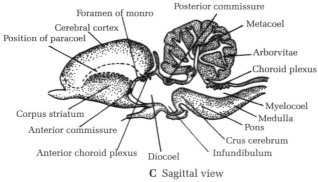

Figure 17.9 C
V.L.S. of brain of pigeon

through the foramina of monro. Posterior commisure separates the diencephalon from the mid brain.

Optic lobes with optocoels are of very large size in correlation with the highly organized visual sensory system necessary in flying and feeding. Cerebellum is far more advanced as compared to the reptilian cerebellum. The surface area is enormously increased by the folding in of the surface in a radiating manner, the folds carry the superficial grey matter deep into the white matter, the whole thing presenting a tree-like branched pattern, called *arbor vitae*. Entire cerebellum consists of a middle portion, the vermis, indistinctly divided into anterior, middle and posterior portions by means of transverse furrows, and the laterally into floccular lobes or flocculi. Large size of the cerebellum is in correlation with control of movements of birds in many planes during flight. By means of many fibrous tracts cerebellum is connected with the cerebral cortex and the spinal cord making coordination between the activities of the cerebral hemispheres and cerebellum and controlling spinal activities. In certain birds, a fibrous tract, called **pons**, develops on the ventral side of the cerebellum and below the medulla oblongata. Pons contains fibres that connect the cerebellum with the cerebral cortex, the fibres decussate as they pass back into the cerebellum. Metacoel is very small.

Medulla oblongata has the same characteristics as in the reptilian medulla. There is a well marked ventral flexure. Posterior choroid plexus is entirely covered by the cerebellum. The dorsal anterior portion of medulla has centres for equilibration and hearing. Myelocoel is connected with diocoel through the iter.

Mammalia: The brain (Figs. 17.10 A, B, C) such as of any one of these animals –squirrel, rabbit and sheep—is of a large size, and its very outlook reflect anatomical complexity.

The forebrain is greatly developed in all mammals the cerebral hemispheres forming the bulk of it and being so enlarged that the olfactory lobes comes to lie below them, while behind, they cover the diencephalon and a considerable part of the mid brain. In squirrel

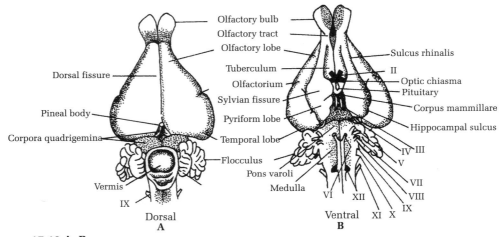

Figures 17.10 A, B
Brain of rabbit   A. Dorsal view   B. Ventral view

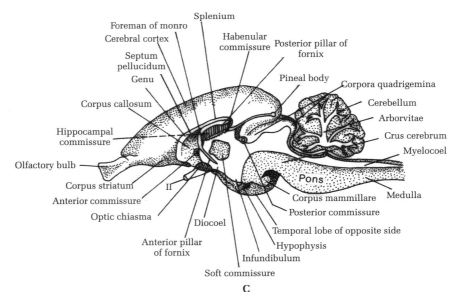

Figure 17.10 C
Sagittal section of the brain of rabbit

and rabbit, the surface of hemisphere is smooth, except at places where there are deep fissures dividing the hemispheres into lobes. In sheep and higher mammals, the surface is folded so as to form wavy ridges, the *gyri* and depressions between them, called *sulci*. Gyri increase the surface area of the cerebral cortex without making the hemisphere proportionately voluminous. The more prominent fissures separating the two hemispheres; and the sylvian fissure, at the side dividing the hemisphere into an anterior frontal and a postero lateral temporal, lobe. The olfactory lobes, more distinctly visible on the ventral side of the hemisphere, consists of two parts, an anterior swollen olfactory bulb, and a posterior stout olfactory tract leading back to a slightly rounded elevation; the tuberculum olfactorium. The hippocampal areas, concerned with olfactory centres in the cerebral hemispheres of reptiles have, in mammals been pushed medially and ventrally to form large hippocampal lobes, separated from the hemispheres by a shallow hippocampal sulcus on each side. Internally, the two lobes are united in a median commissural area called lyra or psalterium. Hippocampal lobes have almost lost their importance as centre of smell and are believed to be related to the production of 'emotional responses' such as searching movements and signs of anxiety and defensive anger. Hippocampi are also concerned with the appetite for food, activity of ductless glands are regulation of blood pressure. Longitudinal grooves, the sulcurhinalis, demark the olfactory tract from the frontal lobes. In correspondence with the trend of olfactory parts taking more ventral positions, the parts of the olfactory lobe of mammals corresponding to a similar part in the brain of lower vertebrates, has been shifted from original lateral placing to ventro-lateral position to be known as the pyriform lobe. The roof of the hemisphere is thickened to form the cerebral cortex.

Mammalian cerebral hemispheres have further been complicated by the appearance, for the first time in vertebrates, of a broad, white mass of medullated fibres, the *corpus callosum* or anterior pallial commissure connecting the cerebral cortex of the two hemispheres. The corpus callosum lies above the roof of the lateral ventricles when viewed in a sagittal section of the brain. Its anterior part is bent slightly downwards to form the **genu** posteriorly, it is bent downwards and forward, constituting the **splenium** which unites with another tract of longitudinally running white modulated fibres, the fornix. The fornix is formed by the union of two strands of fibres called taeniae hippocampi, the ridges are separated both in front and behind by the main body of the fornix, forming the anterior or posterior pillars of fornix. It is with the posterior pillar that the splenium is united. The anterior pillar passes ventralwards connecting the hippocampal lobe with the hypothalamus and mammillary body. Below the corpus callosum, between it and the fornix, is a thin membranous septum the septum pellucidium, separating the cerebral hemispheres on either side of the septum is a venticle. This is not a true ventricle, but is often termed fifth or zero ventricle, pseudocoel is a better term for it.

On the floor of the cerebral hemispheres are corpora striata joined by the anterior commissure. Lateral ventricles or paracoel have paired, anterior and posterior prolongations or horns. Anterior horns are continued into the olfactory bulbs to form the rhinocoels. Cerebral cortex receives information directly or indirectly from nearly all the receptors in the body and every muscles and gland cell is influenced by the impulses originating in it. Almost every single act is influenced by the cerebral cortex.

Diencephalon is almost completely shadowed by the backwards extension of the cerebral hemispheres. Its roof is vascular and non-nervous, forming the anterior choroid plexus, which penetrates into the lateral ventricles through the foramina of monro. The pineal body lies immediately behind the plexus. Side walls of diencephalon, the thalami, are thickened prominently on their inner sides and are fused in the middle of the diocoel to form a median mass of grey matter, called interthalamus, soft or middle commissure or massa intermedia. A rounded outgrowth near the anterior end of the external surface of each thalamus is the geniculate body, which serves to relay impulses from the eyes and ears to the cerebral cortex. The hypothalamus includes:

(i) a large optic tract formed by the union of decussating fibres to the optic nerves that constitute the optic chiasma,

(ii) the thickened floor of diocoel called tuber cinerum, having many autonomic centres,

(iii) the infundibulum, arising as an outgrowth and forming the posterior lobe of the pituitary from its distal parts

(iv) the corpus mammillare appearing externally as a pair of small bodies containing centres for the integration of olfactory sense with related structures of the body

(v) the hypophysis

Optic lobes are small when compared with the respective sizes in other vertebrates. They are almost solid and, by a transverse fissure, are divided into four lobes, forming the corpora quadrigemina. The anterior pair called the superior colliculi, receive some direct fibres from the optic nerves, hence are concerned, to some extent, with the visual sense; posterior pair called (inferior colliculi) receive fibres of auditory nerves from the hearing organs, hence serve to integrate auditory impulses. The posterior commisure connects the superior colliculi, and a habenular commisures connecting the thalami runs across the roof of the diocoel. The crura cerebri (tegmentum) of mammals are far better developed than in the other vertebrates. This area receives fibres from the fore brain and cerebellum and sends fibres to the spinal cord.

Cerebellum is large with a central lobe, the vermis, divided by transverse furrows into anterior, middle and posterior portions. Vermis has bilateral extensions, called lateral lobes or cerebellar hemispheres. To the sides of the posterior portion of vermis are the flocculi or floccular lobes. Surface of cerebellum is thrown into gyri between which are sulci and there is a cerebellar cortex on the dorsal side. The superficial grey matter is carried deep into sulci to form a branching pattern, called arbor vitae. Cerebellum is connected with other parts of brain by three pairs of fibrous tracts, the **cerebellar peduncles**, *anterior* or *superior peduncles* connect cerebellum with the posterior optic lobe; **middle peduncles** connect the cerebellum with the cerebral hemispheres and form a prominent tract on the ventral side of the cerebellum called pons varolii and posterior or inferior peduncles connect it with the medulla and indirectly with the spinal cord. Between the anterior peduncles runs a fibrous tract called **valve of vieusens**.

Cerebellum is largely concerned with neuromuscular mechanism to control the body movements and posture. Metacoel is hardly developed. Medulla oblongata has a thin, non-nervous roof forming the posterior choroid plexus. The sides and floor are thickened and contain the four functional components arranged as in spinal cord. The myelocoel is quite spacious, it presents a median groove, which ends behind, just at the junction of medulla and spinal cord, forming a pointed depression called **calamus scriptorius**. A number of nerves arise from the medulla and bring information from visceral receptors, such as in the heart, lungs etc., and send motor impulses back to their effectors.

Peripheral Nervous System

The brain and spinal cord are connected to various parts of the body by means of afferent and efferent nerves, through which they receive information and send out signals back to organs or structures concerned. These nerves constitute the peripheral nervous system, which is differentiated into two closely related sub systems:

(i) Peripheral central nervous system: comprising the cranial nerves from the brain and spinal nerves from the spinal cord, and composed of somatic sensory and somatic motor fibres and

(ii) Autonomic nervous system: Comprising sympathetic and parasympathetic nerves from some parts of the brain and spinal cord, and composed chiefly of visceral motor fibres.

Cranial nerves

These have the same basic pattern as the spinal nerves, and consists of a dorsal and a ventral root on each side of the brain, in each segment, the roots, however, unite. Segmental plan of cranial nerves is not so clearly visible in the adult vertebrate. The dorsal root ganglia are located over the nerves very close to the brain. In some cases, the dorsal roots and ganglia have been lost, in others. Ventral roots have disappeared. The dorsal root lies posterior to the corresponding ventral root of a segment. It is usual practice to number the cranial nerves in **roman** figures. Three types of nerves are distinguished: Purely sensory, purely motor and mixed. The dorsal root carry both sensory and motor fibres, their motor fibres supply the non-myotomal muscles of the head. Ventral roots have only motor fibres and supply the myotomal muscles and their derivatives.

There are 10 pairs of nerves in the anamniota and 12 pairs in the amniota (Table 17.1).

1.	Olfactory	5.	Trigeminal	9.	Glosso pharyngeal
2.	Optic	6.	Abducens	10.	Vagus
3.	Occulo-motor	7.	Facial	11.	Spinal Acessory
4.	Trochlear	8.	Auditory	12.	Hypo glossal

In addition to these, all vertebrates, except birds, have a sensory nerve not related to the first cranial nerve, yet connected with the olfactory organs. This has been termed *terminal* or *zero* (0) nerve. It is best developed in elasmobranchs, such as *Scoliodon* (Fig. 17.6).

In lower vertebrates, specially the fishes, tailed amphibia and larvae of tailless amphibia, certain nerves are associated with a system of lateral line organs, these are absent in the higher vertebrates.

Cranial Nerves of *Scoliodon*: There are 11 pairs of nerves (Fig. 17.11) inclusive of the terminal nerves. Terminal nerve (0) emerges from the ventral side of cerebrum through the neuropore and passes on into the olfactory mucous membrane. It is sensory nerve and bears a ganglion mid-way from its origin (Fig. 17.6 B).

I – Olfactory (SS) is composed of non-medullated afferent fibres originating from the olfactory lobe of the brain. The cell bodies of these fibres lie in the olfactory epithelium. This nerve is not to be confused with the olfactory tract.

Table 17.1: Cranial Nerves

Number	Name	Superficial Origin	Composition of Functional Components	Point of Origin	Terminal Supply
0	Nervus terminalis	Telencephalon	General somatic afferent	Cerebral hemispheres	Nasal septum
I	Olfactory	Rhinencephalon	Special visceral afferent	Olfactory mucous membrane	Olfactory bulb and lobe
II	Optic	Diencephalon	Special somatic afferent	Retina	Lateral geniculate body: pulvinar: superior colliculus
III	Oculomotor	Mesencephalon	Somatic efferent.	Oculomotor nucleus	Extrinsic eye muscles (except superior oblique and lateral rectus)
			General visceral efferent	Edinger-Westphal nucleus	Ciliary ganglion and intrinsic eye muscles
			General somatic afferent		Proprioceptive from eye muscles (except as above)
IV	Trochlear	Anterior medullary velum	Somatic efferent	Trochlear nucleus	Superior oblique muscle
			General somatic afferent		Proprioceptive from superior oblique muscle
V	Trigeminal	Myelencephalon	General somatic afferent	Gasserian ganglion	Skin and stomodaeal epithelium
			Special visceral efferent	Motor nucleus of V	Motor fibres to muscles of mastication
			General somatic afferent		Proprioceptive from muscles of mastication
VI	Abducens	Myelencephalon	Somatic efferent	Abducens nucleus	Lateral rectus muscle
			General somatic afferent		Proprioceptive from lateral rectus muscle
VII	Facial	Myelencephalon	Special visceral afferent	Geniculate ganglion	Taste-buds in anterior two thirds of tongue:
			General visceral efferent	Superior salivatory nucleus	Submaxillary and sublingual salivary glands
			Special visceral efferent	Motor nucleus of VII	Superficial face and scalp muscles, platysma, posterior belly of digastric stylohyoid muscle
VIII	Acoustic	Myelencephalon	Special somatic afferent	Spiral ganglion	Organ of Corti
			Proprioceptive	Vestibular ganglion	Semicircular canals, utriculus, sacculus
IX	Glosso-pharyngeal	Myelencephalon	General visceral afferent	Petrosal ganglion	Pharynx and posterior third of tongue
			Special visceral afferent	Petrosal ganglion	Taste buds in posterior third of tongue
			General visceral efferent	Inferior salivatory nucleus	Parotid gland
			Special visceral efferent	Nucleus ambiguus	Muscles of pharynx
X	Vagus	Myelencephalon	General somatic afferent	Jugular ganglion	External ear
			General visceral afferent	Nodosal ganglion	Pharynx, larynx, trachea, Oesophagus, thoracic and abdominal viscera
			Special visceral afferent	Nodosal ganglion	Taste buds of epiglottis
			General visceral efferent	Dorsal motor nucleus of X	Thoracic and abdominal viscera via sympathetic ganglia
			Special visceral efferent	Nucleus ambiguus	Striated muscles of pharynx and larynx
XI	Spinal accessory	Myelencephalon accessory	General visceral efferent	Dorsal motor nucleus of X	Thoracic and abdominal viscera via sympathetic ganglia
			Special visceral efferent	A. Nucleus ambiguus	Striated muscles of pharynx and larynx
			Special visceral efferent	B. Anterior gray	Trapezius and column of cord sternocleidomastoid muscles
XII	Hypoglossal	Myelencephalon	Somatic efferent	Hypoglossal nucleus	Musculature of tongue

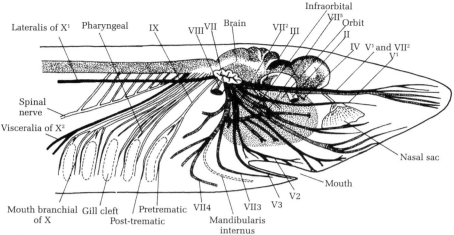

Figure 17.11
Cranial nerves *Scoliodon*

II – Optic (SS) is composed of efferent fibres whose cell bodies lie in the retina of the eye. The two optic nerves cross beneath the diencephalon to form the **optic chiasma**. The origin of optic nerve indicates that it is really not a nerve but a part of the brain, hence the name is usually termed optic tract.

III – Occulomotor (SM) – is the first Ventral nerve root. It leaves the ventral side of the mid brain and supplies four of the six eye muscles, viz-inferior oblique, superior rectus, inferior rectus and internal rectus; all derived from the first myotomal segment.

IV – Trochlear (SM) is the second ventral nerve root. It leaves the mid-brain from its dorsal side. It is the smallest nerve and supplies the superior oblique muscle of the eye derived from the second myotomal segment. A few sensory fibres are present in both III and IV-nerve.

V Trigeminal (SS and SM) is derived from the dorsal roots of the first two segments. It arises from the side of the anterior end of the medulla below the corpora restiformia. The root ganglia fuse to form a large gasserian ganglion located at the origin of the nerve. The main nerve is divided into three branches.

V^1 – Ophthalmic nerve (SS) is again divided into two: (a) Ramus ophthalamicus superficialis, (b) Ramus opthalamicus profundus or deep. The superficialis branch, along with a similarly named branch of the VII, passes along the upper border of the orbit and supplies the skin on the dorsal side of the head and snout.

V^2 – Maxillary nerve (SS) is the main branch of trigeminal, which together with the buccalis branch of **VII**, forms ribbon-shaped infra-orbital nerve running along the floor of the orbit. It finally divides into branches and supplies the skin over the upper jaw, upper lip, lower eye lid and teeth of upper jaw.

V^3 – Mandibular nerve (SS and SM) runs obliquely backwards in the orbit, then goes directly to skin above the lower jaw and muscles of lower jaw.

VI – Abducens (or Abducent) (SM) is the third ventral root arising from ventral side of the medulla near its anterior end and supplies the posterior of external rectus muscle of the eye-ball. Like III and IV nerves, this also has some sensory fibres. A branch of the VI goes to the nictitating membrane.

VII – Facialis is the dorsal root of the third head segment bearing mixed fibres. It leaves the medulla just behind the trigeminal and has several branches.

VII^1 – Superficial ophthalmic nerve, (SS), runs with the V^1 on the dorsal side of the orbit, then into the dorsal surface of the snout, giving off branches to lateral line organs.

VII^2 – Buccalis (SS), accompanies the maxillary branch of trigeminal (V^2), the combined trunk called infra-orbital or bucco-maxillary nerve runs across the floor of the orbit, the buccalis part of the nerve supplies the lateral-line-organs of the maxillary region of the snout. A small branch the ramus opticus, supplies the anterior extremity of the lateral canal.

VII^3 – Mandibularis externus (SS), is a small nerve that enters the orbit along with the VII⁴, then passes backward to supply the lateral line organs in the mandibular region of the head.

VII^4 – Hyomandibularis, (VS, VM and SS), runs above the hyomandibular cartilage, its somatic sensory fibres are distributed to the mandibular group of ampullae. The main nerve then passes forward and towards the midventral line of the throat. After giving off a branch, the mandibularis internus supplies the mucous membrane of the buccal cavity. It terminates in the muscles of the throat as the hyoidius nerve of visceral motor fibres.

VII^5 – Palatinus (VS), arise with the hyomandibularis, but takes a different course along the floor of the orbit. It gives off a small branch to the roof of the pharynx and terminates in the roof of the buccal cavity.

The cell bodies of sensory fibres of facial nerves are aggregated in the geniculate ganglion, located near the base of the hyomandibularis.

VIII – Auditory (SS) arises from the medulla close to the V and VII. The cell bodies of the fibres composing it are located in the acoustic ganglion lying close to the geniculate ganglion of facial nerve. The nerve just after its origin divides into two: (a) vestibular branch, connecting up with part of ear concerned with equilibration, and (b) saccular branch, connecting up with the auditory part of the ear. Vestibular nerve is better developed.

IX – Glossopharyngeal is a dorsal nerve-root composed of mixed fibres. It arises from the ventro-lateral border of the medulla and passes obliquely backwards through the floor of the auditory capsule into the region of the first-gill slit. The root-ganglion, called petrosal ganglion, lies near the base of the nerve. The main nerves divides into a pretrematic branch of visceral sensory fibres passing into the posterior side of the first gill slit, and a post trematic branch of visceral sensory and visceral motor-fibres supplying the hemi branch on the posterior border of first-gill slit. A small pharyngeal branch of visceral sensory fibres goes to the pharynx, where it supplies the taste buds and other receptors.

X-Vagus or pneumogastric is the most important nerve and the one most widely distributed. It represents several dorsal roots combined and arising from the side of medulla very close to the origin of IX. It is composed of both sensory and motor fibres, which are disposed off into three principle branches of the nerve.

X^1 **–Lateralis nerve, (SS)** it leaves the main nerve through a) Lateralis ganglion, and soon passes under the lateral-line canal running the whole length of the body behind the gill slits and between the epaxial and hypaxial muscle; it supplies the lateral line organs situated within the canal.

X^2 **– Branchio visceral trunk** arises through a ganglion called jugular or vagus ganglion and immediately breaks up into two branches.

(a) **Branchialis**, giving off smaller branches to the gill region behind the III visceral arch and the first gill slit. There are four such branches each bearing, at the base a small epibranchial ganglion, from which arise—(i) A small pharyngeal nerve of visceral sensory fibres supplying the roof of the pharynx. (ii) Pre-trematic supplying the anterior hemibranch with visceral sensory fibres (iii) Post-trematic of visceral sensory and visceral motor-fibres, supplying the posterior hemibranch and visceral muscles.

(b) **Visceralis (VS and VM)** supplies the heart and the anterior coelomic viscera.

Cranial Nerves of Frog: The general pattern of cranial nerves in tailed aquatic amphibians, which posses gills and lateral –line system of organs, is the same as that in fish but modifications take place in the tail-less forms, which lose gills and lateral-line organs during metamorphosis. I and II cranial nerves are SS and have the same origin and distribution as in the fish (Fig. 17.12). III nerve arises from the crus cerebrum beneath the optic lobe and close to the median line and supplies somatic motor fibres to 4 of the 6 eye ball muscles as in the fish. IV and VI somatic motor nerves have the same origin and distribution as in the fish.

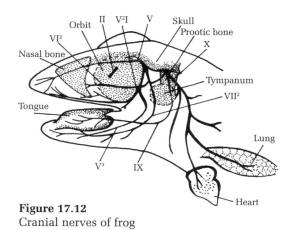

Figure 17.12
Cranial nerves of frog

The **V** is the mixed nerve arising from the side of the medulla, and bearing the gasserian ganglion at the base. it divides into three branches: V^1 – Ophthalmic (SS), which after emerging from the ganglion enters the dorsal side of the orbit, then, through a foramen passes below the nasal bone to supply the skin of the snout and part of the mucous membrane of the nostril. V^2 – maxillary and V^3 – mandibular are flattened like a ribbon and arise from a common stem leaving the root of the trigeminal. After passing from the posterior side of the orbit the two separate; the maxillary passing into the upper jaw from below the orbit divides into the two branches supplying the skin and teeth of the upper jaw and the side of the nostril, a small branch goes to the eye lid. The mandibular nerve courses a little backwards to the angle of jaw, then dips down into the lower jaw from beneath the upper jaw-bone and supplies the skin of lower jaw and throat.

The **VII** arises just behind the **V** and soon joins the gasserian ganglion. It divides into two branches.

VII^1 **– Palatinus (VS)**, a thin nerve that passes to the roof of the buccal cavity, the nerve becomes automatically visible, when the eye-ball is completely removed.

VII² – Hyomandibularis, (VS, SS, VM), is the main branch of VII which passes down into the throat region and supplies the skin and muscles of the lower jaw, the hyoid and throat. **VIII** – (SS) brings impulses from the sensory parts of the ear. The **IX** and **X** cranial nerves of frog, both of mixed fibres, arise together from the side of the medulla, their root-ganglia (petrosal ganglion of IX and vagus ganglion of X) are merged into each other. The IX sends a branch to join the facial, and supplies the mucous membrane of tongue, the taste-buds, wall of mouth-cavity and certain muscles of hyoid. From X-branches go to heart, lungs, larynx and other coelomic viscera.

Cranial Nerves of Rabbit: The cranial nerves of rabbit, rat and squirrel have almost the same plan. There are in all 12-pairs of nerves (17.13 A).

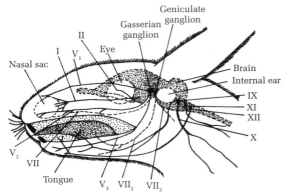

Figure 17.13 A
Cranial nerves of rabbit

I – Olfactory nerve arises from the olfactory epithelium as a number of bundles of non-medullated threads, which pass through the olfactory foramina in the cribriform plate of the ethmoid bone as their way to the brain. Several olfactory axonic fibres synapse with the dendrites of a single cell in the olfactory bulb in a peculiar interwoven manner, the pattern thus formed is called olfactory glomerulus. A branch of olfactory nerve supplies the Jacobson's organ.

II – Optic nerves arise from the retina as usual and form the characteristic optic chiasma. In mammals, such as primates, which have a binocular vision, all the optic-fibres do not cross to the opposite side, only about half the fibres in each optic nerve pass on to the opposite side.

III, IV and VI – Occulomotor, trochlear and **abducens** have the same origin and distribution as in the frog. They supply the muscles of the eye ball. Some fibres of the III pass onto

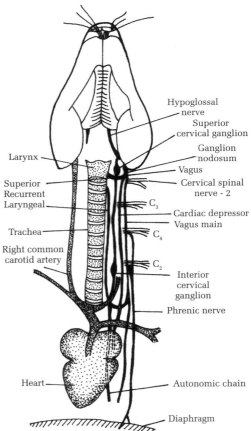

Figure 17.13 B
Neck nerves of squirrel showing distribution

the muscle concerned with the elevation of the upper eyelid, some to ciliary ganglion and ciliary apparatus of the eye, which is concerned with accommodation, and some to the muscles of the iris. A branch of the VI supplies the nictitating membrane.

V – Trigeminal has the three divisions:

V^1 – Ophthalmic, sensory nerve, divides into smaller branches supplying the skin of the upper eyelid, skin of the front of the head and of the inside as well as outside of the nose with a small branch it also supplies the cornea.

V^2 – Maxillary – first pass through the alisphenoid canal of the brain-box, then runs across the floor of the orbit giving branches to the upper jaw and finally, through the infra-orbital foramina reaches the skin of the face.

V^3 – Mandibular, a mixed nerve, leaves the skull by the foramen ovale and distributes its motor-fibres to the muscles of the lower jaw concerned with chewing; the sensory fibres connect with the skin of lower lip, teeth of lower jaw and lower part of the face. A twig of the mandibular supplying the skin is called the buccal branch, another branch, the lingual continues down to supply the receptors for touch and pain on front of the tongue. Another branch, the auriculotemporal, turns backwards and upwards to supply the skin of temporal region of head and external ear.

VII – Facial nerve is a mixed nerve consisting mainly of motor fibres. Its somatic sensory components loses significance with the loss of lateral line system of organs. The main nerve leaves by the stylo-mastoid foramen near the ear and passes through the parotid salivary gland, branching, thereafter to supply the motor-fibres to the muscles of the head, face and neck, including those around the eyes, nose and mouth. The visceral sensory fibres of the facial run through a separate branch which after crossing the tympanic cavity (middle ear) joins the lingual branch of V to form chorda tympani nerve. This supplies the taste buds on front of the tongue.

VIII – Auditory nerve (SS) divides into two branches a vestibular branch transmitting impulses from the equilibratory part of the internal ear, and a cochlear branch from the phonic i.e. hearing part of the ear (cochlea). The root ganglion of VIII is subdivided into vestibular and cochlear parts.

IX, X and XI – Glossopharyngeal, vagus and **spinal accessory** nerves – represent the dorsal roots and jointly form a complex. The IX, soon after its origin, divides into two branches: (i) **Lingual** of visceral – sensory fibres supplying the back of the tongue and, (ii) **pharyngeal** of the visceral motor fibres supplying some muscles of pharynx.

The **vagus** nerve leaves through the jugular foramen. The cell bodies of its sensory fibres are located in a ganglionic swelling, divided into a jugular ganglion within the skull and a *nodosal* ganglion lower down. Past the nodosal ganglion vagus give off (i) *superior* or *anterior laryngeal nerve* to the muscle of larynx, (ii) *cardiac depressor nerve* bringing impulses from the heart and aortic arches. A finer branch forms a plexus in the sinuauricular node. *Vagus then runs* back through the neck and shortly in front of the heart gives off, (iii) *recurrent laryngeal nerve*, which loops around the subclavian artery and passes forward to supply certain muscles of larynx. The main nerve then goes to form a series of plexuses in the visceral organs, such as oesophagus, lung, stomach and small intestine (Fig. 17.13 B).

The **spinal accessory** is a motor dorsal root and has a double origin. It receives some visceral motor-fibres from the vagus, and somatic motor fibres from some of the anterior cervical spinal nerves. The vagal fibres of spinal accessory pass on to the pharynx and larynx accompanying the vagal branches of these organs. The spinal fibres of the nerve supply the sternomastoid and trapezius muscle. XI nerve leaves these skull by jugular foramen.

XII – Hypoglossal nerve is a somatic motor-nerve derived from several posterior ventral spinal roots and supplies the muscles of the tongue and those below it. Some fibres of XII together with fibres of certain spinal nerves form a plexus in the neck called ansa hypoglossi, from which nerves go to muscles that draw the larynx backwards.

Spinal nerves

These nerves arise in segmental pairs, one from each side of the spinal cord, and one pair of nerves to each body somite. Each spinal nerve connects with the spinal cord by means of a dorsal and a ventral root. The roots unite after a short distance in all vertebrates except *Branchiostoma* and cyclostomes to form the trunk of the nerve. The origin and relations of

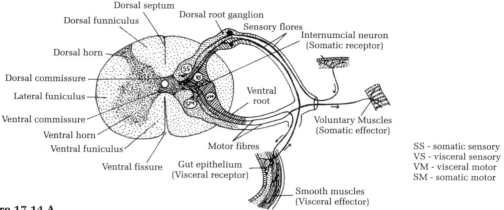

Figure 17.14 A
T.S. Spinal cord showing gross anatomy and functional anatomy with nerve components

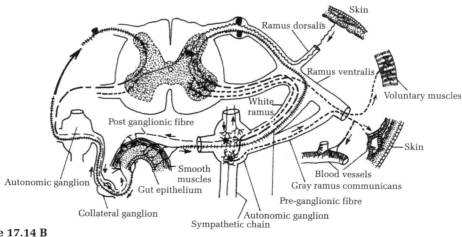

Figure 17.14 B
T.S. Spinal cord

the two roots are different. Dorsal roots are formed of neurons derived from the neural crest. Each such neuron connects with the dorsal horn of the spinal cord by means of its axon and sends out a dendron to a receptor of sensory organ. The cell bodies of all such sensory neurons are collected outside the spinal cord to form a dorsal root ganglion, while their axons and dendrons together form the dorsal root (Fig. 17.14).

Ventral root is formed of motor neurons, which have their cell bodies in the ventral horn of grey matter and their axons coming out to form the root. Fibres in the ventral root connect with effector organs. The dorsal roots are sensory or afferent in function and serve to receive informations both from somatic and visceral receptors by means of somatic and visceral sensory fibres. The ventral roots, on the other hand, are motor or efferent in function serving to carry the impulses away from the spinal cord to the somatic and visceral effectors by means of somatic and visceral motor fibres only in some lower vertebrates, the dorsal roots contain visceral motor-fibres as well.

The roots are enclosed within the meningeal membranes. The trunk of the spinal nerve, formed by the union of the two roots, emerges from the neural canal of the vertebral column through opening known as intervertebral foramina, usually between the arches of the adjacent vertebrae. Soon after, the trunk divides into three branches or rami :

(i) **Ramus dorsalis** to the skin and the epaxial muscles of the dorsal part of the body. It contains medullated somatic sensory fibres.

(ii) **Ramus ventralis** to the ventral (hypaxial) and lateral voluntary muscles of the body. It contains medullated somatic motor fibres, and

(iii) **Ramus communicans** which later join the ganglion of the autonomic system, it contains visceral sensory and visceral motor fibres. Ramus communications really consists of two parts, the white ramus of medullated fibres and the grey ramus of non-medullated fibres.

Generally, at the level of limbs in tetrapoda, the ventral rami of several segments are concentrated to form a complex branching network called plexus. In vertebrates lower than mammals, there are cervico-branchial plexuses, one in connection with each of the forelimbs and lumbo-sacral-plexuses, similarly for the hind limbs. In mammals, there are distinct cervical, lumbar and sacral plexuses.

In lower vertebrates, such as fish, the spinal cord is as long as the vertebral column so that the spinal-nerve-trunks are arranged all along the length of the vertebral column. In higher vertebrates, the spinal cord fails to keep pace with the vertebral column, it is always shorter than the vertebral column specially so in mammals, where spinal cord finishes in the lumbar region. The intervertebral foramina corresponding to the posterior segmental spinal nerves, thus, become distant and the spinal nerves have to travel long distances within the neural canals of the posterior vertebrae to reach their formina. The bunch of such terminal nerves together with the terminal non-nervous part of the spinal cord, appears like a horse tail, hence is called **cauda equina**.

Spinal nerves in chordates

In *Branchiostoma* the spinal nerves are in the form of segmental dorsal and ventral nerve roots, which remain separated hence contain non-medullated fibres.

In **fishes** such as *Scoliodon*, the dorsal and ventral nerve roots are united but not before they have come out of the vertebral column through separate foramina. The dorsal roots come off from the spinal cord a short distance posterior to the ventral roots. The mixed spinal trunks divide into dorsal and ventral rami, the former going into the skin and dorsal muscles, the later supplying the myotomal muscles. In some of the anterior segments, the dorsal roots have disappeared and from the ventral roots arise fibres that compose the hypobranchial nerve, supplying the gill-region. Cervico branchial and lumbor sacral plexuses are developed in connection with pectoral and pelvic fin.

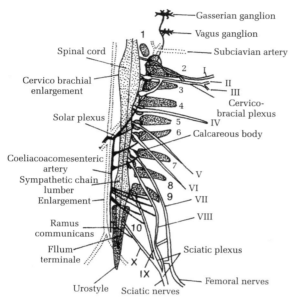

Figure 17.15
1 to X Spinal nerves of frog, 1 to 10 vertebrae and sympathetic chain on left

In **amphibia**, the dorsal and ventral roots unite and the dorsal root ganglia are situated just at the exit for the spinal trunks. The ganglia are protected in tailed amphibia by spongy and fatty masses; in tail less forms such as frog, there are chalk masses instead, called calcareous bodies or **glands of swammerdam**; apparently calcium reserves. The dorsal root arise a little posterior to the corresponding ventral roots (17.15).

In tailed amphibians the number of spinal nerves is large but in tail-less, the number is reduced. *Rana tigrina* has 9 or 10 pairs of spinal nerves corresponding to the number of vertebrae, the first nerve arising from between the first and second vertebrae.

The I spinal nerve has only its ventral root present, the dorsal root having disappeared at the time of metamorphosis. It is motor nerve supplying the floor of buccal cavity and tongue musculature. This nerve represents the hypoglossals of mammals, hence is usually called hypoglossal nerve.

The II pair is very large, it leaves between 2nd and 3rd vertebrae, and is joined by the small third nerve, which leaves between 3rd and 4th vertebrae, the two are joined by a small branch of the first, altogether forming a simple cervico-branchial plexus. Nerves leave the plexus to supply the skin and muscles of the forelimb. Spinal nerves, IV, V and VI leave respectively between 4th and 5th, 5th and 6th, 6th and 7th vertebrae. They pass obliquely backwards and supply the body wall both to skin and muscles. Nerves **VII**, **VIII**, and **IX** leave similarly between vertebrae, all pass backwards and are united with one another by oblique-cross-branches so as to form the lumbosacral or sciatic plexus. Two main nerves, the femoral and sciatic pass back from the sciatic plexus and supply the skin and muscles of the legs. All nerves have rami communicans. Tenth nerve is usually absent, or present only on one side and leaves from the side of the urostyle. Nerves from **II** to last are mixed nerves having both sensory and motor fibres. Last three pairs of spinal nerves and the filum terminate compose the cauda equina.

In **reptiles** the general pattern of spinal nerves is similar to that of amphibians. There are no calcareous bodies. Cervico-brachial and lumbosacral plexuses are present in all reptiles with limbs. Lumbosacral plexus is present in limb-less lizards and snakes as well.

In **birds** there is nothing special about spinal nerves. The lumbosacral plexus is often broken into lumbar and sacral plexuses, the former supplying the thigh and the latter the hind leg. The pudendal plexuses behind the sacral plexus supply the cloacal region and the tail.

In **mammals** the spinal nerves are grouped into regions corresponding to regions of vertebral column. There are cervical, thoracic, lumbar, sacral and coccygeal spinal nerves. The first nerve merges between the occipital bone of the skull and the atlas vertebra. The dorsal and ventral roots unite before emerging out of the intervertebral foramina. All are mixed nerve.

In **rabbit**, there are 37 pair of spinal nerves and they are classified into 8 cervical spinal nerves leaving through foramina between the cervical vertebrae and first thoracic vertebra (Fig. 17.16). Each has typically, a dorsal and a ventral root but no rami-communicans.

1. Cervical spinal nerve leaves in front of the atlas vertebra. It is small and lies at the back of the neck.
2. Cervical spinal is also small but the rest are conspicuous bundles. These supply the muscles and skin of the neck. Fibres from IV, V and VI give rise to the phrenic nerve, supplying the muscles of the diaphragm. The brachial plexus is formed by the branches of the cervical spinal from IV to VIII and the first thoracic spinal nerve. Three main nerves arise from the brachial plexus, these are the radial nerves supplying the flexor muscles of the upper and fore-arm, the ulnal and median nerves, to the muscles and skin of the forearm and hand.

Thoracic spinal nerve, XII or XIII pair form the intercostal nerves, supplying the intercostal muscles between the ribs and the skin of the chest. Next in the series are VII pair of lumbar nerves; first three of them supply the abdominal muscles and skin. The remaining four, together with first three sacral spinal nerves, unite to form the sciatic or lumbosacral plexus which give rise to the femoral, obturator and the skin of the hindlimbs. Last sacral and six coccygeal spinal nerves supply the skin and muscles of the tail.

Peripheral Autonomic System

The organs and structures innervated by the autonomic nerves are not under voluntary control, they send out signals through the dendrites of the visceral sensory neurons whose cell-bodies lie in the dorsal root ganglia of certain spinal and cranial nerves, and receive impulses from the central nervous system by means of visceral motor neurons, whose fibres leave out with certain spinal and cranial nerves (Fig. 17.17). It is these efferent or motor fibres that constitute the autonomic system of nerves. It is obvious that the brain, spinal cord, cranial and spinal nerves and the autonomic nerves form an integrated system, both in structure and function.

The autonomic effectors are supplied by two efferent neurons.

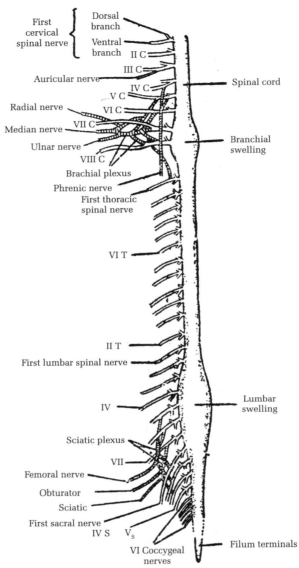

Figure 17.16
Spinal nerves of rabbit (Only left side)

1. Pre-ganglionic whose cell body and dendrites lie in the visceral motor component of the brain or spinal cord and whose medullated axon passes through the white ramus communicans to terminate in the autonomic ganglion.

2. Post ganglionic whose dendrites and cell body lie in the autonomic ganglion and the non-medullated axon proceeds to contact the destination i.e., the effector organ. **Autonomic ganglia**, are therefore, relay centres, and impulses arising through the Pre-ganglionic fibres are relayed through the post ganglionic fibre whose dendrites synapse with terminal branches of the pre-ganglionic axonic fibres within the ganglia.

Based upon the origin and function, the autonomic nerves fall in one of the two categories: (1) **Sympathetic system** (2) **Parasympathetic system.**

Sympathetic nerves are composed of efferent fibres derived from the spinal roots of thoracic and lumbar regions of spinal cord, the parasympathetic fibres are derived from the cranial nerves, III, VII, IX and X and the spinal roots of the sacral region of the spinal cord (Fig. 17.17). Furthermore, the relay centres of sympathetic system are situated, usually, at a distance from the effectors organs, hence their pre-ganglionic fibres are very short. Functionally, the two systems work antagonistically, i.e. if one speeds up a certain activity, the other inhibits or slows it, a compromise between the two leads to the normal functioning of the effector organs. Autonomic system works with an endocrine coordinating system of high complexity. Substances, similar to hormones, are produced on stimulation at the autonomic nerve ending. Majority of post-ganglionic sympathetic fibres produce a substance similar to adrenaline , hence these fibres are often known as adrenergic. Similarly, post-ganglionic parasympathetic and pre-ganglionic sympathetic fibres of sweat glands and uterus produce a substance called acetylcholine and hence called cholinergic fibres.

The sympathetic system comprises:

1. the pre-ganglionic neurons whose cell bodies lie in the spinal cord.,
2. their medullated axonic fibres in the ventral roots of all thoracic segments, and 2nd or 3rd lumber segments of the spinal cord.
3. the white rami communicantes, through which these fibres pass and connect with the sympathetic chain of ganglia,
4. the sympathetic chain ganglia as relay centres, and
5. the post-ganglionic fibres carrying impulses to the effector organs.

The sympathetic post-ganglionic fibres reach the visceral effectors either directly, or often passing through the grey rami communicates, hence accompanying a spinal nerve. Not all pre-ganglionic fibres synapse in sympathetic ganglia, specially, in case of organs of the abdominal viscera, they pass uninterrupted to certain splanchnic or collateral ganglia, where they synapse with post-ganglionic neurons.

A **sympathetic** chain lies on either side of the vertebral column, extending from the foramen magnum and even from farther inside the brain-box, to the sacral region of the backbone. At fairly regular intervals, the sympathetic chain carries segmental sympathetic ganglia. The more anterior of these have fused, forming three main enlargements. Of these the anterior most is the superior cervical ganglion lying dorsal to the carotid artery, roughly, at the place where the latter divides into external and internal carotids. Next in the series, are a small middle cervical and a larger inferior or stellate cervical ganglion. These receive their pre-ganglionic fibres from the white rami communicants of the upper thoracic segments, which ascend through the sympathetic chain for some distance to reach the cervical ganglia. Post-ganglionic fibres from cervical ganglia pass through the grey rami communicants to join certain cranial and cervical spinal nerves to supply the ciliary apparatus concerned with accommodation, the muscles of the iris regulating the size of pupil, roof of mouth-cavity, the salivary glands and glands in the mucous membrane of the nose. Post-ganglionic fibres also go to the heart, larynx, trachea, bronchi and lungs.

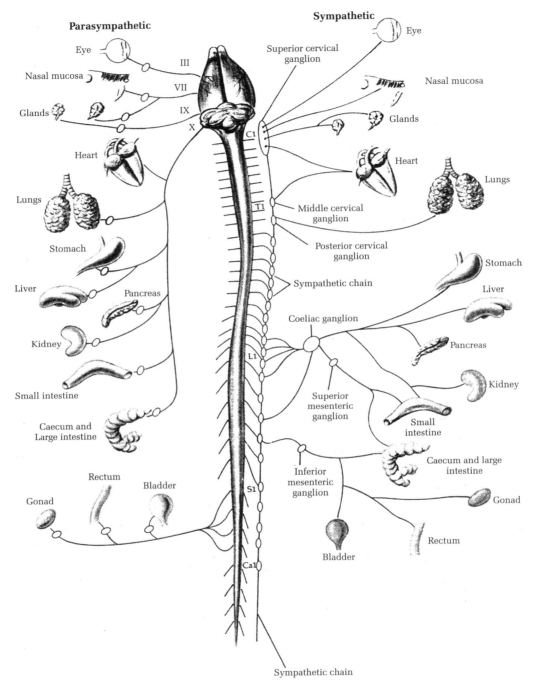

Figure 17.17
Visceral nervous system

In the thoracic and lumbar regions, there is one sympathetic ganglion to each spinal nerve; each ganglion receiving its pre-ganglionic fibres through the white ramus communicans, post-ganglionic fibres of a few anterior thoracic segments leave through the rami communicans to accompany corresponding spinal nerve and supply the heart and part of the respiratory system. In the remaining segments of thoracico-lumbar region the pre-ganglionic fibres pass without synapses through the ganglia of the sympathetic chain and connect with one or more of the splanchnic ganglia. Here they synapse with post-ganglionic neurons. There are three such ganglia, the coeliac, superior or anterior mesenteric and inferior mesenteric. Nerves leading from sympathetic chain ganglia to these three ganglia are called splanchnic nerves.

Post-ganglionic fibres from the coeliac and superior mesenteric ganglia form a complex network of nerves known as **coeliac** or **solar plexus** and supply the oesophagus, stomach, small intestine, anterior part of large intestine, liver, pancreas, and blood vessels of the abdomen. The viscera of the lower abdomen, such as remaining part of large intestine, bladder and urogenital organs, receive post ganglionic fibres from the inferior mesenteric ganglion. The complex network of nerves in this region of the abdomen gives rise to hypogastric plexus and the pelvic plexus further down at the side of rectum.

Post-ganglionic fibres from the sympathetic ganglia pass through the grey rami into the dorsal and ventral roots of certain spinal nerves, which run into the skin of the head, trunk and limbs, innervating the involuntary muscles of the blood-vessels, the arrector pili muscles of the hair and sweat glands.

Parasympathetic system   There is no such thing as a parasympathetic chain of ganglia. The relay centres of the system are situated very close to or within the organs, hence their pre-ganglionic fibres are rather long and post-ganglionic fibres relatively much shorter. Parasympathetic pre-ganglionic fibres are found accompanying cranial nerves III, VII, IX and X. In the sacral region, Pre-ganglionic fibres travel though the white rami communicantes of 2nd, 3rd and 4th sacral spinal nerves and reach their destination directly where they synapse with post-ganglionic neurons.

Pre-ganglionic fibres in the III nerve run to the ciliary ganglion (para-sympathetic relay-centre) in the orbit. Post-ganglionic fibres pass with the short ciliary nerves to supply the sphincter muscles of iris causing the contraction of pupil and ciliary muscles of accommodation. It may be recalled that sympathetic fibres of superior cervical ganglion supply the dilator muscle of iris. Thus, the two systems work antagonistically.

Pre-ganglionic fibres in the VII nerve run a complicated course to reach the submandibular and spheno-palatine ganglia in the head. Post-ganglionic fibres from spheno-palatine ganglion pass on to the lacrimal glands, mucous membrane of nose, palate and the upper part of pharynx. Fibres from the sub-mandibular ganglion pass through the chorda tympani and lingual nerves and control the submandibular and sublingual salivary glands.

Pre-ganglionic fibres in the IX nerve run into the optic ganglion, its post ganglionic fibres control the parotid salivary glands and mucous membrane of the mouth.

Pre-ganglionic fibres in the X nerve run to the ganglia, which lie within the organs and are very small. There are two plexuses of such ganglionic cells, their post-ganglionic fibres form an outer *myenteric* or *Auerbach's plexus* and an inner submucous or *Meissner's plexus*. Post-ganglionic fibres supply the smooth muscles and glands of alimentary canal. Other pre-ganglionic fibres terminate in the heart, larynx, trachea, bronchi, lungs, blood vessels of abdomen, liver, gall bladder and pancreas and synapse with post-ganglionic neurons, whose axons supply the said organs.

Pre-ganglionic fibres accompanying the sacral spinal nerves terminate in ganglia situated within the lower part of large intestine, kidneys, bladder and reproductive organs, and form synapsis with post-ganglionic neurons, whose axons supply the organ mentioned above.

Functions of autonomic system

Stimulation of sympathetic system cause dilation of bronchi, acceleration of heart beat, increased output of blood, constriction of cutaneous blood vessels, contraction of arrector pili muscles causing goose flesh and erection of hair, secretion of sweat, dilation of pupil, reduction in the amount of saliva, relaxation of smooth muscles of digestive tract, relaxation of bladder muscles, contraction of sphincter of bladder, increase in blood sugar and increase in the number of circulating blood cells over all effect of stimulation of this system are found to be such as would be expected for the purpose of preparing the body for action in attack or defence.

The stimulation of parasympathetic system causes the reverse effects, such as contraction of muscles of bronchi, lowering of the heart beat, dilation of blood vessels resulting in a low blood pressure, contraction of pupil, increase in the salivary and gastric secretions, Contraction of smooth muscle of gut, contraction of bladder muscles and relaxation of bladder sphincter, and dilation of blood vessels of external genitalia. Overall results of stimulation of this system are such as to assist in restoring the energies of the body after action. Autonomic system has a 'central control' located in the hypothalamus of the brain.

Comparative anatomy of autonomic system

Only sympathetic system is represented in *Branchiostoma*. The dorsal nerve-roots contain fibres running in the wall of the gut, where they synapse with motor neurons forming a plexus in gut wall. Fibres of the plexuses, control the movements of the gut.

In **elasmobranchs the sympathetic** system consists of an irregular series of ganglia, approximately segmental, lying dorsal to the posterior cardinal sinus and extending back above the kidneys. Pre-ganglionic fibres pass out of the spinal cord through the ventral spinal roots and the rami communicans. Post-ganglionic fibres terminate in the smooth muscles either of the arterial walls or of the viscera. There are no grey rami communicates, consequently no post-ganglionic fibres pass into the spinal nerves. Another peculiarity of the sympathetic system in these fishes is that it is confined to the body behind the head, there being no sympathetic fibres accompanying cranial nerves. **Parasympathetic** system, though present, is not so widely spread as in mammals. There are fibres of this system in the vagus and probably infacial and glossopharyngeal nerves. Parasympathetic fibres supply the heart and gut only. A definite sympathetic chain of ganglia is first found in the bony fish, in which it extends from the level of V cranial nerves backwards. Segmental ganglia correspond, in the head region, to the cranial dorsal roots, but the ganglia receive pre-ganglionic fibres from the trunk segments. In the trunk region, each ganglion receives pre-ganglionic fibres from a corresponding part of the spinal cord through the ventral root. Some post-ganglionic fibres pass back to the spinal nerve via grey ramus communicans to supply the skin and cause the contraction of pigment cells, the melanophores.

Parasympathetic fibres in cranial nerve III supply the muscles of iris, vagal fibres supply heart and stomach. In other cranial nerves, parasympathetic fibres are probably lacking.

Further advancement follows in **amphibia**, specially in frogs and toads. In *Rana tigrina*, the sympathetic and parasympathetic system are distinct and well developed. There is a sympathetic chain of 9 or 10 segmental ganglia on either side of the vertebral column. Each ganglion receives pre-ganglionic fibres that pass through the ventral spinal root of the corresponding segment of spinal cord and enter the ganglion through the white ramus communicans. Post-ganglionic fibres supply the heart, blood vessel-stomach, intestine, kidneys, bladder and reproductive organs. The sympathetic chain is continued within the head where post-ganglionic fibres accompany cranial nerves III, VII, IX and X and supply iris muscles, of roof of the mouth-cavity, laryngo-tracheal chamber and lungs.

Pre-ganglionic fibres of the **parasympathetic** system **accompany VII, IX and X cranial nerves. Similar fibres originate from the sacral region, too.**

Reptiles, birds and mammals, in general, have a typical autonomic system as given in detail.

Chapter 18

Receptor System

Events taking place outside and within the body of an animals are perceived by sense organs and transduced into nerve impulses. Sense organs or receptors, are transducers of energy. They transduce or change, mechanical, electrical, thermal, chemical or radiant energy into nerve impulses in an afferent nerve fibre. The energy constitute a stimulus. Receptors have been classified into three categories (Table 18.1)

1. The **exteroreceptors**
2. The **proprioceptors**
3. The **interoreceptors**

The **exteroreceptors**, respond to environmental stimuli, They are environmental organs dealing with the world outside. They receive mechanical, chemical and radiant stimuli. The **proprioceptors** constitute such receptors which respond to stimuli arising in muscles, joints and tendons and with in semi circular canals and utriculus. These are deeply located and control the musculature and working of body parts. The **interoreceptors** are those receptors which respond to alterations in the internal environment. They are also called visceral receptors. They are connected with the digestive tract where as exteroreceptors and proprioreceptors are somatic receptors.

Exteroreceptors

A) When the external stimuli is **mechanical** i.e. **contact**

    If contact is temporary like touch and pressure—**Tangeoreceptor**

    If it is intensive contact and results in pain—**Algesireceptor**

    If contact is vibratory in nature (air current)—**Phonoreceptor**

    If contact is vibratory in nature (water current)—**Rheoreceptor**

B) When the stimuli is **chemical** i.e.

   Receives gaseous stimuli—**Olfactoreceptors** (for smell)

   Those that receive liquid stimuli—**Gustatoreceptor** (for taste)

   Those that receive stimuli from irritating substances—**Irritatoreceptors** (for irritation)

C) When the stimuli is **radiant** i.e.

   Heat receiving thermal receptors – **Caloreceptors**

   Cold receiving receptors – **Frigidoreceptors**

Sight receiving receptors – **Photoreceptors** or **photicreceptors.**

Table 18.1: Classification of receptors

1. Outside sources ....Exteroceptors
   1. Mechanical stimuli
      a) Temporary contact ....Tangoreceptors   Touch, Pressure
      b) Vibratory contact .... Phonoreceptors   Hearing
      c) Intensive contact .... Algesireceptors   Pain
      d) Currents of water ....Rheoreceptors   Orientation

   2. Chemical stimuli
      a) Gases ....Olfactoreceptors   Smell
      b) Liquids .... Gustoreceptors   Taste
      c) Irritating substances...Irritoreceptors   Irritation

   3. Radiant stimuli
      a) Thermal .... Caloreceptors   Heat
      b) Thermal .... Frigidoreceptors   Cold
      c) Photic................ Photoreceptors   Sight

II. Inside sources
   1. Muscular control .... Proprioceptors   Tonus
   2. Alimentary control .... Interoreceptors   Hunger, Thirst

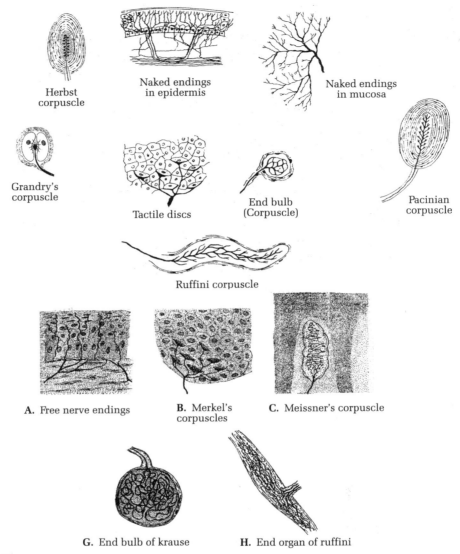

Figure 18.1
Some general receptors are somatic receptors. The naked endings in the mucosa and Pacinian corpuscles are visceral receptors. Corpuscles are encapsulated receptors

For convenience in description receptors may be grouped into:
1. Cutaneous sense organs
2. Special chemical sense organs
3. Gravity sense organs
4. The master senses of hearing and sight
5. Internal sensory mechanisms

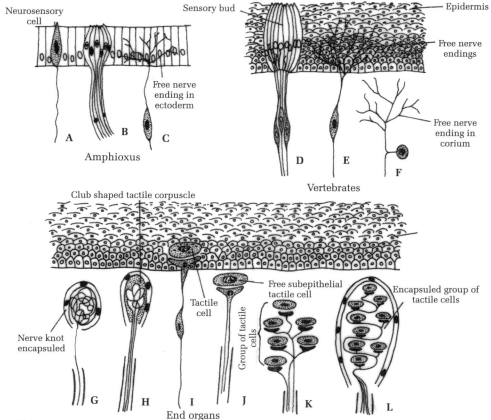

Figure 18.2
Cutaneous sense organs. **A-C** Nerve termination in the skin of Amphioxus. **D-F** Nerve termination in vertetrates. **G-L** Different stages in development of capsuled end organs

Proprioceptors

Striated muscle cells, muscle and tendons, are supplied with sensory endings that get stimulated when striated muscles contract. The sensation evoked by these stimuli is known as **proprioception**. It is also called **kinesthesia** and deep sensibility. Although some of the afferent impulses reach cerebral cortex, many of them are shunted to cerebellum. The result of the reflex arc is coordinated with muscular movements necessary for the maintenance of posture, for locomotion, for grasping, and for performance of skilled activities of the appendage such as using tools and playing a piano.

Interoreceptors

The mucosa of the internal tubes and organs of the body, the cardiac muscles, the smooth muscles including those of the blood vessels, and the capsules, mesentries, and meninges of the viscera are supplied with **interoreceptors.** The visceral receptors are stimulated mechanically by stretching or chemically by the presence of certain substances such as acid in pyloric stomach. They are also stimulated by tactile and thermal stimuli in pharyngeal and oesophageal region.

Somatic Receptors

At least four senses are located in skin. These include pressure or touch, warmth, cold and pain, corresponding to those receptors there exists peripheral nerve terminations or endings.

The endings may be:

(i) **Free nerve endings** in shape of fine branching twig.

(ii) **End organ** made up of specialized cell or group of cells (Fig. 18.1, 18.2).

(i) In **simple tactile** corpuscle the free nerve endings terminate with a cup in which a lenticular tactile cell is located. Such a tactile cell may be epidermal or subepidermal (Fig. 18.1). The free nerve endings may extend between the cells of the epidermis as far as the stratum granulosum e.g. root sheath of hair follicle.

(ii) **Merkel's disks** In mammals, many nerve-endings terminate in localized areas of epidermis forming little disks – merkel's disks e.g., snout of pig, tip of mole's nose (Fig. 18.1). Portions of skin that are without hair carry encapsulated tactile endings located mainly in outermost region of the dermis. These include end-bulb of Krause, Vater-Pacinian corpuscle, Herbst corpuscle, Meissner's corpuscles, grandry's and Ruffini Corpuscles (Fig. 18.1).

End bulb of krause It is small in which nerve ending is club-shaped and encapsulated e.g. mucous membrane of mammalian tongue and lips; mammary glands, conjunctiva of eye, corium of finger tip, external genitalia of man, moist snout of grazing cattle. These bulbs respond to decrease in temperature (Fig. 18.1).

Vater-Pacinian They are largest of the end organs. The terminal branches of the end-organs are surrounded by successive layers of connective tissue. Found in subcutaneous layer of skin, in pleural wall, around larger blood vessel in diaphragm in peritoneum, mesentry of body cavity, also occur in tendons and joints.

Grandry's corpuscles They are seem along the margins of the beaks of birds like ducks. Each of these

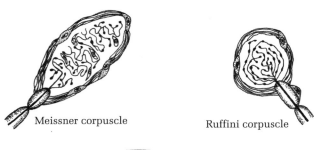

Meissner corpuscle   Ruffini corpuscle

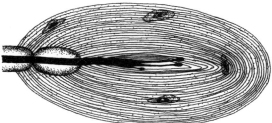

Pacinian corpuscle

Figure 18.3
Encapsulated nerve endings

consists of two large disc of non-nervous stimulator cells, between which the nerve endings of a sensory neuron are sandwiched, surrounded by an envelope of connective tissue.

Herbst corpuscle It is similar to Grandry's corpuscles, are found in mouth parts of birds. Each of these is made up of two rows of non-nervous stimulator cell arranged on either side of a neural core, all surrounded by connective tissue sheath.

Meissner's corpuscle There is a regular pyramid of non-nervous stimulator cell, interlaced by the branching ends of sensory receptor neurons, surrounded by a capsule of connective tissue (18.3). It occurs in friction area of skin of primates such as hand and feet (particularly sensitive to touch).

Ruffini corpuscles The end bulbs of Ruffini corpuscles are flattened, bulb-like endings within a fine, connective tissue network that responds to temperature increase (18.3).

Ophidian pit organs

Pit receptors Pit receptors of reptiles bear a striking morphological resemblance to the external neuromast of fish. In snakes and lizards the receptors are present in form of pits that open to the surface between epidermal scales-most common is apical pits. Encapsulated nerve endings in reptile are buried under dense cornified scales, apical pits provide sites for input of stimuli, probably tactile.

Pit organs in crotalid snakes and in *Boas* – a single pair of more specialized pit receptor is found in the head. (Fig. 18.4)

Loreal pits They are pits of pit vipers and they are located at the posterior end of the loreal scale (location between external nares and eye). These pit organs are so sensitive that they can detect temperature changes below 0°C at a distance of several feet.

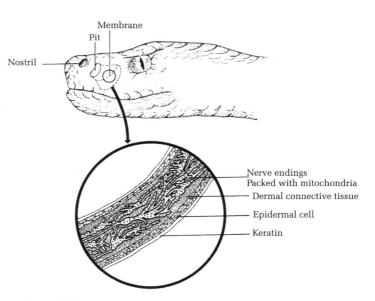

Figure 18.4
Head of rattle snake showing location of pit organ

Labial pits In *Boas*, pits are slit-like and less obvious. Since they are associated with labial scale, they are called labial pits.

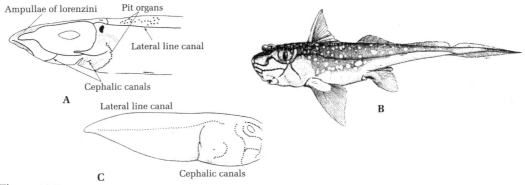

Figure 18.5
Distribution of neuromast organs in aquatic craniates **A.** External openings of pit organs cephalic canal, ampullae of lorenzini *Squalus* **B.** The elevated canal systems of *Chimaera* **C.** Lateral line canal of an anuran tadpole

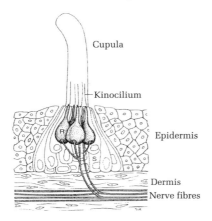

Figure 18.6 A
A neuromast organ in *Necturus*

Lateral line organs (accoustico-lateralis system) found in aganathans, fish and aquatic amphibians are also cutaneous sense organs (Fig. 18.5). A single lateral line organ or neuromast consists of a cluster of elongated epidermal cells the basis of which are connected to the terminal fibrils of afferent nerve fibre. The canal in which neuromast lie is lined by epidermis and filled with mucus.

In **cyclostomes** and **amphibians** each neuromast lie in a separate pit (Fig. 18.6). In dogfish, head has large number of pores, each leading into a deep pit, at bottom of which lie neuromast. These sense organs are called ampullae of Lorenzini (Fig. 18.6 C). Neuromast are stimulated by low frequency vibrations in water or by pressure or possibly by currents in the water.

Mormyomasts In the mormyrids, the modifications of the lateral line are called tuberous organs. They are stimulated by the fish's own electrical potential which are modified by nearby physical objects. It is thought that with this electric "radar" the mormyrid constantly scans for

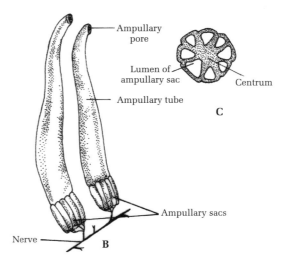

Figures 18.6 B, C
Scoliodon **B.** Ampulla of lorenzini **C.** T.S. of sensory ampullary sac

information about its environment. There is also some indication that they may be used in intraspecific communication.

Special Visceral Receptors

Chemoreceptors The lining epithelium of oral and nasal cavities, in a vertebrate is provided with organs of chemical sense i.e. the organs that are stimulated by chemicals.

Olfactoreceptors olfactory organs are special visceral chemoreceptors which develop from the surface ectoderm. The epithelium lining the organ arises as a pair of ectodermal placodes situated above the stomodaeum. These placode sink into head to form a pair of olfactory (nasal) pit. In fishes the olfactory pits become surrounded by connective tissue and are thereafter known as olfactory sacs. The olfactory organs have a lining of olfactory mucous membrane or schneiderian membrane made up of basal cells, supporting cells and elongated neurosensory cells (Figs. 18.7 A, B). The olfactory epithelium or schneiderian

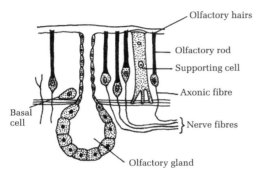

Figure 18.7 A
V.L section of olfactory epithelium of rabbit

Figure 18.7 B
Olfactory organ in *Scoliodon*

membrane is usually extensive on account of numerous schneiderian folds. (Fig. 18.7 B). The olfactory receptor cells are specialized neurosensory cells, spindle shaped and contain Nissil's granules, olfactory rod bearing 5–12 delicate nonmotile olfactory hair at its tip (Fig. 18.9) From other end arises axonic fibre. Olfactory epithelium besides receptor cells, contain supporting and basal cells, and many mucus producing olfactory or Bowman's glands. In mammals, the sense of smell is very acute. They have elongated nasal passage containing folded turbinal bones. The turbinal bones are covered with Schneiderian membrane which not only moisten and warm the inhaled air but is also olfactory in nature. In birds the olfactory apparatus is least developed, that is they have a poor sense of smell, hence called microsmatic.

Vomero-nasal organs (Jacobson's organ) It is meant for smelling food when it is inside the mouth cavity, is well developed in all lizards, snakes and in sphenodon, but is insignificant in aquatic reptiles (Figs. 18.8 A, B). Besides lizards and snakes, it is also found in monotremes, marsupials, insectivores and rodents. But in turtles, crocodiles, birds and many mammals such as primates and cetacea it is found only in embryo.

The organ consists of a pair of blind pocket lined with olfactory epithelium and lying ventral to nasal sacs. Each pocket directly opens into the mouth cavity. The tips of tongue of snakes and some lizard when retracted, are placed in the pockets of Jacobson's organs.

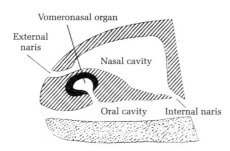

Figure 18.8 A
Parasagittal section showing the vomeronasal organ in a lizard

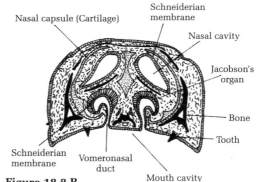

Figure 18.8 B
T.S. head of lizard showing the nasal and jacobon's organs

These apparently, are devices for detecting volatile chemical substances that may adhere to the surface of the tongue when it is thrust out of the mouth.

Organs of taste (Gustatoreceptors)   Most of the flavours of food which are credited to the sense of taste are really sensations derived from nasal (olfactory) sense organs. The oral sense organs are able to give sensation of sweet, sour, bitter and salty substances (Figs. 18.9 A, B).

Taste buds are widely distributed in fishes being found in the mouth, pharynx, branchial cavities and outer surface of the head. In other vertebrates the taste buds are confined to the

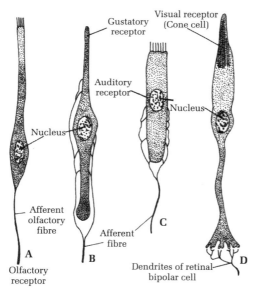

Figures 18.9 A, B, C, D
Receptor cells   **A.** Olfactory receptor for nasal epithelium   **B.** Gustatory receptor from a taste bud   **C.** Auditory cell from organ of corti   **D.** A cone cell of retina

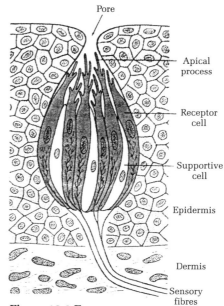

Figure 18.9 E
Receptor cells and their innervation in a taste bud

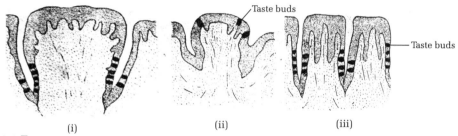

Figure 18.9 F
Position of taste buds (shown in black) on (i) Circumvallate papillae (ii) Fungiform papilla; (iii) Foliate papillae

tongue, oral cavity and pharynx. In frog the tongue has filiform (conical) and fungiform (knob-like) papillae, but taste buds are confined only to the fungiform papillae. In birds, the horny tongue has no taste buds, they are found in lining of mouth and pharynx.

Taste bud is an ovoid cluster of neuroepithelial or neurosensory columnar epithelial cells which bear at its free ends a delicate bristle or "hair" and most peripheral non-sensory supporting cells. The sensory hairs project directly over the surface or, as in the mammals alone, project into a small pitlike tastepore (18.9 E). Taste buds are carried over small elevations called taste-papillae (in cat they are cornified and used for rasping) and are of following types:

A) **Filiform** or **conical papillae**   Confined to anterior 2/3, they do not have taste buds. Cornified (ectodermal) conical papillae are present in the cat.

B) **Fungiform**   Knoblike and rounded, bear few taste buds (ectodermal).

C) **Foliate or leaf like**   They are present near base of the tongue, bear taste buds on either lateral sides (ectodermal).

D) **Circumvallate**   They are largest papillae of rounded from. They have a slightly raised central area and low trench (endodermal in origin). The functional life of a taste-cell is only 200-300 hours or ten days.

Photoreceptors

Also known as or light receptors, are sensitive to radiation in a narrow spectrum. Vertebrates have two sets of photoreceptors.

The lateral (paired) eyes

The median (unpaired) eyes

Lateral eyes   The receptor site of the lateral eye is the **retina**, a membrane rich in nervous tissue and synapses at the rear of a fluid-filled vitreous chamber of the eyeball. Besides this the other two layers of eyeball are outer coat called sclera or sclerotic coat and a middle coat uvea. The sclerotic is composed of tough connective tissue showing in front as the white of an eye. The white part forms a protuberance in front of the lens of eye called cornea. Cornea is transparent and thicker than rest of sclerotic. **Sclerotic** coat gives a shape to eyeball;

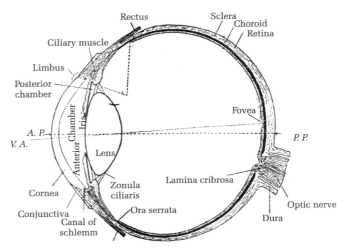

Figure 18.10 A
V.L section through the right eye of man, schematic A.P., anterior pole; P. P., posterior pole; V.A., visual axis

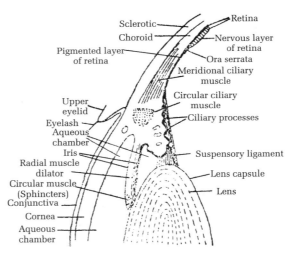

Figure 18.10 B
Section showing the ciliary part of mammalian eye

serves to protect it and provide surface for attachment of eye-muscles. A thin transparent layer of epithelial cells line the inner surface of eyelids and is called the conjunctiva. It is fused with cornea. Fine capillaries of blood belong to conjunctiva. Conjunctiva induces reflexes such as dropping of eyelids and flow of tears helping in removing foreign bodies (Figs. 18.10 A, B).

The **uvea** consists of three parts.

(i) **Choroid** It is vascular, pigmented and is present below sclerotic.

(ii) **Ciliary body** The choroid coat expands in front of eyeball to form ciliary body. It operates the mechanism of accommodation.

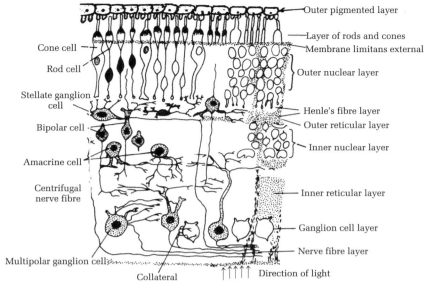

Figure 18.11
Human retina showing layers

(iii) **Iris** Behind cornea, choroid separate from sclerotic and bends sharply inward to form a circular pigmented disc, the iris. Iris is perforated in centre by pupil. Two sets of muscles, circular sphincters and radial dilators are under parasympathetic and sympathetic control. A space between iris and cornea is called anterior chamber of the eye. Similar but narrow posterior chamber, lies behind the iris between it and the lens. The two chambers communicate with each other around the margin of the pupil. Both chambers are filled with a sticky fluid-the aqueous humour, containing about 98 per cent water some proteins and sodium chloride.

Retina It has the same extent as the middle coat and is composed of light sensitive cells, together with a complicated arrangement of conducting nerve cells. The sensitive cells have a limited distribution, they are confined to the part of retina in immediate contact with the choroid. The thin non-sensory part of retina extends further and contributes to the ciliary body and iris, where it terminates at the rim of pupil. Light sensitive part is demarcated from the nonsensory part by an irregular and serrated area known as **ora-serrata or ora terminalis.** The lens of the eye is a transparent, biconvex body just behind the pupil, and held in place by elastic suspensory ligament or zonule, composed of radially arranged fibres. A thin but tough transparent lens capsule surrounds the lens. Shape of the lens depends upon arrangement of fibres. The anterior surface of the lens is less curved than the posterior. The large space between lens and retina is called **vitreous chamber**, this is filled with **vitreous humour**. Shape of the eyeball is determined, largely by the amount of vitreous humour and the pressure it exerts over sclera. Fluids in the eyeball keep moving out through canal of schlemn into veins paralleling outer edge of cornea. Nerve fibres of the retina leave the eyeball as optic nerve that pierces the choroid and sclerotic to reach the brain.

Retina consists of mainly four layers starting from outer side (Fig. 18.11).

(i) **Outer-pigmented epithelial layer**   It is closely applied over the choroid.

(ii) **A layer of visual or photo-receptor cells**   About 125 millions in human retina, visual cells are arranged in two kinds—the rods and cones.

(iii) **A layer of bipolar neurons**   These synapse with the rod fibre of visual cells on one side, and on the other with the cells of the fourth layer.

(iv) **Innermost layer of retina**   It consists of large bipolar neurons and forms the ganglionic cell layer. The point at which axons of the ganglionic layer converge, there are no rods and cones. It is thus called blind spot (incapable of producing impulses).

In man, cones occur mainly in a circular area over the retina, called fovea centralis, which itself is in a shallow depression in a larger area, called macula lutea or area centralis. This area lies almost exactly at the other end of an imaginary optical axis passing through centre of the cornea, pupil and lens. In rest of the retina rods out number cones. In pig, rat and cattle-retina is dominated by rods, hence called colour blind. Cones are responsible for recognition of colours.

Working of eye

The working of eye is almost similar to that of a photographic camera. Rays of light from object are reflected by cornea and light is focussed by convex lens so that a sharp, smaller inverted image is formed on retina. Image is carried by optic nerve to brain, where it is interpreted and the real sensation of sight arises and animal sees the object in an upright way. (However, the image is never re-inverted by brain.)

Type of vision, in higher mammals is called **binocular vision**, the fields of vision of two eyes overlap or even coincide. Thus both eye can see a object by which the animal can also estimate the distance. In (rabbit) lower mammals two eyes have divergent axis of vision as the fields of vision do not overlap. Therefore rabbit can see all around by its two eyes in which each eye covers a different field of vision. This type of vision is called **uniocular vision,** there is a great power of accommodation found in eyes of mammals, firstly because objects to be seen are situated at various distances from eye while the distance between lens and retina is fixed and unchangeable. Secondly the lens is very elastic and is always in a state of compression due to pressure exerted on it. For looking at an object which is at nearer distance the lens convexity becomes more, this change is brought about by ciliary muscle, ciliary process and suspensory ligament which form **accommodation apparatus**. Ciliary muscles contract and carries ciliary processes forward. Thus attachment of suspensory ligament to ciliary process becomes loose and it is brought closer to lens. The pressure on anterior surface of the lens is lessened and it becomes more convex and thicker due to its own elasticity. The thickening of the lens shortens the focal length and thus near objects are focussed. The reverse process occurs for seeing the objects at greater distances. The adjustment of eye to different distance is called accommodation.

Median eyes

The pineal and parapineal organs are ancient photoreceptors, which like lateral eyes, arise as evaginations of the diencephalon. Both pineal and parapineal bodies are light sensitive in lampreys (Figs. 18.12 A, B). Although the pineal eye is not as specialized as the lateral eyes and does not form image, it has been shown to be sensitive to change in light intensity.

In amphibians pineal is considered to be sensory, but not like photoreceptor cells of pineal of lower forms. Parapineal of reptiles referred to as parietal eye is overlaid by translucent tissue and serves as third eye. In larval frog the third eye is called **frontal organ** or **stirnorgan**.

At metamorphosis the photoreceptive part of the pineal complexity regresses, leaving only a glandular component that produced a hormone (melatonin) in larva.

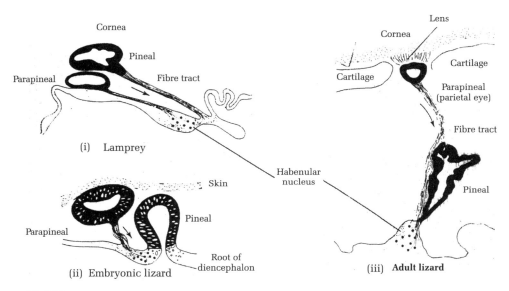

Figure 18.12 A
Epiphyseal complex of lamprey (i) Epiphyseal complex of embryonic and (ii) Adult lizard (iii) The lizard parapineal occupies the parietal foramen of the skull

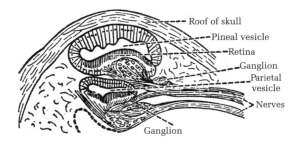

Figure 18.12 B
V.L. section through median eyes of *Petromyzon*

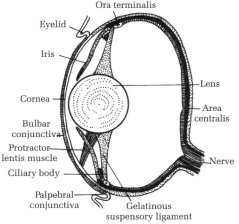

Figure 18.13
V.L.S through the eye of *Scoliodon*

Comparative account of the eye of vertebrates

In *Branchiostoma* there are a series of small light-sensitive and pigmented eye spots along the ventro-lateral sides of the spinal cord. Each spot is a photosensitive cell surrounded by a pigmented capsule. A single nerve-fibre leaves the cell. The long processes sent out by the flagellated cells of the infundibular organ are believed to be photosensitive. The large cephalic **eye spot** of the lancelet is not a photoreceptor.

In **fishes**, typical vertebrate eyes are found, with certain special features of their own. Eyes of elasmobranchs, such as *Scoliodon*, are large-sized and elliptical with the three coats of the eyeball well developed (Fig. 18.13). Eyelids, though present, are mere folds of skin capable of little movement. A third eyelid, called nictitating membrane is present and can be drawn over the entire eyes. The sclerotic is of cartilage and the six-eye muscles are inserted over it. In front the sclerotic forms the transparent cornea with which conjunctiva is fused. The conjunctiva of the cornea is called the **bulbar part** and of the eyelids is called **palpebral part**. The uvea has a pigmented and vascular choroid and the iris; the ciliary body has no or very few ciliary muscles. The pupil is a vertical slit, in other sharks it may be rounded. Iris musculature is not properly developed in *Scoliodon* but in some other sharks it is well developed and even better developed than in bony fishes. Pupil thus remains constant in *Scoliodon*. The lens of fish is spherical and very hard. In sharks the large lens pushes the iris far forward so that it almost touches the cornea. The iris divides the aqueous chamber into an anterior and a posterior chamber, both filled with aqueous humour. The lens is suspended by a suspensory ligament, which is not fibrous but gelatinous.

The inner surface of choroid, in a large number of elasmobranchs contains platelets covered with light – reflecting nitrogenous purines called guanine and hypoxanthine. This layer, called the tapetum lucidum lies behind the retina. Light falling over the plates is reflected back over the retina specially at night, thus, visibility is improved.

In bright light, tapetum is shielded by the retractile pigment of the retina. A thickened, vascular portion of choroid, known as suprachoroid lies between the choriod and sclerotic coats. This is present in fishes which have an optic stalk connecting the eyeball with the orbit; such a stalk is present in *Scoliodon*.

The retina as usual, has sensitive and non-sensory irridial and ciliary portions. Only rods are present, therefore, eye is adapted for dim light vision. A blind spot and an area centralis are present but without fovea. The lens is focussed for distant vision. Accommodation for near vision is partly provided by the protractor lentis muscle on the ventral side of the lens, which upon contraction causes the lens to swing forward. Eyes of bony fishes are a bit more complicated. The eyeball is elliptical and cornea is formed of four layers. The choroid is provided with the tapetum layer, having similar guanine crystals cover the outer surface of the iris, where they form the argenta of iris and in some fishes form the retinal tapetum. Retina has both rods and cones and an area centralis is also present, some possess a fovea too. Fresh water fish have porphyropsin in rods, instead of rhodopsin. Ciliary muscles are not developed, but iris is often provided with sphincters and dilators.

The **amphibian** eyes range from the fish type, found in aquatic amphibians, to the eyes adapted for vision through the air found in amphibia that spend some or all time outside water. For clear vision through air it is essential that the surface of eye be protected from injury and from dust particles and kept moist.

Eyelids and glands have, therefore, gained much significance in salamanders, newts, frogs and toads (Fig. 18.14). The upper eyelid is more or less fixed, but the lower is very mobile and its upper portion is rendered transparent, forming the nictitating membrane a structure specially well-developed in the frogs and toads. Normally, when the eye is open, the nictitating membrane lies folded beneath the remaining opaque portion of the lower eyelid. Closure of the eye, both in *triturus* and *frog*, is achieved by withdrawing the entire eyeball within the orbit by means of the retractor bulbi muscles (derived from external rectus), having their origin at under surface of the parasphenoid bone and insertion on both the upper and lower surfaces of the eyeball. The eyeball then bulges into the buccal cavity. Gradual withdrawal brings the eyelids nearer and the nictitating membrane to be pulled over the cornea. The eye is made to bulge out and open again by means of levator bulbi muscle, lying ventral to the orbit. Nictitating membrane is gradually folded back, this action is aided by the contraction of a small muscle in the lower eyelid, called depressor membrane-nictitans.

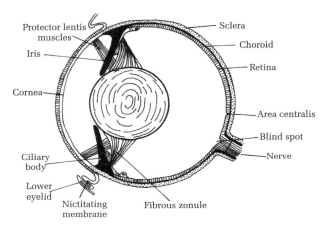

Figure 18.14
V.L.S through eye of frog

Cornea is kept moist and clean from dust, etc., by the secretion of tiny gland in the eyelids of salamanders. Definite harderian glands are found in frogs and toads, where they are located at the inner angles of the eyes. Tear glands are rare. The eyeball of frog *Rana tigrina* is rounded and cornea is bulging. Conjunctiva is fused with the cornea. The lens is flattened very slightly (perfectly rounded in tailed amphibia) and held far back from the cornea by means of fibrous suspensory ligament or zonule, which is joined with the ciliary body at its origin. The sclerotic is made up of cartilage and uvea has the three usual parts. Iris muscles are strongly developed and pupil responds quickly to the changes of illumination. The shining colour of the iris is not due to guanine or other similar crystals. The retina has both rods and cones, the rods have rhodopsin and the cone have iodopsin. An **area centralis** is present in the eyes of both *Rana* and the toad *Bufo*. Recently, it has been shown that certain fibres in the frog's optic nerve respond mainly to blue light. It is suggested that function of the blue sensitive system is to direct the jump of a frightened frog in such a way that it will leap into water to escape its enemies. Accommodation for near vision is brought about by the two protractor lentis-muscles, one dorsal and one ventral to the lens and attached to the zonule. These muscles pull the lens slightly bringing it nearer the cornea, thus, adjusting the focus over the retina. Other muscle known as **muscularis tensor choroidea**, running radially around the lens, probably extend a helping hand to the protractor lentis muscles, and are looked upon as fore runners of ciliary muscles of higher vertebrates. Eyes of limbless amphibia are very small and are used seldom. They are hidden under the transparent skin, which is often fused with the 'Harderian glands' which serve to lubricate the chemotactile **tentacle**.

The **reptilian eye** such as that of *Uromastix* shows important advancements and is fully equipped with glands that keep the cornea and eyeball moist (Fig. 18.15) Both eyelids are movable, the lower eyelid being more so. A real third eyelid, the nictitating membrane, not homologous with a similarly named structure in the frogs, is developed as a separate structure and lies folded beneath the lower eyelid when the eye is open. Nictitating membrane is lubricated by the secretions of the harderian gland. Lacrimal glands are present in many lizards, chelonians and crocodilians. A set of overlapping sclerotic ossicles is present in nearly all reptiles. The iris is capable of contracting or dilating the pupil by means of iris muscles. Ciliary body is better developed than in amphibians. It consists of ciliary muscles and ciliary processes, which are attached to a soft tissue around the lens, the annular pad, along its circumference. The colourless lens is flexible and slightly flattened towards the side of the cornea. A fibrous zonule holds the lens in position. Retina has more cones than rods, an **area centralis** is present in majority of reptiles. Some snakes, lizards, turtles and sphenodon also have a **fovea**. Accommodation in *Uromastix* and other lizards is brought about by the contraction of striated ciliary muscles that in turn cause the ciliary processes to squeeze the lens making its anterior surface more rounded. The arrangement of sclerotic ossicle is such that their concave surface tends to resist the pressure of fluids in the eyeball when the lens is squeezed. A peculiar conical, blackish-brown coloured papilla, known as **conus papillaris** or **pecte**n projects out into the vitreous chamber from the blind spot. This is an ectodermal structure composed of neuroglia and fine blood vessels that bring additional nourishment to the retina, into which it reaches after having first passed on to the viterous humour. Yellow droplets found over the retina serve to protect the retina from excessive exposure to light. No eyelids are present in *geckos* and some limbless lizards.

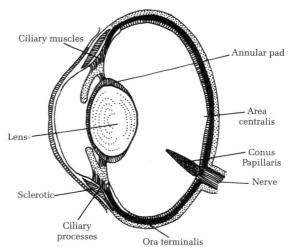

Figure 18.15
Section through the eye of *Uromastix*

The eyes of **snakes** present some special feature of their own. Eyeball is rounded and there are no eyelids or **sclerotic ossicles**. A transparent scale, the **spectacle**, covers the surface of the cornea. A narrow interconjunctval space is left between the spectacle and the cornea and fills the interconjunctval space. A duct drains the same fluid into the nose and the mouth-cavity. The lens of many diurnal snakes is coloured yellow which compensates the loss of yellow droplets over the retina. Accommodation is brought about in a different manner. There are no ciliary muscles. The iris with the help of the certain muscles attached to it, exerts a pressure over the lens causing it to squeeze and become more rounded towards the vitreous chamber. Retina has a large number of cones in diurnal forms and the pecten is of mesodermal origin. A plexus of blood vessels (*hyaloid plexus*), spreading over the retina, brings additional nourishment to the retina.

In **birds** great diversities are found in almost every feature of the eye and the diversities are connected, in some way or the other, with the special habits of the birds, yet the structure of the eye is basically reptilian. Eye of pigeon is of large size and not spherical (Fig. 18.16). No bird in fact, has a spherical eye. The cornea and lens are always bulging out and the shape is maintained by the ring of **sclerotic ossicles**. Besides an upper and a lower movable eyelid, a true nictitating membrane is present; it is lubricated by the secretion of harderian gland. Lacrimal glands serve to keep the cornea and eyeball moist. The lens is soft and accommodation is brought about by the strongly developed ciliary muscles divided into two sets, the, **muscle of Crompton**, which when in action pulls the cornea, and the **muscle of Bruke**, which draws the lens forward into the anterior chamber so that the lens becomes more curved. Thus, the accommodation is two fold, with the lens and with cornea. Pigeon, being a diurnal bird, has more cones than rods. An area centralis is present, but the density of cones is quite high even outside this area, so that every part of the sensitive retina gives a good and detailed picture. The lower part of the pigeon's retina contains yellow, and upper part red droplets that serve as filters increasing the contrasts of blues and greens respectively, as required for vision against the sky in the one case, and the ground in the other. Projecting from the blind spot into the vitreous chamber is a pleated, fan shaped, some what blackish vascular structure called **pecten**. This structure is especially well developed in diurnal birds. Pecten probably increases contrast and allows a clearer detection of movement of objects. It is also suspected that being an erectile structure it exerts, by various degrees of erection, a pressure over the vitreous fluid, that in turn helps in altering the curvature of lens during accommodation. Possibly the pecten also brings additional nourishment to the retina.

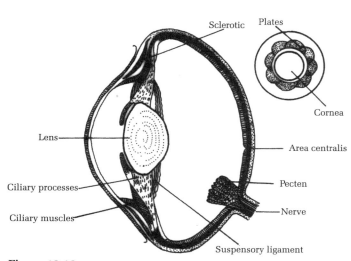

Figure 18.16
V.L.S Eye of pigeon

Phonoreceptors

Inner ear or labyrinth, evolved very early in vertebrate history and, with many variations in configuration but none of basic design and function, has been retained by all vertebrates. The middle ear evolved as tetrapods evolved, and the external ear is scarcely found except in mammals. The auditory apparatus of a mammal consists of three parts (Fig. 18.17), the internal ear or membranous labyrinth, the middle ear or tympanic cavity and the external ear. The main body of the inner ear consists of a sac partially divided by a constriction into

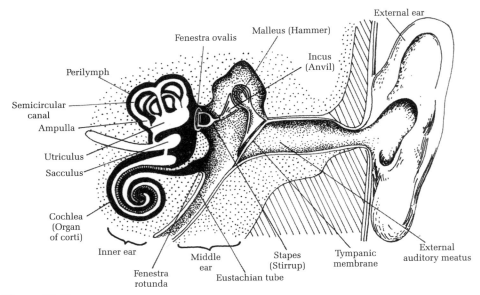

Figure 18.17
Section through human head showing the auditory apparatus

an upper chamber the utriculus and a lower chamber the sacculus (Fig. 18.17 and 18.18). From sacculus arises a narrow tube the ductus endolymphaticus, which runs to the dorsal surface of the skull where it dilates to form the saccus endolymphaticus. This may open by a small pore on the surface of the head in elasmobranchs. In others, it is closed. The utriculus opens into three loops, the semicircular canals of which two are vertical and one end forming an ampulla. The ampullae are at the ventral ends of the two vertical canals and at the anterior end of the horizontal canal. The wall of the inner ear consist of a dense fibrous tissue lined by an epithelium derived from the ectoderm, which at certain places is modified to form the receptor organs of the labyrinth. In elasmobranchs ear there are six such patches, one in each ampulla, one in utriculus and two in sacculus. The receptors of the ampulla are called *cristae* and those of the utriculus and sacculus are called *maculae*.

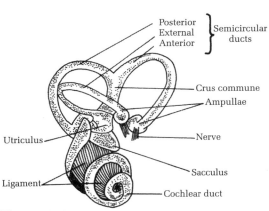

Figure 18.18
Membranous labyrinth of rabbit

The cristae and maculae have similar structures (Figs. 18.19 A, B and 18.20 A, B) each consists of a group of elongated receptor cells, each provided with a tuft of sensory "hairs" and connected below with a fibre, derived from the eight nerve. The membranous labyrinth is filled with a fluid called **endolymph**. In dogfish, otolith particles are suspended in endolymph, but in bony fish there is a single large otolith in relation with each maculae. Membranous labyrinth is surrounded by a fluid the **perilymph**, which is

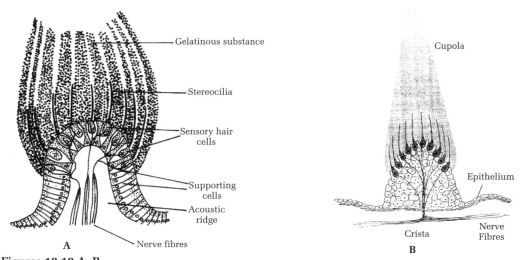

Figures 18.19 A, B
A. Mammalian crista   B. Crista found in ampullae of semi-circular canal

crossed by irregular strands of connective tissue the trabeculae in some cases. In most fish one of the maculae is enlarged and moves into a bud-like outgrowth of the sacculus, the lagena. In carps; the *weberian ossicles* connect the auditory capsule to the anterior end of the air bladder.

Equilibratory function of the labyrinth

a) **Ampullar cristae** are stimulated by the rotational movements of the head, the impulses reach the medulla and set the righting movements of the back, limbs and eye-muscles into action, which compensate for the rotation of the head. A sensory patch in the ampulla is called crista ampullaris, and it is mounted over an acoustic ridge (Figs. 18.19 A, B). The epithelium of the ridge is composed of sensory or receptor cells and non-sensory supporting cells. A receptor cell is produced into a long non-vibratile process or hair (stereocilia) towards the lumen of ampulla and lies embedded in a gelatinous material, the **cupola**, secreted by the supporting cells. Slightest movement in the endolymph of ampulla sets the cupola into motion, which in turn causes distortion of stereocilia.

b) **Maculae** are stimulated in response to gravitational forces regarding the position of body at rest. A macula consists of supporting cells and hair cells, similar to those of ampullary

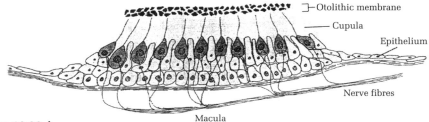

Figure 18.20 A
Neuromast organs of membranous labyrinth of basal craniates. Macula with otolithic membrane

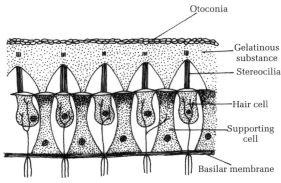

Figure 18.20 B
Mammalian macula

cristae. The hair are not actively motile and are embedded in a gelatinous otolith membrane. Small crystal like bodies, the otoconia of calcium carbonate and proteins are deposited on the outer part of the otolith membrane (Figs. 18.20 A, B). The two maculae are set vertically at right angles to each other. Endolymph fills the sacculus and utriculus. Movements in the endolymph and of the otoonia stimulate the sensory hair and produce impulses that reach the brain through the nerves supplying the utriculus and sacculus.

Thus, both the cristae and maculae send out signals to the brain for the maintenance of posture in all states of the body.

Auditory function of the labyrinth

The cochlear duct is the auditory or acoustic part of the inner ear which responds to the stimuli coming as sound waves (Fig. 18.21 A). Soon after its origin from the sacculus, it turns round to make 2¾ turns in man and 2½ turns in rabbit (Fig. 18.22 B). The duct contains endolymph and is surrounded by the bony cast forming the cochlear canal. The duct and canal together form the cochlea. The cochlea in fact turns round a hollow conical central pillar of bone called **modiolus**, from which a thin osseus lamina of bone projects like a spiral shelf about half way towards the outer wall of the canal. The cochlear duct is shaped like a tube, but is a triangular rather than a round tube. It forms a shelf across the inside of the bony cochlea, but divides the entire perilymph space into two tubes, a lower scala tympani and an upper scala vestibuli, both containing perilymph. The cochlear duct itself is called the scala media containing endolymph. Near the base of scala vestibuli, the wall of the membranous labyrinth comes in contact with fenestra ovalis of the vestibular window. At the lower end of the scala tympani is the fenestra

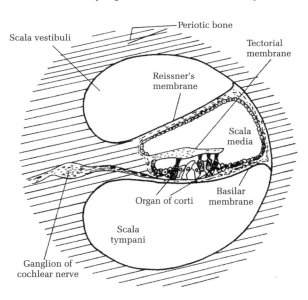

Figure 18.21 A
T.S of mammalian cochlea

rotunda or cochlear window. A narrow strip of the osseus lamina passing down the floor of the scala media is called the basilar membrane. It is over this membrane the receptor

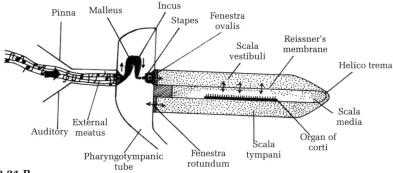

Figure 18.21 B
Mechanism of hearing (arrow indicate movements of sound)

apparatus, known as *organ of corti*, rests. It consists of the cubical epithelium lining the scala media, some supporting cells and some sensory hair-cells. The hair of the hair cells are embedded in an overhanging gelatinous tectorial membrane resting very lightly over them. Axonic fibres from the other end of the hair-cells pass into the basilar membrane and thence into the cochlear branches of the VIII cranial nerve. The thin walled sloping roof of the scala media, separating it from the scala vestibuli, is known as the *Reissener's membrane*. Scala vestibuli communicates, at the tip of cochlear duct, with the scala tympani by a minute pore, the **helicotrema**.

Working of the ear

Sound waves striking the tympanic membrane are transmitted to the membrane covering the fenestra ovalis by means of the three ear-ossicles (Fig. 18.21 B). Vibrations of the membrane of vestibular window are communicated to the perilymph of the scala vestibuli which vibrate in unison. Since the ear is enclosed in bone and the fluid in and around the membranous labyrinth is incompressible, and helicotrema is too small to pass on the vibrations to the scala tympani, the vibrations of perilymph of scala vestibuli set the Reissner's membrane, the spiral lamina, basilar membrane, and fluids in the scala media and tympani into similar vibrations. Every inward movement of the vestibular window is followed by an outward movement of the membrane covering the cochlear window to compensate for the pressure of sound waves causing the perilymph of the scala vestibuli to squeeze in. When the endolymph of scala media and basilar membrane vibrates, the hair of the sensory cell of the organ of corti are distorted, bent or stretched, setting up a train of events that pass into the nerves as sound impulses and interpreted in the medulla of brain. It is interesting fact that although birds lack an organ of corti, they perceive tone and pitch.

Middle ear It is a cavity between ear drum and auditory capsule. It is present in all tetrapods except urodeles, apodans and snakes. It is derived from first gill pouch. It communicates with mouth cavity by narrow eustachian duct which is filled with air. Between the tympanic membrane and the inner ear is a chain of three small bones, the ear ossicles in mammals. In amphibians-one bone the *columella* extends from ear drum to the fenestra vestibuli and transmit vibrations of ear drum caused by sounds to endolymph of the inner ear. Even in reptiles and birds there is a single bone. In mammals-three bones or ossicles i.e. **malleus**, **incus** and **stapes** are present (Fig. 18.21 B).

Malleus (articular)—a hammer shaped bone rests over the ear drum and by the other end connects movably with the **incus** (quadrate) an anvil shaped bone movably linked with the **stapes** (hyomandibular) a stirrup-shaped bone plugging a membrane over a small opening of the inner ear, called fenestra ovalis.

Vibration's of this membrane is transmitted to the perilymph and from there to the endolymph of the inner ear. The middle ear cavity remains in communication with the pharynx throughout life via the auditory (eustachian) tube.

Although conduction of sound via a drum is the predominant route in tetrapods, a drum is not always employed. Urodeles, apodans, a few anurans, and limbless lizards and snakes have no eardrum, no middle ear cavity, and even the columella is often vestigial (of course, aquatic urodeles have a lateral line system, which compensates in part for lack of auditory structures). Since in the absence of a drum the columella may end at the squamosal or quadrate bone, sound conduction via bone in these organisms is enhanced if the floor of the buccal cavity, the lower jaw, or the anterior limbs are in contact with a dense substrate – water or earth. The hyomandibular route to the membranous labyrinth was available long before fish emerged onto land, since the hyomandibular articulates with the otic capsule as a primitive condition. An intermediate stage between fish and tetrapods in the evolution of a sound conducting hyomandibular is seen in primitive urodeles. Some larval amphibians have a long slender rod extending between their lung and otic capsule and performing the same function as weberian ossicles.

Amphibians have a similar rod just under the skin of the lower jaw. These rods are not homologous.

Outer ear   The ear drum is on the surface of the head in amphibians and most reptiles. In crocodilians, birds and mammals, it is deeper in the head, at the end of an air filled passage way, the external auditory meatus. In mammals an appendage, the pinna, collects sound waves and directs them into the canal. External opening of this tube bear the pinna. Pinna is provided with muscles and is a mammalian acquisition. They are large in bats, hares, deer and kangaroos.

The pinna, a mammalian acquisition is a peculiarly molded, flat, skin covered, elastic cartilage in man with a cuplike concha, with centre, surrounding the entrance into the auditory canal. This opening is guarded on either side by two projections, the tragus, and antitragus. The lower end, or lobule of human ear is flesh, pendulous, and without cartilaginous support. In a considerable percentage of human kind the lobule is absent or at least not free from the side of the head. The upper curving edge or helix of the pinna often presents an appearance suggesting animal ancestors. When unrolled at the upper margin and more or less pointed, it is known as the "**satyr ear**". If the projecting part is folded inward it makes the so called **Darwin's Point**.

In mammals the external auditory canal is the passage way that leads to the middle ear. It is about 2.5 cm long in man, slightly bent and larger at either end than in the middle. For the external third of its length its walls are kept rigid by cartilage, continuous with that of the pinna, while for remainder of the way, into the skull the walls become bony, forming a projecting tube from the temporal bone. The auditory canal is lined with the skin, and is supplied with wax glands and outward projecting hairs, both of which are devices that serve not only as dust arresters but also for the discouragement of crawling and flying insects.

Comparative anatomy of ear of vertebrates

In *Branchiostoma*, the sense of equilibration and hearing is not known, According to Young, *infundibular organ* is a gravity or pressure receptor. The cephalic eye-spot is definitely not photoreceptive, probably is an organ for orientation of body movements.

In **fish,** three semicircular ducts, a utriculus and a sacculus, are distinctly differentiated. Small outgrowths of the utriculus and sacculus containing sensory patches, probably, answer to the cochlea of mammals and serve as sound receptors, otherwise the ear of fish is mainly for the control of balance and position (Fig. 18.23)

Ear of *Scoliodon* lies enclosed within the auditory capsule, whose cartilage follows the outlines of the membranous labyrinth and is called the cartilaginous labyrinth. Three semi- circular ducts, the anterior and posterior vertical and the external horizontal spring from the utriculus. Of these, the anterior and horizontal ducts arise dorsally from a common part of the utriculus, called anterior utriculus. The ampullae of anterior and external horizontal ducts jointly open back into the utriculus. The posterior duct arises from a distinct flattened chamber of the utriculus, called posterior utriculus, the posterior ampulla opens back into this chamber. Each ampulla has a sensory crista.

The sacculus lies ventral to the utriculus. Both chambers posses sensory maculae associated with large calcareous bodies, called **ear stones or otoliths**. The hair cells of maculae respond to the change in the position of the otoliths. Otoliths are masses of accumulated otoconia. In addition to the said structures, there is, on the ventral and posterior side of the sacculus, a blunt projection called *pars lagena*, and a conical projection towards the antero-ventral side of the utriculus called *pars neglecta* or *recessus utriculi*. These probably have sensory maculae and serve as phonoreceptors. The sacculus communicates with the outside, at least in the embryo, by means of the invagination canal, representing the original connection of the otic vesicle with the body surface. In the adult, the canal pierces the cartilaginous brain box and lies beneath the overlying skin. Membranous labyrinth is filled with the endolymph, which in all probability, is sea water that passed into the ear through the invagination canal. The sand grains drawn through the same canal with the sea water took part in the formation of otoliths. Ear is innervated by sensory fibres of VIII cranial nerve, which divides into a

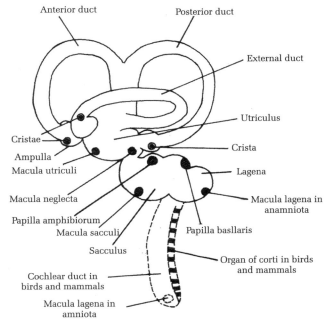

Figure 18.22
Location of the sensory patches in the membranous labyrinth of different vertebrates

vestibular and a saccular nerve. The vestibular nerve receives fibres from the utriculus, ampullae of anterior and external ducts. Saccular nerve receives fibres from the sacculus and the ampulla of posterior canal.

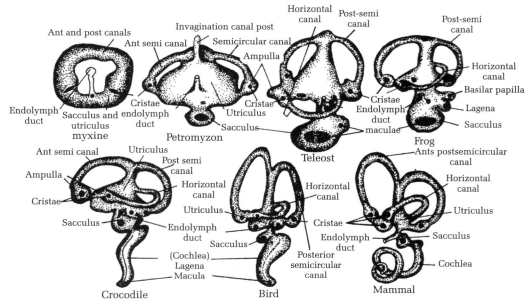

Figure 18.23
The inner ear of different vertebrates

In a group of bony fishes, ostariophysi (Cypriniformes), example *Rohu*, the ear contains three semi circular ducts and three otic chambers, the utriculus, sacculus and lagena, each with an otolith. Many bony fishes can hear sounds with their lagena. The sacculus of each ear of *Rohu* is connected with the transverse canal. A special extension of the perilymph space, the **Sinus impar**, is connected to the air bladder by a set of bones known as **Weberian ossicles**. Changes in the pressure of gas in bladder are transmitted through these ossicles into the perilymph of the ear in the form of vibrations. These functions of the bladder are linked with the equilibratory and auditory functions of the ear.

Amphibians show an advancement over the ear of lower vertebrates by adding the middle ear to their balancing cum hearing inner ear. This advancement is very characteristic of the tail-less amphibian, which must keep them alert on land and in water by means of signals coming as sound waves through the air. Sound is of significance in identifying the voice of a partner and often of a prey. The ear of frog (Fig. 18.23) consists of an inner ear and a middle ear. The inner ear or membranous labyrinth lies within the auditory capsule of the skull. There are three semicircular ducts with ampullae and cristae in them. The utriculus is larger than the sacculus from which it is separated by a constriction, the two chambers have sensory maculae and otoliths. A small outgrowth of sacculus, the lagena projects backwards, it has a sensory macula. The narrow endolymphatic duct extends dorsally from the sacculus and expands into a endolymphatic sinus over the hind brain. Besides these usual structures, in some frogs, there is a small outgrowth of the utriculus, called *pars basilaris*. Macula neglecta and macula or papilla basilaris, are believed to respond to the sound waves. In some frogs the pars basilaris is lacking, but another outgrowth of the sacculus, the *pars amphibiorum* is present containing a sensory macula and having an auditory function.

Endolymph filled membranous labyrinth floats in the perilymph of the auditory capsule. The fenestra ovalis makes a connection with the middle ear. The latter comprises an air-filled tympanic cavity connected with the buccal cavity by the eustachian tube. Single ear-ossicle in the form of a bony rod called **columella auris** and the tympanic membrane a patch of the darkened skin on the surface of the head. The columella communicates with the fenestra ovalis by means of a cartilaginous nodule, the stapedial plate, a detached part of otic capsule (Fig. 18.24). At the other end, columella is fixed over the tympanic membrane by means of a flattened piece of cartilage.

In **reptiles** the structure of the ear is roughly the same as in the amphibians. The semicircular ducts, utriculus and sacculus, show nothing special about them. The only advancement is that there is an increasing trend towards making the ear more sensitive to sound. Correspondingly, the *pars lagena* is larger and the *macula basilaris* is better-developed. In many lizards such as *Uromastix* (Fig. 18.24) and wall lizard *Hemidactylus* the tympanum lies at the back of the jaw, sunk a little below the surface into an ear-hole (Fig. 18.25) The ear ossicle consists of two parts, a basal portion, the bony stapes, plugging the fenestra ovalis, and an extra columella cartilage connecting with the tympanum. Snakes have neither a tympanum, nor an eustachian tube and tympanic cavity, yet are able to receive sound vibrations. The columella auris is attached at its outer end to the quadrate bone. The sound vibrations travelling through the earth, reach the jaw bones of which quadrate is a part, then are carried to the columella auris and its stapedial plate plugging the fenestra ovalis. Snakes can hear air-borne sounds of 100-500 cycles per second (Weaver and Vernon, 1960). Crocodilians have large lagena, spirally coiled to form the cochlea. Eustachian tubes open into the pharynx by a common aperture.

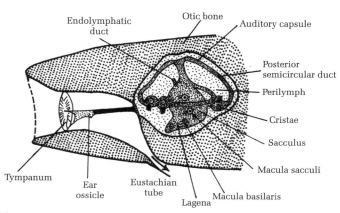

Figure 18.24 Ear of *Uromastix*

In **birds** the complexity of ear almost reaches the mammalian organization of the ear, but there is no external ear. The membranous labyrinth has all parts well developed. It is closely surrounded by a layer of dense ivory like bone forming the bony labyrinth. The *cristae ampullaris* have been assigned special roles, such as providing the bird with a special sense of direction during flight, however, no practical proof is available. A macula neglecta and a macula lagena are present in the inner ear of pigeon (Fig. 18.22 and 18.23) and many other birds, but their significance is largely shadowed by the complexity of lagena, which has become larger and even curved to from the cochlea, and the macula basilaris has been elaborated into the **organ of corti**. This has a basilar membrane supporting the hair cells in contact with the tectorial membrane as in mammals. The persistent macula lagena, at the tip of the cochlear duct, responds to lower notes; the organ of corti receives higher frequencies.

The **middle ear** of pigeon contains the columella auris whose inner portion, the stapes, plugs the fenestra ovalis and the outer portion contacts the tympanum by means of a three-rayed cartilage called extra columella. Eustachian tubes open into the pharynx by a common aperture. Tympanum lies at the bottom of ear hole.

Mammalian ear has been described in detail earlier. Bats are known to emit rapid pulses of sound far above the range of human hearing. These sounds bounce off objects as echoes and are picked up by the bat's ears, giving them a kind of 'sound picture' of surroundings, just as radar gives a radio picture for an aeroplane pilot.

Chapter 19

Endocrine System

Endocrine glands do not form an organ system. They lie in diverse parts of the body, with no distinct interconnections (Fig. 19.1). Moreover, although some are discrete organs (e.g. thyroid, adrenals), others are simply small parts of organs (e.g. islets of langerhans within the pancreas, or endocrine cells within the alimentary canal). Endocrine tissues arise from any of the three germ layers. Their hormones vary chemically according to their embryonic origin: protein, polypeptide, or amino acid hormones are secreted by endocrine glands of endodermal, or ectodermal epithelial origin, and steroids by those of mesodermal origin. However, although not anatomically like an organ system, endocrines do function as a system in controlling and regulating diverse portions of the body. In this they do not work alone. The pituitary, an extreme case, is not only controlled by the hypothalamus but some of its secretions are actually synthesized in the hypothalamus. Many other endocrines are influenced by autonomic innervation

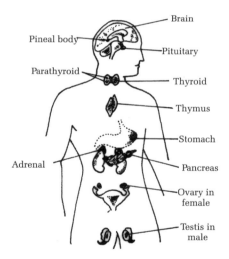

Figure 19.1
The location of various endocrine glands in man

Pituitary Gland or Hypophysis

The pituitary gland, which is located just below and attached to the hypothalamus is structurally and functionally the most complex of endocrine glands. It releases at least nine protein and polypeptide hormones which influence every portion of the body. Although frequently called the **master gland**, the pituitary itself is regulated through feedback mechanisms, by the nervous system and by other endocrines. It would more properly be thought of as a coordinating organ.

Structure

The pituitary has two major components, which develop from separate tissues and grow together (Fig. 19.2). The **neurohypophysis** is formed by a ventral growth from the hypothalamus; the **adenohypophysis**, by a dorsal evagination from the roof of the mouth. This evagination called Rathkes pouch, grows up anterior to the neurohypophysis, adheres to it, and then looses its contact with the oral cavity.

Neurohypophysis

The neurohypophysis includes the pars nervosa and the infundibulum (the stalk connecting pars nervosa and hypothalamus). The cells of the pars nervosa are glial elements called

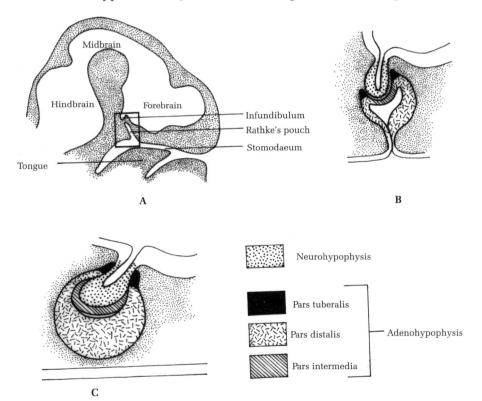

Figures 19.2 A, B, C
Development of the pituitary   A. Sagittal section through the entire head   B. and C. Higher magnification of later stages in pituitary development

pituicytes; they neither synthesize nor secrete hormones (Fig. 19.3). Hormones are synthesized in the cell bodies of the supraoptic and paraventricular nuclei of the hypothalamus. The hormones pass down the axons of these hypothalamic cells which run through the infundibulum in the hypothalamo-hypophyseal (Fig. 19.3) tract. They are then released into the highly vascularized space around the pituicytes, passing into the circulatory system.

In jawed fish, except sarcopterygians, the part of the hypothalamus just behind the infundibulum is a highly folded and vascularized structure of unknown function called the saccus vasculosus. Although frequently described as part of the piscine pituitary, it is truly part of the hypothalamus.

Adenohypophysis

Pars tuberalis In tetrapods, but not in fish, a small portion of the adenohypophysis wraps around the infundibulum and is called the pars tuberalis. It receives autonomic innervation and is highly vascularized, its large cuboidal epithelial cells contain granules. It apparently, produces no hormones.

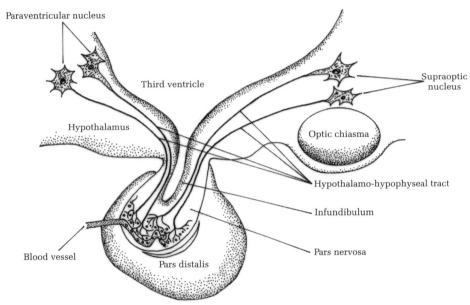

Figure 19.3
Structural and functional connection between the hypothalamus and the pituitary's pars nervosa

Pars distalis This largest and most prominent part of the adenohypophysis is composed of irregular cords of polyhedral epithelial cells separated by vascular sinusoids. There are three types of cells: **acidophils** have granules which stain prominently with acid stains; **basophils** have granules which stain with basic stains and **chromophobes** have granules which stain with neither of the stains.

In tetrapods the pars distalis has no innervation, but is connected with the median eminence (or tuber cineraum) of the hypothalamus by a unique portal vascular system. From a capillary bed in the median eminence venules lead down through the infundibulum to the pars distalis. (Fig. 19.4). These venules form the sinusoids in tetrapods. Specific hormone-releasing factors synthesized by cells in the median eminence are carried by this hypothalamo hypophyseal portal system to the pars distalis. There they regulate its flow of hormones.

A smaller portal system is present in the chondrosteans, holosteans, and elasmobranchs. In teleosts, however, there is nervous innervation. The hormone-releasing factors from the cells of the median eminence travel down the cells axons through the infundibulum and are released in the pars distalis, which in teleosts is densely innervated.

In cyclostomes neither a distinct portal system nor nervous innervation of the pars distalis has been demonstrated. The pars distalis of cyclostomes, teleosts, and chondrichthyeans is divided into rostral and proximal portions, frequently separated by connective tissue septae and having different staining characteristics. In chondrichthyeans there is also a separate ventral lobe of unknown significance.

Pars intermedia That portion of Rathke's pouch which makes contact with the neurohypophysis is called the pars intermedia (Fig. 19.2 A). It is separated from the pars distalis by a small extracellular space, or by an area of mixed pars distalis and pars intermedia cells. The pars intermedia has distinct polyhedral cells with basophilic granules. In at least frogs, elasmobranch and the bowfin, it receives autonomic innervation, with terminals directly on the gland cells. There is no pars intermedia in birds.

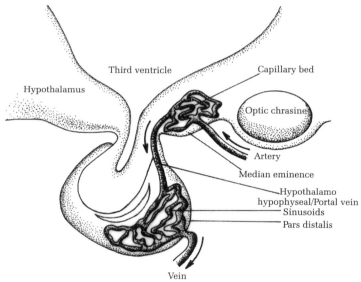

Figure 19.4
Connection between the hypothalamus and the pituitary's pars distalis

Hormones of the pituitary

The pituitary hormones have been most extensively studied in mammals; non-mammals have the same hormones, with but minor differences in amino acid sequences.

Neurohypophysis

Two octapeptide hormones produced in the hypothalamus are stored in and finally released by the neurohypophysis. Both affect the adenohypophysis as well as other target organs.

Oxytocin helps regulate the release of various hormones of the adenohypophysis. In mammals it also causes contraction of the uterine muscles and of the myoepithelial cells surrounding the mammary gland alveoli (causing milk to be "let down" i.e. stored prior to release by suckling).

Vasopressin, the antidiuretic hormone, helps regulate the release of pars distalis hormones. It also has two other important functions: to increase water resorption in the kidney, which helps keep the osmotic concentration of the internal fluids at a constant level and to stimulate contraction of the smooth muscles of blood vessels, which increases blood pressure. In birds, oddly enough, vasopressin causes the smooth muscle of blood vessels to relax.

Adenohypophysis

Pars distalis Six known hormones produced by the pars distalis. (Research has shown which cells produce each and has strongly suggested that no cells produce more than one.) The growth hormone, **somatotrophin**, produced by acidophilic cells, is a branched protein with about 200 amino acids. It is necessary for the normal growth and development of skeletal and muscular tissues.

The lactogenic hormone, **prolactin**, also produced by acidophilic cells, is a protein with a molecular weight of about 25, 000. In mammals it promotes the synthesis of milk, and in birds the production of a proteinaceous secretion of the crop called crop milk. In all vertebrates it affects sexual behaviour and the maintenance of the post ovulatory structures of the ovary.

The **adrenocorticotropic hormone, (ACTH)**, is 39 amino-acid polypeptide produced by chromophobe cells. It stimulates the activity and secretion of the adrenal cortex (inter-renal tissues in anamniotes) and also in most vertebrates affects the dispersion of melanin with in melanocytes.

The **luteinizing hormone**, LH, and the follicle stimulating hormone, FSH, both affect the reproductive organs. They are produced by slightly different basophilic cells, and are both small glycoproteins. In female, FSH, activates the growth of the ovarian follicle, and LH stimulates the interstitial cells of the ovary and the development of corpus luteum. In male, FSH stimulates activity of its interstitial cells. The thyroid—stimulating hormone; TSH, is produced by basophilic cells and is a glycoprotein where molecular weight is approximately 10,000. Its primary function; as its name implies, is to stimulate the secretion of thyroid hormone.

The production and release of these six hormones of pars distalis are promoted by specific releasing factors which are synthesized and secreted by the median eminence of the hypothalamus into the hypothalamo-hypophyseal portal system. Pars distalis activity is also directly influenced by feed back from other endocrines.

Pars intermedia The pars intermedia produces only the melanin stimulating hormone, MSH, polypeptide with 13-18 amino acids. It's action is most pronounced in anamniotes, where it stimulates the synthesis and dispersion of melanin within melanocytes, and the contraction of the purine particles in iridophores.

Comparative account of pituitary gland

Urochordates: The **adneural** and **subneural** glands of tunicates has been homologized with the pituitary gland of vertebrates. According to some the adneural and subneural glands represent the posterior lobe of pituitary.

Cephalochordates: According to recent views the **organ of Muller** of cephalochordates is the forerunner of adenohypophysis of pituitary gland of vertebrates.

Cyclostomes: Pituitary gland in cyclostomes lies between the ventral part of diencephalon and the nasopharyngeal pouch. During embryonic development, a nasohypophyseal stalk appears in close association with the developing olfactory sac at some distance from the dorsal lip and stomodaeum. It extends posteriorly beneath the forebrain. Later on the caudal tip of this stalk buds off certain cells which differentiate to from the intermediate lobe of pituitary. The remaining nasohypohyseal stalk forms the anterior lobe.

In hagfish, the pituitary gland is present between the infundibulum and the nasopharyngeal pouch. Here, it is made of clusters of cells. The neural and stomodaeal components, unlike other forms, are separated by means of layer of connective tissue.

Fish: In case of fish, the pituitary gland is called hypophysis and it remains in the *sella turcica*. In case of Selachian fish, the posterior lobe of pituitary is not much organized and is in the form diffused structure.

In fish like *Polypterus* and *Latimeria* the pituitary gland is found in the most primitive condition. Here, it remains connected with the mouth cavity by means of a hypophyseal cleft or the cavity of Rathke's pouch.

In *haddock* fish, pituitary gland is not present in the sella turcica. It lies closely pressed to the brain and is flattened. This gland is found at the end of very long infundibular stalk, in case of angler fish (*Lophius piscatorius)*, here the pituitary is actually lodged an inch in front of the usual position. There are many other minor variations in the overall constitution of pituitary in fish. However, its fundamental structure is maintained.

Amphibians: The stomodaeal evagination, which corresponds to Rathke's pocket, is a solid outgrowth from the beginning as in case of teleosts also. In **salientian amphibians** e.g. frog, the anterior lobe of pituitary is prominent. However, it lies posterior to the other parts of the gland. The intermediate lobe lies transversely above the anterior lobe and is like a dumb bell. The hypophyseal cleft is absent here. The overall structure of pituitary gland in caudate amphibians is largely similar to the one found in salientians.

Reptiles: The pituitary of reptiles is largely typical and similar to that of other vertebrates. However, in correlation with the unusually long, slender and posteriorly extending infundibular stalk, the pituitary gland is found lodged in a ventroposterior position to the diencephalon. The anterior lobe of the pituitary is rather loosely attached with its sister parts. The pars tuberalis is conspicuously absent in reptiles like snakes and lizards.

Birds: In birds, the pituitary gland is farther displaced from the brain than in other vertebrates. Here, the sella turcica is deeper. The pituitary gland lies in this depression. This depression is specific of the sphenoid bone. The absence of intermediate lobe along with the hypophyseal cleft is the most unusual characteristic of the pituitary gland of birds. It is pertinent to mention that the hypophyseal cleft is traceable during development. However, it gradually disappears. There is no clue available towards the fate of the intermediate lobe during development. It is interesting to note that the hormone intermedin, ordinarily secreted by the intermediate lobe, is now available in the anterior lobe.

Mammals: However, there are minor variations found in mammalian pituitary gland in different species of mammals. For instance, the central cavity of the infundibular stalk dips deeper into the posterior lobe in case of cat. However, in foxes it may be short and extend only to a slight degree into the infundibular stalk. There is found unusual structure projecting into the hypophyseal cleft from the intermediate lobe in case of *ox* and *pig*. This peculiar structure is called the **cone of Wulzen**. It has been seen that the cellular structure of cone of Wulzen is closely similar to that of the anterior lobe of pituitary gland and not to the pars intermedia.

Discrete intermediate lobe is inconspicuous in the pituitary glands of **armadillos, whales** and **Indian elephant**. Further, the anterior and posterior lobes are separated from each other by means of a connective tissue.

Thyroid Gland

Structure

The thyroid is the chief endocrine gland that regulates metabolic rate. In all vertebrates it is composed of spherical or oval follicles within a highly vascularized, loose connective tissue. (Fig. 19.5) The follicles contain a viscous, semigel colloid. The walls of the follicles are composed of a single layer of secretary epithelial cells, resting on a firm connective tissue basement membrane which forms the outer wrapping of the follicle. Against this basement membrane is a highly folded plasma membrane. On their apical surfaces the secretory epithelial cells have microvilli and occasionally flagellae or cilia which extend into the colloid (Fig. 19.6). Between follicular cells there are tight junctions, with no intercellular space. The cytoplasm of the follicular cells contains complex ergastoplasmic membranes and large cistern containing follicular fluid. In addition to the follicular cells there are occasional parafollicular glandular cells, which stain more lightly than do follicular cells and contain granules rather than cisternae. In all vertebrates the thyroid develops as a midline structure, evaginating from the floor of the pharynx. Throughout early development it retains contact with the pharynx by a hollow or solid column of epithelial cells. By the adult stage this contact has been lost and the thyroid is a true endocrine gland, vascularized and ductless.

In the *Lamprey* we can see what the phylogeny of the thyroid may have been. In the ammocoete larva it is called the subpharyngeal gland and is continuous with the pharynx via duct; the gland lies in a groove called the endostyle which is also continuous with the pharynx. After metamorphosis the floor of the pharynx is closed off, and the thyroid tissue becomes endocrine. It is likely, therefore, that the vertebrate thyroid gland evolved from an exocrine gland very early in vertebrate evolution; or perhaps even in prevertebrate chordates.

Despite similarities of development and microscopic structure, the thyroid's gross structure varies greatly among vertebrates. Among bonyfish discrete thyroid glands are found in only a few teleosts; other bony fish and adult lampreys have thyroid follicles dispersed along the ventral aorta below the pharynx. In cartilaginous fish compact thyroid follicles encapsulated by a loose connective tissue lie

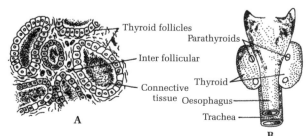

Figure 19.5
Position of thyroid and parathyroid glands

among the hypobranchial musculature in the anterior pharynx, deep to its epithelium. Such isolated thyroid follicles among head and neck tissues are not unusual even in vertebrates that have discrete thyroid glands; in mammals, for instance, they frequently occur at the base of the tongue. These isolated particles are "left overs" from the migrations of the thyroid tissue anlagen.

In amphibians, birds and some reptiles, the evagination from the floor of the pharynx divides, forming paired thyroids on either side of the trachea. In snakes and turtles, however, there is a single, midline thyroid. In adult mammals a bilobed thyroid lies ventrolaterally on either side of the larynx; there is usually a thin strand of thyroid tissue ventrally crossing the larynx and connecting the two lobes. (Fig. 19.7)

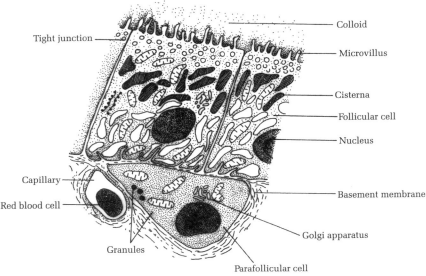

Figure 19.6
Electron microscopic details of both follicular and parafollicular cells of thyroid

Hormones of the thyroid

The follicular and parafollicular cells of the thyroid synthesize and secrete different hormones.

Follicular hormones

The follicular cells concentrate the body's iodine and utilize it in synthesizing two thyroid hormones. The two hormones, *triiodothyronine* and *tetra-iodothyronine* (thyroxine), are iodinated amino acids having similar effects on the body. Triiodothyronine is produced in smaller amounts but is considerably more potent than tetraiodothyronine. The final synthesis of these hormones occurs not within the follicular cells but within the colloid contained in the follicles. A major component of the colloid is a large protein, called thyro-globulin, secreted by the follicle cells. In the colloid, iodine, becomes attached to amino acids of the thyroglobulin. The resulting product splits off as one of the thyroid hormones (which one depends upon whether three or four iodine molecules are involved). The completed hormone must pass back through the thyroid cells to reach the capillaries outside the follicle.

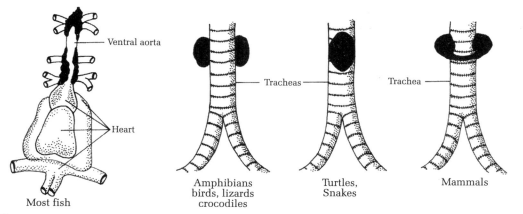

Figure 19.7
Location of thyroid tissue (in black) in diverse groups of vertebrates

The production and release of thyroid hormone are stimulated by TSH, one of the hormones of the pars distalis of the pituitary. In turn, the production and release of TSH are accelerated by releasing factors from the hypothalamus; both production and release of TSH are also limited by the secretion of thyroid hormone and are stimulated by its absence. This is an example of the intricate feedback mechanisms of the endocrine and nervous system.

The primary physiological effect of thyroid hormone is to increase the oxidative respiration of cells and therefore to increase the animal's metabolic rate-particularly important in the homeothermic mechanism of mammals and birds. Thyroid hormone also works in conjunction with the growth hormone from the pituitary. In amphibians it initiates metamorphosis if the gland is removed from amphibian tadpoles they will not metamorphose. In neotenic amphibians such as *Necturus* (mud puppy) the thyroid, although present, is very small.

Parafollicular hormone

The parafollicular cells of the thyroid gland produce quite a different hormone, called *calcitonin* or *thyrocalcitonin*, which is a polypeptide with 32 amino acids. This hormone lowers the level of calcium in the blood by causing more calcium to be deposited in bones; less calcium is therefore, excreted with urine. Calcitonin can also cause calcium to be deposited in the kidney, current evidence indicates that parafollicular cells developed differently than the follicular cells, being derivatives of the ultimobranchial bodies. These ultimo-branchial bodies originate from the corners of the posterior pharyngeal pouches in all vertebrates. They give rise to parafollicular cells of the thyroid and parathyroid only in tetrapods, remaining as ultimo-branchial bodies in fish. Apparently, however, the ultimobranchial bodies themselves secrete both calcitonin and parathyroid hormone in adult fish.

Comparative account of thyroid gland

Protochordates: True thyroid glands are absent. However, in structures like **endostyle**, iodine compounds are elaborated in amphioxus and tunicates. It is pertinent to point out that the above mentioned glandular, ciliated, groove like endostyle found in the ventral wall of pharynx of tunicates and *Amphioxus* has been compared as an homologue to thyroid gland of higher vertebrates. The chemistry of these iodine compounds differs from that of the thyroxin. In any case, these protochordate specific iodine compounds are carried on into the digestive tract with the food materials.

Cyclostomes: Again, ciliated grooves much like those of *Amphioxus* and tunicates is called subpharyngeal gland, which is sometimes referred to as the endostyle. It is believed that this gland elaborates an iodinated compound which is finally carried on to the digestive tract.

In **hagfish,** thyroid gland is made of ovoid capsules embedded in fat. These are median in position. These may also be oriented separately or in groups between the pharynx and the ventral aorta. They lie between the gill pouches on either side.

Fish: The characteristic feature related to thyroid gland is that it is only found in the embryonic stage. During adult sage, its location varies considerably. In **cartilaginous fish** like elasmobranchs, thyroid gland is a single lobed structure. It is found posterior to the mandibular symphysis at the point where the ventral aorta bifurcates. It is bilobed in case of **teleosts** and is found near the first branchial arch on either side. In case of **dipnoi,** it apparently appears to be a single mass. In reality, however, it is composed of two prominent lateral lobes connected by a constricted central portion. It is located under the epithelium tissue covering the tongue just anterior to the muscles of the visceral skeleton and just above the symphysis of the hyoid apparatus. Inadequate supply of iodine in the food of fish leads to goitre.

Amphibians: In **caudate** amphibians, thyroid gland appears early in the developing embryo in the ventral wall of the pharynx. At this stage, it is unpaired. Later on, the single mass divides into two. Thyroid gland lies ventrally on either side of the throat region slightly anterior to the aortic arches in adult salamanders. This gland possesses variable number of follicles. It is believed that the number of follicles is correlated with the size of the animal.

In **anurans,** thyroid gland is paired. It is deeply concealed and small in size. It can be located as an oval shaped structure lying lateral to the hyoid apparatus in the notch formed by the function of the lateral and thyrohyoid process e.g. frog.

Thyroid gland is very crucial for the metamorphosis of the larval stages into adults. In the absence of thyroid gland, tadpole larvae continue to grow in size and acquire lungs and reproductive organs by bypassing metamorphosis. However, if such abnormally large tadpoles are administered thyroid secretion, then these usually experience metamorphosis. The tail is lost, limbs develop and the mouth undergoes pronounced transformation. Interestingly enough, administration of thyroxine into very small tadpoles induces spectacular early metamorphosis giving rise to tiny individuals.

Reptiles: Thyroid gland is found in unpaired condition in reptiles like snakes, turtles, and crocodilians. Lizards possess bilobed thyroid in the young stage but paired in the adult. It is found lying ventral to the trachea approximately halfway along its course in case of lizards. However, in other reptiles, it is found more posteriorly lying in front of the pericardium. Thyroid seems to exercise, at least, in part, control over *ecdysis* in reptiles.

Birds: Here, the paired thyroid gland lies on either side of the trachea at the point of its bifurcation into bronchi. It is believed that excess of thyroxine exercises colour patterns in birds.

Parathyroid Gland

The parathyroid occur as discrete glands only in adult amphibians and amniotes (Fig. 19.5). They are composed of irregular epithelial cords of polyhedral cells, called *chief* cells. Like all endocrine glands they are highly vascularized. Although endodermal derivatives of pharyngeal pouches 2,3 and 4 they may migrate considerable distances: they are usually found in close association with the thyroid gland, often being embedded in its tissues near the cervical and upper thoracic ventral blood vessels.

The hormone of the parathyroid is a polypeptide of 75 amino acids. Its rate of secretion is controlled by the level of calcium and phosphate in the blood. Its target organs are bone, where it mobilizes calcium by causing bone resorption, and kidney, where it causes both the resorption of calcium from urine and the active excretion of phosphates. Thus it's functions are antagonistic to those of calcitonin.

Histologically, parathyroid glands exhibit densely packed masses of polygonal cells arranged in the form of cords (Fig. 19.6). A majority of the cells are pale in colour having a clear cytoplasm and are called **principal cells** which appear to be the essential elements of the gland. The other kind of cells available in small percentage are known as **oxyphilic cells.** These cells are filled with granules and readily stain with acidic dyes.

Parathyroid hormone secreted by parathyroid gland is the only proteinaceous hormone. It plays crucial role in the metabolism of **calcium** and **phosphorus.**

Comparative account of parathyroid gland

Cyclostomes: It is small in size and is of diffused nature. it has not been proved whether buds of small epithelial structures are parathyroid glands in reality.

Fishes: Certain glandular structures in the pharyngeal region may represent parathyroid glands. Their functional identity remains unexplained.

Amphibians: Parathyroid glands assume their true shape for the first time in amphibians. In **caudates,** the parathyroid glands are found lying lateral to the aortic arches and ventral to the thymus gland. Here, 2-3 parathyroid pairs appear on each side. In other forms, only a single pair may be present. These glands are widely separated from the thyroid in caudates. In anurans, these glands are small in size, rounded in appearance and red in colour. There are present two glands on either side of the posterior portion of the hyoid cartilage next to the inner ventral surface of the external jugular vein.

Reptiles: As in amphibians, parathyroid glands in reptiles arise from the ventral ends of 3rd and 4th pharyngeal pouches. These are generally two in number located somewhat posterior and lateral to the thyroid. These glands are advanced nearer the skull in case of snakes.

Birds: These are small, paired bodies lying slightly posterior to the thyroid gland on each side. There may be one or two pairs. In some birds, these glands may lie on the dorsal side of the thyroid.

Mammals: In case of man, each parathyroid gland is of the size of grain of corn (Fig. 19.5). The glands originating from the 3rd pharyngeal pouch are displaced posteriorly during embryonic development and hence are called **inferior parathyroids.** In contrast, the glands derived from 4th pharyngeal pouch are oriented anteriorly and are called **superior parathyroids.**

Abnormal functioning of parathyroid give rise to specific diseases in man. For instance, **tetany** occurs in the total/partial absence of parathyroid glands. Here, calcium decreases while phosphorus increases in the blood stream leading to excessive excitation of muscles and nervous tissue. This condition can be cured by injecting calcium salts along with the oral administration of vitamin D.

Pancreas

The endocrine portion of the pancreas is the *islets of Langerhans*. Three types of cells are found in islet tissue: **alpha cells**, containing alcohol insoluble granules; **beta cells**, the most common, containing alcohol soluble granules; and **D cells** or clear cells, the least common of the three cell types. (Fig. 19.8)

Although islet tissue occurs in all vertebrates, there is some variation in its composition and a great deal of variation in its distribution. In both the larval and adult lamprey aggregates of islet cells, primarily beta cells, are found within the intestinal epithelium. In the hagfish a compact organ of beta cells lies in the intestinal walls. Gnathostomes have all three cell types. In most jawed fish islet tissue occurs diffusely throughout pancreas, but in a few teleosts such as the catfish it is separate from the exocrine portion of the pancreas. In all tetrapods it occurs in separate islets within the exocrine pancreas.

Hormones of the Islets

The alpha cells secrete glucagon, a polypeptide hormone of 29 amino acids; beta cells secrete insulin, a proteinaceous hormone of 51 amino acids. Both regulate carbohydrate metabolism, but in antagonistic ways.

Glucagon mobilizes glycogen. Its target organ is the liver, where it causes stored glycogen to be broken down into glucose molecules which are released into the general circulation. Insulin, on the other hand, facilitates the efficient use of the glucose in the blood and prevents the excessive breakdown of glycogen. When insulin is not produced, as in the metabolic disorder called **diabetes**, most of the blood's glucose passes out with the urine without being used in cell metabolism or stored as glycogen.

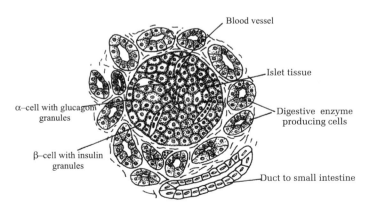

Figure 19.8
T.S. pancreas showing an islet with α–cells and β– cells

Comparative Account of pancreas

Cyclostomes: In *lamprey* a definite pancreas is absent. However, few cell masses found embedded in liver and in the wall of intestine are believed to be comparable to the islets of Langerhans.

Fish: In **cartilaginous fish** like elasmobranchs, the islet cells surround a cavity. In case of **teleosts,** small cluster of islet cells are found scattered in the general gut region. The pancreas is of diffused nature in these fishes. It is interesting to note that the islets of Langerhans are characterised by their large size having a pale colour. These islets of Langerhans are easily discernible with naked eye by virtue of their being thicker and more translucent in comparison to the surrounding exocrine pancreatic alveoli. In other species of teleosts is found a **principal island** which is actually an encapsulated islet of Langerhan of a very large size lying in the region of gall bladder/pyloric caeca/spleen. Its connection with the original pancreas is slight. Principal island is the only glandular tissue present in some teleost fish. In such fish, its removal naturally leads to a marked increase in the blood sugar levels.

Tetrapods: The islets of Langerhans are found normally distributed in the exocrine alveoli of pancreas in all the tetrapods viz., amphibians, reptiles, birds and mammals. In case of mammals, however, there is a tendency on the part of islets of Langerhans to concentrate in that portion of the pancreas which lies in close proximity of spleen. The number of islets of Langerhans may be 7.5 lakhs to 15 lakhs in human beings. In other vertebrates, the number may even exceed. Glucagon is the second hormone secreted by the pancreas. A cells specialize in the production of this hormone. Blood sugar levels are raised by the secretion of glucagon.

Adrenal Glands and Tissues

In mammals, birds, and reptiles the adrenals are compound glands with two types of tissues (Fig. 19.9, Fig. 19.10) **chromaffin tissue**, so named because its cells stain dark brown with chromates, composes the inner portion, or adrenal medulla. **Steroidogenic tissue**, so named because it produces steroid hormones, composes the outer portion, or adrenal cortex. In reptiles and birds there is often some cortical tissue interspersed with the chromaffin tissue of the medulla.

In **anamniotes**, both chromaffin and steroidogenic tissues are present in the region of the kidneys, but they are usually separate from one other at all developmental stages, and no discrete adrenal glands are formed. When steroidogenic tissue is isolated from chromaffin tissue in this way it can also be called inter-renal tissue, it acts as does the steroidogenic tissue of discrete adrenal glands.

In **amphibians** the chromaffin tissue is embedded in the ventral portion of the kidney along with steroidogenic tissue. In actinopterygians, separate, small aggregates of chromaffin and steroidogenic tissue are usually found along the veins in the trunk region and in the anterior portion of the kidney. In chondrichthyeans, chromaffin tissue is in small, paired patches on the medial side of the kidneys, and steroidogenic tissue is in a midline organ lying along the median dorsal blood vessels. In cyclostomes these are isolated, microscopic patches of separate chromaffin and steroidogenic tissue along the walls of the blood vessels and kidneys (Fig. 19.10).

Chromaffin tissues

These tissues are of neural crest origin specifically, their polyhedral cells are modified post ganglionic cells of the sympathetic nervous system. Thus they are innervated by, and their activity is controlled by, preganglionic sympathetic cells. Chromaffin cells, however, secrete large amounts of epinephrine (plus a minimal amount of norepinephrine), while other postganglionic sympathetic cells secrete primarily nor-epinephrine.

Both the epinephrine from chromaffin tissue and the nor epinephrine from the sympathetic system prepare the animal for periods of stress, with some differences. Epinephrine, for instance, dilates arterioles in muscles and liver and thus greatly increases the flow of blood through these organs; norepinephrine has no effect on this blood flow. Epinephrine has a much greater effect on increasing the heart rate than does norepinephrine.

Steroidogenic tissues

The steroidogenic (or inter-renal) tissues of all vertebrates are composed of polyhedral epithelioid cells (derived from mesoderm) containing secretory granules.

These tissues have been most extensively studied in mammals. In the mammalian adrenal cortex

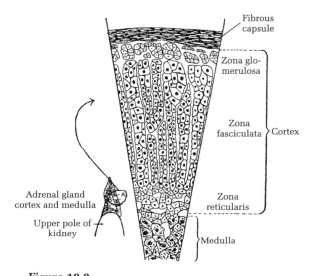

Figure 19.9
Adrenal gland showing differentiation into medulla and cortex, the latter itself being divided into three zones not clearly demarcated from one another. The part of the gland from which the section was taken is shown alongside

the steroidogenic tissues can be divided into zones on the basis of the arrangement of cells and their relationships to the numerous blood sinusoids in the outermost or **zona glomerulosa**, the cells are in irregular aggregates (Fig. 19.9). In a wide zone just deep to this, the **zona fasciculate**, the cells form long parallel cords separated from each other by vascular sinusoids. In the innermost **zona reticularis**, the cells are in irregular cords interspersed with sinusoids. A fourth zone is present in foetal mammals, called **zone X** or the **foetal cortex**, its cells are irregularly placed between the medulla and the zona reticularis. It is evidently an active foetal tissue but is resorbed before birth.

Nearly 50 steroids of the adrenal cortex have been isolated and identified in mammals. Some have no known function while others have very similar functions and can be considered in groups. There are two such groups, found widely in living vertebrates, that are produced only by the adrenal cortex and regulated by ACTH.

The first group, from the zona fasciculata and zona reticularis, regulates the intracellular metabolism of carbohydrates and proteins; cortisone is the most familiar of these. The second group, from the zona glomerulosa, is important in regulating water and electrolyte balance; these steroids act primarily on kidney tubules during urine formation. The most potent of these is aldosterone, which is particularly important for osmotic balance in sea birds and in those fish which can live in either fresh or salt water. In addition to these the adrenal cortex also produces small amounts of all those steroids that are produced in much larger quantities by reproductive organ.

Cyclostomes: In case of *Petromyzon*, there are found 2 distinct series of bodies of which the first consists of small, irregular, lobe-like structures positioned across the postcardinal veins, renal arteries and the arteries running dorsal to the opisthonephric kidneys. In some instances, these may run inside the lumen of vessels. These first series is referred to as **inter-renal or cortical.** The other series is known as the **chromaffin** series extending from the anterior end of the gill region to the tail. It is believed that the second series specializes in **epinephrine** production. In **cyclostomes** like **hagfish**, the first series is absent.

Fishes: The internal bodies are found lying between the posterior ends of opisthonephric kidneys. These may be paired in fishes belonging to family batoidea and unpaired in squali. These bodies are **ochre-yellow** in colour. The chromaffin bodies are paired lying across the segmental branches of the aorta on either side of spinal cord.

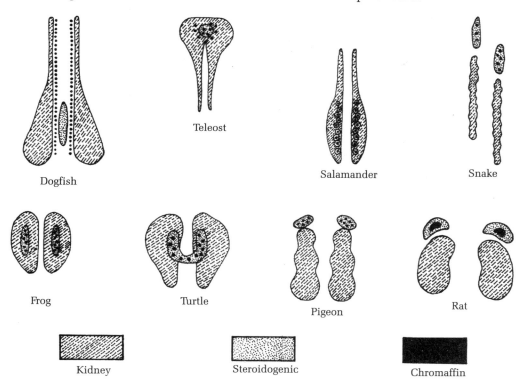

Figure 19.10
The distribution of steroidogenic and chromaffin tissues in relation to the kidneys in different vertebrates

In **teleost fish** again, inter-renal and chromaffin bodies stay separately. Inter-renal bodies are pinkish in colour and may sometimes be embedded in the opisthonephros. The chromaffin bodies are found towards the anterior end of the walls of post cardinal veins.

Inter-renal bodies are absent in **dipnoi** fish. Chromaffin bodies, on the other hand, are found around the intersegmental arteries and in the walls of the anterior part of the post cardinal and right azygos veins.

Amphibians: The cortical and the medullary portions show a marked proximity with each other in amphibians onwards. In these animals are found orange/yellow bands on the ventral surfaces of the opisthonephric kidneys. These bands are made of two types of tissues viz., inter-renal elements and chromaffin cells. The former cells are arranged in cell columns of different shapes and sizes whereas, the latter cells are irregularly distributed throughout the gland.

In **urodeles,** the adrenal glands are less compact. Here, inter-renal and chromaffin bodies are found intermingled in tissue extending the entire length of the opisthonephros. In some instances, these may exceed the anterior borders of the opisthonephros as much as to the subclavian arteries.

Reptiles: Inter-renal and chromaffin bodies are found profusely intermingled. In some reptiles, on the other hand, the chromaffin cells are concentrated towards the dorsal side of the adrenal gland. The arrangement and orientation of these glands is different in **crocodilians** and **chelonians.** In fact, here, adrenals, are akin to the one found in birds.

Birds: Here, adrenal glands lie on either side of vena cava just anterior to the kidneys in close proximity of the reproductive organs. These are ochre-yellow in colour. The characteristic feature of adrenals in birds is that inter-renal and chromaffin bodies are intensely interwoven with each other. Actually, the chromafin tissue comes to lie in the crevices of the inter-renal tissue.

Mammals: Adrenal glands of mammals are typical. These are found lying in close proximity with the cranial ends of the kidneys. In some mammals, there may be found a small space between the adrenal gland and the anterior end of the kidney. The right adrenal is nearer to the vena cava.

The mammalian adrenal gland possesses a true cortex and **medulla.** The medulla portion corresponds to the central region of the gland. Its cells display the typical chromaffin staining reaction. The outer portion of adrenal gland i.e., adrenal cortex is homologous with the interrenal tissue of lower vertebrates.

The adrenal glands of females are larger than those of the males in some species of rats.

There may be found accessory 'cortical' and 'medullary' bodies in the vicinity of adrenal glands. The size and number of these accessory bodies is variable in different species and even in individuals.

The histological constitution of cortex and medulla is species specific. In **prototherians,** these two portions are indistinct. Here, cords of medullary tissue penetrate into the cortical tissue. In higher vertebrates like mouse and bear, cortical and medullary tissues are separated by a connective tissue layer. Cortical portion dominates the medullary portion in mammals like guinea pigs and porcupines. Adrenal glands may be lobulated.

Portion of rat's adrenal gland showing zonation. The gland is shown below in cross section. The wedge above indicates the microscopic appearance of the medulla and the cortical layers.

Adrenals are of bigger size at the time of birth in human beings.

Histologically speaking, adrenal cortex is made of three or more distinct zones viz., **zona glomerulosa, zona fasciculata** and **zona reticularis** (Fig. 19.9). These zones may not be distinct in mammals like **opossum.**

The cells making medulla of adrenal gland are homogenous. Medullary portion is derived from the sympathetic ganglion cells. In some cases, there is present an additional zone called **X zone** or **X layer** between the medulla and zona reticularis of cortex. This layer disappears within one year after birth in man.

Adrenal cortex elaborates more than 30 different steroids and closely allied chemicals. These are known as **adrenocorticoids.** Some important adrenocorticoids include **aldosterone, corticosterone** and **cortisol.** In view of the astonishing revival effects of these adrenocorticiods, adrenal gland is the most important gland next to the pituitary. Based upon their functional role, these hormones are categorized as

1. Glucocorticoids
2. Mineralocorticoids
3. Sex hormones

In the absence of adrenal cortex, there occurs scarcity of sodium and water from the body leading to dehydration and fall in blood pressure. These symptoms are similar to the ones found in patients suffering from **addison's disease.** Dramatic recovery is observed after inoculation of cortisone in patients suffering from **rheumatoid arthritis,** addisons' diseases and certain types of leukemia.

Prominent hormones of adrenal medulla include **epinephrine** and **norepinephrine**. These hormones are in fact associated with fear or anger and are thus employed under conditions of stress and strain. Epinephrine and norepinephrine actually substitute the effect of sympathetic stimulation on visceral organs but many times more than the natural sympathetic effect. There is similarity between these hormones as far as their effects on visceral organs in concerned. However, epinephrine exercises greater effect on cardiac activity, produces lesser constriction of the blood vessels in the muscles. The latter effect is of greater importance because norepinephrine elevates arterial pressure higher than the one raised by epinephrine.

Epinephrine is believed to enable certain fish, amphibians and reptiles to camouflage by blending the colour of the skin pigments with that of immediate environment.

Gastrointestinal Hormones

Three hormones produced by cells in the epithelium of the mammalian stomach or duodenum help to regulate digestion. **Cholecystokinin**, from the duodenum, causes contraction of the gall bladder and release of digestive enzymes from the pancreatic cells. **Secretin**, also from the duodenum, stimulates the flow of pancreatic juice. **Gastrin**, from the stomach, stimulates the fundic glands of the stomach to secrete hydrochloric acid. These three polypeptide hormones have similar amino acid sequences that are also similar to that of glucagon; it is interesting that all form of these hormones are products of the endoderm of a restricted part of the alimentary canal.

Minor Endocrine Glands

Pineal

The tetrapod pineal, an outgrowth from the brain, was long suspected of being an endocrine gland but without experimental proof (Fig. 19.11). Now it has been demonstrated that in at least a few vertebrates its secretions can affect both the gonads and the pigment cells. For instance, in larval, amphibians a known pineal hormone, melatonin, causes the skin's melanophores to contract and lighten the tadpole's colour. However, the significance of the pineal gland as a whole remains obscure.

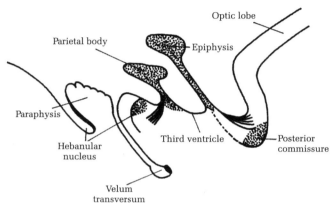

Figure 19.11
Sagittal section showing location of pineal in epithalamus (roof) of the brain

Thymus

Derived from the pharyngeal pouches, the thymus gland is unusual in that it reaches its maximum size in very young animal and then is largely resorbed and replaced by fat and connective tissue as the animal reaches maturity (Fig. 19.12). A very young animal whose thymus is removed will not produce antibodies. If thymus tissue is subsequently injected, or an exogenous thymus gland transplanted, antibody formation will occur. It is thus suggested that the thymus produces perhaps a hormone, which stimulates the lymphoid tissue to produce specific antibodies.

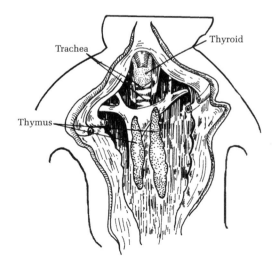

Figure 19.12
Location of thyroid and thymus glands (in one month old baby)

Urophysis

In teleosts, chondrosteans, and elasmobranchs, the posterior part of the spinal cord contains a group of large neurons, whose axons extend into vascular tissue (Fig. 19.13). In teleosts this tissue form a compact vascular organ called the urophysis. In the other fishes the urophysis is more diffuse. These *dahlgren* cells are clearly neurosecretory cells; when appropriately stained their secretion can be observed both in the axons and emptying into the urophysis. It is known that the urophysis releases a protein or polypeptide hormone which elicits rhythmic contraction of smooth musculature. The precise significance of this hormone and why it is found only in certain fish are not yet understood.

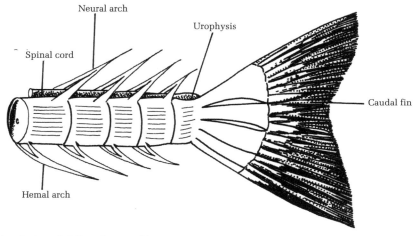

Figure 19.13
Location of urophysis (caudal fin of a perch)

Angiotensin

In mammals, and very likely in other vertebrates, the kidneys, liver, and vascular system have an interesting relationship. The kidney releases a proteolytic enzyme, **renin**, into the blood; the liver releases a globular protein into the blood; the renin partially digests the globular protein, thus producing a ten-amino acid polypeptide called *angiotensin-I*. Further enzymatic action in the blood stream converts this ten-amino acid polypeptide to an-eight-amino acid polypeptide, called *angiotensin-II*. Angiotensin-II in turn causes constriction of the arterioles and thus an increase in blood pressure and stimulates the zona glomerulosa of the adrenal gland to produce aldosterone. Aldosterone causes kidney tubules to resorb water and sodium during urine formation and also inhibits the further production of renin.

Reproductive Hormones

In addition to the reproductive hormones secreted by the pituitary gland the powerful hormones are produced by **ovary, testis** and accessory reproductive organs.

Testes

Testes are gonads specific to males. These are both **cytogenic** (spermatogenesis) and **endocrine** (male hormones) in nature. As far as cytogenic nature of testes is concerned, **spermatozoa** are produced by the process of **spermatogenesis** in the seminiferous tubules of the testes. In lieu of its endocrine nature, testes produce male hormones like **testosterone, androstenedione** etc., which are crucial to the growth of secondary sexual characters associated with growth in puberty and accessory reproductive organs as well. The luteinizing hormone (LH) of the anterior pituitary stimulates the Leydig cells to secrete the androgens.

Histologically speaking, each testis is covered externally with a connective tissue capsule called **tunica albuginea.** Each testis is composed of seminiferous tubules (Figs. 19.3 A, B). These tubules are grouped by partitioning with the help of septa. Each seminiferous tubule consists of cells belonging to different stages of spermatogenesis. In addition, large **sertoli cells** surround these for the purpose of nutrition.

The endocrine part of testes concerns the extratubular tissue in which are embedded **interstitial cells** or Leydig cells (Fig. 19.3 A). It is these cells which elaborate male hormones.

Comparative account of testes

Cephalochordates: There is no scientific proof in support of the presence of endocrine tissue in *Amphioxus*.

***Cyclostomes*:** In case of cyclostomes, urinogenital sinus assumes different shapes in males and females specially during breeding season. One would regard this to be a result of hormonal activity.

Fish: In many fish, sexual dimorphism prevails with regards to the difference in colour patterns, modification of fins, appearance of excrescences on the skin etc. These features are at their climax during breeding season and get diminished outside the breeding period. Simultaneously, many changes appear in kidney tubules, archinephric duct, cloacal glands etc., and these also get diminished outside the breeding period. All these changes have a direct relation with the interstitial cells which lie scattered in between the seminiferous tubules and secrete hormones to induce the above mentioned changes along with several other secondary sexual characters.

Amphibians: Changes similar to the ones described above in fishes are also met within amphibians. For instance, swelling of thumb pads in case of *Rana pipiens*, enlargement of the mental gland in case of *Eurcea bislineata*, colour changes in case of *Triton cristatus and enlargement* of cloacal glands in case of *Ambystoma texanum* are certain examples which appear due to the interference of sex hormones produced by the interstitial cells of the testes. These changes, as do the other reproductive organs, gradually disappear after the breeding period is over. The extent of the breeding period varies in various amphibians. For instance, it may extend from early spring (e.g. *Rana pipens, Eurycea bislineata, Ambystoma texanum*) to early autumn (e.g. *Ambystoma opacum*).

Reptiles: Unlike most vertebrates, majority of reptiles display seasonal reproductive activity and periodicity in appearance and disappearance of secondary sexual characters in a most striking manner. Interestingly enough, unlike lower vertebrates, interstitial cells in reptiles are visible throughout the year. However, these do undergo changes in size and number towards the onset of breeding season. The activity of interstitial cells seems to be synchronized with the appearance of secondary sexual characters. All these changes are under the control of sex hormones secreted by interstitial cells.

Birds: Avian testes also play dual role by being cytogenic as well as endocrine in function. Certain characters like beak colour, plumage changes in males and females may be correlated with breeding cycle. In others, secondary sexual characters prevail constantly throughout the year. e.g. domesticated species of the birds. The extent of the photoperiod has a direct effect on the testes of the birds. It is believed that both cytogenic and endocrine parts are affected. The endocrine activity of testes, in turn, is stimulated by the hormones from the anterior lobe of pituitary (Follicle stimulating hormone or **FSH**, Luteinizing hormone or **LH**).

Mammals: The mammalian testes possess interstitial cells. These interstitial cells (**Leydig cells**) are positioned in the interstices formed by the adjacent seminiferous tubules. Secondary sexual characters are definitely under the control of male hormones. For instance, a cryptorchid individual despite having less or no spermatozoa, does develop secondary sexual characters due to the presence of endocrine activity.

In the absence of male hormones, male individuals of meat providing livestock produce a very good variety of flesh. These commercial livestock animals are castrated to eliminate the effect of masculinity.

Besides testosterone hormone, other male hormones are androsterone, androstenedione. Androgens also produce appropriate changes in the skeleton of males. For instance, the pelvis of a castrated male shows definite differences from that of a normal male. Actually, male hormones are instrumental in determining the appearance/disappearance of certain cartilaginous/bony structures in male individuals. Besides, the larger size of males is due to the interference of these hormones.

Ovaries

Just like male gonads (testes), the female gonadal organs (**ovaries**) are cytogenic as well as endocrine in function. Owing to its cytogenic activity, **eggs** or **ova** are produced by the process of **oogenesis.** The ovaries produce **hormones** due to their endocrine activities. In contrast, to testes, the cytogenic and endocrine tissue of the ovary bear much closer morphological similarity. In general, ovaries mature at puberty and secrete few important hormones during oogenesis. Now, it is fully established that the cells of the **theca interna,** and **graafian follicles** are the sites of female hormones secretion at different times of the female cycle. The all important steroid called **estradiol** is secreted by these cells. Estradiol is the most potent of all the natural **estrogens.** Ovarian estrogens are influential in promoting secondary sexual characters in females. It is pertinent to mention that under the direction of hypothalamus, the anterior pituitary FSH cells secrete FSH (Follicle stimulating hormone) which in turn stimulates ovaries to secretes estrogens.

Comparative account of ovaries

Cephalochordates and Cyclostomes: In **cephalochordates,** not much is known about the endocrine functions of the ovary. In case of cyclostomes like *lampreys*, it has been observed that **urinogenital sinus** forms a large vestibule with a vulva. It is believed that the ovary in these female lampreys elaborates few female hormones which cause the above changes in the urinogenital sinus.

Fish: In fishes, endocrine functions are relatively more pronounced than in cyclostomes. For instance, in case of *Phoxinus laevis,* the attractive red colouration during breeding season is caused due to the secretion of hormones from the ovary. In ovoviviparous **elasmobranchs,** it is believed that **corpora lutea** develop at the sites of follicle rupture in the ovary.

Amphibians: In amphibians, ovary does possess endocrine functions. During breeding season, hormones coming out of the ovary induce many changes. For instance, oviducts, and uteri enlarge greatly and the glandular cells of the lining epithelium begin to secrete actively.

Reptiles: In snakes like *Storeria, Phtamophis, Thamnophis* and *Natrix,* well developed **corpora lutea** are found in the ovaries after ovulation. It is believed that corpora lutea secrete progesterone.

Birds: In most birds, only left ovary is functional. However, in birds of prey both ovaries may persist. Possibly, the follicular cells surrounding the large yolk-filled ova specialize in secreting estrogens. For instance, clinical administration of estrogen results in the activation of the oviduct in case of domestic fowl.

Mammals: A very important hormone i.e. **estradiol** is produced by the cells of the developing *Graafian follicles* of the ovary. One or more of these follicles enlarge each lunar month to burst open with the help of the pressure of the fluid. This releases ova from the follicles by the process of **ovulation.** Further, the ruptured follicles fill up their cavities to be converted into corpora lutea. The corpus luteum becomes endocrine as it elaborates **progesterone** hormone also known as the hormone of **pregnancy.** Corpus luteum, interestingly enough, lasts only up to two weeks if the **egg** is not fertilized. On the other hand, it persists for a greater period of time extending up to 3-4 months if fertilization occurs.

Progesterone is responsible for carrying on the growth of the endometrium, implantation of the fertilized egg in the uterine wall and stimulates the development of mammary glands. **Estrogen** is instrumental for inducing sex changes at puberty, development of accessory sex organs and secondary sexual characters, initiation of the growth changes in the endometrium, vagina and production of estrous or heat period.

Glossary

A

a sexual reproduction A reproductive process like fission, strobilation, budding, gemmule and statoblast formation, that does not involve the union of gametes.

abdomen (L. *abdere.* to hide). That portion of the body cavity containing the greater part of the digestive system.

abduct [L. *ab*, away/*ducere*, to lead] To move an appendage, or part of an appendage away from the body's axis. Opposite of adduct.

abductor L. *abducere,* to draw away). Serving to draw a part away from the axis of the body.

abomasum [L. *ab*, away from + omasum, bullock's tripe] Glandular portion of a ruminant's stomach.

aboral Away from the mouth.

absorption The passage of water and dissolved substances into a cell or an organ.

acanthodii [Gr. akantha, thorn] The subclass of the class Placodermi which is most similar to living bony and cartilaginous fishes.

acclimatize To become accustomed to a new environment.

accommodation [L. *ac*, toward + *com*, with + *modus* measure or manner + *atus*, adapted] Adjustment of the eye's cornea and/or lens to focus images onto the retina.

acelous [L. *a-*, without + *celom*, hollow]. An adjective describing a structure whose surfaces lack concavities; usually referring to the centrum of a vertebra.

acelullar Without cellular organization.

acetylcholine [L. *acetum*, vinegar + *chole*, bile] Acetic acid ester of choline; neurotransmitter of somatic motor nerves to skeletal muscles.

acidophil [L. *acid(us)* sour + Gr. *philos*, loving] A structure, usually a cell, which stains with acid stains.

acinus (L. *acinus*, grape), Referring to a minute saclike structure.

acoelomate Without coelom.

acousticolateralis [Gr. *akouazesthai*, to listen + L. *lateralis*, of the side] Adjective describing vertebrate sense organs with hair cells, that is, the inner ear and the lateral line system.

acrodont [Gr. *acros*, topmost, highest + *odontos*, tooth] Adjective describing sharp-pointed teeth having no roots.

acrosome (Gk. *akron*, tip, + *soma*, body) the anterior extremity of the head of a spermatozoan. Derived from golgi -complex, it contains proteolytic enzymes that, dissolve egg envelopes and egg-plasma membrane during fertilization.

actinopterygii [Gr. *aktinos*, ray + *pterygi*, wing or fin] Subclass of osteichthyes including all the ray-finned fishes.

activation Process of triggering of development in egg, normally achieved by a sperm.

adaptation A phenotypic feature of an individual that contributes to that individual's survival; a feature's form or function and associated biological role with respect to a particular environment.

adduct [L. *adducere*, to lead toward or bring into] To move an appendage, or part of an appendage, toward the body's axis. Opposite of abduct.

adductor (L. *adducere*, to bring toward). Serving to draw a part toward the median line.

adenohypophysis (G. *aden.* gland, + *hypophysis*, an undergrowth). The anterior lobe of the hypophysis (pituitary).

adrenal (L.*ad*, to +*ren*, kidney). An endocrine gland near or upon the kidney.

adrenergic *(adrenalin+* O, *ergon*, work). Relating to nerve fibres that liberate adrenalin.

aerial locomotion Active flapping in flight; volant.

aerobic Using or requiring oxygen.

aestivate (L. *aestivare*, to reside during the summer] To greatly reduce metabolic activity for extended periods of time in response to external stress such as warm weather or lack of food or water. Similar to hibernate, which however is a response to cold.

afferent (L. *ad*, to + *ferre*, to bear). Bringing to or into.

afferent A vessel or structure leading to or towards a given position.

after-birth Extra embryonic membranes that are delivered after the birth of a young one in mammals.

agglutination Cluster formation; the adhesion of spermatozoa to each other during fertilization as a consequence of fertilizin-antifertilizin reaction. These chemicals are produced by ova and spermatozoa respectively.

agnatha [Gr. *a-* without + *gnathos*, jaw] The class of vertebrates lacking true jaws.

agnathan A vertebrate lacking jaws.

akinetic skull A skull lacking cranial kinesis, that is, movable joints between skull bones.

alar plate [L. *ala*, wing] Portion of the central,. nervous system dorsal to the sulcus limitans from which develop the sensory nuclei; in the aggregate these nuclei form a wing shape in the spinal cord.

albicans (L. *albicare*, to be white). White in colour.

albumen (L. *albumen*, white of egg). White of an egg such as that of birds.

aldosterone [NL. *al(cohol) dehyd(rogenatum)*, dehydrogenated alcohol + Gr. *stereos*, hard or solid] A powerful steoid hormone of the adrenal cortex.

alecithal eggs Literally meaning "without yolk", these eggs contain little or no yolk altogether.

allantois (Gk. *allas*, sausage, +*eidos*, appearance), An extraembryonic saclike extension of the hindgut of amniotes (reptiles, birds and mammals serving excretion and respiration).

allantois An extraembryonic extension of the hindgut of amniote embryos that functions in excretion and sometimes in respiration.

allele (Gk. *allelon*, reciprocally). In Mendelian heredity, one of a pair of alternative genetic characters.

alveolus (L. *alveolus*, a small chamber, or cavity). Any small cavity; such as smallest chamber in the lung or the socket for a tooth.

ameloblast [OF. *amel*, enamel + Gr *blastos*, germ] A cell which produces the enamel or enamel-like substance forming the outer part of a tooth or scale.

ammonotelism Excretion of ammonia directly through the kidneys.

amniogenic (Gk. *amnion*, fetal membrane, +*genesis*, origin). Relating to the formation of the amnion.

amnion (Gk. *amnion*, fetal membrane). An inner extra-embryonic membrane surrounding the embryo. It serves as a kind of private pond to the embryo and protects it from mechanical shocks and desiccation; found in reptiles, birds and mammals.

amniote [Gr. *amnos*, Iamb + -*ion*, diminutive suffix] Any vertebrate with extraembryonic membranes that include the amnion: reptiles, birds, and mammals.

amphiarthrosis [Gr. *amphi*, both + *arthrosis*, joint] A skeletal joint with limited movement.

amphibia (Gk. *amphibios*, leading a double life). A class of vertebrates, intermediate in many characters between fishes and reptiles, which live part of the time in water and part on land.

amphibious Capable of living on land as well as in water.

amphiblastula A larval from in the development of *Sycon* sponge.

amphiblastula Double structured blastula as in certain sponges.

amphicelous [Gr. *amphi*, both + *celom*, hollow] An adjective describing a structure both of whose ends are concave; usually pertaining to the centrum of a vertebra.

amphimixis Mixing up of nuclear material of male and female gametes during fertilization

amphiplatyan [Gr. *amphi*, both + *platys*, flat] Adjective describing a structure flat on both sides. Usually refers to the centrum of a vertebra, in which case it is interchangeable with acelous.

amphystyly Jaw suspension via two major attachments: the hyomandibula and the palatoquadrate.

amplexus The nuptial embrace of female frog by a male in pairing or copulation until the eggs are laid by female.

ampulla (L. *ampulla*, a flask). A saccular dilation of a canal.

anaerobic Oxygen independent process or organisms.

analogous Features similar in function but different in origin.

analogy (Gk. *analogos*, comfortable). Bearing reference to similarities in function of two organs or parts in different species of animals.

analogy Features of two or more organisms that perform a similar function; common function.

anamniota (Gk. an without + amniote (which see) Any vertebrate which lacks an amnion although other extraembryonic membrane may be present; all fishes and amphibians.

anapsida [Gr. *an*, without + *apsis*, a loop or mesh] Subclass of reptiles which lacks temporal fenestra: turtles.

anastomosis [Gr. *ana*, again + *stoma*, mouth] A tubular structure, usually a blood vessel, which interconnects two other similar structures.

anatomy [Gr. *ana*, up + *tome*, a cutting] The study of the structure of organisms, usually as revealed by dissection.

anatomy Study of the gross structure of an organism.

andro (Gk. *aner, andros*, man). A prefix meaning "male".

androgen [Gr. *andro*, male + *-gen*, production (of)] A general term for male hormones.

anestrous [Gr. *an*, without + *oistros*, mad desire] An extended period of time between estrous cycles.

angiotensin [Gr. *angeion*, a vessel + L. *tensus*, to stretch] A vasoconstrictive protein produced by the action of renin on a globulin.

angular (L. *angulus*, an angle or corner). Relating to a part occupying an angle or corner.

animal Hemisphere That half of the egg which contains the nucleus and active cytoplasm. It is most active in cell cleavage characterized by less yolk. The other half of the egg is vegetal hemisphere.

animal pole An imaginary point on the surface of the egg or zygote which lies on the centre of the animal hemisphere. It is that region of the egg, where polar bodies are eliminated during maturation phase of oogenesis.

anlage (pl., anlagen). [Ger. set-up, layout; *an*, on + *lage*, position] The embryonic primordium of a structure.

anoxybiotic Same as anaerobic.

antagonist (Gk. *anti*, against + *agonizomai*, I fight). That which opposes or resists the action of another.

antidiuretic hormone [Gr. *anti*, against or opposite + *diouretikos*, from *oureen*, to urinate] Hormone increasing water resorption in the kidney and maintaining constant osmotic level in tissues; also stimulates contraction of smooth muscles in blood vessels. Syn., vasopressin.

antrum (Gk. *antron*, a cavity]. Any closed or nearly closed cavity.

anura (*an-* not, +Gk, *aura*, tail) Amphibia without a tail.

anus (L. *anus*, fundament). The rear opening of the digestive tract.

aorta (Gk. *aeiro*, I lift up). A main trunk of the arterial system.

apatite (Gk. *apate*, deceit). Name given to a group of minerals of the general formula $Ca_{10}F_2(PO_4)_6$.

apnea Temporary cessation of breathing.

apocrine [Gr. *apokrino*, to separate] A secretory process in which a small part of the cell forms the secretion; also an adjectival form designating a specific class of large sweat glands.

apoda (Gk. *apodos*, footless). Any of several different groups of animals that lack limbs or feet. Refers especially to one group of Amphibia.

aponeurosis [Gr. *apo-*, away + *neuron*. sinew or tendon] A fibrous sheet functioning as a tendon.

appendicular (L. *appendere*, to hang upon). Relating to appendages or limbs.

aqueduct (L. *aquaeductus*, a conduit). Relating .to a canal or passageway.

arboreal [L. *arbor*, tree] Adjective designating animals adapted for living in trees.

arboreal Tree dwelling.

arborization [L. *arbor*, tree] A spreading out, as the branches of a tree; used in describing the dendritic pattern of a neuron.

archencephalon (Gk. *arche*, beginning, +*enkephalos*, brain). The anterior division of the primitive brain.

archenteron (Gk. *arche*, beginning, +*enieron*, intestine). The primitive, or embryonic, digestive tube ; found in gastrula and communicating with outside by blastopore.

archeornithes (Gk. *arche*, beginning, +*ornithes*, birds). The original birds.

archetype [Gr. *arche*, first + *typos*, type] The original pattern or model after which a thing in made. In biology this term means an organism, real or hypothetical whose morphology is so general that all other members of its group could be derived from it.

archinephric duct general term for the urogenital duct; alternative names (wolffian duct) are given it at different embryonic stages (pronephric duct, mesonephric duct, opisthonephric duct) or in different functional role (vas deferens).

archinephros [Gr. *arche*, first + *nephros*, kidneys] A type of vertebrate kidney running the entire length of the body with but one tubule per segment. Syn., holonephros.

archipallium (Gk. *archi*, first, +*pallium*, cloak). The olfactory cortex.

archistriatum [(Gr. *archi*, foremost, old + *striatus*, furrow or channel] The ventro-lateral portion of the avian basal ganglia; also, used to designate parts of the basal ganglia of other vertebrates.

archosauria [Gk. *archi*, foremost, old + *sauros*, lizard] Subclass of Reptilia including crocodilians, dinosaurs, pterosaurs, etc.

arcual (L. *arcua*, a bow). Relating to an arch.

arcualia (L. *arcua*, a bow). The primordial elements from which a vertebra is formed.

area opaca A marginal darker zone of the Chick blastoderm.

area Pellucida The central, transparent region of the Chick blastoderm in which the body of the embryo develops.

area Vasculosa The inner part of the area opaca, which develops a network of blood vessels during the second and third days of incubation.

area Vitellina The outer part of the area opaca.

artery (L. from Gk. *arteria*, an air-conveyer; the arteries were once believed to be air-tubes). A blood vessel that transports blood away from the heart.

articular (L. *articulaire*, to connect). Relating to a joint.

articulation [L. *articularis*, pertaining to the joints] A joint; the connection of two skeletal elements.

artificial taxon A group of organisms not corresponding to an actual unit of evolution. Compare with natural taxon.

artiodactyla (Gk. *artios*, even in number, +*dakylos*, finger). A division of the ungulate or hoofed animals, having toes even in number, two or four.

aspiration Drawing in by suction.

aspondyly the condition in which centra are absent from vertebrae.

atlas (Gk. *Atlas*, mythological Titan who supported earth on his shoulders). The first cervical vertebra.

atrium *(L. atrium*, antechamber, entrance hall). Usually refers to heart chamber, but also applied to other organs.

auditory *(L.audire*, to hear). Relating to the perception of sound.

auricle (L. dim. of *auris*, ear). Relating to the external ear or a receiving chamber of the heart.

autolysis A process disintegration of a cell by its own lyosomal enzymes.

autonomic (Gk. *autos*, self, +*nomos*, law). Self-controlling, independent of outside influences.

autonomic nervous system [Gr. *autos*, self + *nomos*, law] The system of all neurons, both central and peripheral, that have general visceral motor functions.

autostyly Jaw suspension in which the jaws articulate directly with the braincase.

aves (L. pl. of *avis*, a bird). The class of birds.

aves [L., *aves*, birds] The class of vertebrates all of whose members are birds.

axial (L. *axis*, axle of a wheel). Relating to the central part of the body as distinguished from the appendages.

axis of the Embryo An imaginary line representing the anteroposterior axis of the future embryo.

axon (Gk. *axon*, axis). The principal process of nerve cell.

azygous (G. *a*, not, +*zygos*, yoke). An unpaired anatomical structure.

B

baculim A bone within the penis.

baculum [L. *baculum*, walking stick or staff] A sesamoid bone found in the penis of many mammals.

baleen [L. *balaena*, a whale] A hard keratinous substance hanging from the roof of the mouth in toothless whales; functionally it acts as a plankton filter.

baleen Keratinized straining plates that arise from the integument in the mouth of some species of whales.

Basal (L. *basis*, footing or base). Relating to a base.

basal ganglia [L. *basis*, basis + Gr. *ganglion*; a swelling or knot) A prominent group of nuclei deep to the base of the telencephalon Syn., corpus striatum.

basal plate [L. *basis*, basis + Gr. *platys*, broad or flat] (1) The fused parachordal cartilages of the neurocranium which form the skeletal floor of the hind-brain. (2) The portion of the central nervous system which develops ventral to the sulcus limitans; this includes the motor nuclei in the spinal cord and brain.

basement membrane [L. *basis*, basis +*membrana*, fine skin] Extremely thin connective tissue network supporting an epithelium.

basi (L. *basis*, base), Prefix pertaining to the base.

basibranchial (L. *basis*, base, +*branchia*, gill). Median-ventral component of a gill-arch.

basihyoid (L. *basis*, + G. *hyoeides*, Y-shaped). Median-central component of the hyoid arch.

basilar membrane [NL. *basilaris*, pertaining to the base, .especially of the skull + L. *membrana*, fine skin] Connective tissue sheet supporting the organ of Corti.

basilar papilla [NL. *basilaris*, pertaining to the base, esp. of the skull + L. *papilla*, nipple] A sensory epithelium for hearing in amphibians, reptiles and birds.

basophil [Gr. *basis*, base + *philos*, loving] A cell readily stained by basic dyes.

batoidea |Gr. *batis*, a ray or skate] The order of Chondrichthyes comprising skates and rays.

betz cell (Named for Vladimir Betz, a nineteenth century Russian anatomist] Large motor cell of the neocortex.

bicuspid (L. *bi*, two, +*cuspis*, point). Having two points or prongs.

bigeminal (L.*bi*, two,+*geminus*, twin). Referring to a doubling or twinning.

bilateral symmetry An anatomical arrangement in which an organism has two equal but opposite right and left halves when it is cut lengthwise in the middle vertical plane.

bile (L. *bilis*, gall). The fluid secreted by the liver.

biogenetic Law The embryos of higher species tend to resemble embryos of lower species in certain respects but they are never like the adults of lower species.

bioluminiscence Emission of light by living organisms.

biped [L. *bi-*, two + *ped*, foot] An animal which locomotes on two legs.

blastema (Gk. *blastema*, a sprout). A primitive cell aggregate from which an-organ develops.

blastocoel (Gk. *blastos*, germ +*koilos*, hollow). The cavity in the blastula.

blastocyst (Gk. *blastos*, germ + *kystis*, bladder) The hollow sphere of early mammalian development.

blastocyst (Gk. *blastos*, germ. +*derma*, skin). The primitive cellular plate at the beginning of embryogeny; living portion of egg from which both embryo and all of its membranes are derived.

blastodisc (Gk. *blastos*, germ + L. *discus*, plate). A plato of cytoplasm at the animal pole of the ovum.

blastomere (Gk. *blastos*, germ +*meros*, part). One of the cells into which the egg divides during the cleavage phase of development; cellular unit of developing egg or early embryo, prior to time of a gastrulation. Smaller blastomeres are micromeres; intermediate ones are mesomeres ; larger ones are macromeres, where there is great disparity in size.

blastopore (Gk. *blastos*, gerrn +poros, opening). The opening of archenteron (gastrocoel) to exterior, occluded by yolk plug in amphibian embryo consisting of a slit like space between elevated nerve of blastoderm and underlying yolk of chick egg; represented in amniota as primitive streak.

blastula (Gk. *blastos*, germ). An early embryo, the cells of which are commonly arranged in the form of a hollow sphere; an embryonic stage between the appearance of distinct blastomeres and the end of cleavage (i.e., the beginning of gastrulation).

blood Islands Pre-vascular groups of mesodermal cells migrated from the postero-lateral margins of the area pellucida into the area opaca. These give rise to the endothelium of the blood vessels and blood corpuscles of the chick embryo.

brachial (L. brachialis, arm). Relating to the arm.

brachiation Arboreal locomotion by means of arm swings and grasping hands, with the body suspended below the underside of branches. Compare with scansorial locomotion.

bradycardia Abnormally slow heart rate. Compare with tachycardia.

branchial (Gk. branchia, gill). Relating to gills.

branchial basket The expanded chordate pharynx that functions in suspension feeding.

branchiomerism (Gk. branchia, gill + meros, part). Segmentation corresponding to the visceral arches.

bronchus (Gk. brounchos, windpipe). One of the two branches given off the trachea and leading to a lung.

bryozoa A phylum of sea-weed like invertebrates: moss animals.

buccal (L. bucca cheek). Relating to the mouth; the surface towards as cheek.

bud In animals, an asexually produced nodule that develops into a new individual by going through blastogenesis.

bulbus (L. bulbus, a bulbous root). Relating to any structure of globular form.

bunodont Pertaining to teeth with peaked cusps. Compare with lophodont and selenodont.

C

caecum (L. *caecus*, blind). Any structure-ending in a cul-de-sac.

calamus [Gr. *kalamas*, reed or stalk] The proximal end of the shaft of a quill leather.

calcification A specific type of mineralization involving calcium carbonates (invertebrates) or calcium phosphates (vertebrates) in the matrix of special connective tissue.

calcitonin *(L.calx,* lime + Gk., *tonos,* tone). Referring to the level of blood plasma calcium.

calcitonin [L, *calx,* lime] Hormone secreted by the parafollicular cells of the thyroid and parathyroid glands.

callosum *(L. callosus,* hard). Relating to something thick or hard.

calyx [Gr. *kalyx*, husk or covering] Dilation of the ureter within the metanephric kidney.

cancellous *(L. canellus,* grating or lattice-work). Noting spongy bone.

canine (L. *caninus*, relating to a dog). Ordinarily refers to a tooth type, but there are other references as well.

cannon bone Hindlimb bone resulting from fusion of metatarsals III and IV (as in horses).

capillary (L. *capillaris*, relating to hair). A microscopic blood vessel, intermediate to arteries and veins.

capsula (L. *capsa*, a chest). A capsule in any sense.

carapace [Sp. *carapacha* carapace] Dorsal half of a turtle or armadillo shell.

carboniferous [L., *carbon*, charcoal + *ferous*, bearing, producing, yielding] Geologic period in the latter Paleozoic era characterized by large deposits of organic carbon from plants.

cardiac (L. *kardia*, heart). Relating to the heart.

cardinal (L, *cardinalis*, hinging, important). Of special importance.

carnassial teeth [F. *carnassier,* flesh-eating] Last upper premolar and first lower molar of carnivores; these teeth are specialized for shearing flesh.

carnassials Sectorial teeth of carnivores, including upper premolars and lower molars.

carnivora (L. *caro; carnis,* flesh +*vorare,* to devour). An order of flesh-eating mammals which possess teeth and claws adapted for attacking and devouring prey.

carotenoid (L. *carota*, carrot, + Gk. *eidos*, like). Referring to a group of pigments having a yellow colour.

carotid (Gk. *karoo*, cause, to sleep profoundly). Relating to blood vessels supplying the head, compression of which produces unconscious state.

carpus (Gk. *karpos* wrist). The collection of bones comprising the wrist.

cartilage (L. *cartilago*, gristle). A gristlelike skeletal tissue.

catadromous Characterizing fishes that hatch in salt water, mature in fresh water, and return to salt water to breed; for example, some eels. Compare with anadromous.

caudal (L. *cauda*, tail). Relating to the tail or rear.

cavernosus (L. *caverna*, cave). Relating to a cavern or cavity.

cavum (L. *cavum*, hollow). Any hollow, hole, or cavity.

cavum arteriosum [L. *cavum*, hole + Gr. *arteria*, windpipe, artery] Portion of the ventricle in the chelonian and lepidosaurian heart which receives blood from the left atrium.

cavum pulmonale [L. *cavum*, hole + *pulmo*, lung] Portion of the chelonian and lepidosaurian heart ventricle from which the pulmonary artery exit.

cavum venosum [L. *cavum*, hole + *venosus*, vein] Portion of the chelonian and lepidosaurian heart ventricle from which the right and left aortae exit.

cbondro- (Gk. *chondros*, cartilage). Combining form meaning cartilaginous.

cell (L. *cella*, a small chamber). The structural and functional unit of plant and animal organization or protoplasm surrounded by a lipoproteinaceous plasma membrane and containing one or more nuclei.

cell lineage Study of origin and fate of specific blastomeres during embryonic development.

cementum [L. *caementum*, rough quarry stones] Layer of dense material holding tooth root into socket.

cenozoic [Gr. *kainos*, new or recent + *zoe*, life] The most recent (and current) geologic era, which began about 70 million years ago.

centriole (Gk. *kentron*, a point). The central granule in a centrosome. It is composed of microtubules and forms mitotic or meiotic spindles during cell division.

centrolecithal Pertaining to eggs with yolk accumulated in centre of egg, e.g., in arthropods.

centromere *(Gk. kentron*, centre + *meros*, part). Region of constriction of chromosomes. It holds sister chromatids together and spindle fibres attach to it during cell division.

centrum (L. *centrum*, center). A central structure of any kind.

cephalic (Gk. *kephale*, head). Relating to the head.

cephalic flexure Ventral bending of embryonic head at the level of midbrain and hindbrain.

cephalization [Gr. *kephale*, the head] The concentration of important organs in the head.

cephalochordata (Gk. *kephale*, head +*chorde*, string chord). A subphylum of Chordata in which the notochord extends into the head.

cerato- (Gk. *Keras*, horn). Prefix denoting composition of horny substance or resembling a horn in shape.

cerebellum (L. dim. *cerebrum*, brain). The brain mass deriving from the roof of the metancephalon.

cerebrum (L. *cerebrum*, brain*)*. The brain mass deriving from the roof of the telencephalon.

ceruminous gland [L. *cera*, wax] Gland of the external auditory meatus secreting ear wax.

cervical (L. *cervix* neck). Relating to the neck.

cetacea (L. *cetus*, whale). An order of completely aquatic mostly marine, mammals of the sub class Eutheria, including the whales, dolphins and porpoises.

chalaza (Gk. *chalaza*, a sty). Twisted opalescent chord of heavy albumen immediately, surrounding the vitelline membrane in a bird's egg.

chelonia [Gr *chelone*, tortoise] The order of Reptilia comprising turtles and tortoises.

chiasma (Gk. *chiasmaa*, figure of X). Describing the crossing of fibres.

cholecystokinin (Gk. *chole*, bile, +*kystis*, bladder, +*kineo*, move). A hormone liberated by the intestinal mucosa causing contraction of the gall-bladder.

cholinergic (Gk.*chole*, bile,+*ergon*, work) Relating to nerve fibres that liberate acetylcholine.

chondrichythyes (Gk. *chondros*, cartilage, +*ichthyes*, fish). Fishes with a cartilaginous skeleton, e.g., sharks.

chondroblast (Gk. *chondros*, cartilage, +*blastos*, germ). An embryonic cartilage forming cell.

chondrocranium (Gk. *chondros*, cartilage, +*kranion*, skull). A cartilaginous skull.

chondrocyte (Gk. *chondros*, cartilage, +*kytos* cell). A cartilage cell.

chondrostei (Gk. *chondros*, cartilage +*osteon*, bone), An order of fishes characterized by a largely cartilaginous skeleton.

chord,: Chorda (L. *chorda*, string or cord). A prefix or suffix meaning cord.

chordata (L. *chorda* fr. G. *chorde*, a string). The phylum of animals with-a-notochord, transient or persistent, *i.e.*, in the adult or at a stage in their development.

chordomesoderm Embryonic notochordal and mesodermal cells.

chordroclast (Gk. *chondros*, cartilage, +*klastos*, broken in pieces). A cell concerned in the destruction of cartilage.

chorio-allantois A common membrane formed by the fusion of the inner wall of the chorion and the outer wall of the allantois (both mesoderm) in the chick embryo.

chorion (Gk. *chorion*, a skin). One of the extra embryonic membranes in reptiles, birds, and mammals ; forms outer cover around embryo and all other membranes, and in mammals contributes to the structure of placenta.

choroid (Gk. *chorion*, skin, +*eidos*, like). Relating to a coat or membrane.

chromaffin tissue [Gr. *chroma*, color + L. *affinis* affinity] Tissue particularly in the adrenal medulla which can be stained deep brown by chromic salts.

chromatophore (Gk. *Chroma*, color, +*phorus*, bearer) A pigment containing cell.

chromosome (Gk. *Chroma*, colour, + *soma*, body). A deeply straining rod shaped or threadlike body in the cell nucleus; bearer of genes.

chyme [Gr. *chymos*, juice] Partially digested food passed from the stomach to the small intestine.

ciliary (L. *ciliariss*, resembling eyelash). Relating to any cilia or hairlike processes.

cingulum [L. *cingere*, to surround] An association tract of the mammalian cerebrum running from the medial cortex to the hippocampus.

cisterna [Gr. *kiste*, chest] Any cavity or enclosed space; particularly, small cavities made by double-walled endoplasmic reticulum.

clade (or Monophyletic Group) One that includes a common ancestor and all descendants

clade A natural evolutionary lineage including an ancestor plus all and only its descendants.

cladogram A branching dendrogram representing the organization and relationships of clades.

clavicle (L. *clavicula*, a small key)-. The collar-bone.

cleavage (Gk. *kleiben*, to cleave). Denoting division or splitting. The series of mitotic cell divisions that transform a single-celled fertilized egg into a multicellular embryonic stage called *blastula*. Syn., segmentation.

cleidoic (Gk. *kleido*, lock up). Referring to eggs, such as those of birds, that are insulated from the environment by albumen, membranes, and shell.

cleidoic egg the shelled container in which the fetus is laid, as in reptiles, birds, and primitive mammals. Compare with egg.

clitoris (Gk. *kleitoris*). An organ composed of erectile tissue, the homologue in the female of the penis.

cloaca (L. *cloaca*, sewer). The combined urogenital and rectal receptacle.

cochlea (L. *cochlea*, snail shell). The spiral canal and its contents in the inner ear.

coel, Coela, Coelo (Gk. koilos, hollow). A suffix or prefix, meaning "cavity".

coelacanthini [Gr. *koilos*, hollow + *acantho*, spiny] The group of sarcopterygian fishes of which *Latimeria* is the only living form.

coeliac (Gk. *koilia*, belly) Relating to the abdomen.

coeloblastula An embryonic stage in the form of a spherical ball consisting of one or several layers of cells arranged around a large, central cavity, the *blastocoel*, e.g., echinoderms, Amphioxus.

coelom (Gk. *koiloma*, a hollow). A cavity bounded by mesodermal epithelium ; subdivided in higher forms into pericardial

colic caecum [Gr. *kolikos colon* + L. caecum, blind .gut] A blind diverticulum occurring at the junction of small and large intestines.

collagen (Gk. *kolla*, glue + *gennao*, I produce). An albuminoid present in connective tissues.

collateral (L. *con.*, together +*lateralis*, relating to the side). Accompanying or running by or from the side of.

collum (L. *collum* gen. *coli*, the neck). Relating to the neck or a collar.

columella [Dim. of L. *columna*, pillar] Middle ear ossicle of amphibians, reptiles and birds.

commensalism Association of individuals of two or more different species to the benefit of one or more and harmful to none.

commissure (L. *commissura*, connection). Passing from one side to the other.

competence Ability of embryonic area to react to stimulus.

concha [L. *concha*, shell] A bone, usually scroll-shaped, covered by mucous membranes, which gives added surface area to the nasal chamber.

concrescence Migration of groups of cells on the two sides of the blastoporal ring during the course of gastrulation until they meet to form a single mass of tissue,

concurrent Flow of adjacent currents in the same direction.

conductor (L. *conducere*, to lead). That which conveys or transmits.

cone [Gr. *konos*, pine cone] Visual receptor cell of the retina which responds to specific wavelength; Thus used in colour vision.

cone, Fertilization Conical projection of cytoplasm from surface of egg to meet spermatozoon which is to invade egg cortex. Cone makes contact and then draws sperm into egg. Not universally demonstrated or seen in frog, but seen in starfish (Chambers) — (Syn. exudition cone).

conjunctiva (L. *fem*, *conjunctives*, to connect). The mucous membrane covering the front surface of the eye and lining the eyelids.

constrictor (L. *constringere*, to draw together). Anything that binds or squeezes a part.

Convergence Movement of groups of embryonic cells towards a single point *e.g.,* towards blastopore or primitive streak.

coprophagy The eating of feces, behavior performed usually to process undigested material again; refection.

copuala (L. *copula*, yoke, joining). Relating to a part connecting two structures.

copulation path The passage marked by a trail of pigment in the egg cytoplasm by the migration of sperm nucleus towards the egg nucleus during fertilization; second portion of sperm migration path through egg toward egg nucleus; when there is any deviation from entrance or penetration path; path of spermatozoon which results in syngamy, *e.g.,* frog.

coracoid (Gk. *korax*, raven, +*eidos*, appearance). Shaped like a crow's beak.

cordis (L. *chorda*, a string). Relating to any cordlike structure.

corium [L. *corium*, skin, hide, leather] The mesodermally derived, dense connective tissue layer of the skin. Syn., dermis.

cornea (L. *corneus*, horny). The transparent front portion of the Eye.

corneum (L. *corneus* horny). Relating to a horny layer.

cornu (L. *cornue*, p. *cornua*, a horn). Resembling a horn in shape.

corona Radiata A layer or follicle cells immediately surrounding the egg of the bird or mammal. These cells are elongated and acquire intercellular canals radiating outwardly from the egg to the surrounding theca. Syn., follicular epithelium.

coronary (L. *coronarius*, a crown; encircling). Denoting various encircling anatomical parts.

coronoid (Gk. *korone*, a crow's +*eidos*, resembling). A process shaped like a crow's beak.

corpora cavernosa [L. pl. of *corpus*, body + *cavernosus*, full of hollows] Two dorsal columns of erectile tissue in the amniote penis.

corpora quadrigemina [L. pi. of *corpus*, body + *quadri*, four + *geminus*, a twin] Roof of the mammalian midbrain, composed of two superior colliculi and two inferior colliculi.

corpus (L. *corpus*, the body). Any body or mass.

corpus callosum [L. *corpus*, body + *callosus*, hard] Neo-cortical commissure of eutherians; the largest tract in most eutherian brains.

corpus luteum [L. *corpus*, body + *luteum*, yellow] Remains of the ovarian follicle after ovulation.

corpus luteum pl. **Corpora lutea**. Progesterone-secreting bodies in vertebrate ovaries formed from remnants of follicles after ovulation. Substance called lutein.

corpus spongiosum [L. *corpus*, body + Gr. *spongos*, a sponge] Ventral medial column of erectile tissue surrounding the urethra in the mammalian penis.

corpus striatum [L. corpus, body + Gr. *striatus*, furrow or channel] Prominent group of ventral medial telencephalic nuclei. Syn., basal ganglia considered collectively.

cortex [L. *cortex*, barkl Outer portion of an organ, such as the kidney; specifically used to designate the outer portion of the central nervous system when it is gray matter rather than white matter.

cosmine An older term designating a derivative of dentin that covers some fish scales; cosmoid scale.

cosmoid scale [Gr. *kosmos*, ornament or form] Heavy bony scales of ancient sarcopterygians which had ornately patterned outer surfaces.

costal (L. *cosia*, rib). Relating to a rib.

cotyledonary (Gk. *kotyledon*, the hollow of a cup). Referring to aggregations of villi on the chorionic surface of a placenta.

cotylosauria [Gr. *kotyle*, cup-shaped + *sauros*, lizard] Stem order of Reptilia.

countercurrent Flow of adjacent currents in opposite directions.

cowper's gland [Named for William Cowper, late seventeenth century British anatomist] A mammalian male secondary genital gland emptying directly into the urethra. Syn., bulbourethral gland.

cranial Flexure Bonding of embryonic forebrain toward the underlying yolk at the level of midbrain in chick embryo, at the fourteen somite stage.

cranial kinesis Movement between the upper jaw and braincase about joints between them; in restricted sense, skulls with a movable joint across the roofing bones. Compare with akinetic skull, prokinesis, mesokinesis, metakinesis.

cranial Relating to the skull or to brain.

cranium (Gk. *kranion*, srkull). The bones of the head collectively.

crescent, Gray Crescentic area between original animal and vegetal pole region on surface of frog's egg, gray in colour because of migration of pigment away from area and toward sperm entrance point (Roux, 1888); region of presumptive chordamesoderm, future blastopore, and anus.

cresent, Yellow A yellow crescentic area on the surface of the Ascidian egg, which will give rise to the mesoderm of the embryo, surrounding the ovum in a mature Graafian follicle of the mammalian ovary.

cretaceous [L. *creta*, chalk] Most recent period of the Mesozoic era.

crista (L. *crista*, a crest). Referring to any ridge or crest projecting from a level surface.

crista [L. *crista*, crest]. The sensory epithelium in the ampulla of the semicircular canal.

crocodilia [L- *kroke*, pebble or gravel + *drilos*, worm] The order of Reptilia including alligators, crocodiles, etc.

crossopterygii [Gr. *krosso*, tassels or fringe + *pterygion*, little wing or fin] The group of sarcopterygian fishes including the coelacanths and rhipidistians.

ctenoid (Gk. *kteis* a-comb, +*eidos*, form). A type of scale featuring toothlike projections.

cumulus Oophorus (Gk. *cumulus*, a heap). A collection or heap of cells.

cupula [L. *cupula*, small tub or dome] Gelatinous material in which the hairs of the hair cells of the ampullary cristae are embedded.

cursorial locomotion Rapid running.

cutaneous (L. *cutaneous*, skin). Relating to the skin.

cuticle (L. *cuticula*, dim. of *cutis*, skin). Outer horny layer of the skin.

cuvier, duct of [Named for Georges Cuvier, French anatomist] The vein which collects from the anterior and posterior cardinal veins and drains into the sinus venosus. Syn., common cardinal vein.

cycloid (Gk. *kyklos*, circle, +*eidos*, form). A type of scale, featuring circular growth rings.

cyclostomata (Gk. *kyklos*, circle, *stoma*, mouth). Fish likeforms, possessing a cartilaginous skeleton, a circular mouth without jaws, and dorsal and caudal fins but no paired fins.

cystic (Gk. *kystis*, bladder or cyst). Relating to a bladder or cyst.

cyto (Gk. *kytos*, cell). Suffix or prefix indicating cell.

cyto architecture [Gr. *kytos*, a hollow vessel (now dually meaning a cell) + L. *architectura*, architecture] The structure of the cytoplasm, i.e., the organization of the organelles within a cell.

cytokinesis (Gk. *kytos*, cell +*kinesis*, motion). Division of the cytoplasm of a cell during cell division.

cytotrophoblast (Gk. *kytos*, cell, +*trophe*, nourishment, +*blastos*, germ). The inner or cellular layer of the trophoblast.

D

dead space The volume of used air not expelled upon exhalation. Compare with tidal volume.

decidua (L. *deciduos*, falling off). The endometrium of the pregnant uterus cast off at the time of parturition.

decussate [L. *decussis*, the number ten] To cross in the form of an X (like the Roman numeral ten).

dedifferentiation Disappearance or loss of specialized or differentiated cell structures reverting them to an undifferentiated state like the embryonic cells; a regressive change toward a more primitive, embryonic, or earlier stage, *e.g.*, a process changing a highly specialized cells to a less specialized cell.

deferens (L. *de*, away, + *ferens*, carrying). Applying to a duct carrying away.

delamination Separation of cell layers by splitting

demibranch [F. *demi*, half or lesser + Gr. *branch*, gill] A gill arch with lamellae on only one of its two surfaces.

dendrite (Gk. *dendrites*, relating to a tree). One of the receiving, branching processes of a nerve cell.

dental (L. *dens*, a tooth). Relating to the teeth.

dentary (L. *dens*, a tooth). A tooth-bearing bone, specifically in the lower jaw.

denticle (L. *denticitlus*, ;i sm;<ll tooth). A projection from a hard surface.

dentine (L. *dens*, a tooth). The substance proper of a tooth.

depressor (L. *drepressus*, to press down). Serving to lower or pull down.

dermatocranium (Gk. *derma*, skin, +*kranion*, skull). Collectively the superficial bones of the skull without cartilaginous precursors.

dermatome (Gk. *derma*, skin, + *tome*, cut, section). One of the embryonic skin segments.

dermis (Gk. *derma*, skin, +*pteron*, wing). Flying lemurs.

desiccate [L. *de-*, .prefix indicating intensity + *siccare*, to dry] To dry thoroughly.

desmosome [Gr. *desmos*, a band + *soma*, body] Tight junctions, usually between epithelial cells.

deuteroencephalon (G. *deuteros*, second, +*enkephalos*, brain). The posterior division of the primitive brain.

deuterostome An animal whose anus forms from or near the embryonic blastopore, the mouth forms at the opposite end of the embryo.

deutoplasm Yolk of egg.

development It is a process by which organisms undergo progressive, orderly and gradual changes in structure as well as in physiology during their entire_life_history.

devonian Geologic period of the Palezoic era characterized by abundance of fish and the first amphibians. So named because strata[1], were first studied in Devonshire, England.

di- (Gk. *dis*, twice, double, two). A prefix meaning "twice" or "double".

diaphragm (Gk. *diaphragma*, a partition). The partition between the abdominal and thoracic cavities.

diaphragmaticus [Gr. *diaphragma*, a partition wall, midriff] Specialized muscle in crocodilians, going from ilium to liver; no structural relationship to the mammalian diaphragm.

diaphysis (Gk. *diaphysis*, a growing through). The shaft of a long bone.

diapsid [Gr. *di-*, two + apsis, a loop or mesh] Referring to a reptile with temporal fenestrae above and below the squamosal-postorbital joint.

diarthrosis [Gr. *di-*, two + *arthron*, a joint] A freely movable joint.

diarthrosis A joint permitting considerable rotation of articulated skeletal elements, and characterized by a joint capsule, synovial membrane, and articular cartilages on joined elements; synovial joint.

diencephalon (Gk. *dia*, through, +*enkephalos*, brain), the posterior of the two subdivisions of the primitive forebrain.

diestrous |Gr. *dia-*. Between + *oisfros*, mad desire] Portion of the estrous cycle characterized by the deterioration of the endometrium after metestrous and before proestrous.

differentiation The events by which cells and other parts of an organism become different from one another and also different from their previous condition. Differentiation is responsible for increasing diversification of form and function, the process which changes a simple to a complex organism.

digastricus [Gr. *di-*, two + *gaster*, belly] The mammalian muscle which lowers the jaw to open the mouth.

digit *(L digitus*, finger), a finger or toe.

digitiform gland [L. digitus, finger + *forma*, form + *glans*, acorn] Salt-secreting gland opening into the large intestine of chondrichthyeans. Syn., rectal gland.

digitigrade A foot posture in which the balls of the feet (middle of the digits) support the weight, as in cats and dogs. Compare with plantigrade and unguligrade.

dikinetic skull A kinetic skull with two joints passing transversely through the braincase. Compare with monokinetic skull.

dilator (L. *dilatare*, to expand). A muscle whose function is to pull open an orifice.

dioecious (Gk. *di*, two +*oikos*, house). A condition in which male and female reproductive systems are found in separate individuals of the same species.

dipleurula A hypothetical invertebrate larva proposed as the common ancestor of echinoderms and hemichordates.

diploid (Gk. *diplous*, double, +*eidos*, resemblance). The full number of chromosomes in the fertilized ovum and in all cells, except the mature germ cells.

diplospondylous [Gr. *diplo-*, double + *spondylos*, vertebra] Condition of having two centra per vertebra.

diplospondyly The condition in which a vertebral segment is composed of two centra. Compare with monospondyly.

dipnoi (Gk. *dipnoos*, with two breathing apertures). A group of, fishes that have, in addition to gills functioning in the usual mariner, a lung or pair of lungs.

discoblastula It is found in sharks, lung fishes, reptiles, birds and egg-laying mammals, discoidal cleavage of telolecithal eggs in these animals results in the formation of a *blastodisc* on top of the undivided yolk.

discoidal (Gk. *diskos*, disc +*eidos*, appearance). Resembling a disc or plate.

distal (L. *distalis*, distant). Farthest from the center or median line.

diverticulum [L. *diverticulum*, a by-road, from *de* + *verto*, to turn aside] A pouch extending from a tubular or saccular organ.

dorsal (L. *dorsalis*, back). Relating to the top or back.

dorsum (L. *dorsum*, [gen. *Dorsi*], back). The upper or back of any part.

ductus (L. *ductus*, to lead). A tubular structure giving exit to or conducting any fluid.

ductus arteriosus [L. *ducere*, to lead + *arteria*, artery] A fetal anastomosis between the aortic arch and the pulmonary arch. Syn., duct of Botallus.

duodenum (L. *duodeni*, twelve). The first division of the small intestine.

E

eccrine glands [Gr. *ekkrinein*, to secrete (from *ex*, out of + *krino*, separate) + L. *glans*, acorn] Small, unbranched, merocrine-secreting sweat glands.

ectoderm (Gk. *ektos*, outside, +*derma*, skin). The outermost of the three primary germ layers.

ectostriatum [Gr. *ektos*, outside + *striatus*, furrow or channel] A portion of the avian cerebral hemisphere receiving visual information.

edentata (L. *dentatus*, toothless). An order of placental mammals without incisor teeth; the other teeth are poorly developed and devoid of enamel.

effector (L. *efficere*, to bring to pass). An organ that reacts by movement, 'secretion, or electrical discharge.

efferent (L. *ex*, out, + *ferre*, to bear). Conducting outward or centrifugally,

egg A fully developed female sexual cell produced by ovary through oogenesis. It is a haploid, immotile, nutrient-filled and metabolically inert cell, which upon union with the male gamete gives rise to a new individual.

egg membranes Covering of the egg such as vitelline membrane. zona radiata, jelly envelopes, chorion and ovarian follicles.

ejaculation Rhythmic, forcible emission of mature spermatozoa from the body of a male after coitus.

elasmobranchii (Gk. *elasmos*, plate, +*branchia*, gills). A group of fishes, comprising the modern sharks and the rays, and their extinct allies.

electric organ A specialized block of muscles producing electrical fields and often high jolts of voltage.

embolomerous (Gk. *embolos*, wedge, +*meros*, part). Referring to vertebrae with two-part centra.

embolomerous vertebra A dispondylous vertebra in which both centra are separate (aspidospondylous) and of about equal size. Compare with stereospondylous vertebra.

emboly (Gk. *emboly*, "throw in" or "thrust in". It is the movement of chorda-mesodermal and endodermal blastomeres from the external surface of the blastula to the interior of the embryo, a morphogenetic movement during gastrulation.

embryo (Gk. *en*, in+*bryei*,to well). The early developmental stage of an organism produced from fertilized egg, juvenile stage of an animals before it comes out from the seed, egg or the body of its mother; a young organism before it becomes capable of leading a self-supporting life.

embryogeny (Gk. *embryon*, embryo, +*gennao*, I produce). The origin and growth of the embryo,

embryology (Gk. *embryon*, embryo, +*logia*, discourse). The study of the origin and development of the embryo.

embryonic Membranes Amnion, chorion, Allantois and the yolk-sac splanchnopleure of the chick embryo.

enamel organ The part of the tooth-forming primordium that is derived from epidermis, becomes associated with the dermal papilla, and differentiates into the ameloblasts that secrete enamel. See dermal papilla.

endocardium. (Gk. *endon*, within, +*kardia*, heart). The lining of the cavities of the heart.

endochondral (Gk. *endon*. within, +*chondros*, cartilage). Within a cartilage or cartilaginous tissue.

endochondral bone formation Embryonic formation of bone preceded by a cartilage precursor that is subsequently ossified; cartilage or replacement bone. Compare with intramembranous bone formation.

endocrine (Gk. *endon*, within, + *krinein*, separate). Denoting internal secretion.

endoderm (Gk. *endon*, within, + *derma* skin). The innermost of the three primary germ layers.

endolymph (Gk. *endon*, within, + L. *lympha* water). The fluid contained within the membraneous labyrinth of the inner ear.

endometrium (Gk. *endon*, within, + *metra*, uterus). The mucous membrane (glandular) lining of the uterus in mammals.

endomysium (Gk. *endon*, within, +*mys*, muscle). The connective tissue sheathing of muscle *fibre*.

endoplasmic reticulum [Gr. *endon*, within + *plasma*, something molded or formed + L. *reticulum*, little net]. The complex of cytoplasmic membranes within a cell.

endoskeleton (Gk. *endon*, within, +*skeletons* dried up). The internal framework of the body.
Endotheliochorial (Gk. *endon*, within, +*thele$_s$* nipple, +*chorion*, skin). Placenta type involving contact between chorionic epithelium and uterine blood vessels.

endostyle [Gr. *endon*, within + *stylos*, pillar] The iodine-concentrating structure of *Amphioxus* and the ammocoete larva.

endothelium (Gk. *endon*, within, +*thele*, nipple, +*chorion*, skin). Placenta type involving contact between chorionio epithelium and uterine blood vessels.

enterocoel (Gk. *enteron*, intestine +*koilos*, hollow). A coelom originally in communication with the lumen of the gut. This type is represented principally by the Echinoderms and the chordates.

enterogastrone (Gk. *enteron*, intestine, +*gaster*, belly). A hormone from intestinal mucosa which inhibits gastric secretion and rnotility.

enteropneusta (Gk. enterion, intestine, +*pneustos*, to breathe). An order of hemichordate "worms."

entorhinal cortex [Gr. *entos*, within + . *rhis (rhin)*, nose + L. *cortex*, bark] Portion of the mammalian cortex intermediate between piriform cortex and hippocampus.

enzyme The protein catalyst that regulates (or catalyzes) biochemical reaction in a living cell.

epaxial [Gr. *epi* upon + L. *axis*, axis] Above or beyond any axis; in anatomy, the portion of the body which develops dorsal to an axis formed by the developing notochord.

ependymal (Gk. *ependyma*, an outer garment). Referring to the layer of cells lining the central canal of the spinal cord.

epi (Gk. *epi*, upon). A prefix denoting above or on top.

epiblast (Gk.*epi*, upon, +*blastos*, germ). The outer or upper layer of a double-layered embryo from which various germ layers maybe derived.

epiboly A method of gastrulation by which the smaller blastomeres at the animal pole of the embryo grow over and enclose the cells of the vegetal hemisphere.

epiboly Morphogenetic movement of cells on the surface of gastrula: growing, spreading, or flowing over, process by which rapidly dividing animal pole cells or micromeres grow over and enclose vegetal pole material. Increase in area extent of ectoderm.

epiboly The spreading of surface cells during embryonic gastrulation.

epicardium (Gk. *epi*, upon, +*kardia*, heart). The visceral peritoneum enveloping the heart.

epicritic [Gr. *epi*, upon + *krino*, separate] Referring to the system of sensory nerves carrying the finest, most discriminative sensations.

epidermis (Gk. *epi*, upon, +*derma*, skin). The outer epithelial portion of the skin.

epididymis (Gk. *epi*, upon, +*didymos*, twin). The first, convoluted, portion of the excretory duct of the testis, passing from above downward along the posterior border of this gland.

epigenesis Caspar Friedrich Wolff postulated that the specialized organs of living things develop from unspecialized tissue. This theory is called epigenesis.

epiglottis (Gk *epi*, upon +*glotta* tongue). The structure that folds back over the aperture (glottis) of the larynx.

epimere (Gk. *epi*, upon, +*meris*, part). The dorsal region of the mesoderm on each side of the neural tube.

epimysium (Gk. *epi*, upon, + *mys*, muscle). The fibrous envelope surrounding a muscle.

epiphysis (Gk. *epi*, upon, +*physis* growth). A separately developed terminal ossification of a long bone; the pineal body.

epithalamus (Gk. *epi*, upon, +*thalamos*, chamber). The dorsal portion of the thalamus of the brain.

epitheliochorial (Gk. *epi*, upon, +*thele*, nipple, +*chorion*, skin). Placenta type involving contact between chorionic epithelium and uterine epithelium.

epithelium (Gk. *epi*, upon, +*thele*, nipple). Cellular layer covering a free surface.

erythrocyte (Gk. *erythros*, red, +*phoros*. carrying). A cell containing red pigment.

esophagus (Gk. *oiso*, 1 shall carry, +*phageton*, food). The portion of the gut between the pharynx and stomach.

estivation A prolonged resting state or hibernation during times of heat or drought that is characterized by lowered metabolic levels and breathing rates.

estrogen (Gk. *oistros*, mad desire, +*gennao*, produce). Substance that produces estrus.

estrous cycle The process of growth and regeneration of the follicle, with its accompanying changes in the ovaries, the vagina, and the uterus is called the estrous cycle.

estrus (Gk. *oistros*, mad desire). Period of sexual excitement in the female.

euryapsid [Gr. *eury*, broad + a*psis*, a loop or mesh] Referring to reptiles with a single temporal fenestra dorsal to the postorbital–squamosal suture.

euryhaline Heaving a wide tolerance to salinity differences.

eutheria (Gk. *eu*, will, +*therion*, **a wild** beast). A subclass of mammalia, embracing the placental mammals.

evagination Growth from a surface outward.

exocrine (Gk. *exo*, out, +*krino*, I separate). Denoting external secretion of a gland.

ex-ova omnia All life comes from the egg (Harvey, 1657).

extensor (L. *extendere*, to stretch out). Serving to straighten a part.

extracolumella [L. *extra*, without, side + *columella*, small column] The outer, usually cartilaginous portion of the auditory ossicle of amphibians, reptiles, and birds.

extra-embryonic membranes Membranes formed of embryonic tissue that do not enter into the formation of the embryo itself. They lie outside the embryo and are (devoted to the care and maintenance of the embryo (protection, respiration, excretion, food supply etc.); include *omnion, chorion allantois* and *yolk sac.*

extrinsic (L. *extrinsecus,* on the outside). Originating outside of the part on which it acts.

F

facet (Fr. *facette,* face). A small smooth area on a bone or other firm structure.

facial (L. *facies,* face). Relating to the face

falciform (L. *falx,* sickle, +form). Having a curved or sickle shape

fascia (L. *fascia,* a band or fillet). A fibrous tissue investing an organ or area.

fasciculus [L. *fasciculus,* little bundle] A bundle of muscle or nerve *fibres* running together.

fate Map Map of blastula or early gastrula stage which indicates prospective significance of various surface areas, based upon previously established studies of normal development aided by means of vital dye marking.

fenestration [L. *fenestra,* window] The evolution of gaps in the dermatocranium usually covered by connective tissue membrane

fertilisation (L. *fertilis,* to bear). Union of ovum and spermatozoon and so of their nuclei, to form a diploid zygote.

fertilizin A chemical substance released by measure eggs of many species, producing temporary agglutination of sperms.

fibroblast [L. *fibra,. fibre* + Gr. *blastos,* bud, sprout] Connective tissue cells which produce connective tissue *fibres.*

fission (L. *fissus,* split) to divide.

flexor (L. *flectere,* to bend). Serving to bend a joint.

foetus (L. *fetus,* pregnant). Prenatal stage of development in man and other mammals, follwing the embryonic stage, roughly from the third month pregnancy to birth.

follicle (L. *folliculus,* a small bag). A vesicular body in the ovary. containing the ovum. Any crypt or circumscribed space.

foramen (L. *foramen,* an aperture). An aperture of perforation through a bone or membrane.

foramen A perforation or hole through a tissue wall.

foramen of Panizza A connecting vessel between the bases of the left and right aortic arches in crocodilians.

foramen ovale the one-way connection between the right and left atria of an embryonic mammal; closes at birth.

fornix [L. *fornix,* arch or vault] A tract of the brain running from the hippocampus to the mammillary bodies and septal nuclei.

fossa (L. *fossa,* trench or ditch). A depression below the level of the surface of a part.

fossorial locomotion Active removal of soil to produce a burrow; digging.

fovea (L. *fovea*, a pit). Referring to any depression or pit.

frondosum (L. *frons*, branch or loaf). In reference to a shaggy surface.

frontal (L. *frons*, brow). Relating to the front or forehead.

fundic [L. *fundus*, bottom] Referring to that portion of the stomach which contains gastric glands.

funiculus (L. dim of *funis*, cord). A small, cordlike structure such as a bundle of nerve fibres.

G

gait The pattern or sequence of foot movements during locomotion.

gamete (Gk. *gametes*, husband *eamete* wife). A haploid germ cell ovum or spermatozoon whose nucleus fuses with its counterpart with the resulting diploid cell (zygote) developing into a new individual.

gametogenesis (Gk. *gametes, gamete,* husband, wife, + *genesis,* origin). The production of gametes inside the gonads.

gametogonium (Gametes, *gamete,* husband, wife, + *gonus,* a begetting). A primordial reproductive cell that produces gametes.

ganglion (Gk. *ganglion,* a swelling under the skin). An aggregation of nerve cells along the course of a nerve.

ganoid scale [Gr. *ganos*, brightness + *-oide*, resembling, like] Heavy scales consisting of bone, cosmine, and many layers of ganoine, giving them a lustrous, metallic color.

ganoine [Gr. *ganos*, brightness) A specific hard, enamel-like protein.

gaster (Gk. *gaster,* belly). General reference to the stomach; the main, body of a muscle.

gastralia (Gk. *gaster,* belly). Special reference to skeletal parts supporting the abdomen.

gastrin (Gk. *gaster,* belly). A hormone produced in pyloric division of the stomach which excites secretion by glands of cardiac stomach.

gastrocoel (Gk. *gaster,* stomach, +*koilos,* hollow). The cavity within the gastrula.

gastrocoel The cavity within the early embryonic gut of the gastrula.

gastrula (L. dim of Gk. *gaster,* belly). The embryo in the stage of development following the blastula, consisting of a sac with a double wall and enclosing a central cavity, the archenteron.

gastrulation Dynamic process involving cell movements which change embryo from a monodermic to either a di-or tridermic form. Generally involves inward movement of cells to form enteric endoderm. Description includes epiboly, concrescence, confluence, involution, invagination, extension and convergence.

gene *(L. gennao,* I produce). A particulate hereditary unit located in a chromosome. By interaction with internal and external environment a gene controls the development of a phenotypic trait; capable of self-replication.

genital *(L.genitalis,* pertaining to birth). Relation to reproduction or generation.

genu [L. *genu,* knee] A general term used to describe any structure shaped like a bent knee.

germ Layer Layer of cells in the embryo *i.e.* , embryonic ectoderm mesoderm and endoderm.

germinal disc The small protoplasmic disc on the surface of the huge yolk mass of avian egg.

germinal epithelium Peritoneal epithelium out of which reproductive cells of both male and female presumably develop (syn., germinal ridges, gonodial ridges).

germinativum (L. *germen,* bud, germ). Relating to multiplication, development or growth.

glans (L. *glans,* acorn). Denoting a mass capping the body of a part.

glenoid (Gk. *glene,* a socket). Resembling a socket.

glia (Gk. *glia,* glue). Referring to supporting and binding substance.

globulin (L. *globulus,* globule). A simple protein present in blood, milk, and muscle.

globus pallidus [L. *giobus,* ball +*pallidus,* pale] Portion of the mammalian basal ganglia just deep to the putamen.

glomerulus *(L. glomus,* a sking). A tuft formed of capillary loops at the beginning of each uriniferous tubule in the kidney.

glomerulus 1. A small bed of capillaries associated with the uriniferous tubule. 2. A small cluster or capillaries on the stomochord of hemichordates.

glossal (Gk. *glossa,* tongue). Relating to the tongue.

glossopharyngeal (Gk. *glossa,* tongue, +*pharyngis,* throat). Relating to the tongue and pharynx, specifically to the ninth cranial nerve.

glottis *(Gk. glottis,* aperture of the larynx). Opening to the larynx and trachea.

glucagon (Gk. *glykys,* sweet). One of the pancreatic hormones concerned with regulating blood sugar.

glycogen [Gr. *glykys* sweet + -*gen,* production of] A branched polysaccharide of glucose; the principal storage carbohydrate of vertebrates.

gnathostomata (Gk. *gnathos,* jaw +*stoma,* mouth): Vertebrates possessing jaws.

gnathostome A vertebrate with jaws.

gonad (Gk. *gone,* seed). A reproductive gland; ovary or testis.

gonadotropic (Gk. *gone,* seed, +*trophe,* nourishment). Referring to a hormone that influences the gonads.

gonopodium [Gr. *gono* reproductive + *podion,* little foot] Modified anal fin of some fishes, used as an intromittent organ.

grade A level or stage of evolutionary attainment; a paraphyletic group.

gradient Gradual change of developmental torces "along an axis.

granulosa (L. *granulosus,* full of grains). In reference to a layer or region having a granular appearance.

gray Crescent A lightly pigmented crescent-shaped region known, because of its (gray) colour, as the gray crescent; located on the dorsal surface of frog's egg, between the animal and vegetal hemispheres.

guanine [Sp. *gmano*. dung] A purine occurring in crystalline form in iridophores.

gubernaculum [L. *gubernaculum*, a helm] A fibrous cord attaching testis to abdominal wall; significant in testicular descent.

gustatory [L. *gusto*, taste] Having to do with taste.

gyrencephalic cortex [Gr. *gyros*, ring + *enkephalos*, brain + L. *cortex*, bark] Convoluted cerebral cortex. Opposite of lissencephalic cortex.

gyrus (Gk. *gyros*, circle). A rounded elevation on the surface of the brain.

H

habenular (L. *habenula*, a strap). Relating particularly to the stalk of the pineal body.

haemal *(Gk. haima*, blood). Relating to the blood or blood vessels.

hair cell A mechanoreceptor cell in the acousticolateralis system, characterized by many stereocilia and usually a single kinocilium on its-free surface.

haploid (Gk. *haplous*, simple + *eidos*, resemblance). An individual or gamete cell (spermatozoon or ovum) containing a single set of chromosomes.

hatching The beginning of the larval life of the amphibian; the process of emergence of the chick embryo from its shell aided by beak tooth.

haversian (from name of English anatomist, Havers). Relating to bone structure described by Havers.

helicotrema [Gr. *helix*, spiral + *trema*, a hole] The continuity at the apex of the cochlea between scala tympani and scala vestibuli.

hemi (Gk, *hemi*, half). A prefix signifying one-half.

hemichordata (G. *hemi*, half, + *chorde* string). A phylum of animals closely related to echinoderms and chordates.

hemipenes [Gr. *hemi*, half + L. *penis*, a tail or a penis] Eversible intromittent organs of snakes and lizards.

hemisphere (Gk. *hemi*, half, + *sphaira*, ball). The lateral half of the cerebellum.

hemochorial (Gk. *haima*, blood, + *chorion*, skin). Placenta type involving contact between chorionic epithelium and maternal blood.

hemocoel Blood-filled channels within connective tissue that lock a continuous endothelial lining.

hemocytoblast [Gr. *hemo*, blood + *kytos*, hollow vessel + *blastos*, bud or sprout] The cell type from which all blood cells differentiate.

hemoglobin (Gk. *haima*, blood, + L. *globus*, globe). The respiratory pigment of the blood.

hemopoiesis [Gr. *hemo*, blood + *poiesis*, a making. The process of blood formation.

hemopoietic tissue Blood-forming tissue. Compare with myeloid tissue and lymphoid tissue.

henle, loop of [Named after Friedrich G.J. Henle, a nineteenth century German anatomist] Long, thin, intermediate segment of the mammalian nephron, characterized by its hairpin turn.

hensen's node Anterior thickened end of the primitive streak of Chick embryo; homologous to the dorsal lip of the blastopore of the amphibia.

hepatic (Gk. *hepar, hepatikos,* relating to the liver). Relating to the liver.

hermaphroditism (Gk. *hermaphroditos,* the son of Hermes, Mercury, + Aphrodite, Venus). The seeming occurrence of both male and female generative organs in the same individual.

heterocelous [Gr. *Hetero,* other, different + *celom,* hollow] Centrum of a vertebra with saddle-shaped ends characteristic of Aves.

heterocercal [Gr. *hetero,* other, different + *kerkos,* tail] Describing a caudal fin where the axis extends dorsally and the fin itself is larger dorsally than ventrally.

heterochrony a time difference.

heterochrony Within an evolutionary lineage, the change in time at which a characteristic appears in the embryo relative to its appearance in a phylogenetic ancestor; usually concerned with the time of onset of sexual maturity relative to somatic development. Compare with paedomorphosis.

heterodont (Gk. *heteros,* different, +*odont,* tooth). Having teeth of varying shapes.

heterodont dentition [Gr. *hetero,* different + *odontos,* tooth] A condition in which tooth morphology varies with the region of the jaw; cf. homodont dentition.

heteroplastic [Gr. *hetero,* different + *plastikos,* formed, molded] Describing a type of bone which-forms within other, already differentiated tissues, such as tendons, ligaments, or heart muscle.

hippocampus (Gk. *hippocampus,* seahorse). An elevation on the floor of the lateral ventricle of the brain.

histogenesis (Gk. *histos,* tissue, + *genesis,* origin). The origin of a tissue; the formation and development of the tissues of the body.

holo- (G. *holos,* entire). A prefix, signifying entire or total.

holoblastic (Gk. *holos,* whole, + *blastos,* germ). Denoting the involvement of the entire egg in cleavage.

holoblastic cleavage Early mitotic planes pass entirely through the cleaving embryo. Compare with meroblastic and discoidal cleavage.

holobranch (Gk. *holos,* complete, +*branchion,* gill). A complete gill with filaments on each side of the supporting arch.

holocephali (Gk. *holos,* entire, + *kep hale,* head). A small group of fishes included in the chondrichthyes. comprising the ratfish

holocrine [Gr. *holo,* entire + *krinein.* To separate] Referring to a gland or its secretion, characterized by the destruction of the gland cell during the release of the secretion.

holonephros (Gk. *holos*, entire, + *nephros*, kidney). Term describing the concept of a single, rather than multiple type kidney.

holonephros A single kidney arising from several regions of the nephric ridge rather than the three types of kidneys (pronephros, mesonephros, metanephros) arising from the nephric ridge.

holospondyly The condition in which the centra and spines of vertebrae are anatomically fused into a single bone. Compare with aspidospondyly.

holostei [Gr. *holos*, entire + *osteon*, bone) Superorder of Actinopterygii with abbreviated heterocercal tail; *e.g.*, *Amia*.

homeostasis [Gr. *Homoios*, like | *stasis*, a standing] The ability to maintain internal integrity unaffected by external environment.

homeostasis The constancy of an organism's internal environment.

homeotherm [Gr. *Homoios*, like + *therme*, heat] An animal capable of maintaining a constant internal body temperature.

homeothermy The condition of maintaining constant body temperature, without regard to the method.

homocercal [Gr. *homos*, one and the same + *kerkos*, tail] Describing a caudal fin which is externally symmetrical although internally asymmetrical, characteristic of Teleostei.

homodont (Gk. *homos*, same, + *odont*, tooth). Having teeth all alike in form.

homodont Dentition in which the teeth are similar in general appearance throughout the mouth

homolecithal egg Egg in which a small amount of yolk is scattered evenly throughout cytoplasm.

homology (Gk. *homos*, same, + *logos*, ratio). Bearing reference to similarity in structure and origin of two organs or parts in different species of animals.

homology Features in two or more organisms derived from common ancestors; common ancestry. Compare with serial homology.

homoplasy Features in two or more organisms that look alike; similar in appearance.

homoplasy similarity due to any cause other than common ancestry.

homotypy (Gk. *homos*, same, + *typos*, type). Correspondence in structure between two parts or organs in one individual.

hormone (Gk. *hormon*, I rouse). A chemical substance formed in one organ or part and carried in the blood to another organ or part which it stimulates to functional activity.

humerus (L. *humerus*, shoulder). The bone of the upper arm.

humor (L. *humor*, fluid). Any fluid or semifluid substance.

hyaline (Gk, *hyalos*, glass). Of a glassy, translucent appearance.

hymen [Gr. *Hymen*, membrane] A thin fold of tissue partly occluding the vaginal orifice.

hyoid (Gk. *hyoiedes*, U-shaped, +L. *mandibula*, jaw). Uppermost segment of the hyoid arch.

hypaxial [Gr. *hypo*, beneath + L. *axis*, axis] Below or under any axis: in anatomy, the portion of the body which develops ventral to an axis formed by the developing notochord.

hyper (Gk. *hyper*, above). Prefix denoting excessive or above normal.

hyperplasia An increase in the number of cells as a result of cell proliferation; usually occurs in response to stress or increased activity. Compare with hypertrophy and metaplasia.

hyperstriatum [Gr. *hyper*, above + *striatus*, furrow or channel] A portion of the avian cerebral hemisphere lying dorsally and superficially.

hypertrophy An increase in the size or density of an organ or part which does not result from cellproliferation. Compare with hyperplasia and atrophy.

hypo- (Gk. *hypo*, under). Prefix denoting beneath something else or a diminution or deficiency.

hypoblast (Gk. *hypo*, under, + *blastos*, germ). The inner or lower layer of a double-layered embryo.

hypodermis |Gr. *hypo*, beneath + *dermis*, skin] The deep, loose, fatty connective tissue of the skin. Svn., subcorium; subcutaneous tissue.

hypoglossal (Gk. *hypo*, beneath, + *glossa*, tongue). Beneath the tongue; usually denoting twelfth cranial nerve.

hypomere (Gk. *hypo*, under, +*meris*, part). The most ventral subdivision of mesoderm.

hypophysis (Gk. *hypophysis*, an undergrowth). An endocrine gland lying at the base of the brain.

hypothalamus (Gk. *hypo*, under, + *thalamos*, chamber). The ventral portion of the thalamus of the brain.

hypothesis A guessed solution of a scientific problem; must be tested by experimentation and, if not validated, must then be discarded.

hypsodont [Gr. *Hypsos*, height + *odontos*, tooth] High-crowned teeth.

I

ichthyopterygia [Gr. *ichthys*, fish + *pterygion*, little wing or fin] Subclass of Reptilia comprised of extinct marine reptiles which externally resemble sharks.

ileum [L. adaptation of Gr. *eileo*, to roll up, twist] The most posterior portion of the mammalian small intestine.

ilium (L. *Ilium*, flank). The dorsal most element of the pelvic girdle.

imago, imaginal (1) An adult insect, (2) Adjective.

impar (L. *impar*, unpaired). Relating to a single or unpaired part.

implantation The attachment of the mammalian blastocyst; the process of adding, or placing a graft within a host without removal of its any part.

incisor (L. *incidere*, to cut into). One of the cutting teeth.

index fossil A fossil animal widely distributed geographically but restricted to one rock layer or time horizon; defining species indicator of a stratum.

induction, inductor (L. *inducere*, to induce). The process in an animal embryo in which one tissue or body part influences the development of certain another tissue or body part. The tissue which is the source of this influence is called the inductor.

infra (L. *infra*, below). Prefix denoting a position below a designated part.

infundibulum (L. *infundibulum*, a funnel). A funnel-shaped structure or passage.

ingestion [L. *ingestus*, poured or thrown into] Process of bringing food into the digestive system.

ingroup The group of organisms actually studied. Compare with outgroup.

innervation [L. *in*, in + *nervus*, nerve + *-ate*, provided with] The distribution of nerves to an organ of the body.

innominate (L. *in*, negative, + *nomen* name). Meaning "nameless" with reference to blood vessels and hip bone.

insectivora (L. *insectum*, insect, + *vorare*, to devour). An order of placental mammals whose chief food consists of insects; it embraces the moles and shrews.

instar Period between consecutive moults in insect development.

insulin (L. *insula*, island). One of the pancreatic hormones concerned with regulation of blood sugar.

integument (L. *integumentum*, covering). The membrane (skin) covering the body.

inter *(L. inter,* between). Prefix conveying the meaning of between or among.

intercalary (L. *intercalare*, to insert). Occurring between two others.

internal capsule [L. *internus*, interior, away from the surface + *capsula*, diminutive of *capsa*, a chest or box] The mammalian *fibre* tract between the dorsal thalamus and the neocortex.

interoceptor A sensory receptor that responds to internal stimuli. Compare with proprioceptor and exteroceptor.

inter-renal [L. *inter,* between + *renes*, kidneys] The steroidogenic tissue of the adrenal gland or its homologue.

interstitial (L. *inter,* between, + *sistere*, to set). Relating to space or structures within parts.

interstitial cells (Leydig cells) These cells form the endocrine tissue of the vertebrate testis. Lying in the interstitial tissue filling the spaces between the seminiferous tubules, they secrete male sex hormones, the androgens.

intervertebral body A pad of cartilage or fibrous connective tissue between articular ends of successive vertebral centra.

intervertebral disk A pad of fibrocartilage in the adult mammal that has a gel-like core derived from the notochord and is located between articular ends or successive vertebral centra. Compare with intervertebral body.

intestine (L. *intestinum*). The digestive tube passing from the stomach to the anus.

intramembranous bone formation Embryonic formation of bone directly from mesenchyme without a cartilage precursor; dermal bone. Compare with endochondral bone formation.

intrinsic (L. *intrinsecus*, on the inside). Belonging entirely to a part.

intromittent [L. *intro*, inwardly + *mittere*, to send] carrying into a body; used to describe the penis and analogous organs.

invagination (L. *in* in, + *vagina*, a sheath), the process of one part being inserted in another; insinking, inpushing or infolding of a layer of cells into a preformed cavity as in one of the processes of gastrulation.

involution (L. *involvere*, to roll up). The turning or rolling in of a part around a margin or edge as in gastrulation of frog and chick embryos; inrolling of a tissue layer underneath another one as in a type of embolic gastrulation.

ipsilateral Occurring on the same side of the body. Compare with contralateral.

iridiophore (Gk. *iris*, rainbow, +phoros, carrying). A cell containing crystals that reflect and disperse light.

iris (Gk. *iris*, rainbow), The pigmented disclike diaphragm of the eye.

ischium {Gk. *ischion*, hip). Posteroventral bone in the pelvic girdle.

iso (Gk. *Isos*, equal); Prefix signifying equal.

isolecithal (Gk. *isos*, equal, + *lekithos*, yolk). Denoting a uniform distribution of yolk within an egg.

isthmus (Gk. *isthmus*, isthmus). A constricted connection or passage between two larger parts of an organ or other anatomical parts.

J

jejunum [L. *jejunus*, empty] Portion of the mammalian small intestine between the duodenum and ileum.

jelly Mucin covering of eggs of sea-urchin, frog. It is secreted by the oviduct and protects the eggs from mechanical injury and from being eaten.

jurassic Middle of the three periods in the Mesozoic era, characterized by great diversity of large reptiles.

juxtaglomerular apparatus [L. *juxtn*, near + *glomerulus* (which see)] A group of myoepithelial cells adjacent to the kidney's glomerulus.

K

karyogamy Fusion of nuclei in process of fertilization.

keratin (Gk. *keras*, horn). A hard, relatively insoluble protein or albuminoid, present largely in cutaneous structures.

kerckring, valves of [L. *valva*, leaf of a folding door; named for Theodor Kerckring, seventeenth century Dutch anatomist] Large folds of the sub-mucosa and mucosa in the small intestine of birds and mammals.

kinetosome Basal body of a cilium or flagellum.

kinocilium [Gr. *kineo*, move + L. *cilium*, an eyelid] Long, narrow cytoplasmic process with internal structure suggesting capability of movement.

kinocilium A modified, rigid cilium of the ear. Compare with microvillus.

kinorhyncha A phylum of marine pseudocoelomates.

L

labium (L. *labium*, a lip). Denoting any lip-shaped structure.

labyrinthodontia [Gr. *labyrinthos*, labyrinth | *odontos*, tooth] An extinct subclass of Amphibia comprising the earliest known fossil amphibians.

lacertilia |L. *lacerta*, lizard] Suborder of Squamata comprised of lizards.

lacrimal *(L. lacrirna,* tear). Relating to the eye and tear ducts.

lacteal [L. *lacteus*, milky | The blind lymphatic capillary of a villus.

lacuna (L. *lacus*, a hollow). A small depression, gap or space.

laeve (L. *levare*, to wash). With reference to a smooth surface.

lagena (L. *lagena*, a flask). An extension from the sacculus of the inner ear; extremity of the cochlear duct.

lagomorpha (Gk. *lagos*, hare, + *morphe*$_t$ form). Group of mammals, including rabbits and hares.

lamella (L. dim. of *lamina*, thin plate). A thin sheet, layer, or scale.

lamina (L, *lamina,* thin plate). A thin place of flat layer.

langerhans, islets of [Named after Paul Langerhans, nineteenth century German anatomist] Endocrine portion of the pancreas.

larva (L. ghost). An immature stage of an animal that is morphologically altogether different from the adult *e.g.,* caterpillar of insect and tadpole of frog.

larva An immature (nonreproductive) stage that is morphologically different from the adult.

larynx (Gk. *larynx*, upper part of windpipe). The box composed of several cartilages, at the upper end of the trachea.

lateral (Gk. *lotus*, side). On the side.

latitudinal cleavage A cleavage furrow appearing transversely above or below the equator.

lecithotrophic Pertaining to the nutrition that the embryo receives from the yolk of the ovum. Compare with matrotrophic.

lemniscus [Gr. *lemniskos*, ribbon or fillet] A flattened tract of nerve fibres in the central nervous system.

lepidosauria [Gr. *lepidos,* scale + *sauros,* lizard] Subclass of reptiles comprised of Rhynchocephalia and Squamata.

lepidotrichia A fan-shaped array of ossified or chondrified dermal rods that internally supports the fin of bony fishes. Compare with ceratotrichia.

lepospondyly A holospondylous vertebra with a husk-shaped centrum usually pierced by a notochordal canal.

leukocyte (Gk. *leukos,* white, + *kytos,* cell). A white blood corpuscle.

levator (L. *levator,* that which lifts). A muscle serving to raise a part.

leydig's gland [L. *glans,* acorn; named after Franz von Leydig, nineteenth century German anatomist] Intermediate portion of chondrichthyean kidney.

lieberkuhn, crypt of [Gr. *kritptos,* hidden; named after Johann Lieberkühn, eighteenth century German anatomist] Deep grooves in the mucosa of the small intestine of birds and mammals.

ligament (L. *Ligamentum,* a band or bandage). A band or sheet of fibrous tissue connecting two or more skeletal parts

ligamentum arteriosum [L. *ligamentum,* band, tie + Gr. *arteria,* windpipe or artery] Ligament that remains after the closing of the ductus arteriosus.

limbic lobe [L. *limbus,* a border + Gr. *iobos,* lobe as of the ear or liver] Medial telencephalic structures of mammals, including hippocampus, septal nuclei, mammillary bodies, and circulate cortex.

linea (L. *linea,* a line). Referring to a line, strip, or streak.

linea alba [L. *linea,* line + *alba,* white] Midventral line of white connective tissue

lines of arrested growth (LAGs) A period when annuli or rings are deposited in bones as a result of cessation or even resorption of bone.

lingual *(L. lingua,* tongue). Relating to the tongue.

lipochrome (Gk. *lipos,* fat, + *chroma* colour). Pigment in the orange-yellow range,

lipophore (Gk. *lipos,* fat, + *phorus,* bearer). A cell containing lipochrome.

lissamphibia [Gr. lissos, smooth | *amphibia,* living a double life] The subclass of Amphibia which includes all the living forms.

lissencephalic cortex [G.r. *lissos,* smooth + *enkaphalos,* brain] L. *cortex,* bark] Smooth cerebral cortex. Opposite of gyrencephalic,

littoral Region of shallow water near the shore between the high and low tide marks.

loop of Henle A region of the mammalian nephron that includes parts of the proximal and distal tubules (thick limbs) and all of the intermediate tubule (thin limb).

lophodont [Gr. *lophos,* a crest + *odontos,* tooth] High-crowned teeth with transverse ridges, *e.g.,* the molars and pre-molars of horses.

lophodont Teeth having broad, ridged cusps useful in grinding plant material. Comprae with bundont and selenodont.

lumbar (L. *lumbare*, apron for loins). Relating to the loins, or the part of the back and sides between the ribs and pelvis.

lumen The space within the core of an organ, especially a tubular organ.

luteum (L. *Lute us*, yellow). Yellow in colour.

lymph (L. *lympha*, clear spring water). Fluid circulating in the lymphatic channels.

lymphocyte (L. *lympha*, + G. *kytos*, cell). A white cell present in the lymph.

lymphoid tissue Blood-forming tissue outside of bone cavities; found, for example, in the spleen and lymph nodes.

lytic (G. *lysis*, loosening). A suffix relating to dissolving or distraction.

M

macrolecithal (Gk. *makros*, large, + *lekythos*, yolk of egg). Referring to large amount of yolk stored in the egg.

macrolecithal Pertaining to eggs with large quantities of stored yolk.

macromere Larger blastomere.

macrophage *(G, macros*, large, + *phago*, I eat). A large blood cell that engulfs and destroys other cells.

macula (L. *macula*, a spot). A coloured spot or patch of distinctive tissue.

macula A mechanoreceptor within the vestibular apparatus of the ear; specialized neuromast organ detecting changes in body posture and acceleration. Compare with crista.

macula densa [L. *macula*, spot + *densus*, thick] Specialized portion of the distal convoluted tubule of the kidney, concerned with controlling osmolarity.

majus (L. comp. of *rnagnus*, great). Denoting large or major.

malleus One of the three middle ear bones in mammals, phylogenetically derived from the articular hone.

mammalia (L. *mamma*, breast). The highest class of living organisms : it includes all the vertebrate animals that suckle their young.

mammary (L. *mamma*, breast). Relating to the mammary glands.

mandibular (L. *mandibula*, a jaw). Relating to the jaws; specifically, the lower jaw.

marrow (Anglo-Saxon, *mearh*}. Any soft or gelatinous material..

marsupialia (Gk. *marsypos*, belly) An order of mammals featuring an abdominal pouch for nurture of young.

masseter (Gk. *maseter*, masticator). One of the muscles for chewing.

mastication [L' *masticatus*, chewed; from *masticare*, to chew] The process of chewing.

matrotrophic Pertaining to the nutrition the embryo receives through the placenta or from uterine secretions. Compare with lecithotrophic.

maxilla (L. *maxilla*, jawbone). Relating to the upper jaw

meatus (L, *meatus*, passage). A canal or channel.

median eminence [L. *medius*, middle + *eminentia*, prominence] Ventral projection from the hypothalamus.

mediastinum (L. *mediastinus*, being in the middle; A septum between two parts of an organ or a cavity.

medulla *(L. medulla,* marrow, pith). The inner, or central, portion of an organ or part.

meibomian gland [L., *glans,* acorn; named after Hendrik Meibom, a seventeenth century German anatomist] Oil-secreting gland of the mammalian eyelids.

meiosis (Gk. *meiosis*, a lessening). The cell divisions of the gametes resulting in halving of the number of chromosomes.

melanin (Gk. *melas melanos*, black), Dark pigment in the skin.

melanophore (Gk. *melanos*, black, + *phorus*, bearer). A cell containing melanin.

membrana granulosa The layer of follicle cells that bound the antrum of the Graafian follicle of the mammal.

menstruation (L. *menstuare*, discharge). Periodic discharge of blood and disintegrated uterine lining through the vagina in human females and apes at the end of menstrual cycle in which fertilization has not occurred.

meridional Cleavage Cleavage passing through the median axis (meridian) of the egg from animal to vegetal pole.

meroblastic (Gk. *meros,* part, + *blastos,* germ). Denoting the involvement of only a restricted cytoplasmic area at the animal pole of the egg in cleavage.

meroblastic cleavage Early mitotic planes that do not complete their passage through the embryo before subsequent division planes form. Compare with holoblastic and discoidal cleavage.

merocrine [Gr. *meros*, a part + *krino*, to separate] Describing a gland or its secretion in which the secretory product is released without damage to the cell.

merycism Remastication together with microbial formation of food in nonruminants. Compare with rumination.

mes-, meso (Gk, *mesos*, middle). Prefix signifying middle.

mesencephalon (Gk. *mesos*, middle, + *enkephalos* brain) The midbrain, the second of the three primitive divisions of the brain.

mesenchyme (Gk. *mesos*, middle, + *enchyma, infusion}.* Embryonic connective tissue.

mesendoderm Mesoderm fused with the endoderm in a newly formed urodele gastrula.

mesentery (L. *mesenterium*, a membranous support)- A double layer of peritoneum enclosing an organ.

mesobronchus [Gr. *mesos,* middle + *bronchos,* windpipe] The primary branch of the trachea in birds.

mesocardium (Gk. *mesos*, middle, + *kardia*, heart). The mesentery supporting the heart.

mesoderm (Gk. *mesos*, middle, + *derma*, skin). The middle layer of the three primary germ layers.

mesokinesis Skull movement via a transverse joint passing through the dermatocranium posterior to the ocular orbit. Compare with prokinesis and metakinesis.

mesolecithal (Gk. *mesos*, middle, + *lekthyos* yolk of egg). Referring to moderate amount of yolk stored in the egg.

mesomere (Gk. *mesos*, middle, + *meris*, segment). The middle region of the mesoderm.

mesometrium Attachment of the uterus to the coelomic wall.

mesonephron (Gk. *mesos*, middle, + *nephros*, kidney). The tubular unit of a mesonephros.

mesonephros (Gk. *mesos*, middle, + *nephros*, kidney). Pertaining to the second stage in the development of the amniote kidney the intermediate kidney, functional kidney in the adult fish and amphibian, but only in the embryonic bird and mammal.

mesorchium Mesentery (mesodermal) which surrounds and supports testis to body wall.

mesothelium Epithelial layers or membranes of mesodermal origin.

mesovarium Mesentery (mesodermal) which suspends ovary from dorsal body wall.

mesozoic [Gr. *mesos*, middle + *zoion*, animal] Geologic era after the Paleozoic and before the Cenozoic, characterized by an abundance of reptiles, and the origin of mammals and birds.

meta (Gk. *meta*, after). Prefix denoting behind or after something else in a series.

metabolism (Gk. *metabola*, change). The sum total of all the chemical reactions taking place in a living system; classified under two broad headings: (1) anabolism and (2) catabolism.

metacarpus (Gk. *meta*, after, + L. *carpus*, wrist). The part of the hand between the wrist and fingers.

metakinesis Skull movement via a transverse hinge that lies posterior between the deep neurocranium and outer dermatocranium. Compare with prokinesis sand mesokinesis.

metamerism (Gk. *meta*, after, + L. *meros*, part). Referring to segmentation, resulting in a series of homologous parts.

metamorphosis (Gk. *metamorphoun*, to transform). Transformation of the larva into adult during which many structural, physiological and behavioural, hormonally, regulated changes take place; the end of the larval period, when growth is temporarily suspended, *e.g.*, amphibia.

metanephron (Gk. *meta*, after, + *nephros* kidney). The tubular unit of a metanephros.

metanephros (Gk. *meta*, after, + *nephros*, kidney). Pertaining to the last stage in the development of the amniote kidney, the permanent kidney of birds and mammals.

metaplasia Change of a tissue from one type to another type. Compare with hypertrophy.

metapterygial fin Basic fin type in which the axis (metapterygial stem) is located posteriorly in the fin. Compare with archipterygial fin.

metapterygial stem The chain of endoskeletal elements within the fish fin that define the major internal supportive axis.

metatarsus (Gk. *meta*, after, + L. *tarsus*, ankle). The part of the foot between the ankle and toes.

metatheria (Gk. *meta*, after, + *therion*, wild beast). A group of mammals co-extensive with Marsupialia.

metencephalon (Gk. *meta*, after, **+** *enkephalos*, brain). The anterior of the two subdivisions of the primitive hindbrain.

metestrous [Cr. *meta*, after + *oistros*, mad desire] Portion of the estrous cycle following ovulation, characterized by the formation of the corpus luteum.

microfilament [Gr. *micros*, small + L. *filum*, thread] Submicroscopic strand within the cytoplasm, found in nerve fibres, muscle fibres, etc.

microlecithal (Gk. *mikros*, small, +*lekythos*, yolk of egg). Referring to small amount of yolk stored in the egg.

micromere. Smaller blastomere.

micropyle An aperture in the egg covering (chorion) of insects through which spermatozoa may enter.

microtubule [Gr. *mikros*, small + L. *tubus*, pipe] Submicroscopic cytoplasmic ductule, such as in nerve fibres.

microvillus [Gr. *mikros*, small + L. *villus*, shaggy hair] Submicroscopic, very narrow cytoplasmic projection from an epithelial cell.

minus *(L. minus, less).* Referring to absence or relative smallness.

mitochondrion (Gk. *minos*, thread, + *chondros* granule). A granule or filament in the cytoplasm of a cell containing enzymes governing metabolism.

modality [L. *modus*, manner] A term to designate any specific form of sensation, such as hearing, taste, etc.

modiolus [L. *nave* of a wheel; literally, small measure; derived from *modus*, manner] The bony core of the cochlea..

molar (L. *molaris*, a mill). A grinding tooth.

molariform A general term describing premolar and molar teeth that appear similar; cheek teeth.

monoecious Refers to female and male gonads within the same individual; hermaphrodite.

monokinetic skull Skull movement via a single transverse joint passing through the braincase.

monophyletic group A clade, all organisms in a lineage plus the ancestor they have in common, therefore a natural group. Compare with paraphyletic and polyphyletic groups.

monophyletic group. See clade.

monospermy Fertilization accomplished by only one sperm.

monospondyly The condition in which a vertebral segment is composed of one centrum. Compare with diplospndyly.

monotremata (Gk. *monos*, single, + *trema*, a hole). Egg-laying mammals**Morphogenesis** (Gk. *morphe*, shape, + *genesis*, production). The origin and development of form, size and other structural features of an organism.

morphogenetic movements Cell or cell area movements concerned with the formation of germ layers *(e.g.*, during gastrulation) or of organ primordia.

morphology [Gr. *morphe*, form + *-logy* a combining form used in names of sciences, etc.] The branch of biology dealing with form structure of organisms

morula Solid ball of cells resulting from cleavage of egg; a solid blastula.

motor end plate The neuromuscular junction; specialized ending through which the axon of a motor neuron makes contact with the muscle it innervates.

moulting Complete or partial shedding of outer covering; in arthropods periodic shedding of the exoskeleton to accommodate an increase in size.

mucosa (L. *mucosus*, mucus). A membrane containing or secreting mucus.

mucus (L. *mucus*, juice). A clear, viscid secretion.

mullerian duct [L. *ductus*, a leading; named after Johannes Müller, a nineteenth century German anatomist] The oviduct including its derivatives.**Myelencephalon** (Gk. *Myelos*, marrow, + *enkephalos*, brain). The posterior of the two subdivisions of the primitive hindbrain.

mutualism Mutually helpful relationship between two organisms of different species.

myelin (Gk. *Myelos* marrow). Fatty sheath of a nerve fibre.

myeloid tissue [Gr. *myel-*, indicating relationship to bone marrow (or spinal cord)] Hemopoietic tissue producing red blood cells and granular white blood cells.

myeloid tissue Blood-forming tissue housed inside bones.

myocardium (Gk. *Myos*, muscle, + *kardia*, heart). The musculature of the heart.

myocoel Temporary cavities within the myotomes of Amphioxus which may have been connected with the coelom; the cavity within which the ovaries of Amphioxus develop.

myocomma (Gk. *Myos*, muscle, + *komma*, separation). A partition of connective tissue separating muscle segments.

myoepithelium [Gr. *mys*, muscle + *epi-*, upon + *thele*, nipple] Group of cells derived from epithelium which surround glands and apparently possess contractile properties.

myofilament [Gr. *mys*, muscle + L. *filum*, thread] Submicroscopic strand of contractile protein in a muscle cell.

myoglobin [Gr. *mys*, muscle + L. *globus*, round body] A globular protein structurally and functionally similar to hemoglobin but found in muscle cells.

myoseptum (Gk. *myos*, muscle, + L. *saeptum*, a barrier). The connective tissue partition between two adjoining muscle segments.

myotome (Gk. *myos*, muscle, + *tomos*, cutting). A prospective or actual muscle segment. Syn., muscle plate.

myotome [Gr. *mys*, muscle + *tomos*, a cut] Portion of the epaxial mesoderm which gives rise to the epaxial musculature.

N

naris (L. *naris*, nostril). A nostril.

naris A nostril.

nasal (L. *nasalis*, nose). Relating to the nose.

neocortex [Gr. *ncos*, new + L. *cortex*, bark] Six-layered mammalian cerebral cortex.

neopallium (Gk. *neos*, new, + L. *pallium*, cloak). The pallium of the brain cortex of recent origin.

neornithes (Gk. *neos*, new, + *ornithes*, birds). All recent and living birds.

neostriatum [Gr. *neos*, new + *striatus*, furrow or channel] A caudal portion of the avian telencephalon.

neoteny (Gk. *neos*, recent, + *teinein, to* stretch). Condition of many urodeles and of experimentally produced (thyroidless) anuran embryos in which larval period is extended or retained, *i.e.*, larvae fail to go through normal metamorphosis. Sexual maturity in larval stage *(e.g.* Axolotl, Necturus).

neoteny Paedomorphosis produced by delayed onset of somatic development that is overtaken by normal sexual maturity.

nephrocoel (Gk. *Nephros,*kidney,+ *koiloma*, a hollow) The cavity of a nephrotome.

nephros (Gk. *nephros*, kidney). The tubular unit of the kidney.

nephrostome (Gk. *nephros*, kidney, + *stoma$_t$* mouth). A funnel-shaped opening by which a kidney tubule communicates with the nephrocoel.

nephrostome [Gr. *nephros*, kidney + *stoma*, mouth) The opening of a kidney tubule to the coelom, found in certain fishes.

nephrotome (Gk. *nephros*, kidney, + *tomos*, a slice). A segment of mesomere, forerunner of a kidney tubule

nerve [L. *nervus*, sinew, tendon] A group of nerve fibres running together in the peripheral nervous system.

neural (Gk. *neuron*, nerve). Relating to the nervous system.

neural crest A paired strip of tissue that separates from the dorsal edges of the neural groove as it forms the neural tube.

neural groove The dorsal longitudinal groove in a vertebrate embryo; enclosed by two neural folds; preceded by neural plate stage and followed by neural tube stage.

neural plate A thickened broad strip of ectoderm along future dorsal side of all vertebrate embryos; later gives rise to the central nervous system.

neurenteric canal A temporary canal, that connects the posterior end of neural canal and archenteron through blastopore.

neurilemma (Gk. *neuron*, nerve, + *lemma* husk). Delicate sheath surrounding the myelin substance of a nerve *fibre* or the axis -of a nonmyelinated fibre.

neuroblast (Gk. *neuron*, nerve, + *blastos*, germ). An embryonic-nerve cell.

neuroblast [(Gr. *neuron*, nerve + *blastos*, germ] Embryonic cell giving rise to neurons.

neurocranium (Gk. *neuron*, nerve, + *kranion*, skull). The part of the skull enclosing the brain.

neuroglia (Gk. *neuron*, nerve, +*glia*, glue). The supporting tissues of the brain and spinal cord.

neuromast organ A mechanoreceptive organ composed of several hair cells, as in the lateral line of the inner ear.

neuron (Gk. *neuron*, nerve). The cellular unit of the nervous system.

neurophyophysis (Gk. *neuron*, nerve, + *hypophysis*, an undergrowth). The posterior lobe of the hypophysis (pituitary).

neuropil (Gr. *neuron*, nerve + *pilos*, felt] Complex interweaving of axons and dendrites within the central nervous system.

neurula The embryonic stage that follows gastrulation and during which neural axis is formed, neural plate is differentiated and basic vertebrate pattern is indicated.

niche [Fr. an-ornamental recess as in a wall, for a statue or similar object] The totality of environmental factors necessary for a species' continuity.

nidamental [L. *nidamentum*, the materials of which a nest is made] The shell gland, formed from a specialized part of the Mullerian duct in Chondrichthyes, Reptilia, Aves, and Prototheria.

nissl substance [Named after Franz Nissl, early twentieth century German neurologist] Large aggregates of ribonucleo-protein in the cytoplasm of neurons.

noradrenaline Hormone and neuro-transmitter, similar in structure and function to adrenaline, which see. Syn., norepinephrine.

notochord (Gk. *notos*, back, + *chorde*, cord, string). A fibrocellular rod constituting the primitive skeletal axis in the embryos of all chordates. In most adult chordates the notochord is replaced by a vertebral column which forms around the notochord.

notochord A long axial rod composed of a fibrous connective tissue wall around cells and/or a fluid-filled space.

nucleus [L. dim. Of *nux*, nut, and hence meaning a little nut, the kernel of a nut, or the inside of something] 1. The portion of a cell which contains the chromosomes and nucleoplasm. 2. An aggregate of nerve cell bodies within the central nervous system.

O

oblique *(L. obliquos,* slanting). Deviation from the perpendicular or horizontal.

octapeptide [L. *octo*, eight + Gr. *pepsis*, digestion] Eight amino acid polypeptide.

oculomotor *(Loculus,* eye, + *motus,* motion). Relating to movements of the eyeball; the third cranial nerve.

odontoblast (Gk. *odont,* tooth, + *blastos,* sprout or bud). One of a layer of cells, lining the pulp cavity of a tooth, which form dentine.

odontoid (Gk. *odont,* tooth, + *eidos,* resemblance). Shaped like a tooth.

olfaction [L. *olfactus,* smell] The sense of smell.

olfactory (L. *olfactus,* to smell). Relating to the sense of smell.

oligo (Gk. *Oligo,* few, small). Prefix meaning 'few' or small.

omasum [L. *omasum,* bullock's tripe] Heavily muscular portion, of a ruminant's stomach.

omentum (L. *omentum,* membrane enclosing the bowels). Mesentery passing from the stomach to another abdominal organ.

on the ventrolateral surface of the telencephalon; it receives olfactory information.

ontogeny (Gk. *on,* being, +*genesis,* origin). The developmental history of an organism, *i.e.,* the sequence of stages in its early development as distinguished from phylogenesis, the evolutionary development of the species.

ontogeny The course of an individual's development from egg to death.

oo (Gk. *oon,* egg). Prefix meaning egg.

oocyte (Gk. *oon,* egg, *kytos,* cell). Presumptive egg cell after initiation of growth phase of maturation. (Sym, ovocyte).

oogamy Sexual fusion in which the gametes of the opposite sex type are unequal, the female gamete being an egg, *i.e.,* nonmotile, the male gamete being a sperm, *i.e.,* motile.

oogenesis *(L. oon,* egg, + *genesis,* origin). Process of maturation of ovum; transformation of oogonium to mature ovum. (Sym. ovogenesis).

oogonia *(Gk.oon,* egg, + *gone,* generation). The primitive ova from which the oocytes are developed; found most frequently in peripheral germinal epithelium.

ooplasm Cytoplasmic substances connected with building rather than reserve material utilized in developmental processes.

opaque (L. *opacus,* shady). With reference to a zone or area not translucent or only slightly so.

operculum (L. *operculum,* cover or lid). Any part resembling a lid or cover.

ophidia [Gr. *ophis,* serpent + *idion,* diminutive suffix] Suborder of Squamata comprised of snakes.

ophthalmic (Gk. *opthalmos,* eye). Relating to the eye.

opistho (Gk. *opisthe,* behind). Prefix denoting behind or to the rear.

opisthocelous [Gr. *opisthen*, behind or in back + L. *celom*, hollow] Referring to a vertebra whose centrum is concave posteriorly and convex anteriorly.

opisthonephros (Gr. *opisthe*, behind, + *nephros*, kidney). The adult kidney in the anamniota.

opisthonephros [Gr. *opisthen*, behind or in back + *nephros*, kidney] The adult kidney in most anamniotes, derived from the intermediate and posterior nephrogenic tissue.

opisthonephros The adult kidney formed from the mesonephros and additional tubules from the posterior region of the nephric ridge. Compare with pronephros, mesonephros, and metanephros.

optic (Gk. *opsis*, sight). Relating to the eye or vision,

optic tectum [Gr. *optikos*, from ops, eye + L. *tectum*, roof (from tego to cover)] The visual portion of the roof of the midbrain; called optic lobes in non-mammalian vertebrates and superior colliculi in mammals.

oral (L. *os*, mouth). Relating to the mouth

orbital (L. *orbita*, a wheel-track). Relating to the eye socket.

ordovician [Named for the Ordovices, Latin name of an ancient Celtic tribe in north Wales] Period of the Paleozoic era in which the earliest vertebrate fossils are found.

organizer An embryonic tissue which organises the entire process of development e.g., the dorsal lip of the blastopore (chordo-mesoderm) in the amphibian which initiates the process of morphogenesis and differentiation by inducing the formation of the neural tube. The later then might induce the development of still other embryonic structures.

organogenesis (Gk. *organon*, organ, + *genesis*, production). The formation of organs.

osmoregulation [Gr. *-osmos*, a pushing or thrusting + L. *regula*, rule or-pattern] Control of the internal concentration of ions.

ossicle (L. *ossiculum* dim. of *os*, bone). A small bone, especially one of the bones of the middle ear.

ossification [L. *os*, bone] The formation of bone.

ossification A specific type of mineralization, unique to vertebrates, wherein hydroxyapatite (calcium phosphate) is deposited on the collagenous matrix leading to bone formation.

ostariophysi [Gr. *ostarion*, a little bone + *physis*, growth] The group of physostomous teleosts having Weberian ossicles; *e.g.*, goldfish, catfish.

osteichthyes (Gk. *osteon*, a bone, + *ichthys*, a fish). Fishes possessing a skeleton composed of bone.

osteoblast (Gk. *osteon*, bone + *blastos*, germ). A bone-forming cell.

osteoclast (Gk. *osteon*, bone, + *klastos*, break). A bone-destroying cell.

osteocyte *(osteon*, bone, + *kytos*, cell). A bone cell.

osteon A highly ordered arrangement of bone cells into concentric rings, with bone matrix surroundings a central canal through which blood vessels and nerves run; the Haversian system.

ostracoderm [Gr. *ostrakon*, shell + *derma*, skin] Order of agnathan fishes comprising the earliest known fossil vertebrates.

otic (Gk. *otikos*, belonging to the ear). Relating to the ear.

otoconia Small calcareous crystals on the maculae of the inner ear; small otoliths.

otocyst (Gk. *otos*, ear, + *kystis*, a bladder). The embryonic inner ear.

otolith (Gk. *otos*, ear, + *lithos*, stone). Calcified body in the succulus inner ear.

otolith A single calcareous mass in the cupula of hair cells.

otolithic membrane [Gr. *ous*, ear + *-lith*, noun ending meaning stone] The gelatinous membrane containing inorganic or Organic crystals, overlying maculae of the inner ear.

ovale (L. *ovum*, egg). Oval or egg-shaped.

ovary (L. *ovarium*, egg-receptacle). One of the reproductive glands in the female, containing the ova or germ cells.

ovi-, ovo (L. *ovum*, egg). Prefix meaning egg.

oviduct (L. *ovum*, egg, + *ductus*, duct). The tube serving or transport and/or house the egg and embryo.

oviparity, oviparous (1) Reproductive pattern in which eggs are released by the female and offspring development occurs outside the maternal body; (2) adjective.

oviposition Process of laying eggs.

ovoviviparous (L. *ovum*, egg, + *vivus*, alive, + *parere*, to bear). Reproductive pattern in which eggs develop within the maternal body; but without nutritive or other metabolic aid by the female parent ; offspring are born as miniature adults (2) adjective.

ovulation Release of an animal egg from its ovary.

ovum (L. *ovum*, egg). The egg or female sexual cell, from which, when fecundated by union with the male element (sperm), a new individual is developed.

oxyhemoglobin [*oxy-*, a combining form of oxygen + Gr. *hemoglobin* (*hemo*, blood + L. *globus*, ball or sphere)] Hemoglobin with loosely bound oxygen.

oxytocin [Gr., *oxys*, sharp, keen + *tokos*, childbirth] Hormone released from the posterior pituitary causing the initiation of labour and the ejection of milk.

P

pacinian corpuscle [L. *Corpusculum*, dim. of *corpus*, body; named for Filippo Pacini, Italian anatomist of the nineteenth century] A pressure-sensitive sense organ located in deep layers of the skin, alimentary canal, and mesenteries.

paedomorphosis Evolution of species that resemble earlier ontogenetic stages of the ancestor.

paedomorphosis the retention of general juvenile features of ancestors in the late developmental stages of descendants. Larval stages of ancestors become the reproductive "adult" stages of descendants. See neoteny and progenesis.

palaeontology Study of ancient life.

palaeoniscoid [Gr. *palaios*, old] Generalized fish of the superorder Chondrostei, particularly those of the Paleozoic era.

palatine (L.*palatinus*, the palate). Relating to the palate,

paleontology [Gr. *palaios*, old + *onto-*, being + *-logy*, science or body of knowledge] The study of fossils.

paleostriatum [Gr. *palaios*, old + *striatus*, furrow or channel] (l) A deep portion of the avian telencephalon. (2) In mammals, the globus pallidus.

paleozoic [Gr. *palaios*, old + *zoion*, animal] The geologic era before the Mesozoic.

pallium *(L. pallium*, cloak). The cerebral cortex with the subjacent white substance.

pancreas (Gk. *pan*, all, + *kreas*, flesh). Abdominal digestive and endocrine gland.

panniculus (L. dim of *pannus*, cloth). A sheet of tissue.

papilla (L. *papilla*, pimple). Any elevation or nipplelike process.

para- (Gk. *para*, alongside). Prefix denoting alongside, near, or departure from normal.

parabronchus (Gk. *para*, alongside, +*bronchus*, windpipe) Component of avian lung.

parachordal [Gr. *para*, beside or near + *chorde*, cord] Paired cartilaginous plates of the neurocranium forming under the hindbrain.

parachuting An airborne fall slowed by the use of canopy-like membranes or body shape that increase drag.

paradaptation The concept that some aspects of a feature may not be adaptive or owe their properties to natural selection.

parafollicular cell [Gr. *para*, beside or near + L. *folliculus*, a small sac] A cell adjoining follicle cells of the thyroid or parathyroid and secreting calcitonin.

paraphyletic group An incomplete clade resulting from removal of one or more component lineages. Compare with monophyletic and polyphyletic groups.

paraphyletic group is one that includes organism which share a common ancestor, but one or more descendants is omitted.

parasagittal plane A sagittal plane parallel with the midsagittal plane.

parasympathetic nervous system [Gr. *para*, beside or near + *syn*, with + *pathetikos*, sensitive] Portion of the autonomic nervous system originating from cranial and sacral regions, having acetylcholine as its neurotransmitter. It increases vegetative functions and inhibits somatic functions.

parathyroid (Gk. *para*, beside, + *thyreos*, oblong shield, + *eidos*,form). An endocrine gland adjacent to the thyroid.

paraxial mesoderm Paired strips or mesodermal populations forming along the neural tube; in the head, it remains as strips of mesoderm called somitomeres, but in the trunk it becomes segmentally arranged as somites.

parenchyma [Gr. *parencheo*, to pour in beside] The distinguishing cells of an organ; e.g., the parenchyma! cells of bone are osteocytes, (those of liver, hepatic cells.

parietal (L. *paries*, wall). Relating to any sidewall.

parotid (Gk.*para*, beside, + *otos*, ear). Denoting several structures near the ear.

parthenogenesis (Gk. *parthenos*, virgin, + *genesis*, origin). Development of an egg without fertilization ; occurs naturally in some animals *(e.g.,* rotifers) and maybe induced artificially in others *(e.g.,* frogs).

parthenogenesis Development of an egg without fertilization.

parturition The act or process of giving birth to a young one.

patagium A stretched fold of skin that forms an airfoil or flight control surface.

path, penetration Initial direction of sperm entrance into egg, often shifting towards egg nucleus along a new copulation path (Sum. entrance path).

pectoral (L. *pectoralis*, breastbone). Relating to the breast or chest.

pelage *A* French word meaning the hair or fur of a mammal.

pellucid (L. *pellucidus* ; *per*, through, + *lucere,* to shine). With reference to any translucent area or zone.

pelvis (L.*pelvis*, a basin). Any basin like or cup-shaped part, as the pelvis of the kidney or the pelvic girdle.

penis (L. *penis).* The male organ of copulation, containing the terminal portion of the sperm duct.

pentidactyl Having five digits per limb; thought to be the basic pattern characteristic of tetrapods but modeled by functional demands.

peramorphosis in contrast to paedomorphosis, results in a descendant that exceeds ancestral development.

peri- (Gk. *peri*, around), "Prefix denoting around or surrounding.

pericardial (Gk. *peri*, around, + *kardia*, heart). Surrounding the heart.

perichondrium (Gk. *peri*, around, + *chondros*, cartilage). The fibrous membrane covering cartilage.

perichondrium The sheet of fibrous connective tissue around cartilage.

perikaryon [Gr. *peri*, around + *karyon*, nut, kernel] The central portion, or cell body, of a neuron.

perimysium (Gk. *peri*, around, + *myos*, muscle). The fibrous sheath enveloping each of the primary bundles of muscle fibres.

periosteum [Gr. *peri*, around + *osteon*, bone] Fine connective tissue coating of a bone.

periosteum The sheet of fibrous connective tissue around bone.

periotic (Gk. *peri*, around, +*otos*, ear). Surrounding the internal ear.

perissodactyla (Gk . *perissos* odd + *daktylos*, finger). An order of nonruminant ungulate mammals that have an odd number of toes.

peristalsis [Gr. *peri*, around + *stalsis*, constriction] The rhythmic contractions of smooth muscles lining the gut.

peristalsis Progressive waves of muscle contractions within the walls of a tubular structure, as within the walls of the digestive tract.

peritoneum (Gk. *periteino*, | stretch over). The lining of the body cavity and covering of organs.

permian [Named for Perm, former province of Russia] Most recent period of the Paleozoic Era.

pessulus [L. *pessulus, a* bolt] Small bone of the avian syrinx, supporting one of the vibrating membranes.

petromyzontia (Gk. *petros*, stone. +*myzon*, suck). The jawless suctorial lampreys.

phagocyte (Gk. *phago*, I eat, + *kytos* cell). A cell possessing the property of ingesting foreign cells and particles.

phagocytosis The intake of solid particles by a cell by forming food-cup and engulfing them.

phalanges (Pl of Gk. *Phalanx*, a line of soldiers). The bones of the fingers or toes.

phallus (Gk. *Phallus*, pertaining to the penis). Shaft of the penis.

pharyngeal slit An elongated opening in the lateral wall of the pharynx.

pharyngula [Gr, *pharynx*, throat] Developmental stage of vertebrates following gastrulation, when the basic body plan is morphologically evident,

pharynx (Gk. *Pharynx*, the throat). Segment of gut between the mouth and esophagus.

pharynx [Gr. *pharynx*, throat] The most anterior portion of the endodermally lined gut; characterized by the laterally extending pharyngeal pouches.

photopic vision Color vision in bright light. Compare with scotopic vision.

photopigment [Gr. *phos*, light + L. *pigmentum*, paint] Molecules reactive to wavelengths of visible light.

phylogeny (Gk. *phyle, phylon*, a tribe, + *genesis*, origin). The evolutionary development of any plant or animal species.

phylogeny [Gr. *phylon*, race, tribe +-*geneia*, origin] The evolutionary history of an organism.

phylogeny Evolutionary history of an organism.

physiology (Gk. Physio, nature + *logos*, a discourse). The study of normal function in living organisms. The term was introduced by the French physician Jean Ferwel in 1552.

physoclistous [Gr. *physa*, bladder + *kleistos*, shut]- Describing those teleost fish which lack a pneumatic duct as adults.

physostomous [Gr. *physa*, bladder + *stoma*, mouth] Describing those teleost fish which retain a pneumatic duct as adults.

pilaster cell [L, *pilastrum*, dim. of *pila*, pillar] Specialized cells lining the blood channels in the secondary lamellae of teleost gills.

pineal (L. *pinea*, pine cone). Relating to an outgrowth from the roofplate of the diencephalon.

pinna (L. *pinna*, wing). The external ear exclusive of the meatus.

piriform cortex |I.. *pirum*, pear + *forma*, form + *cortex*, bark] Surface gray matter

pituicyte [L *pituitarius*. pertaining to or secreting phlegm + Gr. *Kytos*, a hollow vessel]. The parenchymal cells of the pars nervosa of the pituitary.

pituitary (L.*pituita*, slime). Relating to an endocrine gland at the base of the diencephalon.

placenta (L. *placenta*, a cake). The organ of physiological communication between mother and foetus during pregnancy. It is formed in part from the endometrial (uterine) epithelium and in part from the extra-embryonic membranes; develops in most mammals and serves as an organ of nutritive and respiratory exchange between the foetus and the mother.

placenta A composite organ formed of maternal and fetal tissues through which the embryo is nourished.

placode (Gk. *plax*, plate, + *eidos*, like). Denoting any thickened platelike area.

placoderm [Gr. *plax*, anything flat or broad + *derma*, skin] Class of extinct -fish; the first vertebrates to .evolve jaws and paired appendages.

placoid (Gk. *Plax*, plate, + *eidos*, likeness). Relates to a plate.

planktonic Pertaining to a free-floating microscopic plant or animal that is passively carried about by currents and tides. Compare with benthic and pelagic.

plantigrade A foot posture in which the entire sole comes in contact with the ground. Compare with digitigrade and unguligrade.

planula Ciliated free living larva of most coelenterates.

plasma *(Gk. plasma*, anything formed). The fluid portion of blood and lymph.

plasticity [Gr. *plastikos*, formed, molded] In anatomy, the capability of being molded or changing form.

platysma An unspecialized muscle derived from hyoid arch musculature that spreads as a thin subcutaneous sheet into the neck and over the face.

pleura (Gk. *pleura*, side). Relating to the membrane enveloping the lungs.

pleuro (Gk. *pleura*, side). Prefix denoting on the side.

pleurocentrum (G. *pleura*, side, + *kentron*, centre). One of the type of centrum.

pleurodont dentition [Gr. *pleuron*, side + *odontos*, tooth] Condition of having teeth attached to the jaws by one side of the root only, as seen in many lizards.

plexus (L. *plexus*, a brain). A network or inter joining of parts, *e.g.*, blood vessels or nerves.

plumage [L. *pluma*, soft feathers; plural down] All the feathers a bird has at any one time.

pneumatic duct [Gr. *pneuma*, air or gas + L. *ductus*, a leading] Duct connecting the alimentary canal and the gas bladder in physostomous fishes.

podocyte [Gr. *pous*, foot + *kytos*, a hollow vessel] Specialized cell of the inner wall of Bowman's capsule, facilitating filtration.

podocytes Specialized excretory cells associated with blood capillaries of the kidney.

poikilotherm (Gr. *poikilos*, varied + *therme*, heat] Animals not capable of maintaining a constant internal body temperature independent of external temperature.

poikilothermic Cold-blooded.

polar body Relatively minute, discarded cells produced during meiotic divisions of egg cell; synonym; polocyte.

polarity The antero-posterior differentiation in regeneration e.g., Hydra It means that the anterior side of any section of the animal grows a head while the posterior side of the section grows a tail.

poly- (Gk, *polys*, many). A prefix denoting multiplicity.

polydactyly An increase in the number of digits over the basic pentidactylous number. Compare with polyphalangy.

polymorphism Occurrence of several forms within a species.

polyphalangy An increase in the number of phalanges in each digit. Compare with polydactyly.

polyphyletic group An artificial group characterized by features that are not homologous. Compare with monophyletic and paraphyletic groups.

polyphyletic group is a group of organisms that do not share an immediate common ancestor.

polyphyodont A pattern of continuous tooth replacement.

polyphyodont dentition [Gr. *polys*, many + *phyo*, to produce + *odontos*, tooth] Describing the ability to replace lost teeth an .indefinite number of times.

polyploid (Gk. *polys*, many, +*pioos*, equivalent). Term in cytology denoting more than the normal number of chromosomes.

polyspermy (Gk. *polys*, many, + *sperma*, seed). The entrance of more than one spermatozoon into the ovum ; occurring generally in chick egg.

pons (L. *pons*, bridge), Any bridgelike formation connecting two parts; mass in floor of metencephalon.

pontine (L. *pons*, a bridge). Relating to the pons of the brain.

porphyropsin [Gr. *porphyra*, purple + *opsis*, sight] The- photosensitive pigment of rods in fresh-water vertebrates.

portal (L. *porta*, gate). Relating to any porta.

portal system A set of venous vessels beginning and ending in capillary beds or sinuses of the liver.

post- (L. *post*, after). Prefix denoting after or behind.

pre- (L. *prae*, before). Prefix denoting anterior or before, in space and time.

pre- (L. *pro*, before). A prefix denoting before or forward.

preadaptation The concept that features possess the necessary form and function to meet the demands of a particular environment before the organism experiences that particular environment. Compare with paradaptation.

preformation Theory of ontogenetic development that advocated that a living creature existed preformed in the egg or sperm and that development was simply enlargment.

presumptive Predicted outcome of development of a given area based on previous fate map studies.

primates (L. *prims* (*primat-*) chief). The highest order of mammals, including man, monkeys, and lemurs.

primitive groove A narrow furrow running along the middle of the primitive streak bounded by primitive folds and terminated anteriorly by primitive knot and posteriorly by the primitive plate.

primitive streak A thickened dorsal longitudinal strip of blastoderm (ectoderm and mesoderm) formed during gastrulation stage of avian, reptilian and mammalian embryos.

primordial Germ cells Diploid cells which are destined to become germ cells *e.g.*, oogonia and spermatogonia.

procelous [Gr. *pro*, before or in front of | *celom*, hollow] Describing vertebrae whose centra have concave anterior surfaces and convex posterior surfaces.

procoracoid Anterior coracoid (or precoracoid); endochondral bone of the shoulder that first evolved in fishes. Compare with coracoid.

proctodaeum (Gk. *proktos*, anus, + *hodaios*, relating to a way). Terminal portion of rectum formed in the embryo by an ectodermal invagination.

proctodeum The embryonic invagination of surface ectoderm that contributes to the hindgut, usually giving rise to the cloaca.

proestrus (L. *pro*, before, + *oistros*, mad desire). The period immediately preceding heat in female mammals.

progenesis Paedomorphosis produced by precocious onset of sexual maturity in an individual still in the morphologically juvenile stage. Compare with neoteny.

progesterone (Gk. *pro*, in favour of +L. *gestare*, to bear). A hormone of the corpus luteum.

prokinesis Refers to skull movement via a transerse joint that passes through the dermatocranium anterior to the ocular orbit. Compare with mesokinesis and metakinesis.

prolactin (L. *pra*, before, + *lac*, milk). A hormone from the anterior pituitary that stimulates secretion of milk

pronator (L. *pronare*, to bend forward). Serving to rotate a part downward or backward.

pronephron (Gk. *pro-* before, + *nephros*, kidney). Tubular unit of a pronephros.

pronephros (Gk. *pro*, before, + *nephros*, kidney). First stage in the development of the amniote kidney.

pronucleus (Gk. *pro*, before +L. dim. of *nux*, nut or kernel). The nucleus of the spermatozoon or of the ovum prior to fertilization.

proprioception [L. *proprius*, one's own + *capio*, to take] The receiving of sensory information by the nervous system, from end organs located in muscles, ligaments and joints.

Proprioceptor A specialized interoceptor that responds to limb postion, joint angle and state of muscle contraction. Compare with interoceptor and exteroceptor.

prosencephalon (L *pros*, before, + *enkephalos*, brain). The most anterior of the three primitive divisions of the brain.

prostate (L. *pro*, in front, + *stare*, to stand). A glandular body that surrounds the beginning of the urethra in the male.

protopathic [Gr. *protos*, first + *pathos*, suffering] Describing the nerve *fibre*s and cell bodies concerned with the reception and transmission of pain and temperature:

protoplasm (Gk. *protos*, first, + *plasma*, thing formed). Living matter, the substance of which animal and plant tissues are composed.

protostome An animal whose mouth forms from or near the embryonic blastopore.

prototheria (Gk. *protos*, first, + *therion*, a will beast). A primitive group of Mammalia that lay eggs.

protractor (L. *protrahere*, to draw forth). Serving to draw a part forward.

proventriculus [L. *pro*, before + *ventriculus*, dim. of *venter*, belly] Glandular portion of the avian stomach.

proximal (L.*proximus*, next). Nearest the trunk or point of origin.

pseudo (Gk *pseudes*, false). A prefix denoting deceptive resemblance.

pterygoid (Gk *pteryx*, wing, + *eidos*, likeness). Applied to various anatomical parts in the neighbourhood of the sphenoid bone.

pterylae Feather tracks.

ptyalin (Gk. *ptyalon*, saliva). An enzyme in the saliva which converts starch into maltose.

pulmonary (L. *pulmo*, lung). Relating to the lungs.

pupa (L. girl, doll). An intermediate development stage between the larval and adult phases in holometabalous insects; non-feeding, immotile and often enclosed within a cocoon.

purkinje fibre [Named for Johannes E. von Purkinje, Bohemian anatomist

putamen [L. *putamen*, that which falls-off in pruning] The most lateral portion of the basal ganglia of mammals.

pygostyle (Gk. *pygo*, rump, + *stylos* Pillar). The plate of bone forming the posterior end of the vertebral column of birds.

pyloric caecum (pl., caeca). [Gr. *pylorus*, gatekeeper (from *pyle*, gate,+ *ourus*, a warder) + L. *caecus*, blind] Blind diverticulum of the alimentary canal of teleosts, leaving from the small intestine just behind the pyloric sphincter and providing extra surface area for absorption.

Q

quadruped [L. *quattuor*, four 4 + *pes*, foot] Four-footed animal.

quantum evolution Adaptive evolutionary change within a lineage characterized by long periods of little change that are suddenly interrupted by short bursts of rapid change. Compare with punctuated equilibrium.

quaternary [L. *quartus*, fourth] The most recent (and current) period of the Cenozoic Era, only about one million years old.

R

rachis [Gr. *rhachis*, spine, ridge] The portion of the shaft of a contour feather which supports the vane.

radial Cleavage Holoblastic cleavage in which divisions take place in such a manner that all the blastomeres are placed in a radially, symmetrical fashion around the polar axis, *e.g.*, sponges, coelenterates, some echinoderms and Amphioxus.

radial symmetry A regular arrangement of the body about a central axis.

radial symmetry Pattern of an organism where similar parts are arranged about a common centre.

radiata A taxonomic group of radially symmetrical animals.

radius (L. *radius*, spoke), Denoting line from the centre to periphery of a circle. Bone of forearm.

ramus (L. *ramus*, a branch). Denoting branch or division.

raptors Predatory birds that use talons, including hawks, eagles, falcons, and owls.

rathke's Pouch [Named for Martin H. Rathke, nineteenth century German anatomist] Evagination of the stomodeum that forms the anterior pituitary.

recapitulation Theory The successive stages of individual development (ontogeny) correspond with the successive adult ancestors in the line of evolutionary descent (phylogeny): "Ontogeny repeats phylogeny".

receptor (L. *recipere*, to receive). Sensory nerve ending in skin, or a sense organ.

recrudescence Renewal of reproductive interest and readiness of reproductive tracts, usually on a seasonal basis.

rectus (L. *rectus*, straight). Referring to straight.

reflex (L. *re-*, back, + *flectere*, to bend). Denoting a nervous mediated involuntary action.

regeneration The process of repair or replacement of the lost parts of the body is known as regeneration.

regulative eggs There is no predetermination and the fate of various egg portions is usually not fixed until three cleavages have produced eight cells, *e.g.,* the eggs of coelenterates, echinoderms and chordates.

renal (L. *ren*, kidney). Relating to a kidney.

renin [L. *renes*, kidneys] A proteolytic enzyme released by the kidney into the bloodstream.

reptilia (L. *reptilis*, creeping). A class of vertebrates comprising the alligators, crocodiles, lizards, turtles, tortoises, and snakes.

rete (L. *retia*, a mesh). A network of nerve fibres or small vessels.

rete A compact, dense network of capillaries.

rete mirabile [L. *rete*, net + *mirabilis*, marvelous] An extremely tortuous vascular network, usually . involving hairpin turns and the- interruption of .an arterial continuity.

reticulum [L. *reticulum*, little net (from *rete*, net) + *-culum*, dim. suffix] A fine network of connective tissue fibres and cells.

reticulum The second of four chambers in the complex ruminant stomach; a specialized region of the esophagus. Compare with abomasum, omasum, and rumen.

retina (L. *rete*, net). The light-sensitive layer of the eye.

retractor (L. *re*, back, +*trahere*, to draw). Serving to withdraw a part.

retroperitoneal [L. *retro*, backward, behind + Gr. *peritones*, stretched around] Describing the position of an organ in which one surface abuts a serous membrane and the other surface, somatic muscles of the body.

rhachitomous vertebra An aspidospondylous vertebra characteristic of some crossopterygians and early amphibians.

rhipidistii [Gr. *rhipidos*; a fan] Extinct group of sarcopterygian fish which gave rise to the first amphibians

rhodopsin (Gk. *rhodon*, rose, + *ops*, eye). The red pigment in the rods of the retina.

rhombencephalon (Gk. *rhombos*, parallelogram, + *enkephalos*, brain). The posterior of the three primitive brain divisions.

rhomboid (Gk. *rhombos*, a rhomb, + *eidos*, appearance). Having the shape of an oblique parallelogram with unequal sides.

rhynchocephalia [Gr *rhynchos*, snout + *kephalikos*, of the head] An order of Reptilia; the only living member is *Sphenodon*.

rod Photoreceptor of the eye sensitive to black and white.

rodentia (L. *rodere*, to gnaw). An order of placental mammals that possess one or two pairs of long incisor teeth adapted to gnawing.

ruga (L. *ruga*, a wrinkle). A fold, ridge, or crease.

rumen [L.. *rumen*, throat, gullet] Largest chamber of ruminant's stomach, harboring massive numbers of symbiotic bacteria.

rumen The first of four chambers in the complex ruminant stomach; an expanded specialization of the esophagus. Compare with abomasum, omasum, and reticulum.

ruminant A placental mammal with a rumen, a specialized expansion of the digestive tract that processes plant material; Ruminantia.

rumination Remastication together with microbial fermentation in ruminants.

S

sacculus (L.*sacculus*, a little bag). Denoting any small pouch; the smaller of the two sacs in the inner ear.

salivary (I, *saliva*, spittle). Relating to the saliva and the glands producing it.

saltatorial [L. *saltare*, to jump about] Referring to a leaping mode of locomotion.

sarcolemma (Gr. sarx, flesh + *lemma*, shell or husk] Cell membrane of a muscle cell.

sarcoplasm [Gr. *sarx*, flesh + *plasma*, something molded or formed] The cytoplasm of a muscle cell.

sarcopterygii [Gr. *sarx*, flesh + *pterygi*, wing or fin] Subclass of Osteichthyes comprised of the lobe-finned fishes *i.e.*, dipnoans, coelacanths, and rhipidistians.

scala naturae [L. *scala*, ladder + *natura*, nature] The arrangement of living beings into a single linear scale, from simple to complex; now discredited.

scale [Fr. *écaillle*, fish-scale] Overlapping dermal and/or epidermal hardenings, formed in skin folds.

scansorial locomotion Climbing of trees with claws. Compare with brachiation.

schizocoel (Gk. *schizo*, split, + *koilos*, hollow). A coelom formed by the splitting of an originally solid layer of mesoderm.

schizocoel Coelom formed by the splitting of embryonic mesoderm.

schizocoelom the body cavity formed by splitting of the mesoderm. Compare with enterocoelom.

sclera (Gk. *skleros*, hard). The tough outer coat of the eye.

sclerotome *(Gk. Skleros, hard* + *tomos*, cutting). A segment of the skeleton-producing cells derived from a mesodermal somite.

scotopic vision Sensitivity to dim light. Compare with photopic vision.

scrotum (L. *scrotum*, hide). A musculocutaneous sac containing the testes.

sebaceous (L. *sebum*, tallow, grease). Relating to oil or fat.

Secondary cartilage Cartilage that forms after initial bone ossification is complete; formed usually in response to mechanical stress, especially on the margins of intramembranous bone.

secretin (L. *secretus*, to separate). A hormone produced in the intestine which incites pancreatic secretion.

sectorial teeth Teeth with opposing sharp ridges specialized for cutting.

segmentation (L. *segmentum*, to cut). The state or process of being divided into segments.

segmentation Division of a body into more or less similar parts.

selachii [Gr. *selachos*, shark] The order of Elasmobranchii comprised of sharks.

selenodont Teeth with crescent-shaped cusps, as in artiodactyls. Compare with bunodont and lophodont.

self-fertilization The fusion of male (sperm) and female gamete (ovum) produced by a single hermaphrodite organism.

selonodont [Gr. *selene*, moon + *odontos*, tooth] Teeth usually high-crowned with longitudinal crescent-shaped ridges.

semen (L. *seed*). Secretion of male reproductive organs and glands; incorporates spermatozoa and sperm-carrying fluid.

semilunar (L. *semi*, half, + *luna*, moon). Crescentic or half-moon shape.

seminiferous (L. *semen*, seed, + *ferre*, to carry). Carrying or conducting the semen.

septal nuclei [L. *septum*, *a* partition] Anterior ventral medial portion of the telencephalon.

septum (L. *septum*, a partition). A thin wall dividing two cavities or masses of softer tissue.

serial homology Similarity between successively repeated features in the same individual.

serosa (L. fem, of *serosus*, a coat). The peritoneal coat of a visceral organ.

serous [L. *serum*, whey] Relating to or containing serum or a thin, watery, proteinaceous fluid.

sesamoid bone A bone that develops directly in a tendon, for example, the patella (kneecap).

sexual cycle Periodic sequence of changes in the uterine mucosa of the female, the cycle related to ovulation and to endocrine cycle.

sexual reproduction Reproduction involving meiosis and union of haploid gametes, the sperm and ova, to form a diploid zygote.

sinoatrial node [L. *sinus*, sinus + *atrium*, entrance court] The "pacemaker"; the neuromuscular tissue initiating waves of contractions of the heart.

sinus (L. *sinus*, bay or hollow). A space in some organ or tissue.

sinus venosus [L. *sinus*, sinus + *venosus*, vein] The chamber of the heart in fishes, amphibians, and reptiles which receives the systemic venous blood.

sinusoid [L. *sinus*, sinus] Capillary with greatly expanded lumen.

sinusoids Tiny vascular channels that are slightly larger than capillaries and lined or partially lined only by endothelium.

sister group In taxonomy, the particular outgroup most closely related to the ingroup. Compare with ingroup, outgroup.

skeleton (Gk. *skeletos*, dried). The supporting framework of the body.

skull (Early Eng., *skulle*, a bowl). The skeletal famework of the head.

solenocyte A single excretory cell with a projecting circle of microvilli around a central flagellum.

somatic (Gk. *somatikos*, bodily). Relating to body in contrast to germinal cells.

somatopleure (Gk. *somatikos*, bodily, + *plura*, side). The embryonic layer formed by the union of the somatic layer of mesoderm with the ectoderm.

somatotrophin (Gk. *somatikos*, bodily, + *trophe*, nourishment). The growth hormone produced by the adenohypophysis.

somite One of the longitudinal series of segments in segmented animals; especially in an incompletely developed embryonic segment or a part thereof.

spawning Act of expelling eggs from uteri of anamniota (e.g., amphibia).

spermatid *(Gk. sperma,* seed). The rudimentary spermatozoon derived from the division of the spermatocyte.

spermatocyte (Gk. *sperma*, seed, + *kytos*, a hollow cell). A cell-resulting from the division of the spermatogonium, which in turn by division forms the spermatid.

spermatogenesis (Gk. *sperme*, seed, + *genesis*, origin). The formation of spermatozoa.

spermatogonia (Gk. *Sperma*, seed, + *gone*, generation). The primitive sperm cells giving rise, by division, to the spermatocytes.

spermatozoon (Gk. *sperma*, seed, + *zoon*, animal). The male sexual cell.

spermiogenesis The transformation or differentiation of the non-motile, unspecialized spermatid into a motile and specialized mature sperm.

sphenodon [Gr. *sphen*, wedge + *odontos*, tooth] Genus of the only living Rhynchocephalian.;

sphenoid (Gk. *shpen*, wedge, + *eidos* likeness). Relating to any part having wedge shaped form.

sphincter (Gk. *sphinkter*, a band). A muscle that serves to close an opening.

sphincter A band of muscle around a tube or opening that functions to constrict or close it.

spiracle (L. *spiraculum*, breathing hole). An aperture or vent for respiration.

spiracle A reduced gill slit that is first in series.

spiral cleavage Cleavage diagonal to the polar axis of the egg so that resulting blastomeres (generally micromeres) are placed over the junctions between the four lower cells (macromeres), *e.g.* annelids, molluscs and nemerteans.

splanchnic (Gk. *Splanchnon*, viscus). Referring to the viscera.

splanchnic mesoderm The sheet of mesoderm, also called visceral mesoderm, that is applied against the endoderm (of the gut).

splanchnocranium [Gr. *splanchna*, entrails or viscera + *kranion*, skull] Portion of the skull derived from neural crest and forming in relationship to the pharynx.

splanchnocranium That part of the skull arising first to support the pharyngeal slits and later contributing to the jaws and other structures of the head; branchial arches and derivatives; visceral cranium.

splanchno-pleure (Gk. *Splanchnon*, viscus, + *plura*, side). The embryonic layer formed by the union of the visceral layer of mesoderm with the endoderm.

spleen (Gk. *splen*, spleen). A large abdominal gland belonging to the circulatory system.

splenial (Gk. *splenion*, bandage). In anatomy, a structure overlapping another structure. Specifically, the splenial bone.

squamata [L. *squama*, scale] Order of reptiles containing lizards and snakes.

Stage in development when the organism has emerged from its membranes and is able to lead an independent existence, but may not have completed its development. Generally (except in cases of neoteny or paedogenesis) larvae cannot reproduce.

stapes One of the three middle ear bones in mammals, phylogenetically derived from the columella (hyomandibula).

stenohaline Having a narrow tolerance to salinity differences.

stereocilia [Gr. *stereos*, solid + L. *cilium*, eyelid] .Cytoplasmic processes having the shape of kinocilia but lacking their internal structure.

stereocilia Very long microvilli.

stereoscopic vision Denoting the ability to see images in three dimensions.

stereospondylous vertebra A monospondylous vertebra in which the single centrum (an intercentrum) is separate (aspidospondylous). Compare with embolomerous vertebra.

sternum (Gk. *sternon*, the chest). The breast bone.

steroid (Gk. *stereos*, solid, + L. *oleum*, oil, + G, *eidos*, form). Resembling a sterol.

steroidogenic tissue [steroid + Gr. *genesis*, production] Endocrine tissue

stolon A rootlike process of ascidians and other invertebrates that may fragment into pieces that asexually grow into more individuals.

stomach (L. *stomachus*). The portion of the gut between the esophagus and small intestine.

stomodaeum (Gk. *stoma*, mouth, + *hodaios*, relating to a way). Ectodermal invagination forming the mouth cavity.

stratum (L. *stratus*, layer). Any given layer of differentiated tissue.

streptostyly the condition in which the quadrate bone is movable relative to the braincase.

striatum (L. neut. of *striatus*, furrowed). Denoting striping or furrowing.

stroma The mesodermally derived connective tissue supporting the epithelial portions of an organ.

sub- (L. *sub*, under). Prefix signifying beneath, less than normal.

subcorium [L. *sub*, under + *corium*, skin) See-subcutaneous tissue.

subcutaneous tissue [L. *sub*, under + *cutis*, skin] The deep, loose, fatty connective tissue of the skin. Syn., hypodermis, subcorium.

submucosa [L. *sub*, under + *mucosus*, mucus] Dense connective tissue layer of the alimentary canal.

substantia nigra [L. *substantia*, substance + *niger*, black] Nuclear area of the mammalian brain forming the upper part of the cerebral peduncle.

subunguis L. *sub*, under + *unguiculus*, nail or claw] Soft keratin underportion of a claw, nail, or hoof.

sudoriferous gland [L. *sudor*, sweat, + *ferare*, to bring forth] Sweat gland.

sulcus (pl., sulci). [L. *sulcus*, furrow or ditch] A groove on the surface of the cerebral hemisphere.

sulcus limitans [L. *sulcus*, furrow or ditch + *limes*, limit] Groove in the lateral aspect of the developing central canal and ventricular system of the central nervous system, separating dorsal, sensory portions from ventral, motor portions.

superficial Cleavage Cleavage in the cytoplasmic periphery around yolk in centrolecithal eggs.

supinator (L. *supinus*, lying on the back). Serving to turn a part upward or forward.

supra (L. *supra*, above). Prefix denoting a position above the part indicated by the word to which it is joined.

suspension feeding Feeding based on filtering suspended food particles from water; usually involves cilia and secreted mucus; filter feeding or ciliary-mucus feeding.

sustentacular [L. *sustento*, to hold upright] Supporting, as in a sustentacular cell.

swim bladder A gas bladder functioning primarily in buoyancy control.

sym-, Syn- (Gk. *syn*, together, with). Prefix meaning together.

symbiosis Association of two individuals of two different species for mutual benefits.

sympathetic (Gk. *syn*, with, + *pathos*, suffering). Denoting a division of the nervous system.

symphysis (Gk. *syn*, together, + *physis*, growth). A union or meeting point of any two structures.

synapomorphies evidences of shared derived features, rather than overall similatity. (evidences for relationships). i.e. we recognize mammals as those taxa that share the possession of hair, a single bone in the lower jaw, and presence of three ear ossicles.

synapse (Gk. *syn*, together, + *hapstein*, to fasten). The close approximation of the processes of different neurons.

synapsid [Gr. *syn*. with + *apsis*, a loop or mesh] Referring to reptiles with a single temporal fenestra ventral to the postorbital-squamosal suture.

synarthrosis [Gr. *syn*, with + *arthrosis*, articulation] An immovable joint.

synarthrosis A joint through which little or no movement is permitted between articulated skeletal elements.

syncytium (Gk. *syn*, with, + *kytos*, cell). A multi-nucleated protoplasmic mass : aggregation of cells without cell boundaries.

syndesmochorial (Gk. *syndesmo*, ligament, + *chorion*, skin). Placenta type involving contact between chorionic epithelium and connective tissue of uterine mucosa.

synergist (Gk. *syn*, together, + *ergon*, work). That which aids the action.of another.

synergy [Gr. *syn* with + *ergon*, work] Coordinated activity of different parts.

synovial joint A diarthrosis.

syntrophoblast (Gk. *syn* together, + *trophe* nourishment, + *blastos*, germ). The outer syncytial layer of the trophoblast.

syrinx (Gk. *syrinx*, a tube). The vocal organ of birds.

T

talons Specialized bird claws used in striking or catching live prey.

tapetum lucidum [Gr. *tapes*, carpet + L. *lucidus*, from *lux*, light] A portion of the choroid layer of the eye which reflects light, found in various vertebrates.

taxon (pl., taxa). [Derived from taxonomy] Any of the formal categories used in classifying organisms.

taxonomy [Fr. *taxonomic*, from Gr. *taxis*, from *tassein* to arrange or put in order] The branch of biological sciences dealing with the classification of organisms.

taxonomy Study of classification and nomenclature.

tectorial membrane [L. *tectorium*, covering, from *tego*, to cover + *membrane*, fine skin] Gelatinous membrane of the cochlea, onto which the tips of the stereocilia adhere.

tectum [L, *tectum*, roof] The roof of the midbrain, derived from its alar plale.

tegmentum [L. *tegmentum*, cover] The floor of the midbrain, derived from its basal plate.

telencephalon [Gr. *telos*, end + *enkephalos*, brain] Portion of the brain which develops from the anterior end of the central nervous system; cerebral hemispheres and olfactory bulbs.

teleology [Gr. *telos*, end + *-logy*, suffix for branch of knowledge] The study of events or processes which assumes that each has its ultimate goal or purpose.

teleostei (Gk. *teleos*, complete, +*osteon*, bone). The modern bony fishes.

teleostei [Gk. *telos*, end + OSTEON bone] The superorder of Actinopterygii characterized by having a terminal mouth.

telic (Gk. *telikos*, final). Tending to a definite end.

telodendria Gr. *telos*, end + *dendron*, tree] The branching of nerve endings at the end of an axon.

telolecithal (Gk. *telos*, end +*lekithos* yolk). Pertaining to an egg in which the yolk, accumulates in the vegetal half of the cell as in frog egg.

telolecithal Pertaining to eggs in which yolk stores are concentrated at one pole.

temporal (L. *tempora*, temples). Relating to the temple.

tendon (L. *tendo*, a cord). A cord serving to attach a muscle to a bone.

tertiary Membranes The hard calcareous and soft membranous coverings of the chick egg, deposited on the egg within the uterus.

testis, pl. Testes (L. *testis*, testicle). One of the male reproductive glands, producing spermatozoa and male sex hormone.

testosterone Male hormone produced by the interstitial cells of the testis.

tetraodothyronine The primary hormone produced by the thyroid gland, which causes an increase in an animal's metabolic rate. Syn., thyroxine.

tetrapod (Gk. *teira*, four, +*pous*, foot). A vertebrate with four legs.

thalamus (Gk. *thalamos*, a bed). The side walls of the diencephalon.

thalamus [Gr. *thalamos*, a bed or bedroom] That portion of the diencephalon receiving information from the brain stem and sending massive projections to the cerebral hemispheres. Syn., dorsal thalamus.

theca (Gk. *thekey* a box). A sheath.

thecodont dentition [Gr. *theke*, box + *odontos*, tooth] The condition of having teeth whose roots extend into sockets in the bone.

thecondontia (Gk. *theke*, case, +*odous*, tooth). Having teeth inserted in sockets.

theria [Gr. *therion*, wild beast] Subclass of the class Mammalia, which includes both placentals and marsupials.

thermoreceptor A radiation receptor sensitive to infrared energy.

thoracic (Gk. *thorax*, breastplate). Relating to the thorax or chest.

thrombocyte (Gk. *thrombos*, clot, +*kytos*, cell). A small platelike blood cell involved in the clotting mechanism.

thymus (Gk. *thymos*, sweetbread). A ductless gland in the neck.

thyrocalcitonin The hormone produced by parafollicular cells of the thyroid gland. It causes a lowering of blood calcium levels.

thyroglossal (Gk. *thyreos*, a shield, +*glossa*, tongue). Relating to the thyroid gland and tongue.

thyroid (Gk. *thyreos*, on oblong shield, +*eidos*, form). Denoting a gland or cartilage.

tract A collection of nerve fibers coursing together in the brain or spinal cord. Compare with nerve.

triassic [L. *trias*, triad; so named for its three divisions] First period of the Mesozoic era, characterized by the beginning of the great adaptive radiation of reptiles.

triiodothyronine An active form of the thyroid hormone, which contains three iodine atoms whereas thyroxine has four.

tropho,–Troph (Gk. *trophos*, feeder). Prefix or suffix meaning "feeder" or "feeding".

trophoblast {Gk. *trope*, nourishment, +*blastos*, germ). The outer wall of the mammalian blastocyst, concerned with nutrition of embryo within.

trophoblast The outer cellular layer of the mammalian blastocyst.

truncus (L. *truncus*, stem or trunk). A primary nerve or blood vessel before its division.

tuber cinereum [L. *tuber*, bump, swelling + *cinereus*, ashy gray] The region of the hypothalamus surrounding the infundibulum.

tuberculum (L. dim. of *tuber*, a nodule). Relating to a protuberance,

tunica (L. *tunica*, a coat). An enveloping layer.

tusks Specialized, long teeth protruding from the mouth; elongate incisors (elephants), left upper incisor (narwhal), canines (walruses).

tympanic (L. *tympanum*, drum). Relating to eardrum.

tympanum The eardrum or tympanic membrane.

typhosole [Gr. *typhos*, snake] A fold along the lumen of the small intestine of the lamprey.

typology [L. *typus*, type; from Gr. *typos*, blow or impression] An approach to taxonomy which assumes that the structure of all members of the species is similar and individual variation is an abnormality.

U

ultimo (L. *ultimus*, last). Prefix denoting final or last of a series.

ultimobranchial body [L. *ultimus*, last + Gr. *branch*, gill] Endocrine glands of fish producing calcitonin. Also designating the embryonic precursors of the parafollicular cells of the thyroid gland of tetrapods.

umbilical (L. *umbilicus*, navel) Relating to the umbilicus.

undulation [L. undula, dim, of unaa, wave] Movement in a curved, wavy pattern.

unguis [L. *ungiculus*, nail or claw] The hard keratin upper part of a claw, hoof, or nail.

ungulate A hoofed placental mammal belonging to the orders Perissodactyla (horses) and Artiodactyla (cattle, deer, pigs).

unguligrade A foot posture in which the weight is carried on the tips of the toes (as in horses).

unmbilicus The stalk-like connection between the embryo and all extra-embryonic structures of the mammal. The point of connection with the offspring later becomes the naval.

ureotelism Excretion of nitrogen in the form of urea.

ureter (Gk. *oureter*, urinary canal). The tube conducting the urine from the kidney to the bladder.

urethra (Gk. *ourethra*, canal). Duct carrying urine from the urinary bladder to the outside of the amniote body; in the males of most mammals, the urethra also leads sperms to the outside during copulation.

uriniferous (L. *urina*, urine, +*ferre*, to carry). Conveying urine, the tubules of the kidney.

uriniferous tubule the functional unit of the kidney composed of the nephron and collecting tubule.

urochordata (Gk. *oura*, tail, +*chorde*, cord). Chordates with notochord and neural tube well developed in tail of larva.

urodela [Gr. *oura*, tail] Order of Amphibia comprised of salamanders.

urophysis [Gr. *oura*, tail + *phys*, grow] The enlarged caudal tip of the spinal cord of teleosts, chondrosteans, and elasmobranchs, which is apparently an endocrine gland.

uropygeal gland [Gr. *oura*, tail + *pyge*, rump or buttocks] Large, lipid-secreting, holocrine glands found at the base of the pygostyle in birds.

urostyle [Gr. *oura*, tail + *stylos*, pillar] The bone formed from fused caudal vertebrae in Anura.

uterus (L. *uterus*, womb). The womb; the hollow, muscular, expanded organ of the female reproductive tract in which the impregnated ovum develops into the foetus.

utriculus [L. *utriculus*, dim. of *uter*, bag] The largest sac-like-portion of the membranous labyrinth of the inner ear.

uvea (L. *uva*, grape). The middle coat of the eyeball.

V

vagina (L. *vaginae*, sheath). The genital canal in the female extending from the uterus to the vulva ; receives penis of the male during copulation.

valvula [Mod. L. dim. of *valva*, leaf of a folding door] The portion of the teleost cerebellum which extends into the optic ventricle.

vas (L. *vas*, vessel). Denoting a tube or vessel.

vas deferens The tube in a mammalian male conducting sperms from the testes to the urethra.

vasoconstriction The narrowing of a blood vessel; usually resulting from smooth muscle contraction. Compare with vasodilation.

vasodilation The widening of a blood vessel; may be active or passive enlargement. Compare with vasoconstriction.

vasopressin (L. *vas*, vessel, +*pressi*, contriction). The blood pressure-raising hormone of the neurohypophysis.

vegetal (L. *vegetare*, to animate). V. pole, denoting the end of a telolecithal egg containing the yolk.

vein (L. *vena*, a blood vessel). A blood vessel conveying blood toward the heart.

ventilation [L. *ventilare*, to fan] The process of bringing fresh air or water to the respiratory membrane.

ventilation The active movement of water or air across respiratory exchange surfaces. Compare with perfusion.

ventral (L. *venter*, belly). Relating to the bottom or belly.

ventricle (L. *ventriculus*, small cavity). A small cavity especially one in the brain or heart.

vertebra One of several bone or cartilage blocks firmly joined into a backbone that defines the major body axis of vertebrates.

vertebrata (L. *vertebratus*). Having a backbone or spinal column.

vesicle (L. *vesicula*, a small bladder). Denoting any small sac.

vestibular apparatus A sensory organ of the inner ear composed of semicircular canals and associated compartments, such as the sacculus, utriculus, and cochlea (lagena).

vestibule (L. *vestibulum*, an antechamber). Denoting any entry way.

vibrissa (L. *vibrare*, to shake] The stiff tactile hairs of the facial region of mammals; e.g., cat's "whiskers."

villus (L. *villus*, a tuft of hair). A minute projection from the surface of a membrane.

villus A fingerlike projection of a tissue layer, as in the small intestine. Compare with microvillus.

viscera (L. *viscus*, an internal organ). The collective term depicting the internal organs of an animal.

viscus pl. viscera (L. *viscus*, an internal organ). An internal organ, especially one of the large abdominal organs.

vitelline (L. *vitellus*, yolk). Relating to the yolk of an egg, *e.g.*, vitelline artery, vitelline artery, vitelline membrane.

vitreous (L. *vitreus*, glassy). Any structure having a glassy appearance.

viviparity The reproductive pattern of live birth; birth of young not encased in a shell.

viviparous, viviparity (L. *vivus*, alive, +*parere*, to bear). (1) Reproductive pattern in which eggs develop within female body with nutritional and other metabolic aid of maternal parent; offspring are born as miniature adults, (2) adjective.

vomeronasal organ [Named for the vomer and nasal bones] A chemical sensitive sense organ adjacent to the olfactory organ, found in most tetrapods.

vomeronasal organ A chemosensory organ present in the nasal chamber or roof of the mouth of some tetrapods.

vomodor A chemical detected by sensory cells of the vomeronasal organ. Compare with odor.

W

warren An underground maze of excavated passageways used by animals, usually rabbits.

wolffian duct [L. *ductus*, a leading; named for Kaspar F. Wolff, German embryologist in Russia in the eighteenth century] The duct which drains the embryonic kidney in all vertebrates; it has various functions in the adult.

wolffian duct Mesonephric duct.

wulst [Ger. *wulst*, bulge, prominence] A dorsomedial elevated portion of the avian telencephalon.

X

xanthophore [Gr. *xanthos*, yellow + *phoros*, bearing] Pigment cell containing carotenoid or pteridine pigments.

Y

yolk Highly nutrious food material (metaplasm) stored in all except alecithal eggs; the principal components are proteins, phospholipids and fats and globules of fatty material.

yolk plug A plug formed by large yolk cells which are too large to be incorporated immediately in floor of archenteron of amphibian embryo, hence are found protruding slightly from blastopore. Size of plug is often used to determine approximate stage of gastrulation.

yolk sac Extraembryonic splanchnopleure which grows around the yolk mass, remaining connected with the mid gut by the yolk-stalk ; provides extensive surface area for absorption of nourishment.

Z

zona (L. zono, girdle). Any encircling area or structure

zona pellucida A noncellular transparent, secreted layer immediately surrounding the mammalian ovum.

zona radiata Zona pellucida that exhibits radial striations.

zygapophysis [Gr. zygon, yoke + apophysis, offshoot] The articulating process of a tetrapod vertebra.

zygapophysis The projection of a neural arch that articulates with the adjacent neural arch.

zygote The fertilized ovum; the diploid cell formed by the union of two gametes.

Index

A

abducens 354, 523-525, 528
abduction 372, 373
acanthocephala 34, 53, 63
acanthodians 48, 108–111, 277
acanthodii 107, 110
acanthostega 133, 136, 137
accommodation 288, 529, 535, 537, 550, 552, 554–557
acellular level 38
acidophils 569
acipenser 116, 117, 121, 122, 287
acipenseriformes 107, 117
acoelomata 34, 62
acoelous 309, 310, 312
acoustic 524, 526, 559, 560
acrania 85
acromiodeltoid 359, 362, 363
actinopterygii 48, 50, 98, 107, 115, 116, 127, 287, 309, 311
adam's apple 425
addison's disease 583
adduction 372, 373
adenohypophysis 568–571
adenoids 388, 456
adneural 571
adrenergic 535
adrenocorticoids 583
adrenocorticotropic hormone 571
aepyornithiformes 183
afferent branchial 404, 406, 431, 441–443
   spiracular artery 442
age of mammals 223
age of reptiles 166, 478
agnatha 48, 97–101, 103
agranular 427, 428
aldosterone 581, 583, 585

algesireceptor 540, 541
allantoic 149, 196, 197, 200, 467, 468, 469
alpha cells 578
altricial birds 500
alveoli of pancreas 579
ambystomatoidea 145
ameloblasts 239, 240, 245, 383
amiiformes 107, 117
ammocoete 91, 101, 102, 408, 573
ammonotelic 457
amniota 107, 131, 133, 138, 150, 307, 308, 460, 475–480, 511, 523, 563
amniote lineage 132, 139, 140, 155
amphiarthrosis 266
amphibian eyes 554
amphicoela 147
amphicoelous 145, 147, 158, 179–181, 309–312, 316
   centrum 143, 304–307, 309–321, 323, 236, 374, 546
amphiplatyan 309, 310, 317, 319
amphisbaenians 148, 150, 152, 154
amphistylic 277, 278
amplexus 482, 491
ampulla 398, 546, 558, 559, 563, 564
   of vater 398, 400
ampullae of lorenzini 546
ampullar cristae 559
anamniota 460, 475, 477–479, 523, 563
analogies 48, 49
anapophyses 318
anapsid 96, 148, 150, 153, 156, 163, 166, 190, 296, 297
anapsida 150, 153, 156, 162, 297
anapsids 48, 148, 150, 190
anastomoses 429, 430, 449
anastomotic 429, 441, 443
androstenedione 586, 587

androsterone 587
angiotensin 585
angiotensin-II 585
anguiliiformes 119
ankylosis 266
annulus tympanicus 280, 281
ansa hypoglossi 530
antagonistic 577, 578
anterior choroid plexus 506, 509, 514, 516–519, 521
    commissure 194, 196, 200, 506, 509, 510, 514, 515, 516, 517–522, 530, 584
    laryngeal nerve 529
    medullary velum 524
anthracosaurs 133, 139, 140, 150
anthropoid apes 218
anthropoids 206
antiarchiformes 111
antimeres 26
antrum 496
anura 50, 134, 138, 139, 146, 246, 290, 335, 442
anus 8, 20, 25, 27, 28, 30, 39, 57, 52, 55–57, 59, 62, 63, 64, 67–70, 71, 73–77, 80, 87, 93, 95, 99, 102, 115, 141, 196, 201, 235, 236, 330, 390, 396, 469, 487, 489, 496
apatite 262
apatornis 181
aphragmophora 68
aplousobranchia 81
apoda 134, 138–141 143
apodiformes 176, 187
aponeuroses 265
aponeurosis 351
appendicular skeleton 267, 325, 330, 339, 343, 348, 349
appendix epididymis 479–481
    testis 479
apterygiformes 182
aqueduct of sylvius 509
arachnoid villi 504
arbor vitae 519, 522
archaea 43, 45, 46
archaeopteryx lithographica 172, 180
archaeornithes 175, 179
archemetazoan 6-7, 9-12, 20
archinephric 458, 460, 476, 477, 478, 489, 191, 586
    duct 67, 76, 101, 104, 105, 115, 119, 120, 227, 228, 235, 237, 238, 284

archipterygium 126, 332
archisternum 324, 325
archosaurs 48, 151, 152, 299
arcicentrous vertebrae 307
area centralis 552–557
argenta of iris 554
arrectores pilorum 353, 372
    plumarum 353, 372
arthrodiriformes 111
articular 145, 264, 266
    cartilage 99, 108
artiodactyl 346, 347, 392
artiodactyla 214, 215, 302
asterospondylous 309
astragalus (talus) 337, 339
atlas 288
auditory 192, 208
    (SS) 512, 523, 525–529
    nerve 258
auerbach's plexus 537
auricular lobes 515
autodiastyle 227
auto-digestion 393
autonomic ganglia 534
autostylic 127, 129, 278, 279, 287–289
autosystylic 278, 279
axial skeleton 84, 123, 124, 150, 153, 258, 267, 304, 309, 325, 328, 332, 333, 343, 348, 361, 364
axis 267

B

baculum 265, 484
baleen 215, 248, 253, 376
bartholin's glands 497
basalia 328, 331, 333
basapophyses 305, 321
basihyal 275, 276, 278, 280, 281, 286, 301, 302, 304, 356, 365, 379, 380
basibranchial 275, 276, 279, 280, 281, 286, 290
basilar membrane 560, 561, 566
basis cranii 284
basisphenoid bone 271
basophils 569
begging call 500
beloniformes 119
beta cells 578
bicornuate condition 485

bidder's canal 477, 490, 491
bifid glans 484
bilateral symmetry 4, 27, 68
binocular vision 208, 217, 222, 528, 552
bipartite uterus 485
bipolar neurons 552
biradial symmetry 27
blind-sac plan 24-25
blind spot 552, 554–557
blood islands 427, 446
bone stage 306
bony armour 105, 111, 112
brachiopods 42, 52, 55, 58, 60, 61
branchial 267
branchialis 362, 527
branchiomeric 283, 353
    muscles 250
bregmatic bone 273
brontosaurus 165–167
brood patch 500
brood pouches 381, 497
buccalis (SS) 526
buccinator 371, 372, 378, 379
bucco-maxillary 526
bulbar part 554
bulbourethral 487
bursa fabricii 396, 397, 456

C

caecilians 50, 131
calamus scriptorius 522
calcified cartilage 260–262, 264, 314
calcitonin 575, 577
caloreceptors 541
camouflage 583
canal of schlemn 551
cancellous bone 263, 264
canon bone (m3, m4) 347
capitulum 192
caprimulgiformes 176, 187
caput epididymis 494, 495
cardiac depressor nerve bringing impulses 529
carinatae 183
carnassial 211, 385, 387
carotid artery 437
    canals 50, 62
        duct 67, 76, 101, 104, 105, 115, 119, 120, 227, 228, 235, 237, 238, 284

cartilage replacement 263, 264
    stage 6, 8
cartilages 260
casuariformes 182
catarrhini 208
caudata 50, 134, 138–139, 144
cauda equina 531, 533
caudiperyx 174
causes of extinction 168
cavum arteriosum 435–437
    venosum 435–437, 439
cell aggregate plan 24
cell-tissue level 38
cells of Leydig 233, 472
cellular level 38
cementum 383, 385, 387
cephalaspides 99
cephalaspidomorphi 99
cephalic eye spot 554, 563
cephalization 4, 26, 27, 29, 73, 87, 267
cephalochordata 48, 53, 66, 72, 78, 83–86, 89
cephalodiscida 71
ceratohyal cartilage 279, 286
ceratomorpha 214
cerebellar peduncles 522
    ventricle 515, 516
cerebellum 114, 115
cerebral cortex 203, 218
cerebrospinal fluid 503, 504
cervical 192, 194
cervico-branchial plexus 531, 532
cervix 485, 486
chaetognatha 35, 51, 53, 65–67
charadriiformes 176, 186
chassaignac's tubercle 318
chelonia 155–157, 163, 292
chemoreceptors 378, 547
chevron bones 307, 314, 319
chief cells 391, 577
chief endocrine gland 573
chiroptera 204, 205
choanae 50, 122, 133, 138, 289, 302
cholecystokinin 583
cholinergic fibres 535
chondrocranium 194, 269
chondrostei 48, 116, 117, 121, 122, 406
chordocentrous 307, 309, 311
choroid 506
choroidea 506, 509, 555

chromaffin 579
    tissue 573–574
chromophobes 569
chyle 456
chyme 393
cicatrix 493
ciconiiformes 174, 175, 186
ciliary body 550, 551, 553–556
circumvallate 383, 549
clasper organ 360, 483, 484
cladistics 46, 49–51, 152, 153
cladogram 41, 49, 50, 53, 55, 98, 138, 150, 153,155, 177, 191
class 44, 48, 51
clavodeltoid 359, 361, 363, 373
cleft palate 274
cleidoic egg 162, 387
cleidomastoid 362
cleithra 333, 364
cleithrum 136, 364
clitoris 181, 196, 265
cloaca 483, 484
cloacal kiss 483
    thymus 388, 396
clutch 494
cochlear duct 560
    window 560
coeliac or solar plexus 537
coelocanthini 124
coliiformes 177
collateral ganglia 535
collector loop 441, 442
colliculus seminalis 480, 481
colour blind 552
columella auris 565, 566
    communicates 82, 422, 495
column of cord 524
commissure 518, 519
common cardinal veins 432, 446
    carotid artery 444–446
compressor naris 371, 372
conchae 272
cone of Wulzen 573
conical papillae 549
conjunctiva 227, 544
conus arteriosus 103, 113, 115
    papillaris 556
coprodaeum 486

copula 290
coraciiformes 177, 186
corm or fission theory 30
corpora cavarnosa 483
    lutea 552, 588
    quadrigemina 193, 195
    restiformia 515, 517, 521
    striata 506, 513
corpus callosum 193–196
    luteum 475, 493
    mammill 318
    spongiosum 484
corrugator supercilii 371
cortical 472, 474
corticosterone 583
cortisol 583
costocuaneous 370
counter current distribution 405, 464, 474
courtship dance 483
cowper's glands 487, 495–497
cranial dura 503
    kinesis 297
craniostylic 297, 302
cremaster 368, 473
cribriform plate 272, 528
crista 285, 294
    ampullaris 559, 565
    occipitalis 285
cristae 558, 559
    ampullaris 559, 565
crocodilia 153, 156, 160, 164
crop milk 390, 571
cross-trunks 441
crossopterygii 48, 125, 128, 130
cruracerebri 509–510, 515, 517
    tegmentum 509, 522
cryptic 256
cryptobranchoidea 145
cryptorchid 587
cubitals 342
cuccularis 362, 364
cuculiformes 176, 185
cupola 559
curry comb 380
cyclophora 53, 59, 60
cyclomerism theory 7, 29–30
cycloneuralia 52–54
cyclospondylous 309

cyclostomata 91, 97, 100–101
cyclostomes 103–105, 111
cypriniformes 119, 409, 564
cytogenic 586, 587

D

D cells 578
dahlgren cells 585
dasyuromorphia 197
Darwin's point 562
demibranch 275, 403
dental formula 208, 386
depressor membrane-nictitans 555
dermal ribs 321, 322
    bone 311
dermatocranium 267, 272, 367
    osteocranium 272
dermoptera 204
dermoccipitals 272
detrusor 469
deuterostomes 8, 25, 37, 51–52, 54–55, 61, 65–67, 78
deuterostomia 52, 53
developing Graafian 588
dewlap 283
diabetes 578
diaphragmaticus 416–417
diapophyses 305
diapsid 148, 150
diapsida 150, 153–154, 157, 296
diapsids 148, 150–151, 163, 190, 297
diarthrosis 266
diastema 385
didactyla 198
didelphia 196
didelphimorphia 197
diencephalon 127, 505, 509
digastricus 367
digitigrade 211, 348
dinornithiformes 183
dimorphism 494, 586
dinosaur hips 165
dinosaurs 3, 151, 152, 154, 164–166
diocoel 509
diphycercal 97, 103, 122
diphyletic 13, 18, 100, 188
diploblastic 7, 12, 16, 18–19, 38–39, 62

dipneumona 126
dipnoi 48, 50, 107, 115, 124–126, 128
dorsal rib 319–321
diprotodontia 197
division Bilateria 27
domain 44–46
duct of cuvier 432, 447, 448
ductuli aberrantes 480
duplex uterus 485

E

ecdysis 576
echeneiform 121
echiurida 64
ectopic pregnancies 486
ectoprocts 54–55, 60–61
Edinger-Westphal nucleus 524
efferent branchial 406, 441
egg tooth 194, 246, 387
elasmobranchii 48, 107, 114, 127, 306, 403
electric organs 374
electroplax 374
eleuheronis 188
embolomerous 307, 308
enamel organ 245, 383
end bulb of krause 542, 544
end organ 542, 544
endocardial tubes 430
endochondral bone 263
endolymph 558–564
endolymphatic fossa 270
endolymphaticus 558
endomysium 351
endostyle 83–85
endostylophora 82
enterocoel theory 36
enterocoelomata 35
enterocoelous coelomate 65
enterogona 81
enteropneusta 11, 41, 53, 69, 70, 72, 78, 83
enterostome 379, 382
entoglossal 281, 301
    bone 275
entoprocta 53, 59, 60, 63
epicoracoid 325, 336
epigonal organ 489
epimeric 261

epimysium 351
epinephrine 580
epiphyseal apparatus 509
epiphyses 192, 194, 195
epiphysial 270, 517, 518
    apparatus 405–408, 417
episternum 325, 327
epoophoron 480, 492
ergastoplasmic 573
estradiol 587, 588
estrous 487, 588
ethmoid 268, 269
    centres 270–272
    plate 268–270
euryapsid 157, 163
eubacteria 43, 45, 46
eucoelomata 34
eukarya 43, 45, 46
eumetazoa 61, 62
euryapsida 157, 163
eustachian duct 561
eusthenopteron 106, 123, 133, 136, 137, 140, 332
eutheria 189, 191, 192, 195, 196, 199, 200, 203, 223
evolution of tetrapod limbs 136
exoccipitals 222, 272, 288
external glomerulus 459, 460
    intercoastal 320, 358
exteroreceptors 540
extra columella 565, 566
eyes of snakes 556

F

facial 252, 289
    nerve 18, 19
facialis 526
falciform ligament 265
falconiformes 176, 187
false ribs 324
feedback mechanism 568, 575
fenestra ovalis 162, 288
    rotunda 558, 560
fibula 337, 339
filiform 380, 383, 549
filum terminale 312, 511
fimbriae 100, 496
family 44, 49, 86, 197
fin-fold 108, 330, 331

fin-fold hypothesis 108
fin-fold-like origin 331
fissipedia 210, 211

fissure 505, 511
flexure 208, 339, 391, 517
floccular lobes 515, 519, 522
flocculi 519, 522
foliate 382, 383, 549
follicular 475, 573
fontanelles 291
foramen magnum 205, 269
    of magendie 504
    of panizzae 437, 438, 445
foramen triosseum 340, 372
foramina 262, 268, 269
    of monro 506, 509, 514
    transversaria 315, 317, 318
    transversia 323
foreskin 495
fornix 521
fossae 192, 295, 296
four legged trait 166
fovea 552, 554, 556
    centralis 552–557
frenulum 380
frigidoreceptors 541
frontal organ 553
frontalis 371
frontals 272, 287
fungiform 382, 383, 549
funiculi 511
fusion Stage 269

G

gaertner's duct 478–480
galliformes 176, 185
galea aponeurotica 371
ganoid Fish 121, 122, 467
gasserian ganglion 524, 525, 527, 528, 532
gasterocentrous 307, 312
gastralia 321–323, 325, 327
gastric pits 391, 392
gastrin 583
gastro-centrous 307
gastrotricha 13, 34, 53, 63
gaviiformes 175, 186
geniculate body 522, 524

genital ridge 458, 470, 471
genu 520, 521
genus 19, 44, 86, 99, 111, 204, 207, 220, 384
girdles 98, 103, 105
gizzard 265, 391
glands of swammerdam 532
glial elements 568
glorified reptiles 173, 178
glossopharyngeal 289, 526, 538
glucagon 578, 579, 583
glucocorticoids 583
gnathostomata 97–99
goitre 576
gonad 11, 33, 59, 70
gonocoel theory 36, 37
gonopodium 328, 481, 483, 484
grandry's corpuscles 544, 545
granular leukocytes 427, 428, 428
graptozoa 70, 72
gray matter 504, 507
greater omentum 391
grey ramus 531, 539
guanine 408, 409, 554, 555
gruiformes 176, 187
gubernaculum 473, 494, 495
gular pouch 265, 283, 423
gustatoreceptor 541, 548
gymnophiona 50, 129, 134, 139–141, 143
gyri 521, 522

H

habenular commissure 509, 516, 517
haikouella 95
haikouichthys 95, 108
hallux 187, 205, 337
hard palate 302, 379
haversian canal system 262, 263
heat period 588
hedonic glands 234
helicotrema 561
hemibranch 527
hemichordata 35, 41, 48, 51, 53, 54, 65, 66, 69
hemipenes 159, 483, 492
hemipenial sacs 492
hemipenis 492
hemocytoblasts 427
hemolymphatic organ 456
hemopoietic tissue 427

hepatic portal vein 399, 446, 448
    system 6, 10
hesperornithiformes 180
herbst corpuscle 542, 544, 545
hermaphrodite 79, 91, 480, 487
heterocercal 99, 113, 114
heteroplastic 263, 265
    bones 47, 99, 113
heterotopic bones 265
heteronomous metamerism 29
highmore 302
hippocampal 508, 515, 517
    areas 188, 229, 236
    lobes 64, 114, 115
hippomorpha 214
holobranch 404, 441, 442
holocephali 48, 107, 114, 127, 287, 288
holocephalians 127, 287, 288
holonephros 458, 459
holostei 48, 116, 117, 121, 122, 406
holostylic 278, 279
homeobox-containing 108
homeostasis 399, 457
homeothermy 135, 202
homocercal tail 329
homologies 11, 47, 49, 286, 307, 333, 344
homonomous 28, 29
homonomous metamerism 29
homo sapiens 208, 219, 221, 222
humerals 342
husk centra 307
hyaloid plexus 556
hyoid 275, 277
    apparatus 6, 17, 23
hyomandibular 275, 277
    cartilage 99, 112, 115
hyomandibularis 526, 528
hyostylic 277, 278, 286–288
hypapophyses 313, 314, 316, 318
hypaxial muscle 319, 358, 360
hypobranchial 276, 279
hypoglossal 353, 354, 524
    nerve 18, 19, 29
hypophysial cleft 572
    portal system 86, 88, 92
hypophysis 313, 315, 506
hypothalamic D 510
hypothalamus 429, 450, 453
hypotremata 114

hypoxanthine 554
hypurals 309, 329
hyracoidea 212

I

ichthyornis 181, 188, 299
ichthyosaurs 148, 151, 152, 163, 166, 296, 267, 339
ichthyostega 131–134, 136–138, 140, 143
impennae 183
incus 192, 201, 203
    quadrate 155, 156, 158
inferior colliculi 522
    parathyroids 573, 577
    peduncles 522
infra orbital 273, 284, 285
    ridge 160, 238
infratemporal arcade 297, 299, 301, 302
infundibular organ 554, 563
infundibulum 415, 484
inguinal canal 202, 495
innermost layer of retina 552
integumental 204, 353
intercalary discs 352
intercostal nerves 533
intermedin 572
internal glomerulus 460, 461
internuncial 511, 512
interoreceptors 540, 541, 543
interparietal 272, 288, 292
    foramen 156, 158, 161
interrenal 571, 582
    or cortical 581
interstitial cells 472, 571
interthalamus 522
interventricular foramina 509, 514
intervertebral discs 304, 307, 317
    foramina 262, 268, 269
intramembranous ossification 263
intrinsic appendicular 360, 361
    integumental 204, 353
intromittent 360, 483
invertebrate chordates 69, 92
iris 529, 535
irritatoreceptors 541
ischial callosities 208, 248
ischiopubic bar 329, 333
islet tissue 578

islets of langerhans 567, 578
isolecithal 475
isthmus 493
iter 509, 514, 516

J

jugular foramen 529, 530
juxta-glomerular apparatus 464

K

kidney or Leydig's gland 487
kidneys 97, 104, 113
kinesthesia 543
kinetic skull 155, 301
knee cap 265, 266, 339, 347
kingdom 4, 11, 15, 26–28, 37, 38, 44–46
kinorhyncha 34, 53, 63
koilin 392

L

labial pits 545
labyrinthodontia 143, 162, 163, 242, 364
labyrinthodonts 162, 272
lacteal 394, 395
lactogenic hormone 571
lacunae 262, 399
lagena 193, 194, 559
lagomorpha 209
lamellar bone 239, 241
lamina 39, 304, 305
laryngeal skeleton 280, 283, 423
lateral line organs 103, 104, 523
lateral ventricles 506–509
larvacea 66, 82
leaf like 549
left systemic arch 201, 436, 437, 445, 446
lepidosiren 124, 125, 332, 443
lepidosaurs 148, 151, 152, 158
lepospondyli 131, 138, 143
lepospondyls 133, 137, 140
lepto-meninges 503
levator 354, 359
    bulbi muscle 555
    palpebrae superioris 354, 371
    scapulae 335, 361

leydig cells 472, 586, 587
  gland 50, 69, 70
ligamentum arteriosum 440, 444, 446
ligaments 265, 266, 275
linea alba 358, 360, 362, 366
lingual of visceral 529
lissamphibia 133, 138, 139, 142, 143, 355
litters 497
liver 91, 92, 104
longissimus dorsi 357
longitudinal fissure 513, 515, 517
lophiiformes 121
lophophorate coelomata 62, 63
lophoporate 52
lophotrochozoa 54
loreal pits 545
lumbar 165, 203, 312
lumber 535
lumbosacral 532, 533
lumbo-sacral-plexues 531
luminescent 67, 121, 232, 233
luschka 504
luteinizing hormone 571, 586, 587
lymph hearts 554–556
lymphatic organs 456
lymphocytes 453, 454, 456
lyra 521

M

macropodidae 198
macroscelidea 204
macula basilaris 565
macula lutea 552
  neglecta 563–565
  or papilla basilaris 564
maculae 558–560
malleus 192, 201, 203
  articular 145, 279
mammalia 193–197, 199, 201
mammary ridge 238
mammillary 318, 521
  processes 18–20, 39
mandibular 108, 190
  nerve 18, 19, 29
mandibularis externus (SS) 526
manubrium 327, 343
massa intermedia 522
masseter 365, 367

mastacembeliformes 120
master gland 568
maxillary 266, 287
  nerve (SS) 525–527, 529
meantes 145
meckel's cartilage 275, 277
mediastinal 421
medulla oblongata 504, 506
medullary 472, 504
Meibomian glands 236
meissner's corpuscle 544, 545
  plexus 70, 73, 76
melanophores 229, 539, 584
melatonin 553, 584
meninges 503, 505, 543
meninx primitiva 503
mentomeckelians 292
merkel's disks 544
mesaxonic 346
mesectoderm 261
mesencephalon 505, 524
mesenchymal 34, 35, 37, 243
mesenchymal stage 306
mesentries 376, 377, 543
mesolecithal 475
mesometrium 494, 496
mesonephros 458, 459
mesorchium 473, 492
messoptiles 251
mesozoa 61
mesozoic radiation 148, 166
metacarpodigitals 342
metanephros 458, 459
metatheria 189, 191, 195, 196, 198, 199, 200
metasternum 327
metencephalon 505, 507
methystylic 278, 279
microbiotheria 197
micropodiformes 187
mid brain 268, 271
middle 31, 39, 43
middle commissure 509, 522
middle peduncles 522
mimetic muscles 367, 371
mimicry 255
mineralocorticoids 583
mixed 381, 467
modiolus 560
molecular clock 44, 174

molecular techniques 44, 174
monera 44, 45
monimostylic 279
monoliform 307, 309
monophneumona 126
monophyletic 5, 9, 10, 18, 43, 46, 47, 49, 54,
    55, 60, 98, 107, 134, 139, 150, 152, 174,
    190, 223, 188
monophyletic group 43, 46, 47, 49, 98, 107
monopneumona 125
monosynaptic reflex arc. 512
monotremata 193
mormyomasts 546
motor neurons 511, 512, 531
mullerian duct 476, 477
muscle of Bruke 557
muscle of Crompton 557
muscula densa 464
muscularis tensor choroidea 555
musophagiformes 176
myelencephalon 506, 524
myelocoel 509, 514
myeloid tissue 427
myenteric 537
myocardium 430, 431, 437
myoseptae 356–358
myotomes 86, 88, 89
mysticeti 215
myxini 50, 98–101
myxiniformes 100, 101

N

nasals 272, 287, 288
nasohypophyseal stalk 572
nasopharyngeal pouch 378, 572
nemertinea 34, 62, 74
neoceratodus 124, 126, 443, 448
neognathae 174, 175, 183, 184, 299
neopallium 508, 517
neornithes 174, 175, 177, 180
neossoptiles 251
neosternum 325
neotenic 82, 92, 147, 413
nephrocoel theory 37
nephrostome 127, 459
nervus terminal 152
neural 79, 81, 85
    folds 60, 68

neurenteric canal 505
neurocrania 270
neurocranial ossification centres 270
neurocranium 261, 267
neurohypohysis 568, 570
new world monkeys 207–208, 218
nictitating membrane 526, 529, 554–557
nidamental 477, 497
nodosal 524, 529
norepinephrine 580, 583
notochord 31, 65, 66
notoryctemorphia 197
notoryctidae 197
numerical taxonomy 46, 48–49
nutrient 24, 260, 262
    foramina 262, 268, 269

O

obliquus 416, 417
obturator 339, 341, 342
occipital 142, 152
    centres 270–272
occulomotor (SM) 525
ochre-yellow 581, 582
odontoceti 215
old world monkeys 207, 208, 217, 218
olecranon fossa 341, 344, 345
olfactoreceptors 541, 547
olfactory (SS) 523
    capsule 268, 269
    glomerulus 70, 76, 459
    nerve 18, 19, 29
    tract 28, 34, 35
omasum 393
omosternum 325, 336
onychophora 52–54, 64
opercular bones 272, 273, 275
ophiocephaliformes 120
ophthalmic 525–527, 529
    nerve (SS) 525–527, 529
ophthalmosaurus 166
opisthocoela 147
opisthocoelous 145, 147, 309
opisthonephros 460, 462
optic (SS) 525
    capsule 268, 269
    chiasma 388, 509
    lobes 64, 114, 115

nerves 28, 29, 88
ora terminalis 551, 553, 556
oral-aboral axis 26
oral-hood 382
ora-serrata 551
order 11, 31, 38, 44
orbicularis oculi 354, 367
organ of corti 201, 524
    of Muller 476, 477
    of taste 511, 548, 549
origin of paired appendages 328, 330
orinithischia 165
ornithischia 153, 164
ornithodelphia 191, 193
osseous 265
    lamina 304, 305, 356
ostariophysi 119, 564
osteoblasts 263, 505
osteoclast 263
osteons 263
ostracoderms 48, 94, 96–99, 103
otic capsule 268–270
    centres 270–272
    regions 6, 29, 40
otoconia 560, 563
otolith membrane 560
otoliths 563, 564
out-group 49, 50
ovaries 344, 474
ovary 474
oviparous 114, 115, 141
ovisac 490–492
ovoviviparous 490, 497, 588
oxyphilic cells 577
oxytocin 570

P

paired eyes 103, 549
palaezoic radiation 166
palatal bones 274, 279
palatine tonsils 388
palatinus (VS) 526, 527
palato-pterygo-quadrate 289
paleognathae 174, 175, 299, 327, 339, 343
pallium 506
palpebral part 554
pampiniform plexus 474
panniculus carnosus 370

para follicular 388
parabronchi 417, 418, 420
parabronchus 418
parachordal 268, 269, 283
paradidymis 478, 479
parafollicular hormone 575
parapineal 552, 553
paraphyletic group 46, 47, 98, 134, 139, 152
parapophyses 305, 321
parapsid 157, 163
parapsida 157, 296
parasphenoid bone 289, 555
parasternum 322, 324, 325
parasympathetic system 535, 537
paraventricular 568, 569
paraxonic 346, 347
    foot 347
parazoa 61
parenchymula 9
parietal 272, 273
    body 509
    cells 391
    eye 553
paroophoron 480
pars amphibiorum 565
    basilaris 564–566
    distalis 568–571, 575
    intermedia 570–573
    lagena 563
    neglecta 563–565
    nervosa 568–569
    tuberalis 572
parturition 344, 497
patagial 370
patella 265, 266, 339, 340, 343, 345, 347
paucituberculata 197
pecten 556, 557
    projects 556, 560, 564
pectoral girdle sling 361, 362
peduncles 522
pelage 253
pelecaniformes 175, 186
pentastomida 64
peramelemorphia 197
peramelidae 197
perciformes 120
perichondrium 260, 503
perilymph 270, 285, 409, 558
perimysium 351

period of gestation 497
periosteal ossification 263, 264
periosteum 263, 503
perissodactyl 213–215, 346, 400
perissodactyla 213–215
peristalsis 352, 377, 388
petromyzontiformes 100
pessulus 265, 424
petrosal bone 272
    ganglion 504, 511, 513, 514, 523–526
peyers patches 456
pharyngeal chiasma 388
    of the visceral motor fibres 529
pharyngotremy 72
pharyngula stage 42, 258, 259, 261
phlebobranchia 81
pholidota 209
phonoreceptor 540, 541, 557, 563
photic 541
phoronids 52, 55, 59, 60
photo-receptor cells 552
photoreceptors 68, 541, 549, 552
    or photicreceptors 541
phracmophora 68
phrenic nerve 528, 533, 534
phyllospondyli 143
phylum 18, 27, 31, 39, 44
physoclistous 408, 409
physostomous 408
piciformes 177, 185
pigeon milk 390
pilaster cells 404
pillars 521
pikaia 92, 93
pineal 552, 553
    epiphysial foramen 270
    apparatus 518, 529, 535
    body 509–512
pinna 562
pinnipedia 211
pit organs 545, 546
    receptors 545, 547, 548
pituicytes 568, 569
pituitary gland 271, 450, 514, 567
placodermi 98, 107, 110, 111, 410
placoderms 239
plakula theory 8, 14
planaea 9

planctosphaeroidea 70, 72
plantigrade 205, 211, 348
plasma cells 454
platysma 365, 367, 371, 378, 524
plectognathi 121
plesiosaurs 148, 151, 152, 163, 166, 296, 297, 339
pleurapophyses 305, 321, 324
pleurogona 81
pleuronectiformes 121
pleurotremata 114
plumage 251, 253, 587
pneumogastric 526
podicipediformes 175, 186
pogonophora 11, 65, 75, 77
polarity 21–22, 26–27
polylecithal 493
polyphyletic groups 46
polysynaptic 512
polypteriformes 107, 117
polypterus 116–117, 121, 122, 287
pons 519, 520, 522
    varolii 522
portal system 428, 447, 450
porto-abdominal vein 452
post ganglionic 530, 534, 537–539
    trematic efferent 441, 442
posterior 522–527, 530
    choroid plexus 509, 514–522
    commissure 506, 509–511
post-orbital groove 285
prechordal 261, 268, 269, 283, 284
prechordates 74, 77
precocial bird 500
pre-ganglionic 534, 535, 537–539
prepuce 236, 482, 484, 494, 495
pre-trematic efferent 441, 442
priapulida 64
primaries 342
primates 204, 205, 207, 217
primary coelom 34
primitive chordates 69
principal cells 577
proaves 187
proboscidea 212
proboscis 33, 39, 62–65, 69, 73–77
procellariformes 186
procoelous 309, 310, 312–314
procorachohumeralis 363

proctodeum 486
progesterone 588
prolactin 571
promatum 324
pronephros 458–460, 462
proprioception 543
proprioceptors 540, 541, 543
prosencephalon 505
prosimians 204–206, 217
prostate 481, 487, 494, 495, 497
protarchaeopteryx 174, 177
proteida 145
protista 13, 15, 43, 44
protocercal 329
protometazoan 21, 23–24
protoplasmic level 38
protopterus 124, 125, 126, 232, 332, 323, 443
protostomes 25, 37, 41, 51–52, 54, 61–62, 78
prototheria 189, 191–195, 199, 200, 202, 582
protothrix 252
proventriculus 389, 391
psalterium 521
pseudobranch 403, 406, 407, 442, 443
pseudocoelomata 34, 52, 62, 63
pseudocoelomates 37, 63
pseudocoel 521
pseudometamerism 29, 32, 33
pseudosuchia 164, 187, 188
psoas 368
psittaciformes 176, 185
pteraspids 99
pterobranchia 11, 41, 53, 70, 71, 78, 83
pterosauria 153, 164, 187
pterygo-quadrate 273, 286, 289
pudendal plexuses 533
pulsatile bulbils 431
pulvinar 524
purely motor 523
      sensory 523, 525, 526–531
purkinje fibres 431
pyloric caeca 389, 394–396, 579
pyrosomida 81

R

radialia 328, 329, 331–333
radiata 15–16, 18 26, 37, 53–54, 62, 83
rajiformes 114

ramus communicans 530–532, 534, 537, 539
      dorsalis 357, 358, 530, 531
      ophthalamicus superficialis 525
      opthalamicus profundus 525
      opticus 526
      ventralis 531
Rathke's pocket 509, 572
      pouch 568, 570, 572
ratitae 181–182, 188, 324
ray-fins 287
receptors 540–545, 547–549, 552
recessus utriculi 563
recurrent laryngeal nerve 529
red gland 408–409
      muscle fibres 254, 260, 351
reissener's membrane 561
relay centres 510, 534, 535, 537
remiges 249, 250, 342
renal portal system 97, 201, 447, 450–453, 459, 461
renin 464, 585
replacement bone 263, 270, 271, 281
replication-specialization-integration of units 40
restiform bodies 515
rete mirabile 408–409
      testis 471–471, 476
retina 268, 515, 524, 525
retinal tapetum 554
retractor bulbi muscles 555
rhabdopleurida 71
rheoreceptor 540, 541
rheiformes 182
rheumatoid arthritis 583
rhinencephalon 505, 513, 524
rhipidistia 107, 123, 133, 333
rhipidistian 273, 333, 334
rhipidistians 48, 106, 123–125, 128, 140
rhodopsin 554, 555
rhombencephalon 505
rhomboideus 359, 361, 369
rhynchocephalia 74, 153, 155, 158, 164, 295, 357, 415, 416
rhynchocoela 62
ribs 261, 263, 267, 305
right systemic arch 436, 437, 446
rima glottidis 423
ring of sclerotic ossicles 557
rodentia 209

roofing bones 272, 273
rookeries 500
ruffini corpuscles 544, 545
rugae 390
ruling reptiles 164, 187, 299
ruminantia 215
ruminants 215, 389, 480

S

sacculus 390, 524, 558
saccus endolymphaticus 558
    vasculosus 514, 515, 569
sacral 311–314, 316
    plexuses 454, 529, 531–533, 537, 538
salamandroidea 145
salientia 134, 139, 143, 146, 572
sarcopterygii 48, 98, 107, 115, 122, 128, 287
satyr ear 562
saurischia 148, 153, 164, 165, 171
scala media 560, 561
    vestibuli 560, 561
scalene 360, 419
scandentia 204
schizocoel theory 37
schizocoelomata 34
schwann cells 505
sciatic 451, 532
sclerotic ossicles 556, 557
scopeliformes 119
scorpaeniformes 120
sebum 236
secondaries 342
secondary coelom 34, 40
secretin 583
selachii 114
sella turcica 269, 572
sematic 256
seminal vesicle 67, 476, 487
semionotiformes 117, 119
septum pellucidium 521
sertoli cells 472, 473, 586
sesamoid 265, 266, 339, 343, 344, 347
sex hormones 583, 586, 587
sex-cords 471
sexless 472
sexual kidney or Leydig's gland 487
seymouria 149, 162, 163
shell gland 64, 477, 488, 491, 492, 497

shore birds 186
shunts 429, 430
simians 204–206, 217
simplex uterus 485, 486
sinosauropterex 174
sinus impar 409, 564
sinusoids 399, 428
sipunculida 64
sirenian 212, 327, 344, 347
soft palate 203, 379, 380, 388
    spots (fontanels) 273
somatic motor 353, 510, 512, 523
    receptors 512, 513, 515
    sensory 512, 513, 515
somatotrophin 571
sound picture 566
spaces 262, 263, 289
spatial diversity 42
species 16, 19, 23, 33, 39, 42–44
spectacle 160, 556
spermiogenesis 472, 495
sperm-sac 487
sphenethmoid 272, 290, 291, 292
sphenisciformes 175, 183
sphenoid 271, 272, 274
    bone 253, 258, 260
    centres 270–272
sphenodon punctatum 158
spheno-palatine ganglia 537
spinal accessory 529, 530
spinodeltoid 359, 363, 373
spiral lamina 561
    valve 394–396, 403
spirally coiled 196, 200, 565
spiralia 8, 10, 12, 14, 17, 52
splanchnic 227, 259, 261, 353
    nerves 537–539, 553
splanchnocranium 267, 275–277, 284, 353, 379
splenium 520, 521
splint 214, 265, 342, 346, 347
    bones 263, 265–267
squamosal 272–274, 278
squaliformes 114
squamata 48, 50, 150, 153–155, 158, 164, 297
stapes 278–281, 278
    hyomandibular 562
stegocephalia 142, 162
stellate cervical ganglion 535
stenson's duct 381

stereocilia 559, 560
stereospondylous 307, 308, 311
sternebrae 326, 327
sternomastoid 361–363, 365, 367, 368, 530
sternum 261, 263, 267
    proper 265, 288, 289
steroidogenic 579–581
    tissue 579–582, 584
stigma 78, 80, 493
stirnorgan 553
stomochord 69, 73, 76
stones 70, 168, 169, 176, 563
strepsirhini 205
streptostylic 279
strigiformes 176, 187
strobilization 32
struthioniformes 175, 181
stylohyoid process 283
subneural 571
subpharyngeal gland 102, 573, 576
sulci 521, 522
sulcushinalis 521
suina 215
superficial ophthalmic 526
superior 250, 295–297, 346
    colliculi 522
    parathyroids 573, 577
suprascapula 332, 335, 336, 338, 343
suprachoroid 554
supracoracoideus 359, 361, 364, 369, 372
supraoptic 568, 569
supraorbital ridge 285
suspensory ligament 550–555, 557
sylvian 520, 521
symbranchiformes 120
sympathetic system 535, 537–539, 580
symphysis 266, 281, 294
    pubis 266, 322, 325
synapsid 296, 297, 302, 384
synapsida 150, 153, 157, 163, 222
synapsids 148, 150, 151, 189, 190, 222, 384
synarthrosis 266
syncytium 352
syndactyla 198
syngnathiformes 120
synovial fluids 266
    membrane 257, 263–266, 270
synsacrum 177, 316, 318, 319, 323, 341
syrinx 177, 179, 181, 265, 381

T

taeniae hippocampi 521
talons 176, 246
tangeoreceptor 540
tapetum lucidum 554
tardigrada 53, 64
tecti opticum 509
tectorial membrane 560, 561, 566
tectospondylous 309
tegmentum 509, 522
tela choroidea 509
telencephalon 505, 506, 508, 513, 524
teleoptiles 251
teleostei 48, 107, 116–118, 405
telolecithal 475
temnospondyls 131, 133, 139, 140
temporal bone 162, 272, 283, 362, 562
    vacuities 151, 196, 296, 297
    lobe 116, 122, 123
temporal diversity 42
tendons 265, 266, 344, 351
tentacle 56, 555
teres major 359, 362, 363, 369, 373
    minor 283, 359, 361–363
terminal nerve (0) 523
terrestrial birds 183–185
testis 585, 586
testosterone 328, 472, 586, 587
testudinata 156
testudines chelonia 155
tetany 578
tetraodontiformes 121
tetra-iodothyronine 574
tetrapod 123, 131–134
tetrapod ribs 320, 321
thalamencephalon 505
thaliacea 66, 80, 81
theca interna 587
thecodontia 164, 187
third eye 272, 516, 553, 554, 556
thiozoon 13–14
theria 191, 192, 195
thoracico-lumbar 312, 313, 537
three pairs of fibrous tracts 522
thymus 388, 396, 456, 577
    gland 456, 466, 469
thyrocalcitonin 575
thyro-globulin 574

thyroid 83, 85, 104, 282, 283, 356
tibia 334, 335, 337
tinamiformes 175, 183
tissue-organ level 39
tongue-tied 380
tonsils 382, 388, 456
tornaria 8, 70, 74–78
torsion 391
trabeculae cranii 268
traditional approach 46
transducers 540
transversus 358, 360, 368, 416
    muscle 416, 417, 419
trapezius 358, 359, 362, 363, 365
triceratops 166, 167
trigeminal 289, 367, 523–527, 529
    (SS and SM) 525
triiodothyronine 574
triploblasts 18–19
trochlear 344, 354, 523–525, 528
trogoniformes 177
tubal bladders 467, 469
tuber cinerarium 509
    cinereum 516
tuberculin 321
tuberculum olfactorium 520, 521
tube-within-a-tube plan 25
tubulidentata 211
tunica adventitia 428–430
    albuginea 472, 484, 494, 586
    intima 428–430, 454
    media 429–431, 435
    vaginalis 494, 495
turbinal bones 272, 547
tusks 212, 385
two-legged gait 166
tylo-mastoid foramen 529
tylopoda 215
tympanic annulus 291
tympanohyal 282, 283
typhlosole 84, 85, 97, 394
tyrannosaurus 164, 166, 167, 178

U

ultimo-branchial 388, 575
    bodies 575, 577, 581, 582,
unguligrade 214, 348
uniocular vision 552

uremia 466
ureotelic 113, 115, 142, 457
uricotelic 156, 457
urochordates 66, 83, 84, 91, 408, 571
urochordata 44, 48, 53, 66, 72, 78, 79
urodela 134, 139, 143, 144, 498
urodeaum 169, 486, 492
urohyal 282, 301
urophysis 585
uropygial glands 234
utriculus 524, 540, 558
uvea 549, 550, 554, 555

V

vagina masculina 479–481
vagus 289, 421, 431, 439
    nerve 529–535
valve of vieusens 522
    of kerkring 396
valvula cerebelli 515
vasa vasorum 454
vascular system 41, 79, 83, 88, 89, 431
vasopressin 571
vater-pacinian 544
velociraptor 173, 177
velum 378, 379, 382, 524
ventral flexure 517, 520
    rami 532, 533, 535
    rib 305, 311
    rostral 265, 284, 285, 570
vermiform appendix 396
vermis 517, 519, 520, 522
vertebrata 28, 53, 66, 78, 83, 85
vestibule 378, 381
visceral 261, 263, 265, 267
    motor 220, 350–353, 372
    receptors 256, 510, 512
    sensory 252, 299, 351, 379
    skeleton 258, 261, 263
visceralis (VS and VM) 527
vitreous chamber 549, 551, 556, 557
    humour 551, 554, 556
viviparous 115, 151, 196, 200, 202, 490
volkman 262
volutionary conservation 45
vomer 274, 287, 289
vomero-nasal organs 547

W

weberian ossicles 119, 559, 562, 564
wheel organ 87, 89, 379, 382
whiskers 252
white matter 83, 504, 510, 511, 513, 519
    muscle fibres 254, 260, 351, 352, 356
wings 250, 271, 322, 341
wisdom tooth 386
witch's milk 238
wolffian duct 458–462, 468, 470
    vestige 275, 294, 337

X

x layer 583
x zone 583
x1 –Lateralis nerve, (SS) 527
x2 – Branchio visceral trunk 527
xenopus 146, 246
xiphisternal cornua 327
xiphisternum 325–327, 336
x-vagus 526

Y

yunnanozoon 92

Z

zeis glands 236
zero ventricle 521
zona fasciculata 580, 583
    fasciculate 580
    glomerulosa 580, 583, 585
    reticularis 580, 583
zooids 30, 57, 81, 83
zonule 551, 555, 556
zygantra 314
zygomatic arch 297, 301–303
zygosphenes 314